S0-BKY-223

MODERN CELL BIOLOGY

Volume 2

Spatial Organization of Eukaryotic Cells

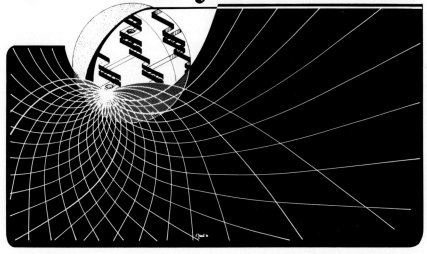

Drawing by Jonathan Izant

SERIES EDITOR

Birgit H. Satir
Department of Anatomy
Albert Einstein College of Medicine
1300 Morris Park Avenue
Bronx, New York 10461

ADVISORY BOARD

Julius Adler
Department of Biochemistry
University of Wisconsin
Madison, Wisconsin 53706

Howard C. Berg
Division of Biology
California Institute of Technology
Pasadena, California 91125

Bill R. Brinkley
Department of Cell Biology
Baylor College of Medicine
1200 Moursund Avenue
Houston, Texas 77030

G. Gerisch
Max-Planck-Institut für Biochemie
D-8033 Martinsreid bei München
Federal Republic of Germany

Anthony R. Means
Department of Cell Biology
Baylor College of Medicine
1200 Moursund Avenue
Houston, Texas 77030

Jean-Paul Revel
Division of Biology
California Institute of Technology
Pasadena, California 91125

Kai Simons
Postfach 10.2209
Meyerhofstrasse 1
6900 Heidelberg
Federal Republic of Germany

Vladimir R. Skulachev
Laboratory of Bioorganic Chemistry
Moscow State University
Moscow 117234, USSR

Joan Steitz
Department of Molecular
    Biophysics and Biochemistry
Yale University
333 Cedar Street
New Haven, Connecticut 06510

Walther Stoeckenius
Cardiovascular Research Institute
University of California
San Francisco, California 94143

Edwin W. Taylor
Department of Biology
The University of Chicago
1103 East 57th Street
Chicago, Illinois 60637

L. Wolpert
Middlesex Hospital Medical School
Mortimer Street
London W1, England

MODERN CELL BIOLOGY

Volume 2

SPATIAL ORGANIZATION OF EUKARYOTIC CELLS

Proceedings of a Symposium held in Honor of
Keith R. Porter,
Boulder, Colorado, April 30–May 2, 1982

Editor

J. Richard McIntosh

University of Colorado
Boulder

Alan R. Liss, Inc., New York

Address all Inquiries to the Publisher
Alan R. Liss, Inc., 150 Fifth Avenue, New York, NY 10011

Copyright © 1983 Alan R. Liss, Inc.

Printed in the United States of America.

Under the conditions stated below the owner of copyright for this book hereby grants permission to users to make photocopy reproductions of any part or all of its contents for personal or internal organizational use, or for personal or internal use of specific clients. This consent is given on the condition that the copier pay the stated per-copy fee through the Copyright Clearance Center, Incorporated, 21 Congress Street, Salem, MA 01970, as listed in the most current issue of "Permissions to Photocopy" (Publisher's Fee List, distributed by CCC, Inc.), for copying beyond that permitted by sections 107 or 108 of the US Copyright Law. This consent does not extend to other kinds of copying, such as copying for general distribution, for advertising or promotional purposes, for creating new collective works, or for resale.

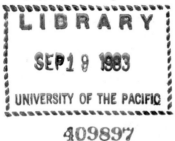

LIBRARY

SEP 1 9 1983

UNIVERSITY OF THE PACIFIC

409897

Library of Congress Cataloging in Publication Data

Main entry under title:

Spatial organization of eukaryotic cells.

  (Modern cell biology ; v. 2)
  Includes bibliographies and index.
  1. Cell compartmentation–Congresses. 2. Cell organelles–Congresses. I. Porter, Keith R. II. McIntosh, J. Richard. III. Series.
QH604.3.S63 1983   574.87′2   83-909
ISBN 0-8451-3301-2
ISSN 0745-3000

Contents

Contributors

M. Beckerle [259]
Department of Molecular, Cellular and Developmental Biology, University of Colorado, Boulder, CO 80309

S. Black [483]
Department of Molecular Biology, University of California, Berkeley, CA 94720

B.R. Brinkley [1]
Department of Cell Biology, Baylor College of Medicine, Texas Medical Center, Houston, TX 77030

David G. Capco [385]
Department of Biology, Massachusetts Institute of Technology, Cambridge, MA 02139

Pradeep Chatterjee [385]
Department of Biology, Massachusetts Institute of Technology, Cambridge, MA 02139

John Condeelis [225]
Department of Anatomy, Albert Einstein College of Medicine, Bronx, NY 10461

William J. Dolan [419]
Department of Cell Biology, New York University Medical Center, New York, NY 10016

Sabine Ermish [385]
Department of Biology, Massachusetts Institute of Technology, Cambridge, MA 02139

Edward G. Fey [385]
Department of Biology, Massachusetts Institute of Technology, Cambridge, MA 02139

J. Gerhart [483]
Department of Molecular Biology, University of California, Berkeley, CA 94720

Bruce L. Granger [143]
Division of Biology, California Institute of Technology, Pasadena, CA 91125

The boldface number in brackets following each contributor's name is the opening page number of that contributor's paper.

Eva B. Griepp [419]
Department of Cell Biology, New York University Medical Center, New York, NY 10016

Mark Groudine [353]
Department of Genetics, Hutchinson Cancer Research Center, Seattle 98104, and Department of Radiation Oncology, University of Washington Hospital, Seattle, WA 98195

Elizabeth D. Hay [509]
Department of Anatomy, Harvard Medical School, Boston, MA 02115

Peter K. Hepler [93]
Botany Department, University of Massachusetts, Amherst, MA 01003

Leroy E. Hood [305]
Division of Biology, California Institute of Technology, Pasadena, CA 91125

Tim Hunkapiller [305]
Division of Biology, California Institute of Technology, Pasadena, CA 91125

Ivan Emanuilov Ivanov [419]
Department of Cell Biology, New York University Medical Center, New York, NY 10016

Ellen Kraig [305]
Division of Biology, California Institute of Technology, Pasadena, CA 91125

Elias Lazarides [143]
Division of Biology, California Institute of Technology, Pasadena, CA 91125

J. Richard McIntosh [xiii, 115]
Department of Molecular, Cellular and Developmental Biology, University of Colorado, Boulder, CO 80309

M. McNiven [259]
Department of Molecular, Cellular and Developmental Biology, University of Colorado, Boulder, CO 80309

Richard Nuccitelli [451]
Zoology Department, University of California, Davis, CA 95616

Stelios Papadopoulos [419]
Department of Cell Biology, New York University Medical Center, New York, NY 10016

Lee D. Peachey [1]
Department of Biology, University of Pennsylvania, Philadelphia, PA 19104

Sheldon Penman [385]
Department of Biology, Massachusetts Institute of Technology, Cambridge, MA 02139

Jeremy D. Pickett-Heaps [241]
Department of Molecular, Cellular and Developmental Biology, University of Colorado, Boulder, CO 80309

Keith R. Porter [259]
Department of Molecular, Cellular and Developmental Biology, University of Colorado, Boulder, CO 80309

Robert O. Poyton [15]
Department of Molecular, Cellular and Developmental Biology, University of Colorado, Boulder, CO 80309

David M. Prescott [329]
Department of Molecular, Cellular and Developmental Biology, University of Colorado, Boulder, CO 80309

Theresa Reiter [385]
Department of Biology, Massachusetts Institute of Technology, Cambridge, MA 02139

Edith S. Robbins [419]
Department of Cell Biology, New York University Medical Center, New York, NY 10016

Enrique J. Rodriguez-Boulan [419]
Department of Pathology, State University of New York, Downstate Medical Center, Brooklyn, NY 11203

Michael J. Rindler [419]
Department of Cell Biology, New York University Medical Center, New York, NY 10016

David D. Sabatini [419]
Department of Cell Biology, New York University Medical Center, New York, NY 10016

Peter Satir [553]
Department of Anatomy, Albert Einstein College of Medicine, Bronx, NY 10461

S. Scharf [483]
Department of Molecular Biology, University of California, Berkeley, CA 94720

L. Andrew Staehelin [73]
Department of Molecular, Cellular and Developmental Biology, University of Colorado, Boulder, CO 80309

Thomas P. Stossel [203]
Hematology-Oncology Unit, Massachussetts General Hospital, and Department of Medicine, Harvard Medical School, Boston, MA 02114

Lewis G. Tilney [163]
Department of Biology, University of Pennsylvania, Philadelphia, PA 19104

Katherine Wan [385]
Department of Biology, Massachusetts Institute of Technology, Cambridge, MA 02139

Harold Weintraub [353]
Department of Genetics, Hutchinson Cancer Research Center, Seattle, WA 98104

Stephen M. Wolniak [93]
Botany Department, University of Massachusetts, Amherst, MA 01003

Conference Participants—Visiting

Peter Andrews
James R. Bamburg
Betty Barbour (Mathews)
Dennis Barrett
David Beer
Gudrun S. Bennett
Ronald W.Berezney
Benjamin G. Bouck
Linda D. Brebnor
Bill R. Brinkley
Mary Bunge
Richard Bunge
Mario H. Burgos
Robert R. Cardell, Jr.
Dona M. Chikaraishi
A. Kent Christensen
John Condeelis
Peter H. Cook
Eugene Copeland
Bill Cunningham
Samuel Dales
Etienne deHarven
Steve Downs
Barry S. Eckert
Mark Ellisman
June Ewing
Clara Franzini-Armstrong
Joseph F. Gennaro, Jr.
John Gerhart
John R. Gibbins
Ursula Goodenough
Dennis Gould

Ed Haskins
Elizabeth Hay
Ronald P. Hathaway
Peter Hepler
Leroy Hood
Terry C. Johnson
Vitauts I. Kalnins
Edward P. Katz
Robert C. King
J.G. LaFontaine
Elias Lazarides
Myron C. Ledbetter
Jon C. Lewis
Katherine Luby-Phelps
Alvin M. Malkinson
Morton D. Maser
Douglas McAbee
William D. Meek
Donald W. Misch
Mary B. Mockus
David T. Moran
Pietro M. Motta
Howard Nornes
Richard Nuccitelli
George E. Palade
Lee D. Peachey
Sheldon Penman
Tilman Prater
Vincent Price
Katherine Pryzwansky
Bill Rainey
John E. Rash

Howard Rickenberg
David H. Robertson
David Sabatini
Birgit Satir
Peter Satir
Manfred Schliwa
Tom Schroeder
Phyllis Schultz
Albert W. Sedar
Jerry Shay
Huntington Sheldon
Emma Shelton
Dorothy Spangenberg
Walther Stoeckenius

Richard Storey
Thomas P. Stossel
George Szabo
Igor Tamm
Stephen Thibodeau
Lewis Tilney
Mukta Webber
Melvyn Weinstock
Harold Weintraub
Jane A. Westfall
John Wolosewick
John P. Wourms
Natille Zimmerman

Conference Participants—Local

The faculty and students of the Department of Molecular, Cellular and Developmental Biology, University of Colorado, Boulder, CO 80309.

Introduction

The retirement of a scientist like Keith Porter from official university service is an occasion worthy of note. To mark this event, Porter's students and colleagues at the University of Colorado, Boulder, organized a symposium on the spatial organization of eukaryotic cells. The subject of the conference was selected in part because it is a topic in which Keith has always taken interest and to which he has richly contributed. In addition it was chosen because it seemed to us a subject of substantial biological breadth and profound importance about which many interesting and interrelated things could be said. We hoped that by assembling scientists who could speak and write about a diversity of the relevant subjects, we would have the chance to look for underlying principles and overlying themes, helping to unify the way one can look at cell structure and function.

The material before you contains much food for thought. Whether there really are significant unities in the subject is for each reader to decide. Satir's concluding remarks address this issue in an important and useful way, but a few words of preamble seem appropriate. Beyond the laws of classical physics and chemistry and the principle of natural selection, "universals" in biology are hard to come by. By any criterion, however, the dictum of Virchow, "Omnis cellula e cellula," must qualify. Much of the science discussed here can be viewed as an effort to understand how a cell organizes its biosynthetic activities so as to construct two copies of itself. Some of the papers on membrane biology develop the idea that existing membrane structures define where newly made membrane components will go; hence they direct their own expansion and proliferation. Analogous phenomena seem to pertain to the growth of parts of the cytoskeleton. Such structural templating is a form of epigenetic information and constitutes a significant relationship between cell organization and inheritance.

The division of cytoplasm into functionally significant compartments is an often mentioned role of cell membranes, and some novel facets of the issue are discussed here. In addition the capacity of membranes to offer a nonaqueous phase for certain chemical reactions is presented as an important feature of the compartment concept for both mitochondria and chloroplasts.

The cytoskeleton is discussed in two sections, one describing important features of each of the well-known fibrous components of cytoplasm and one addressing problems of overall cytoskeletal function. A well-recognized principle that runs through all these papers is that cellular structures are often formed by the self-assembly of proteins whose surfaces include bonding domain that specify the geometry of the resulting polymers. A more novel theme is the limitations of the self-assembly concept in the complex milieu of cytoplasm. Indeed, the most significant insights into the character and mechanisms of cytoskeletal systems seem to come at the beginnings and ends of self-assembly processes where more sophisticated levels of regulation are imposed.

Another theme running through the papers on cytoskeleton is that each of the fibrous components of cytoplasm has a physical and chemical character of its own, based on the structure and the interactions of its subunits. Microtubules and microfilaments are labile, whereas intermediate filaments are generally stable; microtubules are generally rigid whereas microfilaments are usually flexible. Nonetheless, higher levels of organization can confer specific properties on each system. For example, flexible microfilaments can be made stiff by bundling them into a shaft with cross-linking proteins, making them sufficiently rigid to deform the plasma membrane by their outgrowth, a job accomplished elsewhere by microtubules. The components of the cytoskeleton take on the character of the contents of a cytoplasmic tool box. Over the full range of phylogeny, and the course of evolutionary time, different tools have been selected for the same job. The historical component in the makeup of the cells we can study is therefore much in evidence. Finally, two of the papers on cytoskeleton help to remind us how much there is yet to be learned, not only about the complex questions of regulation but even about the very contents of the tool box itself.

The organization of nuclear material seems to have a different character from that of cytoplasm. The structure of DNA dictates nuclear organization at several levels in the organizational hierarchy. Two papers in this section show how nucleotide sequence information defines DNA structural rearrangements that occur over either phylogenetic or ontogenetic time, and how these rearrangements in turn govern aspects of gene expression. Another paper shows how particular DNA sequences are subject to higher levels of organization, not just because they are arranged in supercoils by nucleosomes, but because they are associated with special proteins that promote their transcription. Indeed, the thread running through these treatments of the organization of genetic material is the DNA itself, whose sequence specificity includes information to dictate not only the amino acids and syntax of protein synthesis, but the control of expression and the reorganization of the genome as well. The last paper in the section closes a circle, by discussing ways in which the cytoskeleton may influence the expression of genetic information.

The final part of the book introduces an aspect of structural biology that is still somewhat mysterious: the factors governing global organization and polarity of cells and organisms. How can an organelle know which is the apical and which the basal of a cell? Is the bioelectricity associated with many cells a cause or a result of such polarity? Does the cytoskeleton of embryonic cells help to define the dissimilarity of the daughters at asymmetric cell divisions, or does it merely reflect underlying differences we do not yet known how to visualize? How do embryonic cells specify the anisotropy and heterogeneity of the extracellular matrix material that seems in turn to help define the behavior of other cells as they move through it? These and other questions are defined and discussed so the reader can see the answers that are beginning to emerge from current work. The challenges identified for future investigators are among the most fascinating aspects of the spatial organization of eukaryotic cells.

J. Richard McIntosh

Modern Cell Biology 2:1–12
© 1983 Alan R. Liss, Inc., 150 Fifth Avenue, New York, NY 10011

Scientific Achievements and Contributions of Keith R. Porter to Modern Cell Biology

Lee D. Peachey and B. R. Brinkley

From the Department of Biology, University of Pennsylvania, Philadelphia,
Pennsylvania 19104 (L.D.P.), and the Department of Cell Biology, Baylor
College of Medicine, Texas Medical Center, Houston, Texas 77030 (B.R.B.)

This book is dedicated to Keith Roberts Porter on the occasion of his official retirement as professor of Molecular, Cellular, and Developmental Biology at the University of Colorado, Boulder. The volume honors the career of a unique individual whose life's work helped to change the world's view of the structure and organization of eukaryotic cells. Porter's scientific achievements and contributions are manifold and span a period of 43 years. His remarkable career encompasses the entire era of modern cell biology. Porter, as much as any other single individual, is responsible for the foundation and development of cell biology in America and abroad.

Porter's first major contribution to cell biology was the early observation of cells by electron microscopy in the early 1940s. Claude and Fullman had already examined cell-derived "microsomes" in the new instrument, but had not considered it possible to look at whole cells because it was thought that their thickness would prevent penetration by the 50–60 kV electron beam available in instruments of the day.

Porter, who then was studying cells grown in vitro, realized that some of these might be thin enough to be examined effectively, especially in those regions where the margins of the cells were spread out in a thin layer over the substrate upon which they were growing. Before such a study could be made, however, a number of problems in specimen preparation had to be solved. The cells had to be mounted over the holes in metal specimen screens. Porter did this by growing the cells on cover glasses coated with a thin plastic film, then floating the cells still attached to the film free from the glass support, and picking them up on a fine screen. The cells had to be stabilized by fixation, and for this he chose a method involving osmium tetroxide. This classical fixative from light microscopy had been shown by Strangeways and Canti to produce minimum changes in the patterns of light scattered from a cell as seen in the dark-field light microscope. In addition, osmium tetroxide imparted electron scattering power to the structures present in the cells. The wisdom of this choice is indicated by the fact that osmium tetroxide remains a valued fixative for present-day electron microscopy.

This combination of choosing the right kind of cell and applying to it a set of suitable preparative methods led, in 1944, to the first useful images of the internal structures of cells at resolutions exceeding those obtainable with the light microscope [Porter et al, 1945]. A door was opened through which many would pass, for this also led to the immediate description of a new cellular system, an extensive labyrinth of cytoplasmic vesicles and channels, and a finely divided vacuolar system, which later was named the endoplasmic reticulum by Porter and Kallman [1952].

Porter used the new methods to look at a variety of cultured cells, including tumor cells and virus-transformed cells. An important result during this period was the observation with Helen Thompson that the Bittner milk factor, which transmitted mammary tumors in mice, was very likely associated with a defined particle of uniform size (later identified as a virus) that could be seen in electron micrographs of mammary tumor cells [Porter and Thompson, 1948]. This was the first ultrastructural description of vireons in mammalian cells. Other intracellular and extracellular structures were also explored, some in much greater detail. The studies with Bud Hawn on fibrin [Hawn and Porter, 1947] showed, for the first time, periodicity in a protein fiber, and led Porter into a number of later studies on collagen fibers, especially in relation to their formation by fibroblasts [Porter and Pappas, 1959]. He provided important insights into chondrogenesis [Godman and Porter, 1960] and into the complex problems of the relationships between the organization of extracellular materials and the processes occurring inside the cells themselves [Porter and Machado, 1960a].

In 1950, Porter began a series of studies on muscle, a tissue that was to reappear often in his future work. A study with Charles Ashley on the contraction of isolated myofibrils [Ashley et al, 1951], where the images suffered from excess thickness of the specimen, stimulated his interest in trying to obtain thin sections of embedded tissue. He spent a summer in the laboratory of H. S. Bennett in Seattle, trying to cut thin sections on an ordinary microtome with a steel knife, using thermal effects to provide the fine advance necessary [Bennett and Porter, 1953]. These rather unsatisfactory results convinced him of the need for better methods for thin sectioning, and he devoted considerable effort in the early 1950s toward that goal.

Porter's efforts culminated in the introduction of the Porter-Blum microtome [Porter and Blum, 1953], an instrument that, while not the first, was the most successful and has remained a standard in the field for many years.

At this point, armed with a powerful set of new techniques, Porter and an entire group of biological electron microscopists who had grown up in the decade beginning about 1945, left the study of whole cells grown in culture and focused on the study of thin sections of fixed and embedded cells and tissues. This marked the beginning of a new and important field in cell biology—the study of cell and tissue fine structure. It is difficult today to imagine what cell biology would be like without the knowledge we now have about the fine structure of cells. Keith Porter, as much as any other single person, recognized the importance of this field in modern cell science and contributed to its rapid expansion.

As an example, one of Porter's early papers in this era, with Don Fawcett, reported the fine structure of ciliated epithelia [Fawcett and Porter, 1954], and showed the tremendous potential of the electron microscope and thin-section microtomy for revealing detail of structures, in this case cilia, that were too small to be more than just detectable in the light microscope. The now well-known 9 + 2 pattern within the cilia was described, and with time this elegant structure was worked out in detail. In fact, it was Porter who, with Myron Ledbetter, first showed that there are 13 protofilaments in each microtubule [Ledbetter and Porter, 1964].

Beginning in the early 1950s Porter established an active laboratory at the Rockefeller Institute for Medical Research under the aegis of Herbert Gasser. Many students and young scientists received significant training there. Once the thin-section approach to electron microscopy of biological material had been well established, Porter's interest and that of those who worked with him concentrated in greater detail on a few cell organelles. The endoplasmic reticulum was one of these, and Porter expanded our knowledge of this intracellular membrane system by studies on a variety of specialized cell types [eg, Porter, 1955, 1956; Palade and Porter, 1954; Porter and Palade, 1957; Porter and Machado, 1960b]. A notable case is that of the striated muscle cell, where he and his co-workers named the equivalent organelle the sarcoplasmic reticulum [Bennett and Porter, 1953; Porter and Palade, 1957]. Today we know that the sarcoplasmic reticulum sequesters and re-leases calcium ions in the muscle cell, playing a major role in excitation-contraction coupling. We also know that the transverse tubular system, which originally was described along with the sarcoplasmic reticulum, carries ex-citation into the muscle fiber from its surface [Peachey and Porter, 1959].

The endoplasmic reticulum is well known to play a key role in the synthesis and secretion of a variety of cell products. Porter's seminal work lies at the base of many of these discoveries. One important specialized function of the endoplasmic reticulum is the role of the smooth-surfaced portion of this

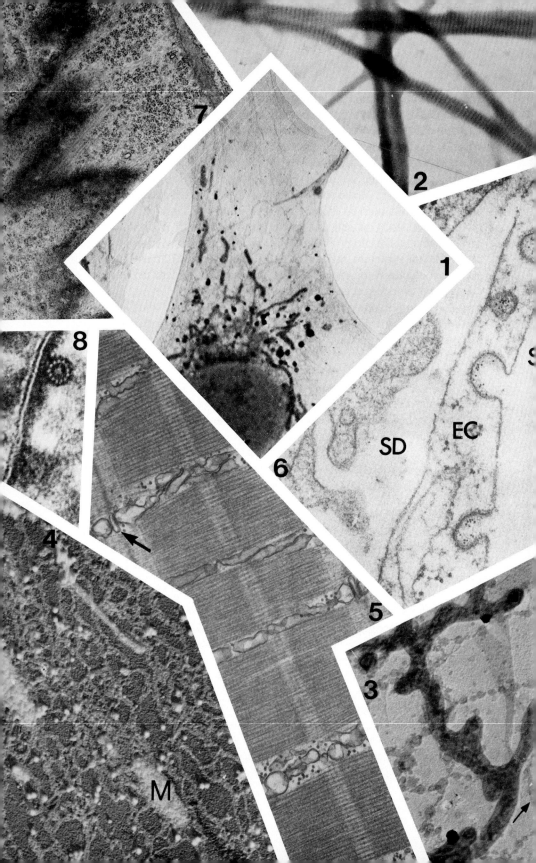

7

2

1

8

SD

EC

S

6

4

5

3

M

organelle in the detoxification of drugs and other toxic compounds. This was first recognized by Porter and Carlo Bruni [1959]. Two additional cellular organelles whose discovery can be attributed to Porter and his colleagues are the autolysosomes and the coated vesicles. The former were found initially in cells of the liver, and it was recognized that they were important organelles in the degradation of cell structures [Ashford and Porter, 1962]. Coated pits and vesicles were first found by Keith Porter and Thomas Roth in mosquito oocytes, and were identified with the selective uptake of proteins into this and other types of cells [Roth and Porter, 1964].

Credit should also be given to Keith Porter for early recognition of microtubules as important and ubiquitous cell organelles [Porter, 1966]. With Myron Ledbetter he found microtubules to be common in plant cells where they were recognized as significant cell structures associated with the assembly of cellulose filaments in the cell wall [Ledbetter and Porter, 1963]. Soon after, they were observed in virtually every eukaryotic cell type studied.

Fig. 1. (1952) Typical early electron micrograph, just before thin sectioning became effective.

Fig. 2. (1947) Bovine fibril polymerized in vitro and stained with phosphotungstic acid. Periodicity 250 Å.

Fig. 3. (1947) Small area of cultured mesothelial cell after fixation over OsO_4 vapors for 12 hours. Image includes mitochondrion and strings of vesicles representing the ER. Arrow points to fold in upper cell membrane.

Fig. 4. (1947) Part of cytoplasm of a mouse mammary tumor cell. The cell was fixed as usual (for that period) with vapors of OsO_4 and dried from water. It was then shadowed with gold to enhance contrast. The small white granules are viruslike particles interpreted to be the Bittner milk factor known then to transmit this tumor. The lacelike structure in the ground substance is the ER. M = mitochondria.

Fig. 5. (1963) Thin section of frog skeletal muscle that shows the sarcoplasmic reticulum and the triad arrangement of SR and transverse tubule at Z line (arrow).

Fig. 6. (1962) Thin section of rat liver including part of sinusoid (S) and limiting endothelial cell (EC) with coated pits. The animal's circulation was injected with ferritin 2 min before sacrifice. The ferritin has been taken up selectively by the coated pits. The space of Disse (SD) is limited by endothelial liver cells.

Fig. 7. (1963) Micrograph includes, in oblique section, the primary wall between two plant cells. It illustrates the cortical disposition of microtubules in these cells during stages of wall formation. (With Ledbetter.)

Fig. 8. (1964) Cross section of plant cell microtubules showing in negative image the 13 protofilaments making up the wall of the tubule.

Porter's work, especially with Lewis Tilney, suggested that microtubules are centrally involved in cytoplasmic movements and in aiding or influencing alterations of cell shape in morphogenesis [Byers and Porter, 1964; Gibbins et al, 1969]. These concepts have been explored by an enthusiastic group of workers, many of whom were inspired by collaboration with Porter while he was at Harvard University [eg, Byers and Porter, 1964; Tilney and Porter, 1965, 1967; Gibbins et al, 1966, 1969; Bikle et al, 1966; McIntosh and Porter, 1967; Junqueira and Porter, 1969].

Porter's interest in the cell surface also emerged during his collaborative studies on the special differentiations of the muscle cell surface. The discovery with Clara Franzini-Armstrong of the real nature of the inward extensions of the muscle cell surface, known as the transverse tubules, opened up an important area of structural and functional studies on muscle cells concerned with the inward spread of contraction activation [Franzini-Armstrong and Porter, 1964]. Again, much of his work was carried on in the careers of younger people who, as students or associates of his laboratory, acquired the Porter enthusiasm for morphological studies. This period in Porter's career, extending from the early 1950s to the late 1960s, produced over 100 publications in his name, and countless others by students and collaborators who followed his pattern and applied it to subjects of their own interest.

In the early 1970s, Porter moved into scanning electron microscopy. These instruments had been available, especially in England, for more than a decade, but had been put to use by very few people except for obvious specimens such as the external parts of insects and specialized skin surfaces of amphibians and reptiles. It seems, once again, to have taken the Porter combination of wisdom and insight to realize that most soft tissues would need special preparative methods to bring them under study by this instrument. He resurrected the critical-point drying method, discovered originally by Thomas Anderson in the 1950s, and applied it both to tissue culture cells and to cells freed from organs and tissues by chemical and enzymatic treatment [Porter et al, 1972; Vial and Porter, 1975]. His study on cell-surface changes during the cell cycle with Prescott and Fry [Porter et al, 1973a] has become a classic. Porter also used the scanning electron microscope to show dramatic cell surface changes during transformation to malignancy [Porter et al, 1973b].

Another parallel story can be told with respect to the high-voltage electron microscope. Again these instruments had been available, in a limited way to be sure, for several years, but had not been applied extensively to biological studies. Keith Porter was instrumental in convincing the biological community and funding agencies that high-voltage electron microscopes should be explored more seriously in biology. He became the director of one of the first two NIH-supported high-voltage electron microscope facilities set up for biological research. Under his direction, this facility has clearly demonstrated that the hoped-for usefulness of the instrument is real, and that a new tool for determining the three-dimensional aspects of cell structure is now available [eg, Wolosewick and Porter, 1975, 1976, 1977a, b, 1979; Ellisman and Porter, 1980; Schliwa et al, 1981].

Porter's laboratory contributed significantly to our early understanding of the ultrastructure of tumor cells. In the late 1960s and early 1970s, Porter was among the first to recognize that cell transformation to malignancy involved major loss or rearrangement of structural organization of the cytoplasm. Specifically, he and Theodore Puck discovered that morphologically transformed cells could be induced to undergo "reverse transformation" when treated by $3'5'$-dibutyryl cyclic AMP [Porter et al, 1974]. Malignant Chinese hamster ovary cells with rounded morphology could be induced to take on "normal" fibroblastic shapes and growth characteristics in culture when exposed to cAMP. Moreover, such "reverse transformation" required the assembly and arrangement of microtubules. Porter and Ted Puck were the first to propose that malignant transformation involved major changes in the cytoskeleton. Moreover, since the shape of the transformed phenotype could be influenced by cAMP, it was concluded that a protein kinase might somehow be involved in the transformation, a theory that continues to be supported in current research. To a large extent, Porter's early interest in microtubules and other formed elements of the cytoplasm laid the foundation for a major thrust among cell biologists of the 1970s and 1980s to chart the remaining structures of the cytoskeleton and to define more precisely the structural and functional organization of the cytoplasm.

A new chapter in the cell biology career of Keith Porter is unfolding at the present time. Once again, Porter has seen what no one else had noticed. Since the thin-sectioning approach was first developed, attention has been focused on the formed organelles in the cytoplasm, especially those delimited by membranes. What lies between these has been referred to as the cytoplasmic matrix, or ground substance, and it has been thought to have little structure and to be of even less interest. Porter has recently studied the cytoplasmic matrix by high-voltage electron microscopy, and found it to be rich in structure [Wolosewick and Porter, 1975, 1976, 1979; Ellisman and Porter, 1980]. He refers to this structure as the microtrabecular lattice, and points out that it forms specific associations with most of the organelles in the cell. He suggests that the microtrabeculae play an important role in control of the positions and movements of organelles in the cell, including those seen during cell differentiation. At the moment, these ideas are properly classified as speculation. If history repeats itself, however, Porter will turn out to be correct, and another new and important domain of cell biology will have been opened.

Although Keith Porter's direct contributions to science through his own research are extensive, diverse, and important, he also has made another significant and lasting contribution through the people he has trained or

otherwise influenced. His former students are found throughout the world, and many hold important positions at leading universities and medical centers. At least 30 postdoctoral associates have received training with Porter. His style and curiosity have been imprinted permanently on their careers. Through his publications, he has set standards and patterns that uncountable workers have followed. Not to be neglected in this regard are the atlases of fine structure he has coauthored and which have introduced the wonders of cell structure to literally thousands of students [Porter and Bonneville, 1968; Ledbetter and Porter, 1970; Motta et al, 1977].

Porter was instrumental in, and often the main driving force behind, the founding of the Tissue Culture Society, the American Society for Cell Bi-

---

Fig. 9.   (1952) First effective model of microtome. The chuck for holding embedded tissue is at left (asterisk). It is continuous with a horizontal bar which is held in a gimble at arrow. The heat from the reading lamp expanded the bar and provided gradual advancement of specimen to knife.

Fig. 10.   (1953) Micrograph of cross section of cilia showing early image of $9 + 2$. (With Fawcett.)

Fig. 11.   (1947) Margin of cultured cell showing characteristic image of ER following 12-hour fixation with OsO_4.

Fig. 12.   (1951) Chick fibroblast engaged in synthesis and disposition of collagen. Numerous fine collagen fibrils extend across the surface of the cell. (With Pappas.)

Fig. 13.   (1971) Scanning image of a small population of rat sarcoma (4337) cells. The SEM provides an impressive image of numerous lamellopodia and microvilli, which cover surfaces of these cells.

Fig. 14.   (1971) After exposure to dbcAMP, the sarcoma cell adopts a spindle shape with the assembly of numerous microtubules oriented parallel to the long axis of these cells.

Fig. 15.   (1964) Micrograph of section through cortex of early mosquito oocyte showing clathrin-coated pits loaded with yolk protein. (With Roth.)

Fig. 16.   (1979) Thin section showing small area of human fibroblast (WI-38) including three elements of cytoskeleton: microfilaments (Mf), intermediate filaments (If), and microtubules (Mt). (With Wolosewick.)

Fig. 17.   (1981) Micrograph depicting small area of cytomatrix in cultured BSC cell (African green monkey epithelial). Besides microtubules (Mt), the image includes numerous microtrabeculae comprising the microtrabecular lattice (MTL). Whole cell was dried by critical-point method.

Fig. 18.   (1982) Part of Holocentrus erythrophore showing pigment dispersed.

Fig. 19.   (1982) Micrograph depicting pigment granules in relation to microtubules and MTL.

ology, the Journal of Cell Biology, and, most recently, the International Federation for Cell Biology. In addition, he has expended considerable energy and affection on the Marine Biological Laboratory in Woods Hole, helping it to make difficult transitions into the realities of modern times.

One could, perhaps, characterize Keith Porter as a scientist who had a habit of being in the right place at the right time, and who knew what to do when he was there. The implication of an element of luck, however, is not consistent with the large number and great diversity of the contributions he has made. Alternatively one might characterize his career as one of opening a new area, staying for a time, and then moving on. Again there is an incorrect implication, this time one of superficiality, that is not truly characteristic of Porter's science, though most certainly Porter's career has had more than an average amount of variety. Probably the best characterization would be to extract positive aspects from both of the previous statements and to say that Porter has shown an unusual ability to recognize opportunities for advancement in knowledge and to set a pattern in a new area. He then finds himself drawn away to a new challenge. This is part of the style of the man. We are happy to have this opportunity to honor the scientific career of Keith R. Porter and to acknowledge the magnitude, breadth, significance, and variety of his contributions to cell biology.

REFERENCES

Ashford TP, Porter KR (1962): Cytoplasmic components in hepatic cell lysosomes. J Cell Biol 12:198–202.
Ashley CA, Porter KR, Philpott DE, Hass GM (1951): Observations by electron microscopy on contraction of skeletal myofibrils induced with adenosinetriphosphate. J Exp Med 49:9–20.
Bennett HS, Porter KR (1953): An electron microscope study of sectioned breast muscle of the domestic fowl. Am J Anat 93:61–105.
Bikle D, Tilney LG, Porter KR (1966): Microtubules and pigment migration in the melanophores of Fundulus heteroclitus L. Protoplasma 61:322–345.
Byers B, Porter KR (1964): Oriented microtubules in elongating cells of the developing lens rudiment after induction. Proc Natl Acad Sci USA 52:1019–1099.
Ellisman MH, Porter KR (1980): Microtrabecular structure of the axoplasmic matrix: Visualization of cross-linking structures and their distribution. J Cell Biol 87:464–479.
Fawcett DW, Porter KR (1954): A study of the fine structure of ciliated epithelia. J Morphol 94:221–281.
Franzini-Armstrong C, Porter KR (1964): Sarcolemmal invaginations constituting the T system in fish muscle fibers. J Cell Biol 22:675–696.
Gibbins JR, Tilney LG, Porter KR (1966): Microtubules in primary mesenchyme cells of sea urchin embryos. Anat Rec 154:347a (abstract).
Gibbins JR, Tilney LG, Porter KR (1969): Microtubules in the formation and development of the primary mesenchyme in Arbacia punctulata. I. The distribution of microtubules. J Cell Biol 42:201–226.
Godman G, Porter KR (1960): Chondrogenesis, studied with the electron microscope. J Biophys Biochem Cytol 8:719–760.

Hawn CVZ, Porter KR (1947): The culture of tissue cells in clots formed from purified bovine fibrinogen and thrombin. Proc Soc Exp Biol Med 65:309–314.

Junqueira L, Porter KR (1969): Pigment migration in Fundulus melanophores. Biophys J 9:FAM-C5 (abstract).

Ledbetter MC, Porter KR (1963): A "microtubule" in plant cell fine structure. J Cell Biol 19:239–250.

Ledbetter MC, Porter KR (1964): The fine structure of microtubules in plant cells. Science 144:872–874.

Ledbetter JC, Porter KR (1970): "Introduction to the Fine Structure of Plant Cells." Berlin: Springer-Verlag, 188 pp

McIntosh JR, Porter KR (1967): Microtubules in the spermatids of the domestic fowl. J Cell Biol 35:153–173.

Motta P, Andrews PM, Porter KR (1977): "Microanatomy of Cell and Tissue Surfaces: An Atlas of Scanning Electron Microscopy." Philadelphia: Lea and Febiger.

Palade GE, Porter KR (1954): Studies on the endoplasmic reticulum. I. Its identification in cells in situ. J Exp Med 100:641–656.

Peachey LD, Porter KR (1959): Intracellular impulse conduction in muscle cells. Science 129:721–722.

Porter KR (1955): The fine structure of cells. Fed Proc 14:673–682.

Porter KR (1956): The sarcoplasmic reticulum in muscle cells of Amblystoma larvae. J Biophys Biochem Cytol 2:163–170.

Porter KR (1966): Cytoplasmic microtubules and their functions. In Wolstenholme GEW, O'Connor M (eds): "Principles of Biomolecular Organization." London: Churchill, pp 308–345.

Porter KR, Blum J (1953): A study in microtomy for electron microscopy. Anat Rec 117:685–710.

Porter KR, Bonneville MA (1968): "An Introduction to the Fine Structure of Cells and Tissues," Ed 3. Philadelphia: Lea and Febiger.

Porter KR, Bruni C (1959):An electron microscope study of the early effects of 3'-Me-DAB on rat liver cells. Cancer Res 19:997–1010.

Porter KR, Kallman FL (1952): Significance of cell particulates as seen by electron microscopy. Ann NY Acad Sci 54:882–891.

Porter KR, Machado RD (1960a): The endoplasmic reticulum and the formation of plant cell walls. In "Proceedings of the European Regional Conference on Electron Microscopy, Delft 1960." Delft: Nederlandse Vereniging voor Electronenmicroscopie, pp 754–758.

Porter KR, Machado RD (1960b): Studies on the endoplasmic reticulum. IV. Its form and distribution in striated muscle cells. J Biophys Biochem Cytol 3:269–300.

Porter KR, Palade GE (1957): Studies on the endoplasmic reticulum. III. Its form and distribution in striated mucle cells. J Biophys Biochem Cytol 3:269–300.

Porter KR, Pappas GD (1959): Collagen formation by fibroblasts of the chick embryodermis. J Biophys Biochem Cytol 5:153–166.

Porter KR, Thompson HP (1948): A particulate body associated with epithelial cells cultured from mammary carcinomas of mice of a milk-factor strain. J Exp Med 88:15–24.

Porter KR, Claude A, Fullam EF (1945): A study of tissue culture cells by electron microscopy J Exp Med 81:233–246.

Porter KR, Prescott D, Frye J (1973a): Changes in surface morphology of Chinese hamster ovary cells during the cell cycle. J Cell Biol 57:815–836.

Porter KR, Todaro GJ, Fonte VG (1973b): A scanning electron microscope study of surface features of viral and spontaneous transformations of mouse Balb/3T3 cells. J Cell Biol 59:633–642.

Porter KR, Puck TT, Hsie AW, Kelley D (1974): An electron microscope study of the effects of dibutyryl cyclic AMP on Chinese hamster ovary cells. Cell 2:145–162.

Porter KR, Kelley D, Andrews PM (1972): The preparation of cultured cells and soft tissues for scanning electron microscopy. In "Proceedings of the Fifth Annual Steroscan Colloquium," pp 1–19.

Roth TF, Porter KL (1964): Yolk protein uptake in the oocyte of the mosquito Aedes aegypti L. J Cell Biol 20:313–332.

Schliwa M, Weber K, Porter KR (1981): Localization and organization of actin in melanophores. J Cell Biol 89:267–275.

Tilney LG, Porter KR (1965): Studies on microtubules in Heliozoa. I. The fine structure of Actinosphaerium nucleofilum (Barrett), with particular reference to the axial rod structure. Protoplasma 60:317–344.

Tilney LG, Porter KR (1967): Studies on the microtubules in heliozoa. II. The effect of low temperature on these structures in the formation and maintenance of the axopodia. J Cell Biol 34:327–343.

Vial J, Porter KR (1975): Scanning microscopy of dissociated tissue cells. J Cell Biol 67:345–360.

Wolosewick JJ, Porter KR (1975): High voltage electron microscopy of WI-38 cells. Anat Rec 181:511–512.

Wolosewick JJ, Porter KR (1976): Stereo high-voltage electron microscopy of whole cells of the human diploid line, WI-38. Am J Anat 147:303–324.

Wolosewick JJ, Porter KR (1977a): Observations on the morphological heterogeneity of WI-38 cells. Am J Anat 197–226.

Wolosewick JJ, Porter KR (1977b): Effect of low temperature on the ground substance of cultured cells. J Cell Biol 75:275a (abstract).

Wolosewick JJ, Porter KR (1979): Microtrabecular lattice of the cytoplasmic ground substance: Artifact or reality. J Cell Biol 82:114–139.

Organization of Cell Membranes and the Compartmentalization of Cytoplasm

Modern Cell Biology 2:15–72
© 1983 Alan R. Liss, Inc., 150 Fifth Avenue, New York, NY 10011

Memory and Membranes: The Expression of Genetic and Spatial Memory During the Assembly of Organelle Macrocompartments

Robert O. Poyton

From the Department of Molecular, Cellular and Developmental Biology,
University of Colorado, Boulder, Colorado 80309

I. INTRODUCTION

One of the principal findings of eucaryotic cell biology to date is that cellular constituents and metabolic pathways are not randomly dispersed in cells but, rather, are localized to specific sites within the three-dimensional fabric of the cellular interior. Studies with a variety of different eucaryotic cell types have revealed that nearly all eucaryotic cells, irrespective of their size, genomic complexity, and morphology are characterized by a "cytoplasm" which is partitioned by membrane-bounded organelles. It is indeed striking that "simple" unicellular lower eucaryotes (eg, yeast) whose genomic content exceeds that of bacteria by only a few-fold have the same set of organelles that are characteristic of "complex" multicellular higher eucaryotes whose genomic content may exceed that of bacteria some ten to twenty thousand fold [Holliday, 1970; Carlile, 1980]. The presence of membrane-bounded organelles in cells from both ends of this spectrum suggests that intracellular membrane systems are indispensible for the evolutionary success of the eucaryotic cell type and raises many questions regarding the origin, function, and assembly of intracellular membrane-bounded compartments [Frederick, 1981].

The early efforts of Porter and Palade to recognize and assign functions to intracellular membrane systems [Porter et al, 1945; Hogeboom et al, 1948; Palade and Porter, 1954; Palade, 1955; Porter, 1956; Palade and Siekevitz, 1956], together with the later efforts of many cell biologists and biochemists who followed their lead, have made it clear that membrane-bounded organelles effectively serve to compartmentalize both cellular metabolism [Hess, 1980] and gene expression [Parthier, 1980; Nover, 1980]. From these studies it is also clear that the spatial organization of metabolic pathways is nearly (but not completely) invariant from one eucaryotic cell type to another and that the enzymic composition of cellular organelles and their membranes is constant from one generation to the next.

This latter property of organelles, when considered with the fact that membranes do not form de novo [Palade, 1978; Sitte, 1980] has led to the proposal [Palade, 1978] that cell membranes and their constituent proteins have *temporal continuity* and, as such, may account for at least part of the epigenetic spatial memory which is present in cells. How does this spatial memory interface with the genetic memory encoded by the genome? One promising approach to this question is through an analysis of the early steps in organelle biogenesis, when newly synthesized constituents are sequestered by the organellar membrane or compartment in which they are to reside. This review addresses this aspect of the spatial organization of eucaryotic cells. After a brief consideration of the principles which underlie the physiological compartmentation of cells, it will focus on the biogenesis of organelles from the perspective of their oligomeric membrane proteins. It will then present a new hypothesis which attempts to provide a molecular basis for memory in membranes.

II. PRINCIPLES OF PHYSIOLOGICAL COMPARTMENTATION

A. Physiological Compartmentation Per Se Does Not Require the Presence of Membranes

It has been recognized for some time that cellular metabolism is compartmentalized at all levels of cellular organization [for reviews see Srere and Mosbach, 1974; Srere and Estabrook, 1978; Nover et al, 1980]. The smallest known metabolically significant compartments make up the active sites of enzymes. These compartments are in the Ångstrom size range and serve to concentrate and sequester substrates and cofactors. At their simplest these compartments are formed by the tertiary structure of a single (ie, monomeric) polypeptide chain. These compartments probably were among the first to evolve [Haldane, 1967]; in their absence, enzymic catalysis and life itself would have been impossible. The next highest level of compartmentation is found amongst oligomeric and multifunctional enzymes. Homo-oligomeric proteins, composed of identical subunits, may have evolved by gene duplication, whereas hetero-oligomeric proteins, composed of nonidentical subunits, are thought to have evolved either by a combination of gene duplication and mutation or by the evolution of catalytic and regulatory polypeptides with sterically compatible surfaces [Gutfreund, 1981]. The aggregate states of both homo- and hetero-oligomeric proteins probably evolved initially because they were energetically more stable than the monomeric state [Schulz and Schirmer, 1979]. However, as interactions developed across protomeric faces, the aggregate state offered the additional advantage of cooperative interactions between active sites, or between regulatory and catalytic subunits. These interactions form the basis for the allosteric regulation of key steps in cellular metabolism and for the functional integration of different metabolic pathways [Monod et al, 1963]. Unlike oligomeric enzymes, which catalyze a single reaction, multifunctional enzymes or multienzyme complexes catalyze two or more steps in a metabolic sequence [Reed, 1974]. As a result of their multiple active sites these proteins serve to concentrate and sequester not only substrates and cofactors but also reaction products of one reaction to be used as substrate for the next. This sequestration of intermediates in a metabolic pathway is referred to as "metabolic channeling." Its role in the control of biosynthetic pathways has been reviewed recently by Stebbing [1980].

The next highest level of compartmentation is found in supramolecular structures such as ribosomes and membranes. Like multifunctional enzymes, ribosomes represent a series of catalysts organized in an aggregate which permits the sequestration of substrates (ie, amino acids), cofactors (ie, ATP and GTP), and reaction products (ie, elongating polypeptide chains) away from the bulk phase of the cytosol. Other examples of compartmentation at the supramolecular level can be found amongst metabolic chains which lie along and/or across biological membranes. For example, the mitochondrial electron transport chain consists of four major oligomeric protein complexes which are composed of more than 25 different polypeptides, cofactors, and a large number of electrons carrying prosthetic groups (Fig. 1). These components act in series to catalyze the transfer of electrons from NADH or succinate to O_2. Once electrons enter this chain they are effectively sequestered away from the aqueous environment on either side of the inner mitochondrial membrane until they emerge from cytochrome c oxidase and combine with protons and oxygen to form water (Fig. 1).

At present, it is not clear if "soluble" enzymes or the multienzyme complexes which constitute metabolic pathways in the cytosol or organellar soluble phases are organized in vivo into aggregates which are larger than those that can be isolated by commonly used cell and biochemical fractionation procedures. Such aggregates have been reported for the enzymes of glycolysis [Mowbray and Moses, 1976; Fossel and Solomon, 1978] and the Kreb's cycle [Srere and Henslee, 1980]. They are presumably stabilized by weak and transient protein-protein interactions. If the enzymes of "soluble" metabolic pathways do exist as aggregates, possibly as part of the microtrabecular lattice [Wolosewick and Porter, 1976; Porter, this volume], they will represent yet another example of compartmentation at the supramolecular level.

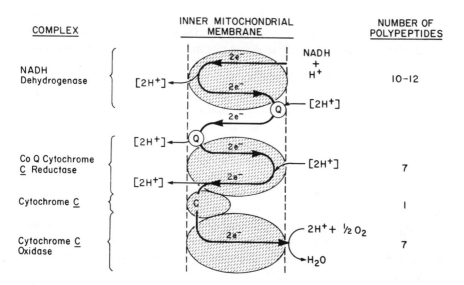

Fig. 1. The mitochondrial electron transport chain as a supramolecular compartment. The mitochondrial electron transport chain is composed of four enzyme complexes (NADH dehydrogenase, coenzyme Q cytochrome c reductase, cytochrome c oxidase, and succinate dehydrogenase—not shown), cytochrome c, and the ubiquinone derivative coenzyme Q. Together these components consist of at least 25 different polypeptides, five different kinds of cytochromes (cytochromes b, c, c_1, a, and a_3), and a number of different noncytochrome electron carrying prosthetic groups. Once electrons enter this chain from NADH or succinate they are effectively sequestered away from the aqueous environment on either side of the membrane until they emerge from cytochrome c oxidase and combine with protons and oxygen to form water.

All of the above described compartments represent *microcompartments* in the hierarchy described recently by Hess [1980]. They are easily distinguished from membrane-bounded organellar *macrocompartments* by their size, dynamic properties, and ability to sequester water-soluble solutes without making use of the permeability barrier imposed by the lipid bilayer of a membrane. Macrocompartments are themselves microcompartmentalized. A particularly striking example of this is the presence of the enzymes of glycolysis and gluconeogenesis, converse biochemical pathways, in the same cellular compartment. Both of these pathways are operative in the cytosol, and both make use of the same enzymes and reaction intermediates. Their physiological separation is brought about not by a membrane partition but by the specificities of enzymes in each pathway for different anomeric forms of the sugars involved [Koerner et al, 1977], by substrate availability [Hess, 1973], and by metabolic channeling [Hess, 1973]. This example serves to illustrate that membrane-bounded compartments are not prerequisites for the compartmentation of metabolism. Why, then, do membrane bounded compartments exist?

B. Advantages of Membrane-Bounded Compartments

Like microcompartments, membrane-bounded macrocompartments also sequester and concentrate reaction substrates, intermediates, and reaction products; they are therefore capable of preventing two metabolic pathways which may use the same substrates and cofactors from competing with one another. Membrane-bounded macrocompartments also have the ability to store potentially harmful substances, such as acids (in lysosomes and vacuoles) and hydrogen peroxide (in peroxisomes). However, the functional importance of membrane-bounded macrocompartments goes well beyond their ability to sequester and concentrate and lies in their ability 1) to contain *complete* metabolic pathways, and 2) to generate and maintain transmembranous concentration gradients as a source of chemical potential energy to drive reactions.

Haldane [1967] has proposed that "the critical event that may best be called the origin of life was the enclosure of several different self-reproducing polymers within a semi-permeable membrane." This limiting membrane not only would have provided for the *containment* of cellular constituents but also would have led to an increase in the concentration of reaction intermediates and end products available to the entrapped enzymes. It has been proposed recently [Stebbing, 1980] that the confinement of reaction intermediates, especially those that are labile, to the vicinity in which they are to be used not only allows for their more efficient utilization but also permits them to enter new routes of metabolism. Stebbing [1980] has argued that the evolution of the metabolic pathways that now exist was favored by two events in the evolution of a cell: 1) the development of a limiting cellular membrane; and 2) the recruitment of enzymes that had a high level of substrate ambiguity [Jensen, 1976]. The confinement of substrates and reaction products by a diffusion limited membrane-bounded macrocompartment is, of course, as important for modern-day cells as it was during the evolution of their progenitors. Indeed, the behavior of any reversible enzymatic reaction depends on the concentration of its substrates, end products, and cofactors. In principle, the rate of any enzymatic reaction may be controlled by varying the aqueous volume of the membrane-bounded compartment in which it resides.

Perhaps the most important role of membranes, and their constituent proteins, in cell metabolism lies in their ability to generate and maintain transmembranous concentration gradients [Wilson and Lin, 1980]. All transmembranous gradients are ultimately driven by energy made available by the oxidation of metabolized substrates or by light energy captured during photosynthesis. Some membranes (ie, mitochondrial inner membrane, chloroplast thylakoid membrane, and bacterial cytoplasmic membrane) are capable of capturing and conserving this energy directly by virtue of their possession of electron transport chains which generate chemiosmotic gradients that can be used to generate ATP or drive transport processes. Other membranes are incapable of capturing this energy directly, but instead make use of ATP hydrolysis to produce gradients. In either case, the gradients produced play key roles in regulating metabolism and in integrating the metabolic activities of one compartment with those of another. Some of the more obvious roles played by these gradients are 1) the efficient conservation of energy made available during respiration or photosynthesis [Nicholls, 1982]; 2) the regulation of macrocompartment volumes [Wilson and Lin, 1980]; 3) the coupling of interrelated metabolic pathways, such as glycolysis and respiration [Erecinska and Wilson, 1978]; and 4) the integration of intracellular energy requirements with the uptake of metabolizable substrates [Elbrink and Bihler, 1975].

C. Membrane-Bounded Compartments Communicate Via Oligomeric Membrane Proteins

Membrane-bounded compartments are functionally interconnected by the permeases, pumps, and other transport proteins which span their membranes and permit the selective flow on ions, electrons, protons, metabolites, and other solutes across them. It is now clear [Klingenberg, 1981; Hobbs and Albers, 1980] that these integral membrane proteins are, nearly without exception, oligomeric and exist in situ either as homo-oligomers or hetero-oligomers (Table I). These proteins are quite diverse in structure and in size. They range from relatively small dimers, composed of identical subunits, to relatively large hetero-oligomers, composed of many nonidentical subunits. Recent studies have revealed that both types of oligomers may interact with peripheral proteins and/or cytosolic proteins in a number of different ways (Table II). The most intimate associations between integral and peripheral polypeptides are those that take place in enzymes where peripheral and integral polypeptides are subunits which play catalytic or regulatory roles in function. These associations are mediated by electrostatic and hydrophobic forces and are strong enough to withstand the disruptive conditions used during membrane fractionation. All other associations between integral and peripheral proteins are transient or weak. They are generally unable to withstand the forces used during standard cell fractionation procedures. It seems clear that even homo-oligomeric integral proteins may exist as hetero-oligomers when associated, however transiently, with peripheral or cytosolic proteins in situ. For example, although the band 3 protein of the erythrocyte

TABLE I. Representative Examples of Transmembrane Protein Oligomers

Protein	Source	Type	Molecular weight Subunit	Molecular weight Oligomer	Reference
Homo-oligomers					
ADP/ATP carrier	Mitochondrial inner membrane	Dimer (α_2)	30,000	60,000	1
Anion transporter (band III)	Erythrocyte plasma membrane	Dimer (α_2)	90,000	180,000	2
Bacteriorhodopsin	Halobacterium cytoplasmic membrane	Trimer (α_3)	26,000	78,000	3–5
Ca^{++}-ATPase	Rabbit sarcoplasmic reticulum membrane	Dimer (α_2) (or)	100,000	200,000	6
		Tetramer (α_4)	100,000	400,000	
M_{13} (f_1) phage coat protein	Bacteriophage M_{13} (f_1) and bacterial host membrane	Polymer* $(\alpha_{2,500})$	5,000	1.25×10^7	7
Hetero-oligomers					
$(Na^+ + K^+$ ATPase	Mammalian plasma membrane	$\left(\begin{smallmatrix}\alpha\\\beta_2\end{smallmatrix}\right)$ 1 or 2	95,000 55,000	150,000	6, 8
Cytochrome c oxidase	Yeast and mammalian inner mitochondrial membrane	Polymer I_1 II_1	55,000 31,000		
		III_1 IV_1 V_1 VI_1 VII_1	30,000 15,000 13,000 10,000 5,000	159,000	9, 10
Oligomycin sensitive ATPase	Yeast inner mitochondrial membrane	Polymer 1_3 2_3 3_1 4_2 5_1 6_4 7_1 8_2 9_4	56,000 52,000 32,000 25,000 24,000 22,000 14,000 10,000 8,000	584,000	11, 12

*The polymer number given is that for the number of subunits in an assembled phage particle. The polymer numbers for subunits in the host membrane is not known.

References: 1, Hacklenberg and Klingenberg [1980]; 2, Yu and Steck [1975]; 3, Henderson and Unwin [1975]; 4, Michel et al [1980]; 5, Engelman et al [1980]; 6, Hobbs and Albers [1980]; 7, Wickner et al [1980]; 8, Kyte [1975]; 9, Wikstrom et al [1981]; 10, George-Nascimento and Poyton [1981]; 11, Todd and Douglas [1981]; 12, Todd et al [1980].

TABLE II. Examples of Associations Peripheral Polypeptides Can Make With Integral Membrane Polypeptides

Type of association	Peripheral polypeptide	Integral polypeptide	Reference
Subunit	F_1 ATPase subunits	Oligomycin-sensitive ATPase (CF_0 subunits)	1
	Subunits IV and VI	Yeast cytochrome c oxidase	2
Substrate donor	Cytochrome c	Cytochrome c oxidase	3
	Creatine kinase	Na^+/K^+ ATPase	4
Structure stabilizer	Ankyrin, spectrin	Erythrocyte anion transporter (band 3)	5, 6

References: 1, Futai and Kanazama [1980]; 2, George-Nascimento and Poyton [1981]; 3, Wikstrom et al [1981]; 4, Saks et al [1977]; 5, Branton et al [1981]; 6, Sakaki et al [1982].

membrane is active in anion transport as a dimeric homo-oligomer, it may take on different properties as a hetero-oligomer when it is associated with cytosolic hemoglobin [Salhany and Shaklai, 1979; Shaklai and Abrahami, 1980], glycolytic enzymes [Strapazon and Steck, 1977; Karadsheh and Uyeda, 1977; Salhany and Gaines, 1981], or cytoskeletal proteins [Branton et al, 1981].

The properties of the oligomeric state and the advantages it poses for transmembranous integral proteins has been considered elsewhere [Klingenberg, 1981]. Briefly stated, the oligomeric state of integral proteins may be expected 1) to increase the stability of hydrophobic proteins by maximizing hydrophilic interactions across subunit faces and hydrophobic interactions with the fatty acyl chains of phospholipids; 2) to facilitate the formation of aqueous pores or channels through the lipid bilayer; and 3) to allow for both negative and positive cooperativity vis-à-vis protein function. This latter allosteric property may be especially important for integrating the transport functions of an integral protein with the metabolic activities of the compartments it interconnects.

III. THE ASSEMBLY OF MEMBRANE-BOUNDED MACROCOMPARTMENTS: GENERAL CONSIDERATIONS

Since cellular membranes do not form de novo [see reviews by Palade, 1978; Sitte, 1980] and since the continued maintenance of transmembranous gradients is essential for cellular viability [Sitte, 1980] it follows that the biogenesis of intracellular membranes must involve the addition of *new* constituents to preexisting compartments under conditions where the functional integrity of their membranes remains intact. Before considering how membranes are assembled it is therefore necessary to consider some aspects of the spatial organization of integral proteins in the plane of a membrane and the organization of membranes within a cell.

A. Spatial Organization of Integral Proteins in the Plane of the Membrane

Although the "fluid mosaic" model of Singer and Nicolson [1972] proposes that integral proteins and phospholipids are free to diffuse in the plane of the membrane, and although there is ample evidence to support this concept [see reviews by Cherry, 1979; Jardetsky, 1982], it is becoming apparent that the lateral mobility of integral membrane proteins in situ may in fact be restricted [Edidin, 1982] by a variety of factors (Fig. 2). Among these are crystalline or nonbilayer phospholipid microdomains present in the plane of the membrane [Karnovsky et al, 1982; Freire and Snyder, 1982; de Kruijff et al, 1982] and the associations which integral proteins make with peripheral, cytoskeletal, or soluble proteins [Mooseker and Tilney, 1975; Branton et al, 1982]. In addition, the mobility of an integral protein may be limited by collisional interactions with other membrane proteins [Edidin, 1982] in a way that permits a "random walk" on only a rather limited region of membrane surface. Each of these restrictions on the movement of membrane proteins may limit, in a statistical fashion, an individual membrane protein to a circumscribed region on the membrane surface. Thus, although highly dynamic, membranes may be composed of "unit areas" of structure which are statistically defined. This concept is important from the viewpoint of membrane assembly since it raises the possibility that not all molecules of a particular protein in a membrane have equal access to all other proteins and that the insertion of new proteins into a membrane may not be random but may be determined instead by the availability of an *open* "unit area" with unoccupied peripheral or integral protein receptors. We will return to this concept in sections VI A and VII.

B. Organellar Membrane Families

Organellar membranes may be subdivided into two groups based on their ability to recognize and fuse with one another. In one group are the organelles of the secretory pathway (nucleus, endoplasmic reticulum, Golgi apparatus, condensing vacuoles, plasma membrane), lysosomes, and peroxisomes. Membranes of these organelles are physically indiscrete insofar as they are either physically continuous (eg, nuclear membrane, rough endoplasmic reticulum membrane, and smooth endoplasmic reticulum membrane) or combine with one another via vesicle intermediates. Their lipid bilayers are both physically and temporally continuous [Palade, 1978] and therefore provide proteins and phospholipids with potential pathways for flowing from one organelle to another [Morré, 1980]. Of course, this process of "membrane

A. Random collisions

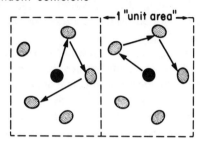

B. Crystalline phospholipid microdomains

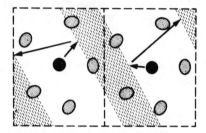

C. Associations with peripheral proteins

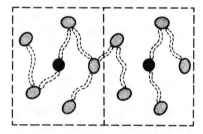

Fig. 2. Factors that may restrict the movement of integral membrane proteins. The mobility of integral membrane proteins may be restricted to small regions on the membrane surface by A) random collisions with other integral proteins [Edidin, 1982]; B) crystalline phospholipid microdomains [Karnovsky et al, 1982] or phospholipids in hexagonal phase [De Kruijff et al, 1982]; and C) associations with cytoskeletal and peripheral proteins [Branton et al, 1981]. Each of these restrictions may limit the mobility of any integral protein to a "random walk" over a limited region of membrane surface and may therefore lead to a loosely organized liquid crystalline array of membrane proteins. We have termed the repeating unit of this array a "unit area," and view it as a region that can be defined statistically by its time averaged composition. It is not meant to imply that integral membrane proteins are fixed absolutely in a particular position along the plane of the membrane.

flow" does not go unchecked. Indeed, each organelle in this group has a characteristic protein and lipid composition. Thus, it would appear that both proteins and lipids are "sorted" as they traverse the secretory pathway [Palade, 1978; Sabatini et al, 1982].

In the other group of organelles are mitochondria and chloroplasts. The inner membranes of these organelles are discrete in that they do not fuse with other cellular membranes. It is not clear at present if the outer membranes of mitochondria and chloroplasts fuse with other cellular membranes. It is also not clear to what extent the inner membrane of one mitochondrion (or chloroplast) fuses with the inner membrane of another mitochondrion (or chloroplast) in the same cell. However, the high level of genetic recombination between mitochondrial genome populations in yeast cells during somatic growth [Birky, 1978] suggests that fusion of mitochondrial inner membranes, and hence the mixing of DNA containing matrix compartments, may be a frequent event.

C. Sites of Synthesis of Membrane Phospholipids

The major intracellular site of fatty-acid and phospholipid synthesis is the endoplasmic reticulum [Bloch and Vance, 1977; Bell and Coleman, 1980]. To some extent mitochondria and, to a greater extent, chloroplasts are capable of synthesizing some of their phospholipids and galactolipid constituents [Bell and Coleman, 1980; Stumpf, 1981; Douce and Joyard, 1981]. From the viewpoint of membrane biogenesis, the growth of the lipid bilayer in the family of organelles from the secretory pathway can be thought of as a two step process involving 1) the insertion of new phospholipid molecules into the lipid bilayer of the endoplasmic reticulum itself, and 2) the diffusion and distribution of phospholipids by membrane flow. The growth of the lipid bilayers in mitochondria and chloroplasts is likely to be more complex, involving 1) the syntheses of new phospholipids and their constituents at both the endoplasmic reticulum and the organellar surface, 2) the delivery of new phospholipids synthesized at the endoplasmic reticulum to the mitochondrial or chloroplast surface, and 3) the insertion of these phospholipids into either the organellar inner or outer membranes.

D. Sites of Synthesis of Organelle and Membrane Proteins

There are four well-documented ribosomal sites for protein synthesis in eucaryotic cells (Fig. 3). At two of these, ribosomes translate nuclear-encoded messenger RNAs and reside either in the cytosol as "free" polysomes or at the surface of the endoplasmic reticulum as bound polysomes. The "free" polysomes synthesize soluble cytosolic proteins as well as proteins destined for the membranous and soluble phases of mitochondria and chloroplasts and the lumen of peroxisomes (see below). Although "free" from membranes,

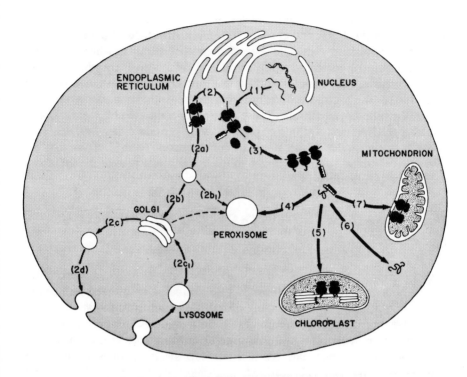

Fig. 3. Nuclear gene products follow multiple pathways to cellular organelles. The flow of genetic information guiding proteins from their genes in the nuclear genome to their positions in organelle compartments may be represented as a series of decisions. These decisions are 1) to transcribe and process a messenger RNA (possibility 1) for translation on soluble or endoplasmic reticulum bound ribosomes; 2) to be inserted into the endoplasmic reticulum membrane *cotranslationally* (possibility 2); or 3) to be elongated and terminated on soluble ribosomes (possibility 3). Once secretory pathway organelle proteins have been inserted into the endoplasmic reticulum they must decide to remain in the endoplasmic reticulum (possibility 2a) or be "sorted" to the Golgi apparatus (possibility 2b) or peroxisomal membrane (possibility $2b_1$); and to pass to the plasma membrane (possibilities 2c and 2d) or lysosome (possibility $2c_1$). In addition, proteins that are translated on soluble ribosomes must decide whether 4) to enter the peroxisomal lumen (possibility 4); 5) to enter the chloroplast (possibility 5); 6) to stay "soluble" (possibility 6); or 7) to enter the mitochondrion (possibility 7).

these ribosomes may reside at specific positions in a cell, fixed by their associations with elements of the microtrabecular lattice [see Penman, this volume]. Polysomes bound to the endoplasmic reticulum synthesize either secretory proteins or membrane proteins which are to reside in one of the secretory pathway organelles (see below). A third ribosomal site for protein synthesis lies inside mitochondria. These polysomes translate mRNA encoded by mitochondrial DNA and are bound to the matrix face of the inner mitochondrial membrane; they synthesize integral polypeptides which find their way into three oligomeric protein complexes of the inner membrane or, in some organisms, the small subunit of the mitochondrial ribosome [Poyton et al, 1982]. The fourth ribosomal site of protein synthesis in eucaryotic cells lies in the chloroplast stroma. These ribosomes translate chloroplast DNA-encoded mRNAs and synthesize proteins which reside either in the stroma itself or in the thylakoid membrane [Von Wettstein, 1981]. They may be either "free" in the stroma or bound to the cytoplasmic face of the thylakoid membrane [Chua et al, 1976].

IV. PATHWAYS OF PROTEIN TRANSPORT TO ORGANELLAR MEMBRANES

A. Nuclear Gene Products

Nuclear-encoded proteins find their way into virtually every known cellular macrocompartment. The synthesis of these proteins is initiated on soluble cytosolic ribosomes and, depending on their intracellular destination, is terminated either on soluble ribosomes or on ribosomes associated with the endoplasmic reticulum (Fig. 3). In general, polypeptides that terminate on membrane-bound ribosomes are inserted into the membrane while in the process of elongation. This *cotranslational* mode of insertion is followed by many, if not all, of the nuclear gene products which are secreted through, or reside in, membranes of the secretory pathway family of organelles (Fig. 3, pathway 2). This mode of insertion is also followed by many proteins which are secreted through or into the cytoplasmic membrane of bacterial cells [Inouye and Halegoua, 1980; Osborne and Wu, 1980; Tai and Davis, 1982; for exceptions see Wickner, 1980] and for membrane proteins encoded by chloroplast and mitochondrial genes [Sabatini et al, 1982] (see below). Polypeptides that terminate on soluble cytosolic ribosomes can be inserted into membranes *posttranslationally* subsequent to their release from the ribosome. *Posttranslational* insertion pathways are followed by the nuclear-coded polypeptides that reside in all compartments of mitochondria and chloroplasts, as well as the luminal compartment of peroxisomes and glyoxysomes (Table III).

1. Topological signals. When one considers both the plethora of cellular compartments between which newly synthesized nuclear-encoded polypeptides must choose in order to be properly sequestered, and the number of steps that characterize each targeting pathway shown in Figure 3, as well as the apparent accuracy with which polypeptides are targeted to and sequestered by organelle macrocompartments, it is obvious that each nuclear gene product

TABLE III. Nuclear Gene Products Targeted to Their Organellar Destination Posttranslationally

Organelle	Protein (s)	Reference
Mitochondrion*	Proteins that reside in all four mitochondria subcompartments*	1–3
Chloroplast	Ribulose 1,5-biphosphate carboxylase Chlorophyll a/b protein Plastocyanin Ferredoxin-NADP$^+$ oxidoreductase Fructose-1,6-bisphosphatase	4–6
Peroxisome	Catalase Urate Oxidase	7, 8
Glyoxisome	Malate Synthase Isocitrate Lyase Malate Dehydrogenase	9, 10

*See Table V for a complete list of proteins whose biosynthetic pathways have been studied. References: 1, Chua and Schmidt [1979]; 2, Neupert and Schatz [1981]; 3, Poyton et al [1982]; 4, Dobberstein et al [1977]; 5, Schmidt et al [1979] 6, Grossman et al [1981]; 7, Robbi and Lazarow [1978]; 8, Goldman and Blobel [1978]; 9, Kindl [1982]; 10, Becker et al [1982].

must have encoded within it a variety of topological signals. For secretory proteins and proteins which reside in the membranes of secretory pathway organelles (ie, those that follow pathway 2 in Fig. 3) these signals fall into at least three categories: 1) those that are operative in binding the ribosome to the membrane during translational elongation; 2) those that permit the protein to become lodged in the membrane or to pass through it completely; and 3) those that are operative in "sorting" membrane and secretory proteins as they transverse the organelles of the secretory pathway (Fig. 4, steps a–d of pathway 2). For proteins that follow *posttranslational* pathways during their sequestration into organellar compartments, these topological signals are probably even more diverse. They may include 1) signals that permit polypeptide translation to proceed on free polysomes; 2) signals that are operative in bringing proteins to the surface of mitochondria (Fig. 3, pathway 7), chloroplasts (Fig. 3, pathway 5), and peroxisomes (Fig. 3, pathway 4); 3) signals that allow those proteins which reside in the lumen of peroxisomes to traverse the peroxisomal membrane; 4) signals that direct both chloroplast and mitochondrial proteins to their correct organellar compartment; and 5) signals that allow cytosolic proteins to remain "soluble."

2. Signal-receptor interactions: An interface between genetic and spatial memory? Although the schemes shown in Figure 3 are undoubtedly over-

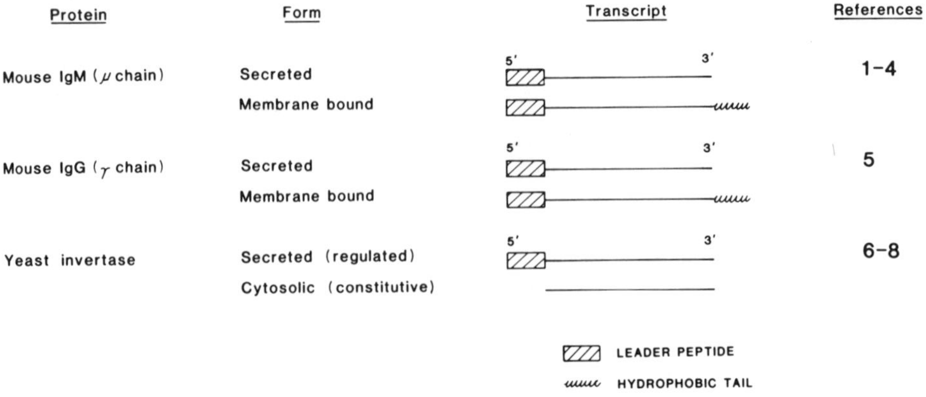

Protein	Form	Transcript	References
Mouse IgM (μ chain)	Secreted	5' ▨─────── 3'	1-4
	Membrane bound	▨────────────ᴗᴗᴗ	
Mouse IgG (γ chain)	Secreted	5' ▨─────── 3'	5
	Membrane bound	▨────────────ᴗᴗᴗ	
Yeast invertase	Secreted (regulated)	5' ▨─────── 3'	6-8
	Cytosolic (constitutive)	───────────	

▨ LEADER PEPTIDE

ᴗᴗᴗ HYDROPHOBIC TAIL

Fig. 4. Alternative transcripts from a single gene can direct gene products to different cellular locations. By differential processing of transcripts (heavy chains of mouse IgM or IgG) or by use of different promoters to produce different transcripts (yeast invertase) protein products of the same gene can be targeted to different intracellular destinations. Those regions of the transcripts that encode leader peptides and hydrophobic polypeptide tails are designated. References: 1, Alt et al, 1980; 2, Early et al, 1980; 3, Singer et al, 1980; 4, Rogers et al, 1980; 5, Rogers et al, 1981; 6, Perlman and Halvorson, 1981; 7, Carlson and Botstein, 1982; 8, Perlman et al, 1982.

simplifications of the in situ situation and ignore the possibility that ribosomes and organelles may be linked to one another in the cytoplasmic infrastructure (or microtrabecular lattice), they do raise a number of questions regarding the interaction between genetic and spatial memory in the targeting of proteins to organellar surfaces. Among these: 1) How do newly synthesized proteins select the organellar surface with which they are to interact? 2) What determines the site on the membrane where they will interact?

These questions can be approached experimentally by adopting the hypothesis that the *topological signals* on membrane-directed proteins recognize organellar surface receptors [eg, see Blobel, 1980], that these receptors are proteins which are unique to a particular organellar surface, and that the insertion event occurs only at sites of the organellar surface where receptors are unoccupied. Conceptually, this hypothesis implies that although topological signals are ultimately derived from the cell's genetic memory, the organellar surface receptors that recognize these signals constitute a different kind of memory based on the spatial organization of preexisting membranes. From the viewpoint of the temporal continuity of membranes, this hypothesis, as stated here, proposes that membrane-bounded compartments are perpetuated from one cell generation to the next because their surface membranes carry receptors for those newly synthesized proteins which are to reside within them from one generation to the next. This hypothesis also proposes that the site at which membrane-directed proteins are inserted into or through the lipid bilayer may be defined very precisely by the spatial memory encoded along the two-dimensional surfaces of organellar membranes. As will be discussed in section VII, this spatial memory may ultimately be encoded by statistically defined "unit areas" (Fig. 2).

There is no a priori reason to assume that the spatial memory which encodes receptors for proteins that are transported across the membrane is identical to that which encodes receptors for proteins that are inserted into the membrane. Indeed, since the insertion of new proteins into a membrane must ultimately lead to the lateral expansion of the membrane surface, it is very likely that membrane and secreted proteins do not use the same memory. Furthermore, integral membrane proteins exist as oligomers (see above) whereas secreted proteins usually do not. It is therefore possible that spatial memory along the surface of a membrane is used also for the assembly of oligomeric membrane proteins. As will be discussed in greater detail below (sections VI A and VII), this memory may be expected to target and assemble subunits of hetero-oligomeric membrane proteins to specific regions of the membrane surface.

3. Genetic regulation of pathways. Other questions raised by a consideration of the multitude of pathways followed by proteins to different organelles (Fig. 3) concern the mechanisms by which the flow of genetic information along each pathway is regulated. Are the nuclear genes for organellar proteins regulated or are they constitutively "on" at all times? If they are regulated, how does each gene "sense" the intracellular concentration of its protein product, especially when its product is sequestered inside of a membrane bounded macrocompartment or lodged in a membrane? At what level—transcription, transcript processing, transport of transcripts out of the nucleus, translation, posttranslational modification, or transport into organelles—is the expression of nuclear-coded organellar protein genes regulated?

Although these questions are largely unanswered, a number of recent studies have demonstrated that the intracellular concentration of many nuclear-coded organellar proteins is regulated. Proteins in this category include 1) secretory proteins produced by mammalian exocrine cells [Jamieson, 1972; Campbell, 1979] and by yeast cells [Bostian et al, 1980; Carlson and Botstein, 1982; Perlman et al, 1982]; 2) cytochrome P_{450} and other microsomal enzymes of the mixed function oxidase system in mammalian liver cells [Gonzalez and Kasper, 1982; Walz et al, 1982; for review see De Pierre et al, 1981]; 3) catalase and other peroxisomal proteins in yeast [Cross and Rius, 1978; Ruis, 1979; Tanaka et al, 1982]; and 4) proteins of both the mitochondrial and chloroplast electron transport chains [Kirk, 1971; Linnane et al, 1972; Poyton, 1980]. Although it is not yet known how the genes for membrane or organellar proteins sense either the total or organellar concentration of their protein products, it is likely that at least part of this regulation is mediated by metabolites and other small molecules in ways which are similar to those employed for regulating the expression of "soluble" cytosolic proteins [Darnell, 1982]. It is clear that for many organelle proteins this regulation is exerted at the transcriptional level. Organellar proteins for which there is evidence of transcriptional control include 1) yeast mitochondrial cytochrome c [Zitomer et al, 1979], RNA polymerase [Lustig et al, 1982a], and cytochrome c oxidase subunits IV–VII [Lustig et al, 1982b]; 2) yeast peroxisomal catalase [Ammerer et al, 1981]; 3) yeast microsomal (ie, secretory) invertase [Carlson and Botstein, 1982]; 4) acid phosphatase [Bostian et al, 1980]; 5) liver microsomal cytochrome P_{450} [Gonzalez and Kasper, 1982]; and 6) egg yolk microsomal (ie, secretory proteins) ovalbumin [McKnight and Palmiter, 1979; Swaneck et al, 1979] and conalbumin [McKnight and Palmiter, 1979].

Recently, Blobel and his co-workers [Walter and Blobel, 1981a–c; Anderson et al, 1982a] and Meyer et al [1982] have provided evidence that the synthesis of some secretory proteins and at least one plasma membrane protein may be controlled at the translational level. These studies have led to the identification of a "signal recognition protein" and a "docking protein" complex which are required for the successful translation of β-lactoglobulin, IgG light chain, and the δ subunit of the acetylcholine receptor. At present, it is not clear how the large signal-recognition protein complex (11S) operates in translation. Nor is it clear which of its components (RNA or polypeptide) is responsible for its activity. Nevertheless, it is clear that either part or all of this complex is required for the complete elongation of some, perhaps all, proteins that are to be inserted *cotranslationally* into the endoplasmic reticulum.

One of the most striking recent observations to bear on the genetic regulation of targeting pathways is that single genes can be transcribed into alternative transcripts which direct their products to different cellular locations (Fig. 4). The genes for the μ chain of mouse IgM or the γ chain of mouse IgG produce two alternative transcripts: one that translates a protein with a hydrophobic carboxyl terminus and which is membrane-bound, and another that translates a protein with a hydrophilic carboxyl terminus and which is secreted [Alt et al, 1980; Early et al, 1980; Singer et al, 1980; Rogers et al, 1980, 1981]. These alternative transcripts result from different messenger RNA splicing pathways [see Hood et al, this volume]. The presence of the hydrophobic carboxy terminus is thought to anchor these otherwise soluble proteins to the plasma membrane. The yeast invertase gene is another example of a single gene which produces two transcripts that target proteins to different intracellular locations [Carlson and Botstein, 1982; Perlman et al, 1982]. However, unlike the IgM and IgG light chain transcripts, the invertase transcripts differ at their 5' ends and probably result from the use of two different promoters, rather than differential transcript processing. It is intriguing that the transcript whose protein product is located in the cytosol is produced constitutively whereas the transcript whose product is secreted is regulated and seems to use a promoter, which permits the reading of a "leader peptide" [Carlson and Botstein, 1982; Perlman et al, 1982]. In all three cases individual cells can produce both transcripts. Except for yeast invertase it is not yet clear which intracellular conditions determine the ratios of the alternative transcripts. It seems likely that additional examples of this phenomenon will be uncovered as the transcripts for more proteins which may be present in two different cellular locations are identified [for additional probable candidates see Kitutani et al, 1981; and Hopper et al, 1982].

B. Organellar Gene Products

Proteins coded by extranuclear organellar genomes are directed to far fewer cellular locations than those coded in the nucleus. The simplest situation exists in the mitochondrion (Fig. 5), since all known mitochondrial gene products reside in the inner mitochondrial membrane as integral membrane proteins and, in some organisms, in the mitochondrial ribosomes attached to it [Poyton et al, 1982]. Thus, mitochondrial gene products are associated with only one of the four compartments (ie, outer membrane, intermembrane space, inner membrane, matrix) of the mitochondrion.

Gene products encoded by the chloroplast genome have so far been shown to be located in only two of the six possible compartments (ie, outer envelope membrane, interenvelope membrane space, inner envelope membrane, stroma, thylakoid membrane, thylakoid space) of the chloroplast (Fig. 5). Some are directed to the thylakoid membrane (Fig. 5, pathway 1) where they exist as integral membrane proteins while others are released into the stroma (Fig. 5, pathway 2) as soluble proteins [Von Wettstein, 1981].

V. INSERTION OF PROTEINS INTO CELLULAR MACROCOMPARTMENTS

In considering the chronology of signal-receptor interactions which take place as proteins are targeted to their intracellular destinations (Figs. 3, 5), it is obvious that one of the earliest signals to be expressed is that which determines whether a protein will associate with a membrane-bound receptor during or after translation. A *cotranslational* interaction obviously takes place before all of the genetic information in a gene is transformed into its protein product. In contrast, *posttranslational* interactions can theoretically make use of all of the genetic information encoded in a gene.

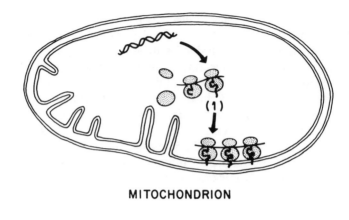

MITOCHONDRION

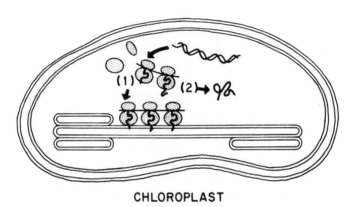

CHLOROPLAST

Fig. 5. Organelle gene products follow fewer pathways than nuclear gene products. Proteins encoded on mitochondrial or chloroplast genomes are targeted to fewer cellular locations than nuclear gene products (Fig. 3). All identified mitochondrial gene products reside at or in the inner mitochondrial membrane. They therefore follow one pathway (1). Chloroplast gene products may reside in either the thylakoid membrane (1) or stroma (2).

A. Cotranslational Insertion of Proteins Into Organellar Membranes

1. Signals. Early interest in the *cotranslational* insertion of proteins into organellar membranes focussed on the proteins of secretory pathway organelles. Studies by Palade, Sabatini, and Blobel [Siekevitz and Palade, 1960; Redman et al, 1966; Redman and Sabatini, 1966; Blobel and Sabatini, 1970] established that these proteins appear in cellular microsomal fractions concomitant with their synthesis and are sequestered in the cisternal space of the endoplasmic reticulum. The early studies [reviewed extensively by Palade, 1975; Sabatini et al, 1982] suggested that the information required to target these proteins to the endoplasmic reticulum is expressed early in their synthesis and lead to formulation of the "signal hypothesis" [Blobel and Sabatini, 1971; Blobel and Dobberstein, 1975a, b]. This hypothesis proposed that the genetic information which directs newly synthesized secretory proteins to the surface of the endoplasmic reticulum, rather than to some other organellar surface, lies encoded in a transient "leader peptide"[1] at their amino terminus. Furthermore, this hypothesis as formulated by Blobel and Dobberstein [1975a, b] proposed that the leader peptide is removed during or after transport to the luminal face of the endoplasmic reticulum membrane, and that the nascent secretory polypeptide traverses the membrane through a protein pore, or channel.

The "signal hypothesis" has received widespread support from studies with both eukaryotic and prokaryotic secretory and membrane proteins [see Sabatini et al, 1982]. Indeed, leader peptides have been found on many, but not all, secretory and membrane proteins (see Table IV for exceptions) which are targeted to membranes *cotranslationally* in both prokaryotic and eukaryotic cells. Although these leader peptides vary in length and amino acid sequence [Austen, 1979; Steiner et al, 1980; Kreibich et al, 1980; Kreil, 1981], they do appear to have some features in common. Most leader peptides have a central region rich in hydrophobic amino acids [Von Heijne, 1982] and a sequence near the cleavage site which has a high β-turn-forming potential [Austen, 1979; Steiner et al, 1980; Kreil, 1981]. In addition, many leader peptides have a charged amino acid near their amino terminus. The hydrophobic nature of leader peptides has been postulated to be sufficient to direct it to the lipid bilayer of a membrane [DiRienzo et al, 1978; Engelman and Steitz, 1981]. However, the recent discovery of "signal recognition" and "docking" proteins [Walter and Blobel, 1981c; Meyer et al, 1982] suggests that membrane recognition by leader peptides is mediated by protein-protein interactions rather than protein-lipid interactions.

Of course, the presence of a leader peptide on a membrane-directed protein does not prove a priori that it will be inserted into a membrane *cotranslationally*. Indeed, some proteins that are inserted into membranes *posttranslationally* also have leader peptides [Wickner, 1980] (see below). Nor does the presence of a leader peptide on a secretory protein demonstrate that it is the *signal* which is responsible for targeting that protein to the surface of the endoplasmic reticulum, especially since it is now clear that other proteins (ie, "signal recognition protein" complex, "docking protein," ribophorins) are also involved [Walter and Blobel, 1981c; Meyer et al, 1982; Kriebich

---

[1]A "leader peptide" is defined here as a *transient* amino-terminal extension on a polypeptide chain. The term is not meant to imply a function for this extension.

TABLE IV. Representative Examples of Organelle Proteins Which Lack Transient Amino-Terminal, "Leader Peptides"

Targeting pathway	Protein	Organelle location	Reference
Cotranslational	Cytochrome P_{450}	Microsomal membrane	1
	Ovalbumin	Microsomal lumen	2–4
	Erythrocyte anion transporter (band 3)	Erythrocyte plasma membrane	5, 6
	Beef cytochrome oxidase subunit II	Inner mitochondrial membrane	7, 8
	Yeast ATPase subunit IX	Inner mitochondrial membrane	9, 10
Posttranslational	Catalase	Peroxisomal lumen	11, 12
	Cytochrome b_5	Microsomal membrane	13, 14
	Adenine nucleotide translocase	Inner mitochondrial membrane	15–18
	Cytochrome c	Mitochondrial intermembrane space	19–22

References: 1, Bar-Nun et al [1980]; 2, Palmiter et al [1978]; 3, Lingappa et al [1979]; 4, Braell and Lodish [1982a]; 5, Braell and Lodish [1981]; 6, Braell and Lodish [1982b]; 7, Steffans and Buse [1979]; 8, Anderson et al [1982b]; 9, Macino and Tzagoloff [1979]; 10, Sebald et al [1979]; 11, Robbi and Lazarow [1982]; 12, Lazarow et al [1982]; 13, Rachubinski et al [1980]; 14, Kreiter and Shires [1980]; 15, Nelson and Schatz [1979a]; 16, O'Malley et al [1982]; 17, Zimmerman et al [1979]; 18, Zimmerman et al [1980]; 19, Montgomery et al [1978]; 20, Korb and Neupert [1978]; 21, Matsuura et al [1981]; 22, Smith et al [1979].

et al, 1980, 1982]. It is also clear that many membrane-directed proteins that cross membranes *cotranslationally* lack transient leader peptides altogether (Table IV), and that the leader peptide by itself is not sufficient to insure the transport of secretory proteins across bacterial cytoplasmic membranes [Moreno et al, 1980; Koshland and Botstein, 1980]. What then is the function of leader peptides on membrane directed proteins? At present, it seems likely that the leader peptide is only a part of the signal which is operative early to bring proteins to their correct organellar surface. It is not yet clear how the leader peptide functions in this regard, nor is it apparent that the leader peptide functions solely as a part of a signal. Indeed, other possible functions of leader peptides on membrane proteins will be considered below (section VI, B).

2. **Mechanism.** Implicit in the "signal hypothesis" is the assumption that the energy for insertion of a protein into or through a membrane is derived

from the elongational steps of protein synthesis. Recently, however, it has been suggested on energetic and structural grounds that leader peptides are capable of forming a helical hairpin [Engelman and Steitz, 1981] or loop [Inouye and Halegoua, 1980], which, upon folding, permits the leader peptide and its attached polypeptide to partition spontaneously into and across the lipid bilayer of the membrane [Von Heijne and Blomberg, 1979]. Although both of these models are feasible on energetic grounds [Von Heijne and Blomberg, 1979], they do not take into account the recently discovered "signal recognition proteins" [Walter and Blobel, 1981a–c] or "docking proteins" [Meyer et al, 1982], and ignore the fact that biological membranes are not energetically static but have an enormous potential to conserve and generate energy via their pumps and transport proteins (see above). In this later regard it is striking that all of those intracellular membranes across which proteins are transported *cotranslationally* have ATPase proton pumps that pump protons in the same direction as that followed by proteins during their insertion (Fig. 6). Are these proton pumps, or the antiport systems that are coupled to them, used to drive proteins across membranes *cotranslationally?*

B. Posttranslational Insertion of Proteins Into Organellar Membranes

At present, little is known about the molecular mechanisms that underlie the sequestration of cytoplasmically translated proteins by mitochondria, chloroplasts, and peroxisomes. For convenience, it is useful to view the transport of these proteins into organelles as a two-event process [Poyton et al, 1982]: 1) the targeting of the protein to the organelle surface; and 2) the internalization of the protein and its subsequent sequestration in the appropriate organellar subcompartment. The first event most likely involves the interaction between a membrane-specific selection sequence on the protein and a receptor on the organellar surface. Conceptually, this type of signal-receptor interaction is analogous to that which operates during *cotranslational* targeting. However, since posttranslational targeting is not initiated until after a protein has been completely translated, it is not necessary to predict that the signal is a leader peptide at its amino terminus. It would seem a priori that internal regions of the primary sequences of these proteins are as likely to be operative as signals as either of their two termini. Other possible candidates for signals on proteins that are targeted to membranes *posttranslationally* would include sites on their surface determined by the tertiary structure of the folded polypeptide or ancillary "pilot proteins" [Poyton et al, 1980] with which the protein becomes associated in the cytosol.

Attempts to identify potential signals on proteins which are targeted to membranes *posttranslationally* have focused so far on the identification of larger precursor molecules which are processed by proteolysis either during

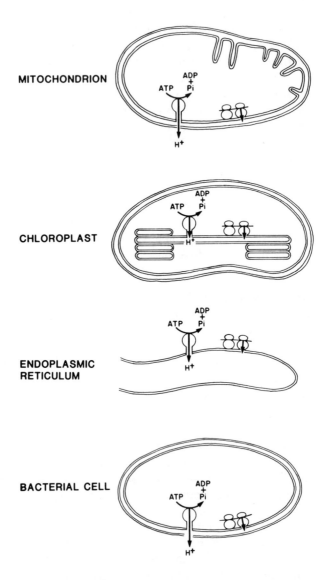

Fig. 6. Topology of proton gradients and cotranslational insertion into biological membranes. Membranes into which proteins are inserted *cotranslationally* have ATPase proton pumps that are capable of pumping protons in the same direction as the elongating polypeptide chain.

or after passing into the organelle. These studies have demonstrated that many, though not all (Table IV), of the proteins which follow *posttranslational* pathways to their organellar surface are initially synthesized as larger molecules. It is not yet clear if these larger molecules have amino-terminal leader peptides. Indeed, of all of the precursors identified so far only one—the small subunit of chloroplast ribulose-1,5-biphosphate carboxylase—has been shown unambiguously to have an amino-terminal "leader peptide" [Schmidt et al, 1979].

The second event in the *posttranslational* internalization of proteins into organellar compartments is also poorly understood. Aside from the finding that proteins that are sequestered into the mitochondrial inner membrane or matrix compartments require an energized inner membrane [Nelson and Schatz, 1979a, b; Zimmerman and Neupert, 1980] and that chloroplast stromal and thylakoid membrane proteins require energy to be incorporated into the chloroplast [Grossman et al, 1980], very little is known about these processes.

The most intensively studied organellar proteins that follow posttranslational pathways are those that are targeted to mitochondria. A consideration of the biosynthetic pathways followed by these proteins suggests that a multitude of signals might be employed for one, or both, events in their sequestration. From Table V it is apparent that 1) many proteins are synthesized as larger precursor molecules; 2) adenine nucleotide translocase, an integral protein of the inner membrane, and cytochrome c, a peripheral protein of the outer surface of the inner membrane, are not synthesized as larger precursors; 3) precursor molecules vary in size from 500 daltons (approx. 4 amino acids) to 10,000 daltons (approx. 80 amino acids) larger than the "mature" protein; 4) the weights of the "extra" amino acid sequences present in the precursor molecules vary from 3% (for rat liver OMM 35) to 50% (for Neurospora ATPase proteolipid) of the weight of the mature protein. Furthermore, it is also apparent from some of the studies summarized in Table V that precursor molecules to some nuclear-coded mitochondrial proteins (ie, yeast cytochrome c oxidase polyprotein, yeast cytochrome b, and rat liver ornithine transcarbamylase) are processed through multiple steps during conversion to their mature size. At present, it is not known if the additional genetic information present in these precursors molecules resides at their NH_2 termini (as "leader" peptides) at their COOH termini (as "trailing" peptides) or at both termini. It is also not clear to what extent other, more subtle, posttranslational modifications (eg, acetylation, methylation, or phosphorylation) of internal amino acids are important for targeting these proteins to the mitochondrial surface or for sequestering them into the correct mitochondrial subcompartment.

From Table V it is apparent that two different types of precursor pathways have been proposed for subunits IV, V, VI, and VII of cytochrome c oxidase.

TABLE V. Biosynthesis and Posttranslational Transport of Nuclear-Coded Polypeptides Into Mitochondria

Protein	Subunit	Size of mature polypeptide	Additional size of precursor	Site of synthesis	Reference
Outer membrane					
Rat liver OMM35	—	35,000	~500?	Free polysomes	1
Intermembrane space					
Yeast ⎫	—	12,300	0	—	2, 3
Neurospora ⎬ Cytochrome c	—	12,500	0	Free polysomes	4
Rat liver ⎭	—	12,500	0	Free polysomes	5, 6
Rat liver sulfite oxidase	—	55,000	~4,000	Free polysomes	7
Yeast cytochrome c peroxidase	—	33,500	~6,000	—	8
Yeast cytochrome b_2	—	58,000	~10,000	—	9
Inner membrane					
Yeast ⎱ Adenine nucleotide	—	32,000	0	—	10, 11
Neurospora ⎰ translocase	—	32,000	0	Free polysomes	12, 13
Neurospora ATPase proteolipid	—	8,000	~4,000	—	14
Yeast cytochrome bc_1 complex	V	25,000	~2,000	—	15, 16
	c_1	31,000	~4,000	—	9, 16
			Polyprotein Pre-proteins		
Yeast cytochrome	IV	14,300	~2,000	—	
oxidase	V	14,000	~10,000 ~2,000	—	17–22
	VI	9,400	~6,000	—	
	VII	4,500	0	—	
			Polyprotein Pre-protein		
Rat liver cytochrome c	IV	16,500	~3,000	—	
oxidase	V	13,200	—	—	23, 24
	VI	10,000	~3,000 —	—	
	VII	6,900	—	—	
Chick liver aminolevulinate synthetase	—	65,000	~10,000	—	25
Rat liver β-hydroxybutyrate dehydrogenase	—	32,000	~5,000	—	7
Matrix					
Yeast F_1 ATPase	α	58,000	~6,000	—	
	β	54,000	~2,000	—	10, 26
	γ	34,000	~6,000	—	
Chicken heart aspartate aminotransferase		44,500	~3,000	Free polysomes	27–29

continued next page

TABLE V. Biosynthesis and Posttranslational Transport of Nuclear-Coded Polypeptides Into Mitochondria—Continued

Protein	Subunit	Size of mature polypeptide	Additional size of precursor	Site of synthesis	Reference
Neurospora citrate synthetase	—	45,000	~2,000	—	30
Rat liver carbamyl PO₄ synthase	—	160,000	~5,000	Free polysomes	31–36
Rat liver ornithine transcarbamylase	—	39,000	~4,000	—	36–40
Rat liver glutamate dehydrogenase	—	60,000	~6,000	Free polysomes	7
Rat liver malate dehydrogenase	—	37,000	~1,000	Free polysomes	7
Yeast superoxide dismutase	—	24,000	~2,000	—	41

References: 1, Shore et al [1981]; 2, Montgomery et al [1978]; 3, Smith et al [1979]; 4, Korb and Neupert [1978]; 5, Matsuura et al [1981]; 6, Scarpulla et al [1981]; 7, Mihara et al [1982]; 8, Maccecchini et al [1979a]; 9, Gasser et al [1982]; 10, Nelson and Schatz [1979a]; 11, O'Malley et al [1982]; 12, Zimmerman et al [1979]; 13, Zimmerman et al [1980]; 14, Michel et al [1979]; 15, Cote et al [1979]; 16, Nelson and Schatz [1979b]; 17, George-Nascimento and Poyton [1981]; 18, Poyton and McKemmie [1979a]; 19, Poyton and McKemmie [1979b]; 20, Lewin et al [1980]; 21, Mihara and Blobel [1980]; 22, Lustig et al [1982b]; 23, Ries et al [1978]; 24, Schmelzer and Heinrich [1980]; 25, Ades and Harpe [1981]; 26, Maccecchini et al [1979b]; 27, Sonderegger et al [1980]; 28, Sonderegger et al [1982]; 29, Sakakibara et al [1980]; 30, Harmey and Neupert [1979]; 31, Shore et al [1979]; 32, Raymond and Shore [1979]; 33, Raymond and Shore [1981]; 34, Mori et al [1979a]; 35, Mori et al [1979b]; 36, Mori et al [1981a]; 37, Conboy et al [1979]; 38, Kraus et al [1981]; 39, Conboy and Rosenberg [1981]; 40, Mori et al [1981b]; 41, Autor [1982].

One of these pathways is thought to involve a common polyprotein precursor to all four of these subunit polypeptides [Poyton and McKemmie, 1979a, b; Ries et al, 1978]. It has been proposed for both yeast and rat liver cells that this precursor is proteolytically processed only after its insertion into the inner mitochondrial membrane [Poyton and McKemmie, 1979b]. This precursor has been found in growing yeast cells both in cytosolic and mitochondrial inner membrane fractions. It has a molecular weight of 55,000 daltons and therefore contains approximately 10,000 daltons (approx. 80 amino acids) worth of nonsubunit genetic information. The other pathway is thought to involve individual, pre-protein, precursors to all subunits except VII [Lewin et al, 1980; Mihara and Blobel, 1980; Lustig et al, 1982b] of yeast cytochrome c oxidase. These precursors cannot be found in vivo during

HYPOTHESIS

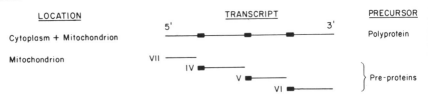

Fig. 7. Nuclear-coded subunits of cytochrome c oxidase may follow two pathways to the mitochondrion. One way to explain the apparent existence of two types of precursor processing pathways (ie, common polyprotein and individual pre-proteins) for subunits IV, V, VI, and VII of cytochrome c oxidase is to propose that the genes for subunits IV, V, VI, and VII are contiguous on the genome and that they may be transcribed in two ways: 1) to give a continuous polyprotein transcript; or 2) to give individual pre-protein transcripts. Since the polyprotein is observed in vivo, both in cell cytosolic and inner mitochondrial membrane fractions [Reis et al, 1978; Poyton and McKemmie, 1979a, b], and pre-proteins have not been observed in growing cells [Mihara and Blobel, 1980; Lewin et al, 1980; Lustig et al, 1982b], it is conceivable that the polyprotein is on a "slow" pathway whereas pre-proteins are on "fast" pathways. The type of pathway followed (ie, transcript made) may ultimately be decided by the state of mitochondrial gene expression, since the three remaining subunits (I, II, and III) of the enzyme are mitochondrial gene products. Perhaps a polyprotein "reserve" of subunits is an important source of subunits for assembly when the mitochondrial genes for subunits I, II, and III are turned on before the nuclear genes for subunits IV–VII. The blackened boxes on the long transcript would correspond to nonsubunit "spacer" polypeptides whereas those on the short transcripts would correspond to leader peptides.

growth but can be found in metabolically shocked yeast spheroplasts and among the polypeptides synthesized in reticulocyte lysates programmed with the appropriate mRNA. In view of these properties it appears that the polyprotein precursor is rather long-lived compared with the pre-protein precursors. Despite the fact that there is a great deal of experimental support for the existence of these two pathways it is not clear that both are operative in vivo. Indeed, their existence is currently controversial. To evaluate properly the existence of these two pathways and determine their relative importance, it will be necessary to identify the genes and transcripts for these polypeptides. It is conceivable at present that both pathways are operative and that a polyprotein gene can be transcribed in two different ways (Fig. 7)—one producing a single long transcript initiated from an external promoter, and another producing four shorter transcripts read from internal promoters. If correct, this scheme would represent another example of one gene read as alternative transcripts [for details, see Hood et al, this volume], although in this case both types of protein products would be targeted to the same intracellular destination, but at different rates.

VI. THE SYNTHESIS AND ASSEMBLY OF OLIGOMERIC MEMBRANE PROTEINS

A. Spatial Memory and the Assembly of Oligomers

Since transmembranous proteins are oligomeric in situ (Table I) and since their oligomeric state is apparently indispensible for the functional communication between cellular macrocompartments, the interaction between spatial and genetic memory may extend beyond the early steps in the synthesis and intracellular targeting of these proteins to their membrane surface (see section IV A). It may include their assembly into oligomers. In view of the fact that the lateral diffusion of integral proteins in the plane of the membrane is restricted (see above) [Edidin, 1982] and may be limited to "unit areas" (Fig. 2), it is intriguing to consider the possibility that spatial memory, encoded in "unit areas" along the surface of a membrane, may be used to determine not only which intracellular membrane a newly synthesized membrane protein selects but also where on that membrane's surface the protein is inserted.

In considering the assembly of membrane proteins it is useful to recall a fundamental difference between those which are homo-oligomers and those which are hetero-oligomers. In theory, homo-oligomers may be derived from polypeptides that are translated from the same polysome (Fig. 8). Subunits of homo-oligomers have identical asymmetric topologies vis-à-vis the two faces of the membrane [Wickner, 1980; Klingenberg, 1981]. Therefore, for homo-oligomers whose subunit polypeptides are inserted into a membrane *cotranslationally*, assembly may be affected by the close juxtaposition of new and identical polypeptides in the plane of the membrane.

The assembly of hetero-oligomeric membrane proteins can be expected to be considerably more complex (Fig. 8) than the scheme depicted above for a homo-oligomer. First, polypeptide subunits have to be made available in the correct stoichiometry. Second, they have to find one another in the plane of the membrane. And third, like homo-oligomers, they have to be assembled into an energetically favorable conformation. In view of the local restrictions on integral protein diffusion, it is attractive to view the insertion of subunit polypeptides of hetero-oligomers into the membrane and their assembly into the hetero-oligomeric structure as coupled processes (Fig. 8). In this model, assembly is viewed as a sequential process in which the first subunit inserted into the membrane serves as a membrane-bound receptor (or template) for the subsequent insertion of other subunits of the oligomer. Although the scheme shown in Figure 8 illustrates this process for polypeptide subunits which enter a membrane *cotranslationally*, the model is equally applicable for integral polypeptide subunits that enter the membrane *posttranslationally*. One of the attractive features of this simple model is that subunits of hetero-oligomers enter the plane of the membrane at a site where

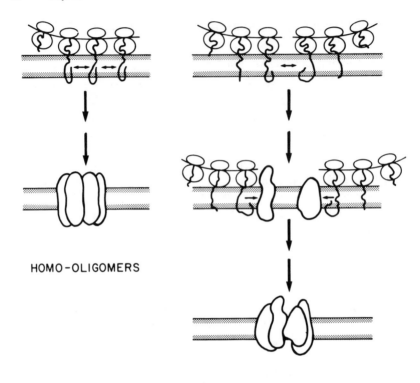

HOMO-OLIGOMERS

HETERO-OLIGOMERS

Fig. 8.   Assembly of homo-oligomeric and hetero-oligomeric membrane proteins viewed from the perspective of the number and types of polysomes involved in the synthesis of their subunits. In theory, the assembly of a homo-oligomer may be fairly simple and may result from the assembly of a set of subunits translated on the same polysome. In contrast, the assembly of a hetero-oligomer is likely to be more complex since its polypeptide subunits are synthesized on different polysomes. The model shown for hetero-oligomers implies that one subunit of the oligomer can serve as a receptor which is recognized by the elongating polypeptide chain of another subunit. Equally plausible, however, for both types of oligomers, is that their polypeptide chains are assembled after their release from the polysome.

they are needed (ie, to complete the assembly of a partially assembled oligomer) rather than at a random site, from which they have to diffuse in order to find, and be assembled with, the other subunits of the hetero-oligomer. How is the synthesis of these subunits regulated? Do leader peptides function as signals, or in other ways, in this sort of coupled insertion assembly? And if so, how? In the discussion which follows we will consider these questions and propose some new functions for leader peptides on integral membrane proteins.

B. "Leader Peptides" and the Biogenesis of Oligomeric Membrane Proteins

From the studies reviewed in section V it is obvious that much remains to be learned about the signal-receptor interactions that become operative as proteins are targeted to intracellular membranes. From both biochemical and genetic studies [reviewed by Kreil, 1981] it is equally obvious that leader peptides are likely, yet still presumptive, signal components for proteins that follow a *cotranslational* route to their intracellular destination. Despite the general acceptance of the "signal hypothesis," a number of questions concerning how leader peptide function during *cotranslational targeting* remain to be answered. These include the following: Do all proteins that are targeted to the same membrane have leader peptide signals that recognize the same receptor? What is the nature of the membrane-bound receptors for secreted proteins and for membrane proteins? How many soluble proteins ("signal recognition proteins," etc) are required for targeting and insertion of membrane-directed proteins? Do leader peptides have functions not previously recognized?

1. Leader peptides on mitochondrial gene products. Particularly promising for addressing these questions are proteins encoded by mitochondrial DNA. These proteins are all synthesized on ribosomes bound to the matrix face of the inner mitochondrial membrane and are inserted *cotranslationally* [for review see Poyton et al, 1982] into the inner membrane where they reside as integral polypeptide subunits of the oligomeric protein complexes (Fig. 9) cytochrome c oxidase, CoQ cytochrome c reductase, and oligomycin-sensitive ATPase. Since the structural genes for these proteins have been identified and sequenced, since some of the proteins themselves have been sequenced, and since all steps in the expression of these genes (including the insertion of their protein products into the inner membrane) are likely to be regulated by nuclear genes [for review see Poyton, 1980], this well-defined set of proteins allows an analysis of signal-receptor interactions at a level of resolution which is afforded by few other membrane-directed proteins. For example, by analyzing the amino acid sequences of these proteins from several species, it should be possible to identify regions of homology (if they exist) which might serve as signals for targeting, insertion, or assembly. It should also be possible to determine 1) if all mitochondrial gene products make use of the same signal-receptor system; 2) if gene products which reside in the same enzyme complex (eg, subunits I, II, and III of cytochrome c oxidase) use the same signal-receptor system; or 3) if the signals and receptors are different for gene products which reside in each protein complex. In addition, it should be possible to determine what role, if any, leader peptides play in targeting mitochondrial gene products to the inner membrane, and, by use of appropriate mutations in their structural genes, determine if they

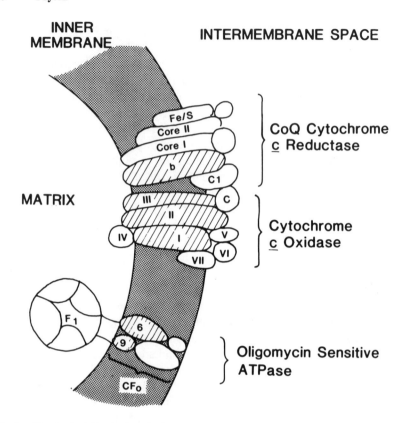

Fig. 9. Topological distribution of mitochondrial gene products across the inner mitochondrial membrane. Note that mitochondrial gene products (cross-hatched) are subunits of genetically mosaic enzyme complexes which contain nuclear gene products (clear) as well. Taken from Poyton et al [1982].

have other functions in the biogenesis of these membrane proteins. Finally, by genetic complementation analysis of nuclear mutants that are defective in the insertion of one or all mitochondrial gene products into the inner membrane, it should be possible to determine the number of different polypeptides required for targeting and inserting these proteins into the inner membrane.

It has recently been demonstrated that some mitochondrial gene products in the yeast, Saccharomyces cerevisiae, and in Neurospora crassa have leader peptides at their amino termini while others do not. In Saccharomyces cerevisiae, only one of the seven known polypeptides encoded by the mitochondrial genome has been shown unambiguously to be synthesized as a primary translation product with an amino-terminal leader peptide. In vitro experiments [Sevarino and Poyton, 1980; DeRonde et al, 1980; Sevarino et al, 1983] and DNA sequence data [Fox, 1979; Coruzzi and Tzagoloff, 1979] have revealed that subunit II of cytochrome c oxidase is synthesized with a 15 amino acid leader peptide (Fig. 10). Sevarino and Poyton [1980] have shown that this leader peptide is normally removed from the mature polypeptide *cotranslationally* but that it can be processed *posttranslationally* as well under some conditions. Previously, we have evaluated the properties of this leader peptide [Poyton et al, 1982; Sevarino et al, 1983] and have pointed out that it has properties in common with leader peptides on proteins that are targeted *cotranslationally* to other cellular membranes.

Are other yeast mitochondrial gene products synthesized with leader peptides? At present, it is possible to state with certainty that ATPase subunit 9 is not preceded by a "leader peptide" during its insertion into the inner membrane since both the polypeptide and its gene have been sequenced [Macino and Tzagoloff, 1979; Sebald et al, 1979]. Definitive data of this nature regarding the presence, or absence, of "leader peptides" on other yeast mitochondrial gene products are not yet available. However, the findings that some mature yeast mitochondrial translation products can be radiolabeled in vivo with formate, and hence probably retain formylmethionine at their NH_2 termini [Mahler et al, 1972; Feldman and Mahler, 1974], and that cytochrome c oxidase subunit II is the only mitochondrial gene product synthesized with a "leader peptide" in an aurintricarboxylic acid–inhibited "read-out" system [Sevarino and Poyton, 1980] suggest that subunit II may be the only yeast mitochondrial gene product that possesses a leader peptide.

It is also not yet clear if some mitochondrial gene products have uncleaved NH_2-terminal segments which function as "leader peptides," as has been proposed recently for cytochrome P_{450} [Bar-Nun et al, 1980] and sindbis virus glycoprotein pE [Bonatti and Blobel, 1979]. However, a comparison of the sequence of the leader peptide (II') on cytochrome c oxidase subunit II from yeast with the NH_2-terminal sequence of mature yeast ATPase subunit 9, a gene product known to be synthesized without a leader peptide (see above), and the NH_2-terminal sequences of all other yeast mitochondrial gene products (as predicted from their DNA sequences) makes this possibility unlikely. Indeed, from a consideration of these sequences (Fig. 10) it is clear that there is little sequence homology between leader peptide II' and all other gene products (with the possible exception of ATPase 6). Furthermore, a comparison of the predicted secondary structures of these NH_2-terminal sequences (Fig. 11) reveals that there is no easily recognizable structure common to the leader peptide II' and the other mature mitochondrial gene products. Even though there are some similarities in the predicted secondary structures of II' and the NH_2 termini of ATPase subunit 6 and ATPase subunit

	Hydrophobicity Index (Residues 1–15)
Leader Peptide	
Cytochrome Oxidase II' — [fMet-Leu-[Asp]-Leu-Leu-[Arg]-Leu-Gln-Leu-Thr-Thr-Phe-Ile-Met-Asn	2.06
Mature Polypeptides	
Cytochrome Oxidase II — [Asp]-Val-Pro-Thr-Pro-Tyr-Ala-Cys-Tyr-Phe-Gln-[Asp]-Ser-Ala-Thr-Pro-Asn-Gln-Glu-Gly	1.43
Cytochrome Oxidase I — fMet-Val-Gln-[Arg]-Trp-Leu-Tyr-Ser-Thr-Asn-Ala-[Lys]-[Asp]-Ile-Ala-Val-Leu-Tyr-Phe-Met	1.63
Cytochrome Oxidase III — fMet-Thr-[His]-Leu-Glu-[Arg]-Ser-[Arg]-[His]-Gln-Gln-[His]-Pro-Phe-[His]-Met-Val-Met-Pro-Ser	0.77
Cytochrome b — fMet-Ala-Phe-[Arg]-Lys-Ser-Asn-Val-Tyr-Leu-Ser-Leu-Val-[Asn]-Ser-Tyr-Ile-Ile-[Asp]-Ser	1.43
ATPase 9 — fMet-Gln-Leu-Val-Leu-Ala-Ala-[Lys]-Tyr-Ile-Gly-Ala-Gly-Ile-Ser-Thr-Ile-Gly-Leu-Leu	1.90
ATPase 6 — fMet-Phe-Asn-Leu-Asn-Thr-Tyr-Ile-Thr-Ser-Pro-Leu-[Asp]-Gln-Phe-Glu-Ile-[Arg]-Thr	1.43
	1 5 10 15 20

Fig. 10. Amino-terminal sequences of yeast mitochondrial gene products. The sequences shown for cytochrome c oxidase subunit II and its "leader peptide," II', and for ATPase subunit 9 are derived from both the amino acid sequence of the polypeptides and the nucleotide sequences of their genes. The sequences for all other mitochondrial gene produced have been deduced from the nucleotide sequences of their genes only. Residue numbers of the mature polypeptides are indicated below the sequences. Charged amino acids are shown in boxes. Hydrophobic indices were determined as described by Poyton et al. (1982).

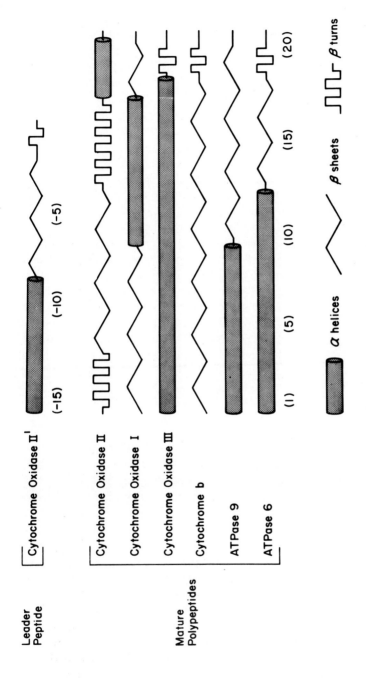

Fig. 11. Predicted secondary structures of the amino-terminal regions of yeast mitochondrial gene products. Probabilities for α helix, β sheet, and reverse turn conformations were determined by the rules of Chou and Fasman [1974, 1978]. Residue numbers of the mature polypeptides are indicated below the predicted secondary structures.

9, the positions of charged amino acids in these sequences, as well as their hydrophobicities (Fig. 10), distinguish them from II'. Although these considerations argue against the presence of an uncleaved leader peptide on ATPase subunit 9 (and other yeast mitochondrial gene products), they do not rule out the possibility that this polypeptide (or other mitochondrial gene products that lack transient leader peptides) has internal peptide segments that perform leader peptide functions (whatever they might be!).

It has recently become clear that even the leader peptide II' is not present on cytochrome c oxidase subunit II in all species of eukaryotes. From an analysis of the sequence of the subunit and its gene from Saccharomyces cerevisiae [Sevarino et al, 1983; Coruzzi and Tzagoloff, 1979], Neurospora crassa [Machleidt and Werner, 1979; Macino, personal communication], human [Chomyn et al, 1981; Anderson et al, 1981], and bovine [Steffans and Buse, 1979; Anderson et al, 1982b], it is obvious that Neurospora crassa subunit II, like that from Saccharomyces cerevisiae, has a leader peptide but that human and bovine subunit II have no leader peptides (Fig. 12). This result is striking when one considers that there is a great deal of sequence homology between the mature subunit II polypeptides from all four species (Fig. 12) and that all four polypeptides are integral polypeptides with the same topological disposition in the inner mitochondrial membrane [Sevarino et al, 1983]. These results imply that the leader peptides on subunit II polypeptides in Saccharomyces cerevisiae and Neurospora crassa may not be required for the signaling or the insertion of these polypeptides into the inner mitochondrial membrane (since subunit II from other species lack it), and suggest that leader peptides in general may perform as yet unrecognized functions in addition to, or instead of, their putative [Blobel and Sabatini, 1971; Blobel and Dobberstein, 1975a, b] membrane recognition signal function.

2. Possible roles of "leader peptides" in the synthesis and assembly of membrane proteins. As discussed earlier, leader peptides are present on polypeptides that follow both *cotranslational* and *posttranslational* routes to cellular membranes. Since there is no a priori reason to assume that a targeting signal has to be at the NH_2 terminus of a membrane-directed protein that follows a *posttranslational* route, it is conceivable that leader peptides perform more than one type of function in the biogenesis of membrane proteins. Aside from their presumed signal functions, at least two types of roles may be envisioned for leader peptides. On the one hand they may serve to regulate the translation of the polypeptide chain which they preceed; on the other hand, they may play a role in the assembly of oligomeric membrane proteins. As discussed below, both types of function may have an underlying role in targeting proteins to specific regions (eg, "unit areas"—Fig. 2) on a membrane's surface, and may therefore be difficult to distinguish experimentally from the signal function commonly attributed to leader peptides.

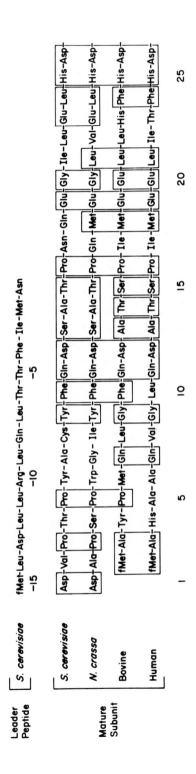

Fig. 12. "Leader peptides" do not occur on cytochrome c oxidase subunit II in all eukaryotic species. From an analysis of the amino acid sequences of subunit II polypeptides and nucleic acid sequences of their genes it is apparent that subunit II from Saccharomyces cerevisiae and Neurospora crassa are translated with amino-terminal "leader peptides" whereas bovine and human subunit II are not. Subunit II polypeptide from the latter two species is formylated. Mature subunit II sequences from all four species have a high degree of homology. Residue numbers of the mature polypeptides are given below the sequences.

a. Synthesis: Cotranslational attenuation. In trying to explain why cytochrome c oxidase subunit II is synthesized with a leader peptide in some species but not in others, it is of interest to note that those species which have a leader peptide are unicellular organisms that are capable of regulating their intracellular levels of cytochrome c oxidase through a fairly wide range of concentrations, whereas those species that lack the leader peptide are not. It is thus possible that the leader peptide on subunit II in Saccharomyces cerevisiae and Neurospora crassa regulates the translation of mature subunit II which in turn determines the intracellular level of holocytochrome c oxidase.

How might the "leader peptide" on subunit II and other membrane-directed proteins which follow cotranslational routes regulate translation? This may occur by a mechanism that is analogous to transcriptional attenuation [Yanofsky, 1981; Platt, 1981]. This mechanism, here termed "cotranslational attenuation,"[2] proposes that the leader peptide, once translated to a length that extends it beyond the ribosomal exit domain [Bernabeu and Lake, 1982], combines with ancillary components (eg, "signal-recognition proteins" and possibly other soluble proteins such as elongation factors involved in translational elongation) and searches for an available receptor on the surface of the membrane (Fig. 13). During the time it takes the leader peptide to combine with the "signal recognition" protein and its receptor (eg, "docking protein") [Meyer et al, 1982], translational elongation ceases. Upon finding an unoccupied receptor translation continues and the protein is inserted into the membrane. If the leader peptide fails to find an unoccupied receptor, translation would be terminated prematurely. This model is analogous to one proposed recently by Walter and Blobel [1981c] and Meyer et al [1982], and receives additional experimental support from the observation that procollagen peptides may control the synthesis of collagen at the translational elongation level [Horlein et al, 1981]. It is derived in concept from transcriptional attenuation in bacteria [Yanofsky, 1981; Platt, 1981]. Both types of attenuation are at the interface between coupled processes—transcription-translation for the former and translation-insertion for the latter—and both types of attenuation use information synthesized in the first process to dictate the continuation or cessation of the second. It will be interesting to see how the leader peptide affects translational elongation and to look for further analogies at the molecular level between cotranslational and transcriptional attenuation. It will also be important to test further the hypothesis that a

---

[2]"Cotranslational attenuation" is distinguishable from "translational attenuation" described recently by Gryczan et al [1980] and Hahn et al [1982]. In the former, the "leader peptide" is proposed to have a role in membrane-receptor recognition and to regulate translation positively when bound to the receptor and negatively when not bound to it. In the latter, the structure of leader region mRNA is thought to regulate translation.

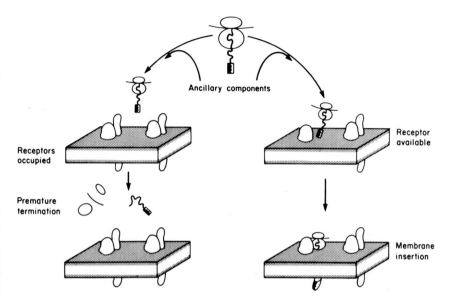

Fig. 13. "Leader peptide"–mediated cotranslational attenuation. By way of analogy to transcriptional attenuation [Yanofsky, 1981], it is proposed here that "leader peptides" play a role in the attenuation of translation for proteins that enter membranes *cotranslationally*. According to this model, translational elongation and normal termination of these proteins are possible only if unoccupied receptors, which are recognized by the "leader peptide," are available on the membrane surface. The leader peptide is shown as a black and white box.

leader peptide can affect translational elongation. A critical test of this model will be achieved most easily through genetics and through demonstration that amino acid substitutions in the leader peptide (at positions which are important for translational inhibition) lead to the overproduction of its attached protein.

b. Assembly: Are "leader peptides" analogous to viral scaffolding proteins? The assembly of hetero-oligomeric membrane proteins is complex and likely to be composed of many sequential events. For reasons that have been discussed earlier in this section, it is attractive to consider the hypothesis that the insertion of subunit polypeptides of hetero-oligomeric protein complexes into the membrane is coupled to their assembly into the oligomeric structure itself. With such a model in mind it is easy to imagine that the leader peptide on subunit polypeptides might serve to catalyze the formation of a receptor for its attached protein from two (or more) different polypeptide subunits that are to reside in the assembled oligomer (Fig. 14).

This *assembly* function of leader peptides would accomplish two things. First, it would target the *new* subunit to a region of the membrane surface

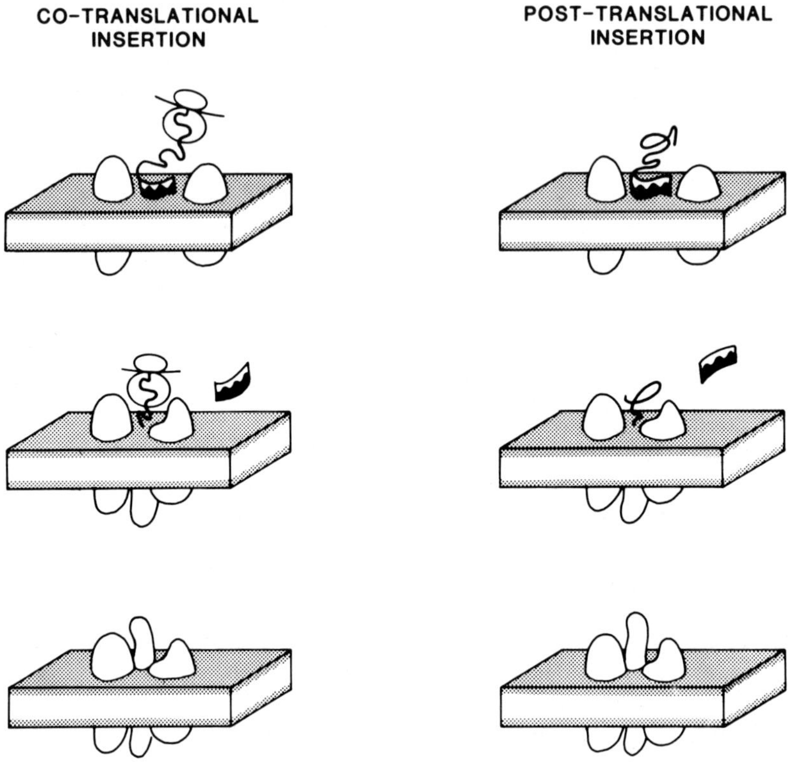

CO-TRANSLATIONAL
INSERTION

POST-TRANSLATIONAL
INSERTION

Fig. 14. A possible role for "leader peptides" in the assembly of oligomeric membrane proteins. Here it is proposed that "leader peptides" have functions analogous to those of viral scaffolding proteins [Wood and King, 1979]. As such, they may serve to catalyze the assembly of membrane oligomers for polypeptide subunits which are inserted either *cotranslationally* or *posttranslationally* into the membrane. The "leader peptide" is designated by a black and white box.

where there exist other subunits of the oligomer. Second, it would catalyze the formation of a partially assembled oligomer, either by bringing subunits that are already in the membrane closer together or by changing their conformation to allow for the subsequent addition of new polypeptide subunits to the oligomer. Since the leader peptide is not found on mature assembled polypeptides of hetero-oligomeric membrane proteins, this assembly function of a leader peptide is transient and would be directly analogous to roles played by scaffolding proteins during the assembly of bacteriophage [Wood and King, 1979; Kunzler and Berger, 1981]. As shown in Figure 14, this putative function of a leader peptide is, unlike the cotranslation attenuation function discussed above, not restricted to proteins that follow *cotranslational* routes to their membrane receptor. It is equally feasible for proteins that follow *posttranslational* pathways. As was true for the putative *cotranslational* attenuation function of leader peptides, this presumptive function can be critically evaluated by looking for defective assembly of membrane oligomers in mutants that have modified leader peptides.

In considering both of the above hypotheses it is important to note that the proposed synthesis and assembly function of leader peptides would be operative to target the attached polypeptide to a specific site on the membrane surface. They therefore do not replace, but rather extend, the presumed signal function of leader peptides.

C. Sequential Aspects of the Assembly of Oligomeric Membrane Proteins

Currently, it is assumed that the assembly of membrane protein oligomers is sequential, with subunit polypeptides entering into the oligomeric state in a partially assembled oligomer one after another [for examples see Poyton, 1980; Cox et al, 1981]. However, very little is known about the sequence or types of subunit-subunit interactions that take place during the assembly of any membrane-protein oligomer. Moreover, it is not yet clear how the insertion of subunit polypeptides into a membrane and their posttranslational modification (by proteolytic processing, glycosylation, phosphorylation, etc) is related to their assembly into the oligomeric state. Nor is it clear if the synthesis and assembly of new oligomer molecules is in any way related to the turnover of preexisting (ie, old) oligomer molecules.

Although generally ignored during considerations of membrane biogenesis, the turnover of membrane protein oligomers may be very important for both the synthesis of new subunit polypeptides and their assembly into the oligomeric state. For example, the dissociation of oligomers into subunits may play a role in "turning on" the synthesis of new subunits and, at the same time, provide old subunit partners with which new subunits assemble. Although the experimental data relevant to these two aspects of membrane assembly are meager, it is interesting that unassembled pools of subunits of some membrane oligomers (eg, cytochrome c oxidase [see review by Poyton, 1980]) have been identified and that the different subunits of membrane hetero-oligomers may turn over at very different rates [see Hare and Hodges, 1982]. Both of these observations are consistent with the interesting possibility (see below) that subunits of old oligomers may assemble with new subunits to make hybrid oligomers.

VII. MEMORY IN MEMBRANES: THE OLIGOMER TEMPLATE HYPOTHESIS

In the above sections of this review it has been proposed that genetic memory and spatial membrane memory interact whenever new proteins are inserted into membranes and that, insofar as membranes do not form de novo, it is the preexisting spatial memory encoded in a membrane that brings new proteins to its surface. It has been proposed also that the molecular basis for this interaction is at the level of signal-receptor recognition; where the signal on a newly synthesized membrane protein is encoded by genetic memory, and the membrane receptor which it recognizes is encoded by spatial memory. Yet to be considered is how this spatial memory is encoded and transmitted during cell growth and division as membrane surfaces expand.

A molecular explanation of these latter two aspects of membrane memory requires some of the same considerations that were necessary to understand the molecular basis of genetic memory and its expression via the genetic code. Realizing that genetic memory is one-dimensional, along a DNA molecule, whereas spatial memory is likely to be two-dimensional, along membrane surfaces, and three-dimensional within the cellular interior, it is probable that spatial memory is more complicated and diverse than genetic memory. Nevertheless, as was the case for genetic memory, in order to understand spatial memory in membranes it is essential 1) to define the unit of memory; 2) to ascertain how it is encoded along a membrane surface; and 3) to determine how it is replicated and transmitted during membrane growth.

In the following discussion we will speculate about each of these aspects of membrane memory in the context of current knowledge concerning the oligomeric structure of integral membrane proteins, the organization of membrane proteins in the plane of the membrane, and the biosynthesis of membrane proteins. To do this we have developed the "oligomer template hypothesis" (Fig. 15). This hypothesis takes note of the fact that the functional specificity of a membrane lies in its enzymes and proteins and proposes therefore that membrane memory is encoded in membrane proteins, not phospholipids, and is determined by protein-protein interactions across the polypeptide subunit (protomer) faces of hetero-oligomeric membrane proteins. It assumes that the *units of memory* are hetero-oligomeric membrane proteins but doesn't specify whether one, or more than one, type of hetero-oligomer is involved. To explain the differentiation of intracellular membranes this hypothesis also assumes that memory-encoding hetero-oligomeric membrane proteins are different from one cellular membrane to the next. Although it is conceivable that the memory of each intracellular membrane is encoded in only one type of hetero-oligomeric protein, it is also possible that all of the hetero-oligomeric proteins which are present in a membrane participate.

In considering how units of memory (ie, hetero-oligomers) might be encoded along the surface of a membrane it is useful to recall the concept of a "unit area" (section III A; Fig. 2). This concept has evolved from the fact that under steady state in vivo conditions the mobility of a membrane protein

I. One "Unit Area"

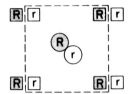

2. Expansion due to insertion of phospholipids

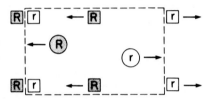

3. Dissociation of oligomers into subunit templates

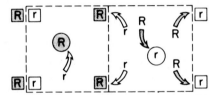

4. Targeting of new subunits to available template receptors

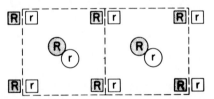

5. Two "Unit Areas"

Fig. 15. The "oligomer template hypothesis" for the transmission of membrane spatial memory during membrane surface expansion. Here, it is proposed that membrane memory is encoded in oligomeric membrane proteins and that this unit of memory is replicated *semiconservatively* with polypeptide subunits of "old" oligomers serving as templates for assembly with newly synthesized polypeptide subunits. In order to simplify visualization of this hypothesis, we have shown the surface expansion of a membrane in one dimension only and have shown only dimeric hetero-oligomers (one whose subunits are designated by circles and another whose subunits are designated by squares). The hypothesis is equally applicable for membrane expansion in two dimensions and for more complex hetero-oligomers. According to this hypothesis each subunit of "old" oligomers is operative as a template receptor for new complementing subunits of the oligomer.

in the plane of a membrane can be restricted by a number of interactions (ie, collisions with other membrane proteins; interactions and collisions with phospholipids organized in crystalline microdomains or in a nonbilayer, non-lamellar, hexagonal phase; and associations with peripheral, soluble, or cytoskeletal proteins) to a "random walk" over a limited region of membrane surface. This region, a "unit area," is viewed here as a repeating unit of membrane surface which may be defined statistically by its time-averaged composition. It is not an area in which the position of proteins are fixed absolutely. At present, there is no definitive experimental data in favor of the existence of "unit areas." Indeed, to the contrary, there is a large body of data that suggest that membrane proteins can diffuse over great distances in the plane of the membrane [reviewed by Cherry, 1979; Edidin, 1982]. However, when viewing this data it is important to note that "unit areas" which may exist in vivo in growing membranes would be easily perturbed in vitro by the introduction of exogenous phospholipids or sterols into the membrane [Schneider et al, 1980a, b; Cherry et al, 1980; Schneider et al, 1982] or by changes in membrane protein-protein interactions brought about by electrophoretic displacement [Sowers and Hackenbrock, 1981], or by changes in salt concentrations, pH, or temperature [see Elgsaeter and Branton, 1974; Hochli and Hackenbrock, 1977]. It is also important to note that the lateral diffusion, or flow, of membrane proteins that is observed during endocytosis and capping [Edidin and Weiss, 1972; Silverstein et al, 1977] may reflect a change in "unit area" structure that ultimately coincides with and is necessary for the endocytotic vesiculation of areas of plasma membrane surface. Indeed, the transformation of plasma membrane, with one type of "unit area" structure, into endocytic vesicle membranes, with another type of "unit area" structure (ie, memory), may be possible only by the rearrangement of molecules (from one type of memory to another) along the surface of the plasma membrane.

Finally, at the heart of the "oligomer template hypothesis" is a mechanism by which memory is passed from one cell generation to the next during the expansion of intracellular membrane surfaces. As seen from Figure 15 this hypothesis proposes a three-step process for the replication and transmission of the membrane hetero-oligomer *unit of memory*. First, the surface area of a membrane increases owing to the insertion of new phospholipid molecules into the bilayer. Second, the increased lipid to protein ratio, resulting from the presence of new phospholipids in the bilayer, leads to the dissociation of membrane hetero-oligomers into their subunit polypeptides. And third, subunits from "old" (preexisting) oligomers serve as templates for newly synthesized subunits which enter the membrane at sites determined by "old" subunit receptors.

In summary, the "oligomer template hypothesis" proposes that 1) the *unit of memory* in biological membranes is encoded in one, or more, type(s) of hetero-oligomeric membrane proteins; 2) each intracellular membrane has (a) unique and diagnostic hetero-oligomeric protein(s) as its unit of memory; 3) units of memory may be encoded along membrane surfaces in quasistable "unit areas"; and 4) memory is transmitted during membrane expansion by hetero-oligomer dissociation and the reassociation of "old" subunits, from preexisting membrane memory, with "new" subunits, from genetic memory, to give hybrid oligomers. Although this hypothesis can only be speculative at present it does make a number of experimentally testable predictions. First, it proposes that some, perhaps all, hetero-oligomeric membrane proteins dissociate during lipid bilayer expansion. Second, it suggests that oligomer polypeptide subunits serve as template receptors for "new" complementary subunits and that "new" and "old" subunits assemble into hybrid molecules. And third, it predicts that the insertion of new subunits of hetero-oligomeric membrane proteins into a membrane is coupled to their assembly into an oligomer. These predictions should permit the "oligomer template hypothesis" to be verified, refuted, or modified.

VIII. CONCLUDING REMARKS

Thanks to the creative energy, interest, and persistence of Keith Porter and to studies by his many colleagues, eukaryotic cell biology now stands poised on the threshold of a future which promises to be as exciting as it is challenging. Among the many challenges to be faced in the future are the acquisition of a better appreciation for the molecular and functional inter-relationships which exist between cellular components, and a clearer understanding of ways in which *genetic memory* is transformed into and mixed with *spatial memory* as information encoded in one-dimension, along a DNA strand, is decoded into the multitude of three-dimensional structures found in the eukaryotic cell.

This article considers, in overview, these two aspects of the spatial organization of the eukaryotic cell. It has sought to identify principles which underlie the functional organization of cellular membranes and compartments and to explain how spatial and functional features of organelle membranes are passed on from one generation to the next. The following are the major conclusions that have been reached.

1. The functional compartmentation of cells occurs at all levels of molecular organization and does not require the presence of membrane-bounded compartments.

2. Membrane-bounded macrocompartments (ie, organellar compartments) have two principal advantages over non-membrane-bounded microcompartments. These are a) the ability to *contain* complete sets of metabolic pathways; and b) the use of the thermodynamic barrier imposed by a membrane lipid bilayer to *conserve* energy in the form of gradients across it.

3. Membrane-bounded compartments communicate with one another via membrane protein oligomers.

4. Membrane proteins may be organized in the plane of the membrane in quasistable liquid crystalline "unit areas." Therefore, an individual membrane protein molecule may not diffuse over an entire membrane surface but may, in fact, be restricted in its distribution to a fairly small region of membrane surface.

5. Membrane proteins may enter a membrane *cotranslationally* or *posttranslationally*. Both routes are followed by nuclear gene products. Proteins of secretory organelles follow a *cotranslational* route whereas proteins of all compartments of mitochondria and chloroplasts and the peroxisomal lumen follow *posttranslational* routes. Mitochondrial and chloroplast gene products that are membrane proteins follow *cotranslational* pathways to the intraorganellar membrane in which they reside.

6. Many, but not all, membrane-directed proteins possess NH_2-terminal leader peptides. Although leader peptides have been proposed to act as signals which target their attached protein to a specific intracellular membrane they may have other functions as well. For example, they may participate in the synthesis of membrane protein by a mechanism we've called *cotranslational attenuation*. They may also play a role in the assembly of membrane protein oligomers.

Finally, in order to visualize how spatial and genetic memory interact and explain how memory may be encoded in membranes, we have proposed the "oligomer template hypothesis." This hypothesis assumes that spatial and genetic memory interact when newly synthesized membrane proteins are inserted into a membrane. It proposes that hetero-oligomeric membrane protein function as the *unit(s) of memory* in membranes and that intracellular membranes are differentiated with respect to one another because they possess different hetero-oligomeric proteins and hence different memories. This hypothesis also proposes that each *unit of memory* is replicated *semiconservatively,* and that polypeptide subunits of "old" oligomers serve as templates for assembly with new polypeptide subunits.

ACKNOWLEDGMENTS

I would like to thank the members of my laboratory and Dr J. R. McIntosh for many stimulating discussions.

Some of the research carrried out in the author's laboratory and reviewed here was supported by grants GM 29838 and GM 30228 from the National Institutes of Health. The author is an Established Investigator of the American Heart Association.

IX. REFERENCES

Ades IZ, Harpe KG (1981): Biogenesis of mitochondrial proteins. Identification of mature and precursor forms of δ-aminolevulinate synthase from embryonic chick liver. J Biol Chem 256:9329–9333.

Alt FW, Bothwell ALM, Knapp M, Siden E, Mather E, Koshland M, Baltimore D (1980): Synthesis of secreted and membrane-bound immunoglobulin Mu heavy chains is directed by mRNAs that differ at their 3' ends. Cell 20:293–301.

Ammerer G, Richter K, Hartler E, Ruis H (1981): Synthesis of Saccharomyces cerevisiae Catalase A in vitro. Eur J Biochem 113:327–331.

Anderson S, Bankier AT, Barrell BG, De Bruijn MHL, Coulson AR, Drouin J, Eperon IC, Nierlich DP, Roe BA, Sanger F, Schreier PH, Smith AJH, Staden R, Young IG (1981): Sequence and organization of the human mitochondrial genome. Nature 290:457–465.

Anderson DJ, Walter P, Blobel G (1982a): Signal recognition protein is required for the integration of acetylcholine receptor δ subunit, a transmembrane glycoprotein, into the endoplasmic reticulum. J Cell Biol 93:501–506.

Anderson S, De Bruijn MHL, Coulson AR, Eperon IC, Sanger F, Young IG (1982b): Complete sequence of bovine mitochondrial DNA. Conserved features of the mammalian mitochondrial genome. J Mol Biol 156:683–717.

Austen BM (1979): Predicted secondary structures of amino-terminal extension sequences of secreted proteins. FEBS Lett 103:308–313.

Autor AP (1982): Biosynthesis of mitochondrial manganese superoxide dismutase in Saccharomyces cerevisiae. J Biol Chem 257:2713–2718.

Bar-Nun S, Kreibich G, Adesnik M, Alterman L, Negishi M, Sabatini DD (1980): Synthesis and insertion of cytochrome P-450 into endoplasmic reticulum membranes. Proc Natl Acad Sci USA 77:965–969.

Becker WM, Reizman H, Weir EM, Titus DE, Leaver CJ (1982): In vitro synthesis and compartmentalization of glyoxysomal enzymes for cucumber. Ann NY Acad Sci 386:329–348.

Bell RM, Coleman RA (1980): Enzymes of glycerolipid synthesis in eucaryotes. Ann Rev Biochem 49:459–487.

Bernabeu C, Lake JA (1982): Nascent polypeptide chains emerge from the exit domain of the large ribosomal subunit. Immune mapping of the nascent chain. Proc Natl Acad Sci USA 79:3111–3115.

Birky W (1978): Transmission genetics of mitochondria and chloroplasts. Ann Rev Genet 12:471–512.

Blobel G (1980): Intracellular protein topogenesis. Proc Natl Acad Sci USA 77:1496–1500.

Blobel G, Dobberstein B (1975a): Transfer of proteins across membranes. I. Presence of proteolytically processed and unprocessed nascent immunoglobulin light chains on membrane-bound ribosomes of murine myeloma. J Cell Biol 67:835–851.

Blobel G, Dobberstein B (1975b): Transfer of protein across membranes. II. Reconstitution of rough microsomes from heterologous components. J Cell Biol 67:852–862.

Blobel G, Sabatini DD (1970): Controlled proteolysis of nascent polypeptides in rat liver cell fractions. I. Location of the polypeptides in rough microsomes. J Cell Biol 45:146–157.

Blobel G, Sabatini DD (1971): Ribosome-membrane interaction in eukaryotic cells. In Manson LA (ed): "Biomembranes." New York: Plenum, Vol 2, pp 193–195.

Bloch K, Vance D (1977): Control mechanisms in the synthesis of saturated fatty acids. Ann Rev Biochem 46:263–298.

Bonatti S, Blobel G (1979): Absence of a cleavable signal sequence in sindbis virus glycoprotein pE$_2$. J Biol Chem 254:12261–12264.

62   Poyton

Bostian KA, Lemire JM, Cannon LE, Halvorson HO (1980): In vitro synthesis of repressible yeast acid phosphatase: Identification of multiple mRNAs and products. Proc Natl Acad Sci USA 77:4504–4508.
Braell WA, Lodish HF (1981): Biosynthesis of the erythrocyte anion transport protein. J Biol Chem 256:11337–11344.
Braell WA, Lodish HF (1982a) Ovalbumin utilizes an NH_2-terminal signal sequence. J Biol Chem 257:4578–4582.
Braell WA, Lodish HF (1982b): The erythrocyte anion transport protein is co-translationally inserted in microsomes. Cell 28:23–31.
Branton D, Cohen CM, Tyler J (1981): Interactions of cytoskeletal proteins on the human erythrocyte membrane. Cell 24:24–32.
Campbell PN (1979): Post-translational events associated with the biogenesis of secretory proteins. In Bull AT (ed): "Companion to Biochemistry." London: Longman, Vol 2, pp 229–244.
Carlile MJ (1980): From prokaryote to eukaryote: Gains and losses. In "The Eukaryotic Microbial Cell," Symp Soc Gen Microbiol Vol 30. Cambridge: Cambridge University Press, pp 1–40.
Carlson M, Botstein D (1982): Two differentially regulated mRNAs with different 5' ends encode secreted and intracellular forms of yeast invertase. Cell 28:145–154.
Cherry RJ (1979): Rotational and lateral diffusion of membrane proteins. Biochim Biophys Acta 559:289–327.
Cherry FJ, Müller U, Holenstein C, Heyn MP (1980): Lateral segregation of proteins induced by cholesterol in bacteriorhodopsin phospholipid vesicles. Biochim Biophys Acta 596:145–151.
Chomyn A., Hunkapiller MW, Attardi G (1981): Alignment of the amino terminal amino acid sequence of human cytochrome c oxidase subunits I and II with the sequence of their putative mRNAs. Nucl Acids Res 9:867–877.
Chou PY, Fasman GD (1974): Prediction of protein conformation. Biochemistry 13:222–245.
Chou PY, Fasman GD (1978): Prediction of the secondary structure of proteins from their amino acid sequence. Adv Enzymol Related Areas Mol Biol 47:45–148.
Chua N-H, Schmidt GW (1979): Transport of proteins into mitochondria and chloroplasts. J Cell Biol 81:461–483.
Chua N-H, Blobel G, Siekevitz P, Palade GE (1976): Periodic variations in the ratio of free to thylakoid-bound chloroplast ribosomes during the cell cycle in Chlamydomonas reinhardtii. J Cell Biol 71:497–514.
Conboy JG, Kalousek F, Rosenberg LE (1979): In vitro synthesis of a putative precursor to mitochondrial ornithine transcarbamoylase. Proc Natl Acad Sci USA 76:5724–5727.
Conboy JG, Rosenberg LE (1981): Posttranslational uptake and processing of in vitro synthesized ornithine transcarbamoylase precursor by isolated rat liver mitochondria. Proc Natl Acad Sci USA 78:3073–3077.
Coruzzi G, Tzagoloff A (1979): Assembly of the mitochondrial membrane system. DNA sequence of subunit 2 of yeast cytochrome c oxidase. J Biol Chem 254:9324–9330.
Cote C, Solioz M, Schatz G (1979): Biogenesis of the cytochrome bc_1 complex of yeast mitochondria. J Biol Chem 254:1437–1439.
Cox GB, Downie JA, Langman L, Senior AE, Ash G, Fayle DRH, Gibson F (1981): Assembly of adenosine triphosphatase complex in Escherichia coli. Assembly of F_0 is dependent on the formation of specific F_1 subunits. J Bacteriol 148:30–42.
Cross HS, Ruis H (1978): Regulation of catalase synthesis in Saccharomyces cerevisiae by carbon catabolite repression. Mol Gen Genet 166:37–43.

Darnell JE (1982): Variety in the level of gene control in eukaryotic cells. Nature (Lond) 297:365–371.

De Kruijff B, Cullis PR, Verkleij AJ (1982): Structural and functional aspects of nonbilayer lipids. In Martonosi A (ed): "Membranes and Transport." New York: Plenum, Vol 1, pp 43–49.

De Pierre JW, Seidegard J, Morgenstern R, Balk L, Meijer J, Astrom A (1981): Induction of drug-metabolizing enzymes: A status report. In Lee CP, Schatz G, Dallner G (eds): "Mitochondria and Microsomes." Reading, Massachusetts: Addison-Wesley, pp 585–610.

De Ronde A, Van Loon APGM, Grivell LA, Kohli J (1980): In vitro suppression of UGA codons in a mitochondrial mRNA. Nature (Lond) 287:361–363.

Di Renzio JM, Nakamura K, Inouye M (1978): The outer membrane proteins of gram-negative bacteria: Biosynthesis, assembly and functions. Ann Rev Biochem 47:481–532.

Dobberstein B, Blobel G, Chua N-H (1977): In vitro synthesis and processing of a putative precursor for the small subunit of ribulose-1,5-biphosphate carboxylase of Chlamydomonas reinhardtii. Proc Natl Acad Sci USA 74:1082–1085.

Douce R, Joyard J (1981): Does the plastid envelope derive from the endoplasmic reticulum? Trends Biochem Sci 6:237–239.

Early P, Rogers J, Davis M, Calame K, Bond M, Wall R, Hood L (1980): Two mRNAs can be produced from a single immunoglobulin μ gene by alternative processing pathways. Cell 20:313–319.

Edidin M (1982): Lateral diffusion of membrane proteins. In Martonosi A (ed): "Membranes and Transport." New York: Plenum, Vol 1, pp 141–144.

Edidin M, Weiss M (1972): Antigen cap formation in cultured fibroblasts: A reflection of membrane fluidity and cell mobility. Proc Natl Acad Sci USA 69:2456–2459.

Elbrink J, Bihler I (1975): Membrane transport: Its relation to cellular metabolic rates. Science 188:1177–1184.

Elgsaeter A, Branton D (1974): Intramembrane particle aggregation in erythrocyte ghosts. I. The effects of protein removal. J Cell Biol 63:1018–1036.

Engelman DM, Henderson R, McLachlan AD, Wallace BA (1980): Path of the polypeptide in bacteriorhodopsin. Proc Natl Acad Sci USA 77:2023–2027.

Engelman DM, Steitz TA (1981): The spontaneous insertion of proteins into and across membranes: The helical hairpin hypothesis. Cell 23:411–422.

Erecinska M, Wilson DF (1978): Homeostatic regulation of cellular energy metabolism. Trends Biochem Sci 3:219–222.

Feldman F, Mahler HR (1974): Mitochondrial biogenesis: Retention of terminal formyl methionine in membrane proteins and regulation of their synthesis. J Biol Chem 249:3702–3709.

Fossel ET, Solomon AK (1978): Ouabain-sensitive interaction between human red blood cell membrane and glycolytic enzyme complex in the cytosol. Biochim Biophys Acta 570:99–111.

Fox TD (1979): Five TGA "stop" codons occur within the translated sequence of the yeast mitochondrial gene for cytochrome c oxidase subunit II. Proc Natl Acad Sci USA 76:6534–6538.

Frederick JF (1981): Origins and evolution of eukaryotic intracellular organelles. Ann NY Acad Sci 361.

Freire E, Snyder B (1982): Compositional domain structure of lipid membranes. In Martonosi A (ed): "Membranes and Transport." New York: Plenum, Vol 1, pp 37–41.

Futai M, Kanazawa H (1980): Role of subunits in proton-translocating ATPase (F_0–F_1). Curr Top Bioenerg 10:181–215.

Gasser SM, Ohashi A, Daum G, Boehni PC, Gibson J, Reid GA, Yonetani T, Schatz G (1982): Imported mitochondrial cytochromes b_2 and c_1 are processed in two steps. Proc Natl Acad Sci USA 79:267–271.

George-Nascimento C, Poyton RO (1981): Further analysis of the polypeptide subunits of yeast cytochrome c oxidase. Isolation and characterization of subunits III, V, and VII. J Biol Chem 256:9363–9370.

Goldman BM, Blobel G (1978): Biogenesis of peroxisomes: Intracellular site of synthesis of catalase and uricase. Proc Natl Acad Sci USA 75:5066–5070.

Gonzalez FJ, Kasper CB (1982): Cloning of DNA complementary to rat liver NADPH-cyto-chrome c (P-450) oxidoreductase and cytochrome P-450b mRNAs. J Biol Chem 257:5962–5968.

Grossman A, Bartlett S, Chua N-H (1980): Energy-dependent uptake of cytoplasmically-synthesized polypeptides by chloroplasts. Nature (Lond) 285:625–628.

Grossman A, Bartlett S, Schmidt GW, Mullet JE, Chua N-H (1982): Optimal conditions for post-translational uptake of proteins by isolated chloroplasts. In vitro synthesis and transport of plastocyanin, ferredoxin-NADP$^+$ oxidoreductase, and fructose-1,6-bis-phosphatase. J Biol Chem 257:1558–1563.

Gryczan TJ, Grandi G, Hahn J, Grandi R, Dubnau D (1980): Conformational alteration of mRNA structure and post-transcriptional regulation of erythromycin-induced drug re-sistance. Nucl Acids Res 8:6081–6097.

Gutfreund H (1981): Some problems in molecular evolution. In Gutfreund H (ed): "Biochemical Evolution." Cambridge: Cambridge University Press, pp 1–14.

Hacklenberg H, Klingenberg M (1980): Molecular weight and hydrodynamic parameters of the adenosine-5'-diphosphate-adenosine-5'-triphosphate carrier in Triton X-100. Bio-chemistry 19:548–555.

Hahn J, Grandi G, Gryczan TJ, Dubnau D (1982): Translational attenuation of erm C: A deletion analysis. Mol Gen Genet 186:204–216.

Haldane JBC (1967): "The Origin of Life." Reprinted in Bernal JD: "The Origin of Life." New York: World.

Hare JF, Hodges R (1982): Turnover of mitochondrial inner membrane proteins in hepatoma monolayer cultures. J Biol Chem 257:3575–3580.

Harmey MA, Neupert W (1979): Biosynthesis of mitochondrial citrate synthase in Neurospora crassa. FEBS Lett 108:385–389.

Henderson R, Unwin PNT (1975): Three dimensional model of purple membrane obtained by electron microscopy. Nature (Lond) 257:28–32.

Hess B (1973): Organization of glycolysis: Oscillatory and stationary control. In "Rate and Control of Biological Processes," Symp Soc Exp Biol, Vol 27. Cambridge: Cambridge University Press, pp 105–131.

Hess B (1980): Organization of biochemical reactions: From microspace to macroscopic struc-tures. In Nover L, Lynen F, Mothes K (eds): "Cell Compartmentation and Metabolic Channeling." Jena and Amsterdam: VEB Gustav Fischer Verlag and Elsevier/North-Holland, pp 75–92.

Hobbs AS, Albers RW (1980): The structure of proteins involved in active membrane transport. Ann Rev Biophys Bioeng 9:259–291.

Höchli M, Hackenbrock CR (1977): Thermotropic lateral translational motion of intramembrane particles in the inner mitochondrial membrane and its inhibition by artificial peripheral proteins. J Cell Biol 72:278–291.

Hogeboom GH, Schneider WC, Palade GE (1948): Cytochemical studies of mammalian tissues. I. Isolation of intact mitochondria from rat liver; some biochemical properties of mi-tochondria and submicroscopic particulate material. J Biol Chem 172:619–629.

Holliday R (1970): The organization of DNA in eucaryotic chrmosomes. In "Organization and Control in Procaryotic and Eucaryotic Cells," Symp Soc Gen Microbiol, Vol 20. Cambridge: Cambridge University Press, pp 359–379.

Hopper AK, Furukawa AH, Pham HD, Martin NC (1982): Defects in modification of cytoplasmic and mitochondrial transfer RNAs caused by single nuclear mutations. Cell 28:543–550.

Hörlein D, McPherson J. Goh SH, Bornstein P (1981): Regulation of protein synthesis: Translational control by procollagen-derived fragments. Proc Natl Acad Sci USA 78:6163–6167.

Inouye M, Halegoua S (1980): Secretion and membrane localization of proteins in Escherichia coli. CRC Crit Rev Biochem 7:339–371.

Jamieson JD (1972): Transport and discharge of exportable proteins in pancreatic exocrine cells: In vitro studies. Curr Top Membr Transport 3:273–338.

Jardetsky O (1982): NMR in the study of membranes. In Martonosi A (ed): "Membranes and Transport." New York: Plenum, Vol 1, pp 109–113.

Jensen RA (1976): Enzyme recruitment of evolution of new function. Ann Rev Microbiol 30:409–425.

Karadsheh NS, Uyeda K (1977): Changes in allosteric properties of phosphofructokinase bound to erythrocyte membranes. J Biol Chem 252:7418–7420.

Karnovsky MJ, Kleinfeld AM, Hoover RL, Klausner RD (1982): The concept of lipid domains in membranes. J Cell Biol 94:1–6.

Kindl H (1982): Glyoxysome biogenesis via cytosolic pools in cucumber. Ann NY Acad Sci 386:314–326.

Kirk JTO (1971): Chloroplast structure and biogenesis. Ann Rev Biochem 40:161–196.

Kitutani H, Sitia R, Good RA, Stavnezer J (1981): Synthesis and processing of the α heavy chains of secreted and membrane bound IgA. Proc Natl Acad Sci USA 78:6436–6440.

Klingenberg M (1981): Membrane protein oligomeric structure and transport functions. Nature 290:449–454.

Koerner TA, Voll RS, Younathan E (1977): A proposed model for the regulation of phosphofructokinase and fructose 1,6-bisphosphatase based on their reciprocal anomeric specifications. FEBS Lett 84:207–213.

Korb H, Neupert W (1978): Biogenesis of cytochrome c in Neurospora crassa. Synthesis of apocytochrome c, transfer to mitochondria, and conversion to holocytochrome c. Eur J Biochem 91:609–620.

Koshland D, Botstein D (1980): Secretion of beta-lactamase requires the carboxy end of the protein. Cell 20:749–760.

Kraus JP, Conboy JG, Rosenberg LE (1981): Pre-ornithine transcarbamylase. Properties of the cytoplasmic precursor of a mitochondrial matrix enzyme. J Biol Chem 256:10739–10742.

Kreibich G, Czako-Graham M, Grebenau RC, Sabatini DD (1980): Functional and structural characteristics of endoplasmic reticulum proteins associated with ribosome binding sites. Ann NY Acad Sci 343:17–33.

Kreibich G, Ojakian G, Rodriguez-Boulan E, Sabatini DD (1982): Recovery of ribophorins and ribosomes in "inverted rough" vesicles derived from rat liver rough microsomes. J Cell Biol 93:111–121.

Kreil G (1981): Transfer of proteins across membranes. Ann Rev Biochem 50:317–348.

Kreiter PA, Shires TK (1980): Cell-free translation of apocytochrome b_5 messenger RNA and analysis of the peptide products. Biochem Biophys Res Commun 94:606–611.

Kunzler P, Berger H (1981): Structure of the scaffold in bacteriophage lambda preheads: Removal of the scaffold leads to a change of prehead shell. J Mol Biol 153:961–978.

Kyte J (1975): Structural studies of sodium and potassium ion-activated adenosine triphosphatase. The relationship between molecular structure and the mechanism of active transport. J Biol Chem 260:7443–7449.

Lazarow PB, Robbi M, Fujiki Y, Wong L (1982): Biogenesis of peroxisomal proteins in vivo and in vitro. Ann NY Acad Sci 386:285–297.

Lewin AS, Gregor I, Mason TL, Nelson N, Schatz G (1980): Cytoplasmically-made subunits of yeast mitochondrial F_1 ATPase and cytochrome c oxidase are synthesized as individual precursors, not as polyproteins. Proc Natl Acad Sci USA 77:3998–4002.

Linnane AL, Haslam JM, Lukins HB, Nagley P (1972): The biogenesis of mitochondria in microorganisms. Ann Rev Microbiol 26:163–198.

Lingappa VR, Lingappa JR, Blobel G (1979): Chicken ovalbumin contains an internal signal sequence. Nature (Lond) 281:117–121.

Lustig A, Levens D, Rabinowitz M (1982a): The biogenesis and regulation of yeast mitochondrial RNA polymerase. J Biol Chem 257:5800–5808.

Lustig A, Padmanaban G, Rabinowitz M (1982b): Regulation of nuclear-coded peptides of yeast cytochrome c oxidase. Biochemistry 21:309–316.

Maccecchini ML, Rudin Y, Schatz G (1979a): Transport of proteins across the mitochondrial outer membrane. J Biol Chem 254:7468–7471.

Maccecchini ML, Rudin Y, Blobel G, Schatz G (1979b): Import of proteins into mitochondria: Precursor forms of the extra-mitochondrially made F_1 ATPase subunits in yeast. Proc Natl Acad Sci USA 76:343–347.

Machleidt W, Werner S (1979): Is the mitochondrially made subunit 2 of cytochrome c oxidase synthesized as a precursor in Neurospora crassa. FEBS Lett 107:327–330.

Macino G. Tzagoloff A (1979): Assembly of the mitochondrial membrane system. The DNA sequence of a mitochondrial ATPase gene in Saccharomyces cerevisiae. J Biol Chem 254:4617–4623.

Mahler HR, Dawidowicz K, Feldman F (1972): Formate as a specific label for mitochondrial translation products. J Biol Chem 247:7439–7442.

Matsuura M, Arpin M, Hannum C, Margoliash E, Sabatini DD, Morimoto T (1981): In vitro synthesis and post-translational uptake of cytochrome c into isolated mitochondria. Role of a specific addressing signal in the apocytochrome. Proc Natl Acad Sci USA 78:4368–4372.

McKnight GS, Palmiter RD (1979): Transcriptional regulation of the ovalbumin and conalbumin genes by steroid hormones in chick oviduct. J Biol Chem 254:9050–9058.

Meyer DI, Krause E, Dobberstein B (1982): Secretory protein translocation across membranes—The role of the "docking protein." Nature (Lond) 297:647–650.

Michel H, Oesterhelt D, Henderson R (1980): Orthorhombic two-dimensional crystal form of the purple membrane. Proc Natl Acad Sci USA 77:338–342.

Michel R, Wachter E, Sebald W (1979): Synthesis of a larger precursor for the proteolipid of the mitochondrial ATPase complex of Neurospora crassa in a cell-free wheat germ system. FEBS Lett 101:373–376.

Mihara K, Blobel G (1980): The four cytoplasmically-made subunits of yeast mitochondrial cytochrome c oxidase are synthesized individually and not as a polyprotein. Proc Natl Acad Sci USA 77:4160–4164.

Mihara K, Omura T. Harano T, Brenner S, Fleischer S, Rajagopalen KU, Blobel G (1982): Rat liver L-glutamate dehydrogenase, malate dehydrogenase D-β-hydroxybutyrate dehydrogenase, and sulfite oxidase are each synthesized as larger precursors by cytoplasmic polysomes. J Biol Chem 257:3355–3358.

Monod J, Changeux JP, Jacob F (1963): Allosteric proteins and cellular control systems. J Mol Biol 6:306–329.

Montgomery DL, Hall BD, Gillam S, Smith M (1978): Identification and isolation of the yeast cytochrome c gene. Cell 14:673–680.

Mooseker MS, Tilney LG (1975): Organization of an actin filament–membrane complex. Filament polarity and membrane attachment in the microvilli of intestinal epithelial cells. J Cell Biol 67:725–743.

Moreno F, Fowler AV, Hall M, Silhavy TJ, Zabin I, Schwartz M (1980): A signal sequence is not sufficient to lead β-galactosidase out of the cytoplasm. Nature (Lond) 286:356–359.

Mori M, Miura S, Tatibana M, Cohen PP (1979a): Cell-free synthesis and processing of a putative precursor for mitochondrial carbamyl phosphate synthetase I of rat liver. Proc Natl Acad Sci USA 76:5071–5075.

Mori M, Morris SM, Cohen PP (1979b): Cell-free translation and thyroxine induction of carbamyl phosphate synthetase I messenger RNA in tadpole liver. Proc Natl Acad Sci USA 76:3179–3183.

Mori M, Morita T, Ikeda F, Amaya Y, Tatibana M, Cohen PP (1981a): Synthesis, extracellular transport, and processing of the precursors for mitochondrial ornithine transcarbamylase and carbamyl phosphate synthetase in isolated hepatocytes. Proc Natl Acad Sci USA 78:6056–6060.

Mori M, Morita T, Miura S, Tatibana M (1981b): Uptake and processing of the precursor for rat liver ornithine transcarbamylase in isolated mitochondria. Inhibition by uncouplers. J Biol Chem 256:8263–8266.

Morré DJ (1980): Flow-differentiation of membranes: Pathways and mechanisms. In Nover L, Lynen F, Mothes K (eds): "Cell Compartmentation and Metabolic Channeling." Jena and Amsterdam: VEB Gustav Fischer Verlag and Elsevier/North-Holland, pp 47–61.

Mowbray J, Moses V (1976): The tentative identification in Escherichia coli of a multienzyme complex with glycolytic activity. Eur J Biochem 66:25–36.

Nelson N, Schatz G (1979a): Energy-dependent processing of cytoplasmically-made precursors to mitochondrial proteins. In Lee CP, Schatz G, Ernster L (eds): "Membrane Bioenergetics." Reading, Massachusetts: Addison-Wesley, pp 133–152.

Nelson N, Schatz G (1979b): Energy-dependent processing of cytoplasmically-made precursors to mitochondrial proteins. Proc Natl Acad Sci USA 76:4365–4369.

Neupert W, Schatz G (1981): How proteins are transported into mitochondria. Trends Biochem Sci 6:1–4.

Nicholls DG: "Bioenergetics. An Introduction to the Chemiosmotic Theory." New York: Academic Press.

Nover L (1980): Structural and functional organization of differential gene expression in procaryotic and eucaryotic cells. In Nover L, Lynen F, Mothes K (eds): "Cell Compartmentation and Metabolic Channeling." Jena and Amsterdam: VEB Gustav Fischer Verlag and Elsevier/North-Holland, pp 367–387.

Nover L, Lynen F, Mothes K (eds) (1980): "Cell Compartmentation and Metabolic Channeling." Jena and Amsterdam: VEB Gustav Fischer Verlag and Elsevier/North-Holland.

O'Malley K, Pratt P, Robertson J, Lilly M, Douglas MG (1982): Selection of the nuclear gene for the mitochondrial adenine nucleotide translocator by genetic complementation of the op_1 mutation in yeast. J Biol Chem 257:2097–2103.

Osborn MJ, Wu H (1980): Proteins of the outer/membrane of gram-negative bacteria. Ann Rev Microbiol 34:369–422.

Palade GE (1955): A small particulate component of the cytoplasm. J Biophys Biochem Cytol 1:59–68.

Palade GE (1975): Intracellular aspects of the process of protein synthesis. Science 189:347–358.

Palade GE (1978): Membrane biogenesis. In Solomon AK, Karnovsky M (eds): "Molecular Specialization and Symmetry In Membrane Function." Cambridge, Massachusetts: Harvard University Press, pp 3–30.

Palade GE, Porter KR (1954): Studies on the endoplasmic reticulum. I. Its identification in cells in situ. J Exp Med 100:641–656.

Palade GE, Siekevitz P (1956): Liver microsomes. An integrated morphological and biochemical study. J Biophys Biochem Cytol 2:171–200.

Palmiter RD, Gagnon J, Walsh KA (1978): Ovalbumin: A secreted protein without a transient hydrophobic leader sequence. Proc Natl Acad Sci USA 75:94–98.

Parthier B (1980): Evolutionary aspects of gene expression organization in macrocompartments. In Nover L, Lynen F, Mothes K (eds): "Cell Compartmentation and Metabolic Channeling." Jena and Amsterdam: VEB Gustav-Fischer Verlag and Elsevier/North-Holland, pp 107–121.

Perlman D, Halvorson HO (1981): Distinct repressible mRNAs for cytoplasmic and secreted yeast invertase are encoded by a single gene. Cell 25:525–536.

Perlman D, Halvorson HO, Cannon LE (1982): Presecretory and cytoplasmic invertase polypeptides encoded by distinct mRNAs derived from the same structural gene differ by a signal sequence. Proc Natl Acad Sci USA 79:781–785.

Platt T (1981): Termination of transcription and its regulation in the tryptophan operon of E coli. Cell 24:10–23.

Porter KR (1956): The sarcoplasmic reticulum in muscle cells of Amblystoma larvae. J Biophys Biochem Cytol 2:163–170.

Porter KR, Claude A, Fullam EF (1945): A study of tissue culture cells by electron microscopy. J Exp Med 81:233–246.

Poyton RO (1980): Cooperative interactions between mitochondrial and nuclear genomes: Cytochrome c oxidase assembly as a model. Curr Top Cell Regul 17:231–295.

Poyton RO, McKemmie E (1979a): A polyprotein precursor to all four cytoplasmically-translated subunits of cytochrome c oxidase from Saccharomyces cerevisiae. J Biol Chem 254:6763–6771.

Poyton RO, McKemmie E (1979b): Post-translational processing and transport of the polyprotein precursor to subunits IV to VII of yeast cytochrome c oxidase. J Biol Chem 254:6772–6780.

Poyton RO, Sevarino KA, George-Nascimento C, Power SD (1980): Protein precursors in the assembly of yeast cytochrome c oxidase, a transmembrane oligomer of the inner mitochondrial membrane. Ann NY Acad Sci 343:275–292.

Poyton RO, Bellus G, Kerner AL (1982): Biosynthesis of mitochondrial membrane proteins. In Martonosi A (ed): "Membranes and Transport." New York: Plenum, Vol 1, pp 237–247.

Rachubinski RA, Verma DPS, Bergeron JJM (1980): Synthesis of rat liver microsomal cytochrome b₅ by free ribosomes. J Cell Biol 84:705–716.

Raymond Y, Shore GC (1979): The precursor for carbamyl phosphate synthetase is transported to mitochondria via a cytosolic route. J Biol Chem 254:9335–9338.

Raymond Y, Shore GC (1981): Processing of the precursor for the mitochondrial enzyme, carbamyl phosphate synthetase. Inhibition by p-aminobenzamide leads to a very rapid degradation (clearing) of the precursor. J Biol Chem 256:2087–2090.

Redman CM, Sabatini DD (1966): Vectorial discharge of peptides released by puromycin from attached ribosomes. Proc Natl Acad Sci USA 56:608–615.

Redman CM, Siekevitz P, Palade GE (1966): Synthesis and transfer of amylase in pigeon pancreatic microsomes. J Biol Chem 241:1150–1158.

Reed LJ (1974): Multienzyme complexes. Accounts Chem Res 7:40–46.

Ries G, Hundt E, Kadenbach B (1978): Immunoprecipitation of a cytoplasmic precursor to rat liver cytochrome oxidase. Eur J Biochem 91:179–191.

Robbi M. Lazarow PB (1978): Synthesis of catalase in two cell-free protein-synthesizing systems and in rat liver. Proc Natl Acad Sci USA 75:4344–4348.

Robbi M, Lazarow PB (1982): Peptide mapping of peroxisomal catalase and its precursor. Comparison to the primary wheat germ translation product. J Biol Chem 257:964–970.

Rogers J, Early P, Carter C, Calame K, Bond M, Hood L, Wall R (1980): Two mRNAs with different 3′ ends encode membrane-bound and secreted forms of immunoglobin μ chain. Cell 20:303–312.

Rogers J, Choi E, Souza L, Carter C, Word C, Kuehl M, Eisenberg D, Wall R (1981): Gene segments encoding transmembrane carboxyl termini of immunoglobulin γ chains. Cell 26:19–27.

Ruis H (1979): The biosynthesis of catalase. Can J Biochem 57:1122–1130.

Sabatini DD, Kreibich G, Morimoto T, Adesnik M (1982): Mechanisms for the incorporation of proteins into membranes and organelles. J Cell Biol 92:1–22.

Sakaki T, Akehiko T, Chang CH, Ohnishi S-I (1982): Rotational mobility of an erythrocyte membrane integral protein band 3 in dimyristoylphosphatidylcholine reconstituted vesicles and effect of binding of cytoskeletal peripheral proteins. Biochemistry 21:2366–2372.

Sakakibara R, Huynh QK, Nishida Y, Watanabe T, Wada H (1980): In vitro synthesis of glutarate oxaloacetic transaminase isozymes of rat liver. Biochem Biophys Res Commun 95:1781–1788.

Saks VA, Lipina NA, Sharov VG, Smirnov VN, Chazov E, Grosse R (1977): The localization of the MM isozyme of creatine phosphokinase on the surface membrane of myocardial cells and its functional coupling to ouabain inhibited Na^+/K^+ ATPase. Biochim Biophys Acta 465:550–558.

Salhany JM, Gaines KC (1981): Connections between cytoplasmic proteins and the erythrocyte membrane. Trends Biochem Sci 6:13–15.

Salhany JM, Shaklai N (1979): Functional properties of human hemoglobin bound to the erythrocyte membrane. Biochemistry 18:893–899.

Scarpulla RC, Agne KM, Wu R (1981): Isolation and structure of a rat cytochrome c gene. J Biol Chem 256:6480–6486.

Schmelzer E, Heinrich PC (1980): Synthesis of a larger precursor for the subunit IV of rat liver cytochrome c oxidase in a cell free wheat germ system, J Biol Chem 255:7503–7506.

Schmidt GW, Devillers-Thiery A, Desruisseaux H, Blobel G, Chua N-H (1979): NH_2-terminal amino acid sequences of precursor and mature forms of the ribulose 1,5-biphosphate carboxylase small subunits from Chalmydomonas reinhardtii. J Cell Biol 83:615–622.

Schneider H, Lemasters JJ, Höchli M, Hackenbrock CR (1980a): Liposome-mitochondrial inner membrane fusion. Lateral diffusion of integral electron transfer components. J Biol Chem 255:3748–3756.

Schneider H, Lemasters JJ, Höchli M. Hackenbrock CR (1980b): Fusion of liposomes with mitochondrial inner membranes. Proc Natl Acad Sci USA 77:442–446.

Schneider H, Höchli M, Hackenbrock CR (1982): Relationships between the density distribution of intramembrane particles and electron transfer in the mitochondrial inner membrane as revealed by cholesterol incorporation. J Cell Biol 94:387–393.

Schulz GE, Schirmer RH (1979): "Principles of Protein Structure." New York: Springer-Verlag.

Sebald W, Hoppe J, Wachter E (1979): Amino acid sequence of the ATPase proteolipid from mitochondria, chloroplasts, and bacteria (wild type and mutants). In Quagliarello E (ed): "Function and Molecular Aspects of Biomembrane Transport." Amsterdam: Elsevier/North-Holland, pp 63–75.

Sevarino KA, Poyton RO (1980): Mitochondrial membrane biogenesis: Identification of a precursor to yeast cytochrome c oxidase subunit II, an integral polypeptide. Proc Natl Acad Sci USA 77:142–146.

Sevarino KA, Bellus G, McKee EE, Power SD, Poyton RO (1983): Leader peptides occur on cytochrome c oxidase subunit II molecules from some, but not all, eucaryotic species. Cell (in preparation).

Shaklai N Abrahimi H (1980): The interaction of deoxyhemoglobin with the red cell membrane. Biochem Biophys Res Commun 95:1105–1112.

Shore GC, Power F, Bendayan M, Carignan P (1981): Biogenesis of a 35-kilodalton protein associated with the outer mitochondrial membrane in rat liver. J Biol Chem 256:8761–8766.

Shore GC, Carignan P, Raymond Y (1979): In vitro synthesis of a putative precursor to the mitochondrial enzyme, carbamyl phosphate synthase. J Biol Chem 254:3141–3144.

Siekevitz P, Palade GE (1960): A cytochemical study on the pancreas of the guinea pig. V. In vivo incorporation of leucine-1-^{14}C into the chymotrypsinogen of various cell fractions. J Biophys Biochem Cytol 7:619–630.

Silverstein SC, Steinman RM, Cohn ZA (1977): Endocytosis. Ann Rev Biochem 46:669–772.

Singer SJ, Nicolson G (1972): The fluid mosaic model of the structure of cell membranes. Science 175:720–731.

Singer PA, Singer HH, Williamson AR (1980): Different species of messenger RNA encode receptor and secretory IgMu chains differing at their carboxy termini. Nature (Lond) 285:292–298.

Sitte P (1980): General principles of cellular compartmentation. In Nover L, Lynen F, Mothes K (eds): "Cell Compartmentation and Metabolic Channeling." Jena and Amsterdam: VEB Gustav-Fischer Verlag and Elsevier/North-Holland, pp 17–32.

Smith M, Leung DW, Gillam S, Astell CR, Montgomery DL, Hall BD (1979): Sequence of the gene for iso-1-cytochrome c in Saccharomyces cerevisiae. Cell 16:753–761.

Sowers AE, Hackenbrock CR (1981): Rate of lateral diffusion of intramembrane particles: Measurement by electrophoretic displacement and re-randomization. Proc Natl Acad Sci USA 78:6246–6250.

Sonderegger P, Jaussi R, Christen P (1980): Cell-free synthesis of a putative precursor of mitochondrial aspartate aminotransferase with a higher molecular weight. Biochem Biophys Res Commun 94:1256–1260.

Sonderegger P, Jaussi R, Christen P, Gehring H (1982): Biosynthesis of aspartate aminotransferases. Both the higher molecular weight precursor of mitochondrial aspartate aminotransferase and the cytosolic isoenzyme are synthesized on free polysomes. J Biol Chem 257:3339–3345.

Srere PA, Estabrook RW (eds) (1978): "Microenvironments and Metabolic Compartmentation." New York: Academic Press.

Srere PA, Henslee JG (1980): Is there an infrastructure in the mitochondrial matrix? In Nover L, Lynen F, Mothes K (eds): "Cell Compartmentation and Metabolic Channeling." Jena and Amsterdam: VEB Gustav-Fischer Verlag and Elsevier/North-Holland, pp 159–168.

Srere PA, Mosbach K (1974): Metabolic compartmentation: Symbiotic, organellar multienzymic, and microenvironmental. Ann Rev Microbiol 28:61–83.

Stebbing N (1980): Evolution of compartmentation, metabolic channeling, and control of biosynthetic pathways. In Nover L, Lynen F, Mothes K (eds): "Cell Compartmentation and Metabolic Channeling." Jena and Amsterdam: VEB Gustav-Fischer Verlag and Elsevier/North Holland, pp 93–105.

Steffans GJ, Buse G (1979): Studies on cytochrome c oxidase. IV. Primary structure and function of subunit II. Hoppe-Seyler's Z Physiol Chem 360:613–619.

Steiner DF, Quinn PS, Chan SJ, Marsh J, Tager HS (1980): Processing mechanisms in the biosynthesis of proteins. Ann NY Acad Sci 343:1–16.

Strapazon E, Steck TL (1977): Interaction of the aldolase and the membrane of human erythrocytes. Biochemistry 16:2966–2971.

Stumpf PK (1981): Plants, fatty acids, compartments. Trends Biochem Sci 6:173–176.

Swaneck GE, Nordstrom JL, Kreuzaler F, Tsai MJ, O'Malley BW (1979): Effect of estrogen on gene expression in chicken oviduct: Evidence for transcriptional control of ovalbumin gene. Proc Natl Acad Sci USA 76:1049–1053.

Tai PC, Davis BD (1982): Direct demonstration of cotranslational secretion of proteins in bacteria. In Martonosi A (ed): "Membranes and Transport." New York: Plenum, Vol 1, pp 307–313.

Tanaka A, Osumi M, Fukui S (1982): Peroxisomes in alkane-grown yeast: Fundamental and practical aspects. Ann NY Acad Sci 386:183–198.

Todd RD, Douglas MG (1981): A model for the structure of yeast mitochondrial adenosine triphosphatase complex. J Biol Chem 256:6984–6989.

Todd RD, Griesenbeck T, Douglas MG (1980): The yeast mitochondrial adenosine triphosphatase complex. Subunit stoichiometry and physical characterization. J Biol Chem 255:5461–5467.

Von Heijne G (1982): Signal sequences are not uniformly hydrophobic. J Mol Biol 159:537–541.

Von Heijne G, Blomberg E (1979): Transmembrane translocation of proteins. Eur J Biochem 97:175–181.

Von Wettstein D (1981): Chloroplast and nucleus: Concerted interplay between genomes of different cell organelles. In Schweiger HG (ed): "International Cell Biology 1980–1981." New York: Springer-Verlag, pp 250–272.

Walter P, Blobel G (1981a): Translocation of proteins across the endoplasmic reticulum. I. Signal recognition protein (SRP) binds to in vitro–assembled polysomes synthesizing secretory protein. J Cell Biol 91:545–550.

Walter P, Blobel G (1981b): Translocation of proteins across the endoplasmic reticulum. II. Signal recognition protein (SRP) mediates the selective binding to microsomal membranes of in vitro–assembled polysomes synthesizing secretory protein. J Cell Biol 91:551–556.

Walter P, Blobel G (1981c): Translocation of proteins across the endoplasmic reticulum. III. Signal recognition protein (SRP) causes signal sequence-dependent and site specific arrest of chain elongation that is released by microsomal membranes. J Cell Biol 91:557–561.

Wickner W (1980): Assembly of proteins into membranes. Science 210:861–868.

Wickner W, Ito K, Mandel G, Bates M, Nokelianen M, Zwizinski C (1980): The three lifes of M_{13} coat protein: A virion capsid, an integral membrane protein, and a soluble cytoplasmic proprotein. Ann NY Acad Sci 343:384–390.

Wickstrom M, Krab K, Saraste M (1981): "Cytochrome Oxidase." New York: Academic Press.

Wilson TH, Lin ECC (1980): Evolution of membrane bioenergetics. J. Supramol Struct 13:421–446.

Walz FG, Vlasnik GP, Omecinski CJ, Bresnick E, Thomas PE, Ryan DE, Levin W (1982): Multiple, immunoidentical forms of phenobarbital-induced rat liver cytochromes P-450 are encoded by different mRNAs. J Biol Chem 257:4023–4026.

Wolosewick JJ, Porter KR (1976): Stereo high-voltage electron microscopy of whole cells of the human diploid line, WI-38. Am J Anat 147:303–324.

Wood WB, King J (1979): Genetic control of complex bacteriophage assembly. In Fraenkel-Conrat H, Wagner R (eds): "Comprehensive Virology." New York: Plenum, Vol 13, pp 581–633.

Yanofsky C (1981): Attenuation in the control of expression of bacterial operons. Nature (Lond) 289:751–758.

Yu J, Steck TL (1975): Associations of band 3, the predominant polypeptide of the human erythrocyte membrane. J Biol Chem 250:9176–9184.

Zimmerman R, Paluck U, Sprinzl M, Neupert W (1979): Cell free synthesis of the mitochondrial ADP/ATP carrier protein of Neurospora crassa. Eur J Biochem 99:247–252.

Zimmerman R, Neupert W (1980): Transport of proteins into mitochondria post-translationally. Transfer of ADP/ATP carrier into mitochondria in vitro. Eur J Biochem 109:217–229.

Zitomer RS, Montgomery DL, Nichols DL, Hall BD (1979): Transcriptional regulation of the yeast cytochrome c gene. Proc Natl Acad Sci USA 76:3627–3631.

Modern Cell Biology 2:73-92
© 1983 Alan R. Liss, Inc., 150 Fifth Avenue, New York, NY 10011

Control and Regulation of the Spatial Organization of Membrane Components by Membrane-Membrane Interactions

L. Andrew Staehelin

From the Department of Molecular, Cellular and Developmental Biology, University of Colorado, Boulder, Colorado 80309

I. INTRODUCTION

One of the most fundamental aspects of cellular membranes is their ability to influence the spatial organization of cellular components. This is most evident in the establishment and maintenance of transmembrane gradients of ions and other molecules, but equally important is their ability to organize membrane-associated components both in the plane of the membrane and across the lipid bilayer continuum. In recent years considerable progress has been made in elucidating both the asymmetric organization of membrane components across the bilayer and how this asymmetry is produced. Less is known about parameters that control and regulate the spatial organization of membranes parallel to the bilayer.

At present five basic mechanisms can be envisaged for controlling the distribution of components within the plane of a membrane: 1) formation of crystalline domains of membrane proteins and/or membrane lipids [Staehelin, 1968; Gulik-Krzywicki, 1975]; 2) cross-linking of membrane components through salt bridges, electrostatic interactions, disulfide bonds, and covalent bonds; 3) tethering of membrane components to structures associated with the cytoskeleton (examples: the interaction of band 3 protein with the spectrin network of red blood cells [Branton et al, 1981]; the interaction of cell surface components in lymphocytes during ligand-induced capping [Sheterline and Hopkins, 1981]); 4) formation of links between adjacent membranes—ie, membrane-membrane interactions as found in intercellular junctions [Staehelin, 1974]; and 5) formation of membrane barriers such as tight junctions that inhibit the free movement of components from one membrane area to another, for instance the translational movement of components from the apical to basolateral membrane regions of epithelial cells [Staehelin, 1974]. In the context of this chapter it is interesting to note that the spatial organization of the tight junctional elements seems to result from a combination of membrane-membrane interactions and interactions with cytoskeletal tethering elements [Hull and Staehelin, 1979].

This report describes how membrane-membrane interactions produce a nonrandom organization of membrane enzymes in chloroplast membranes, how phosphorylation of the membrane adhesion factor regulates the amount of membrane adhesion and thereby membrane spatial organization, and how the latter changes, in turn, affect membrane functional activity.

II. BASIC ARCHITECTURE OF CHLOROPLAST MEMBRANES

The photosynthetic membranes of chloroplasts are known as thylakoids. Thylakoids contain all the pigments and enzymes needed for trapping light energy and transducing it into chemical energy. The thylakoids of higher plants and green algae are spatially differentiated into unstacked (stroma) regions and stacked (grana) regions. All stroma and grana thylakoids are interconnected such that their lumens are part of a single anastomosing internal chamber [Kirk and Tilney-Bassett, 1978].

As shown in Figure 1, the grana regions are formed by closely appressed or stacked thylakoids that are linked by single, unstacked stroma thylakoids. The actual spatial relationship of the two types of thylakoid membranes is shown more clearly in Figure 2, which demonstrates how a given stroma thylakoid can spiral up and around a grana stack and connect with each

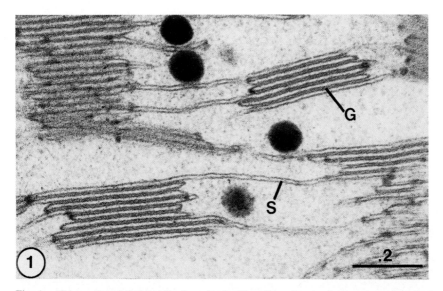

Fig. 1. Thin-sectioned thylakoids of a spinach chloroplast, showing interconnected stacked (grana; G) and unstacked (stroma; S) membranes. × 95,000.

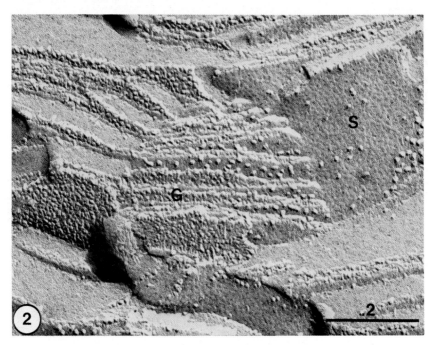

Fig. 2. Freeze-fracture micrograph illustrating the spatial relationship between grana and stroma thylakoids. Note how the stroma (S) membrane on the right is connected to all the membranes of the grana (G) stack in the center. On the left, the cross-fractured stroma membranes show that several such membranes attach to a grana stack, each of these being associated with the grana membranes as illustrated on the right. × 123,000.

thylakoid of the stack. In this manner each grana stack is surrounded and interconnected to several parallel, spiraling stroma thylakoids.

Freeze-fractured thylakoids of green algae and of higher plants reveal four types of fracture faces (Figs. 3, 4) [Goodenough and Staehelin, 1971; Staehelin, 1976]—two associated with stacked regions (referred to as EFs and

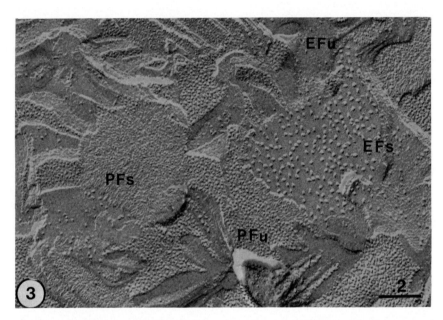

Fig. 3. Freeze-fractured isolated thylakoids from a pea plant illustrating the four types of fracture faces typical for such specimens. The faces EFs and PFs belong to stacked membrane regions, the faces EFu and PFu to unstacked ones. × 62,000.

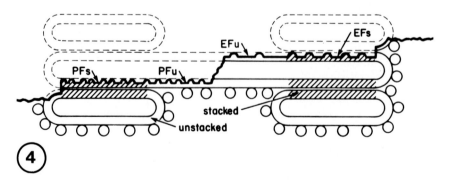

Fig. 4. Illustration of how the membrane faces EFs, EFu, PFs and PFu seen in Figure 3 arise during the fracturing of thylakoid membranes.

PFs), and two with unstacked membrane regions (Efu and PFu). Each type of fracture face carries a distinct complement of intramembrane particles which can be quantitatively characterized by means of particle size histograms (Fig. 5) and particle-density measurements. Table I illustrates that the structural differences between stacked and unstacked thylakoid membranes are correlated with differences in the enzyme composition of the two regions. In recent years considerable effort has been devoted to correlating specific classes of freeze-fracture particles with specific functional units. On current evidence it is generally accepted that the EF particles are structural equivalents of protein-pigment complexes composed of one PSII reaction center surrounded by aggregates of the chlorophyll a/b light harvesting complex (chl a/b LHC) [Armond et al, 1977; Wollman et al, 1980]. The different size classes of EF particles (~ 10.5 nm, ~ 13 nm, and ~ 16 nm in diameter; Fig. 5) seem to be formed by an ~ 8 nm photosystem II (PSII) core particle (PSII reaction center complex) surrounded by varying amounts of chl a/b LHC. The chl a/b LHC is a structural unit consisting of several polypeptides in the 24–29 kilodalton (kd) size range and bound chlorophylls a and b in an a/b ratio of 1.2 [Burke et al, 1979].

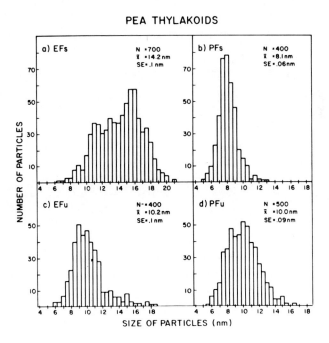

PEA THYLAKOIDS

Fig. 5.   Histograms of particle sizes found on the four fracture faces of pea thyalkoids. Number of particles (N), average diameter ($\overline{X}$), standard error of the mean (SE).

TABLE I. Spatial Distribution of Chloroplast Membrane
Components

Component	Stacked (grana) membranes	Unstacked (stroma) membranes
PSII*	~85%	~15%
PSI*	<15%	>85%
ATP synthetase†	0	100%
NADP reductase*	0	100%
CYT F/B$_6$‡	~50%	~50%
CHL A/B LHC‡	Variable (>50%)	Variable (<50%)

*Anderson [1981].
†Miller [1976].
‡Anderson [1982].

The analysis of PF particles is less advanced, but indirect evidence from studies of mutant thylakoid membranes and of purified membrane protein complexes reconstituted into artificial bilayer membranes suggests the following tentative picture for the relationships between size classes of PF particles and functional membrane units: photosystem I (PSI) complexes and 10 to 11 nm particles [Miller, 1980; Mullet et al, 1980]; cyt f/b$_6$ complexes and 8 to 9 nm particles [Sprague et al, 1981]; CF_0–CF_1 complexes and 9 to 10 nm particles [Mullet et al, 1981b] (Mörschel et al, unpublished results); and free chl a/b LHC complexes and 8 to 9 nm particles [Staehelin and Arntzen, 1979; Simpson, 1979; McDonnel and Staehelin, 1980; Mullet and Arntzen, 1980]. These are still very tentative correlations, and further studies are needed to confirm that the above listed complexes indeed correspond to particles of the indicated size ranges in normal thylakoid membranes. If these data are substantiated, however, the particle size histograms of the four fracture faces would be fully consistent with the distribution of functional membrane complexes shown in Table I. The table is based on morphological as well as recent biochemical fractionation studies in which the grana and stroma regions of purified thylakoids were mechanically sheared apart in a Yeda press and then separated by means of polyethylene glycol-dextran phase partitioning methods [Anderson, 1981, 1982; Miller and Staehelin, 1976].

III. EXPERIMENTAL MANIPULATION OF THE SPATIAL ORGANIZATION OF THYLAKOID MEMBRANES

In 1966 Izawa and Good demonstrated that isolated thylakoids could be experimentally unstacked in low salt media (eg, 50 mM Tricine) and restacked by the addition of $MgCl_2$ or NaCl (Figs. 6, 7). Figure 8 illustrates the relationship between concentration of NaCl and percent restacked membranes, and demonstrates that maximal stacking requires >150 mM NaCl. In MgCl$_2$ an equivalent amount of stacking is achieved with 3–5 mM salt. These salt effects on thylakoid membrane structure, as well as mathematical modeling of the physical force fields associated with membrane-membrane interactions, have led to the theory that thylakoid membrane stacking is largely mediated by a combination of Van der Waals attractive forces and electrostatic repulsive forces between the membranes [Barber, 1980, 1982]. However, more specific protein-protein interactions probably stabilize the links once they have formed [Sculley et al, 1980].

Low-salt-induced unstacking of isolated thylakoids leads not only to the separation of the thylakoids in the stacked membrane regions but also to a complete intermixing and randomization of the membrane particles (Fig. 6) [Goodenough and Staehelin, 1971; Ojakian and Satir, 1974; Staehelin, 1976]. Addition of >150 mM NaCl or >2 mM MgCl$_2$ to a suspension of experimentally unstacked thylakoids produces both stacked membrane regions and a complete resegregation of all membrane components within one hour at 4°C (Fig. 7) [Staehelin, 1976; Miller and Staehelin, 1976; Staehelin and Arntzen, 1979].

IV. IDENTIFICATION OF THE CHLOROPLAST MEMBRANE ADHESION FACTOR

Although it was suspected for many years that thylakoid membrane stacking could be related to the presence of certain chlorophyll-containing light-harvesting complexes [see reviews by Anderson, 1975; Barber, 1976], it was only in the second half of the 1970s that a direct correlation between these two parameters could be demonstrated. Using greening chloroplasts from plants grown first under intermittent light conditions, which allows for only partial development of the thylakoids, Armond et al [1976, 1977] and Argyroudi-Akoyunoglou and Akoyunoglou [1977] showed that the development of grana strictly paralleled the appearance of the chl a/b LHC and the cation regulation of energy distribution between the two photosystems. Other experiments revealed that, when experimentally unstacked thylakoids were treated for 2 minutes with 10 μg trypsin/mg of chlorophyll-containing membrane, the membranes lose their ability to restack and to resegregate their particles after addition of 5 mM MgCl$_2$ [Carter and Staehelin, 1979a, b; Steinback et al, 1979]. Analysis by gel electrophoresis of the proteins of such trypsin-treated, restacking-incompetent thylakoids demonstrated the selective loss of an approximately 1,000-dalton segment from the major chl a/b LHC peptide.

To obtain direct evidence for the involvement of the chl a/b LHC in thylakoid membrane adhesion, we isolated the complex from Triton X-100-

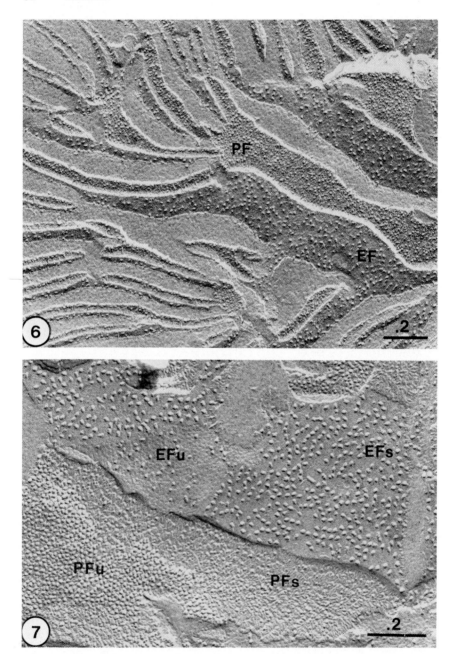

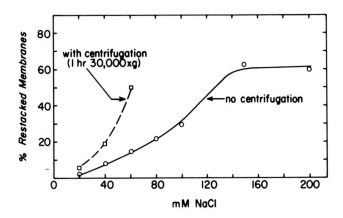

Fig. 8. Diagram illustrating the salt dependency of stacking of thylakoid membranes. Maximal stacking (adhesion) is observed in the presence of > 150 mM NaCl.

solubilized membranes according to the method of Bose et al [1977] and reconstituted it into lipid bilayer vesicles [McDonnel and Staehelin, 1980]. When run on nondenaturing SDS gels, the purified complex gives rise to two green bands CPII (apparent MW approx 30 kd) and CPII* (apparent MW approx 80 kd). Upon rerunning of the material eluted from the two green bands under denaturing conditions, both CPII and CPII* give rise to three polypeptide bands on Coomassie blue–stained gels, one major band at approx 28 kd and two minor ones between 25 and 27 kd. This finding suggests that CPII* is an oligomer of CPII. Whether CPII* corresponds to a natural aggregate of chl a/b LHC molecules has yet to be determined. Liposomes reconstituted with purified chl a/b LHC exist as single vesicles in low salt media, but form aggregates in the presence of $\geq$ 2 mM $MgCl_2$ or $\geq$ 60 mM NaCl. Freeze-fracture micrographs of such aggregates (Fig. 9) reveal ~ 8 nm particles concentrated in the membrane regions involved in membrane adhesion. In most instances these particles form hexagonal arrays that are aligned in adjacent membranes, thereby confirming the adhesive properties of the purified chl a/b LHC particles. Brief treatment of the chl a/b LHC liposomes with trypsin removes an approximately 1,000 kd fragment from the C terminus of the 28 kd chl a/b LHC polypeptide, and abolishes their

Fig. 6. Freeze-fractured pea thylakoids allowed to unstack and unfold for 1 h in a 50 mM Tricine, 0.3 M sucrose solution. Note the sheetlike nature of the thylakoids, and the random distribution of both EF and PF particles. × 63,000.

Fig. 7. Fracture faces of pea thylakoids that were first allowed to unstack in a low salt Tricine buffer as in Figure 6, and then were restacked in vitro by the addition of 10 mM mgCl₂. Each of the fracture faces (EFs, EFu, PFs, PFu) exhibits a virtually normal complement of intra-membrane particles. × 83,000.

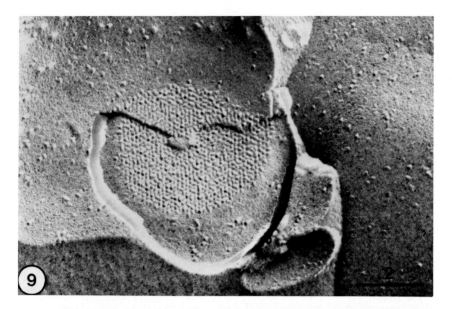

Fig. 9. Lipid vesicles reconstituted with purified chl a/b LHC in the presence of 100 mM NaCl. Note the aggregation of the ~ 8 nm chl a/b LHC particles into aligned hexagonal arrays in the area of contact between the vesicles on the left. In the surrounding nonadhering membrane regions, the particles are randomly distributed. × 95,000.

ability to adhere to each other in the presence of 5 mM $MgCl_2$, as is the case for similarly treated thylakoids. The approx 1,000 kd C-terminus peptide that is released by trypsin possesses the following amino acid sequence as determined by Mullet et al [1981a]: (Lys or Arg)-Ser-Ala-Thr-Thr-Lys-Lys. How this sequence participates in thylakoid stacking is still unknown, but the excess of positively charged groups suggests that electrostatic interactions could play an important role.

V. PHOSPHORYLATION OF THE MEMBRANE ADHESION FACTOR REGULATES MEMBRANE STACKING AND THEREBY MEMBRANE FUNCTION

In most biochemical textbooks the light reactions of oxygenic photosynthesis are depicted in the form of the Z-scheme (Fig. 10). One of the main elements of the Z scheme is the flow of electrons from PSII to PSI. Until recently, most researchers in photosynthesis assumed that efficient transfer of electrons from PSII to PSI, and therefore efficient photosynthesis, would require physical proximity of the two photosystems in thylakoid membranes. Improved thylakoid fractionation techniques have revealed, however, that

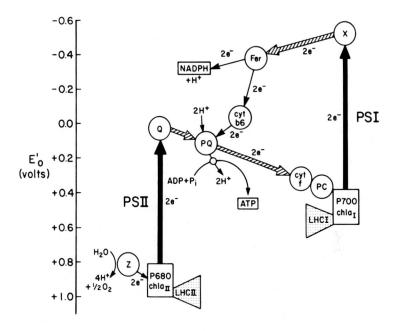

Fig. 10. The Z scheme for electron flow in photosynthesis. The two bold arrows represent the two light reactions associated with PSII and PSI (photosystems II and I); all others are dark reactions. Flow of electrons from H_2O to $NADP^+$ (nicotinamide adenine dinucleotide phosphate) is referred to as "noncyclic" electron flow, and the electron flow from Fer (ferredoxin) through cyt b_6 (cytochrome b_6) to PQ (plastoquinone) and back to PSI is known as "cyclic" electron flow. E'_0: oxidation-reduction potential at pH 7, in volts; Z: electron donor system for PSII; P680: reaction center chlorophyll and energy trap of PSI; P700: reaction center chlorophyll and energy trap of PSI; LHC II and LHC I: light-harvesting complexes associated with PSII and PSI; Q: electron acceptor of PSII; cyt f: cytochrome f; PC: plastocyanin.

this is not the case—ie, that PSII and PSI are not juxtaposed in thylakoid membranes [Anderson, 1981, 1982]. To the contrary, as already pointed out in Table I, the formation of stacked membrane regions seems to be correlated with the physical separation of PSII and PSI reaction centers in chloroplast membranes of higher plants and green algae. Thus > 85% of the PSII units appears to be associated with stacked membrane regions and > 85% of the PSI units with unstacked regions. Based on this finding one has to postulate that transfer of electrons from PSII to PSI is mediated by mobile electron carriers. The most probable candidate for this type of transfer is the pool of plastoquinone carriers, which has a time constant for electron transfer of about 20 milliseconds, compared to fractions of milliseconds and nanoseconds for most other electron transfer steps of the Z scheme [Junge, 1977].

What is the functional significance of this spatial organization of chloroplast membranes of higher plants and green algae? The emerging answer to this question is that the regulation of the functional activity of thylakoid membranes is dependent on the spatial organization of their enzymatic units.

As postulated by the Z scheme, maximal efficiency of noncyclic electron flow requires PSII and PSI to turn over at equal rates. If one photosystem is excited more than the other, the excess excitation energy is dissipated, with a consequent loss of photosynthetic quantum efficiency. Owing to differences in composition of the light-harvesting complexes that trap energy for PSI and PSII, they do not have identical absorption spectra, and they can therefore be excited by different wavelengths of light. Although these differences are beneficial in terms of allowing plants and algae to utilize a greater segment of the light spectrum, they also create problems for optimizing the turnover rates of the two photosystems. These problems can be further compounded by the fact that within a canopy of leaves or within an aqueous environment, the pigments in the topmost leaves or the topmost photosynthetic organisms will selectively remove certain wavelengths of light from the light spectrum, thus forcing the chloroplast in the shaded areas to operate under conditions in which the intensity is reduced and the spectral quality of the incident light is changed. For instance, although the ratio of sunlight energy available at 700 nm vs 680 nm (max absorption of PSI and PSII reaction center chlorophylls) is close to 1 on the top of a maple leaf, the ratio changes to about 5 after having passed through one leaf [Staehelin and Arntzen, 1982].

Fluorescence-emission studies of plants and algae have revealed that the optimization of noncyclic electron flow is achieved by regulating the distribution of absorbed excitation energy to the PSI and PSII reaction centers [Myers, 1971]. During sample illumination with light absorbed preferentially by PSI light (near 710 nm), relatively slow (minutes) changes in the chloroplasts result in an increased proportion of total absorbed excitation energy being transferred to PSII relative to PSI. This condition is referred to as state 1, and can also be produced by prolonged incubation in the dark. Alternatively, when the algae are exposed to light preferentially absorbed by PSII (near 650 nm), they adjust to the unbalance by allowing some of the excitation to be redirected to PSI. This process is often referred to as "spillover" of excitation energy from PSII to PSI, and the spillover state is known as state 2.

A possible biochemical mechanism for the fine tuning of the photosynthetic apparatus by means of state 1–state 2 transitions has emerged recently [Bennett et al, 1980; Allen et al, 1981; Horton and Black, 1981]. These workers have established that the regulation of energy distribution between PSI and

PSII is paralleled by the phosphorylation/dephosphorylation of proteins of the chl a/b LHC (Fig. 11). Phosphorylation of the chl a/b LHC, which occurs on a threonine residue of the peptide that is removed by trypsin [Mullet et al, 1981a], by a thylakoid-bound protein kinase increases the amount of excitation energy that is delivered to PSI (state 2), whereas dephosphorylation catalized by a phosphatase produces the reverse effect. The activity of the protein kinase appears to be regulated by the redox state of plastoquinone, which, being located between PSII and PSI (Fig. 10), is in an ideal position to detect any imbalance of excitation of the two photosystems [Bennett et al, 1980].

As pointed out above, thylakoid stacking seems to be mediated by the chl a/b LHC and the extent of stacking by the balance between attractive Van der Waals interactions and repulsive electrostatic forces. Thus we postulated that incorporation of additional, negatively charged phosphate groups into the membrane-bound chl a/b LHC particles could affect the amount of stacking and thereby the general structural organization of thylakoid membranes. This prediction has been confirmed by means of quantitative electron microscopy of pea thylakoids phosphorylated and dephosphorylated in vitro. Phosphorylation of the chl a/b LHC, which increases the repulsive electrostatic charge between the membranes, reduced stacking by 15%, from 67% to 52%, in the presence of 10 mM $MgCl_2$; dephosphorylation of the chl a/b LHC reversed the change by the same amount. By examining the phosphorylated/dephosphorylated thylakoids in terms of change in density and size of their intramembrane particles in stacked and unstacked membrane regions, we have found that the 15% reduction in stacked membranes associated with the phosphorylation of the chl a/b LHC is brought about by the reversible migration of $\sim$ 8 nm PFs particles to PFu regions (Figs. 12, 13). Several lines of indirect evidence suggest that these $\sim$ 8 nm particles correspond to chl a/b LHC particles [Staehelin and Arntzen, 1979; Simpson, 1979; McDonnel and Staehelin, 1980].

VI. CONCLUDING REMARKS

We can therefore conclude that membrane stacking in chloroplasts of higher plants and green algae serves as a mechanism for controlling the spatial distribution of membrane components and thereby regulating the distribution of excitation energy between PSII and PSI. Of crucial importance for this regulation seems to be the reversible migration of a subset of chl a/b LHC particles between stacked and unstacked membrane areas (Fig. 14). Because the efficiency of resonance transfer of excitation energy from light-harvesting antennae to photosynthetic reation centers drops with the sixth

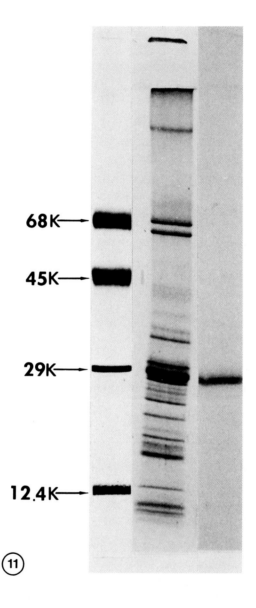

Fig. 11. Light-induced phosphorylation of the chl a/b LHC peptides. Isolated washed barley thyalkoids were illuminated in the presence of 200 M [γ-^{32}P] ATP for 5 min. Thylakoid proteins were precipitated with acetone and fractionated by sodium dodecyl sulfate/polyacrylamide gel electrophoresis. Left lane, standards; middle lane stained with Coomassie blue; right lane, autoradiogram. The major ^{32}P-labeled band corresponds to the location of the chl a/b LHC peptides. Photograph courtesy of P. Haworth, D.J. Kyle, and C.J. Arntzen.

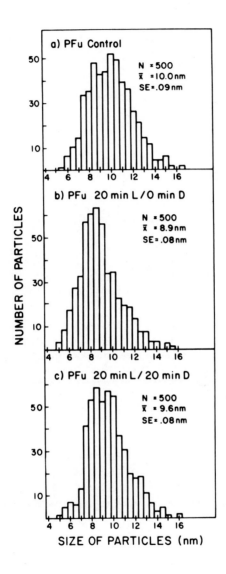

Fig. 12. Particle size histograms showing changes in the size distribution of PFu particles as a result of light-induced phosphorylation, and dark-mediated dephosphorylation of the chl a/b LHC. Note the proportional increase in ~ 8 nm particles as a result of the phosphorylation of the chl a/b LHC. This increase in ~ 8 nm particles appears to be caused by a net migration of such particles from stacked (PFs) to unstacked (PFu) membrane regions. The relatively uniform size distribution of PFs particles can be seen in Figure 5.

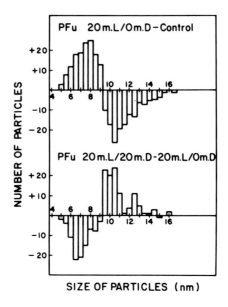

Fig. 13. Difference histograms prepared from the histograms shown in Figure 12. In the upper diagram, the net increase in 6 to 9 nm particles and the dilution dependent decrease in > 9 nm particles caused by the light-dependent phosphorylation of the chl a/b LHC is clearly seen. The lower diagram shows that dark-dependent dephosphorylation of the chl a/b LHC reverses the changes.

power of the distance [Clayton, 1980], the reversible migration of chl a/b LHC particles between PSII-rich stacked and PSI-rich stacked and PSI-rich unstacked membrane regions allows for the delivery of variable amounts of excitation energy to PSII and PSI reaction centers. Confirmation for this theory has come from experiments with thylakoids experimentally unstacked in a low salt medium. Normal phosphorylation of the chl a/b LHC in such unstacked thylakoids can be achieved in the light, and dephosphorylation in the dark; however, no changes in the distribution of excitation energy between PSII and PSI are observed, presumably because all thylakoid components remain randomly dispersed irrespective of the state of phosphorylation of the chl a/b LHC [Barber, 1976, 1980; Staehelin and Arntzen, 1979; Bennett et al, 1980]. Thus membrane-membrane interactions and the resulting spatial organization of membrane components are an essential part of the regulatory mechanism for energy distribution between the two photosystems in chloroplast membranes.

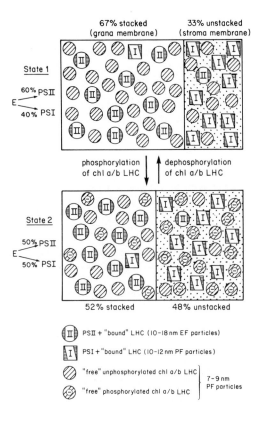

67% stacked
(grana membrane)

33% unstacked
(stroma membrane)

State 1

60% PSII
E
40% PSI

phosphorylation
of chl a/b LHC

dephosphorylation
of chl a/b LHC

State 2

50% PSII
E
50% PSI

52% stacked   48% unstacked

PSII + "bound" LHC (10-18 nm EF particles)

PSI + "bound" LHC (10-12 nm PF particles)

"free" unphosphorylated chl a/b LHC
"free" phosphorylated chl a/b LHC
7-9 nm
PF particles

Fig. 14.   Diagram illustrating how phosphorylation of the chl a/b LHC particles, the "adhesion factor" particles of thylakoid membranes (Fig. 9), leads to a net migration of such particles from PSII-rich stacked, grana membranes to PSI-rich unstacked, stroma membranes, and concomitantly to a reduction in the amount of membrane stacking. Since the chl a/b LHC serves also as light-harvesting antenna, the phosphorylation-dependent increase of such antennae in the PSI-rich membrane regions leads to more of the trapped light energy being transferred to PSI reaction centers. Dephosphorylation of the chl a/b LHC reverses the effect.

ACKNOWLEDGMENTS

Thanks are due Dr Sallie Sprague for reviewing the manuscript, and to Marcia DeWit for excellent technical assistance. Supported by grant GM22912 from the National Institute of General Medical Sciences.

VII. REFERENCES

Allen JF, Bennett J, Steinback KE, Arntzen CJ (1981): Chloroplast protein phosphorylation couples plastoquinone redox state to distribution of excitation energy between photosystems. Nature 291:25–29.

Anderson JM (1975): The molecular organization of chloroplast thylakoids. Biochim Biophys Acta 416:191–235.

Anderson JM (1981): Consequences of spatial separation of photosystem 1 and 2 in thylakoid membranes of higher plant chloroplasts. FEBS Lett 124:1–10.

Anderson JM (1982): Distribution of the cytochromes of spinach chloroplasts between the appressed membranes of grana stacks and stroma-exposed thylakoid regions. FEBS Lett 138:62–66.

Argyroudi-Akoyunoglou JH, Akoyunoglou G (1977): Correlation between cation-induced formation of heavy subchloroplast fractions and cation-induced increases in chlorophyll a fluorescence yield in tricine-washed chloroplasts. Arch Biochem Biophys 179:370–377.

Armond PA, Arntzen CJ, Briantais JM, Vernotte C (1976): Differentiation of chloroplast lamellae. I. Light harvesting efficiency and grana development. Arch Biochem Biophys 175:54–63.

Armond PA, Staehelin LA, Arntzen CJ (1977): Spatial relationship of photosystem I, photosystem II and the light harvesting complex in chloroplast membranes. J Cell Biol 73:400–418.

Barber J (1976): Ionic regulation of intact chloroplasts and its effect on primary photosynthetic processes. In Barber J (ed): "The Intact Chloroplast." Amsterdam: Elseiver/North Holland, pp 89–134.

Barber J (1980): An explanation for the relationship between salt-induced thyalkoid stacking and the chlorophyll fluorescence changes associated with changes in spillover of energy from photosystem II to photosystem I. FEBS Lett 118:1–10.

Barber J (1982): Influence of surface charges on thylakoid structure and function. Ann Rev Plant Physiol 33:261–295.

Bennett J, Steinback KE, Arntzen CJ (1980): Chloroplast phosphoproteins: Regulation of excitation energy transfer by phosphorylation of thylakoid membrane polypeptides. Proc Natl Acad Sci USA 77:5253–5257.

Bose S, Burke JJ, Arntzen CJ (1977): Cation induced microstructural changes in chloroplast membranes: Effects on photosystem II activity. In Packer L (ed): "Bioenergetics of membranes." Amsterdam: Elsevier/North Holland, pp. 89–134.

Branton D, Cohen CM, Tyler J (1981): Interaction of cytoskeletal proteins on the human erythrocyte membrane. Cell 24:24–32.

Burke JJ, Steinback KE, Arntzen CJ (1979): Analysis of the light-harvesting pigment-protein complex of wild type and a chlorophyll-6-less mutant of barley. Plant Physiol 63:237–243.

Carter DP, Staehelin LA (1979a): Proteolysis of chloroplast thyalkoid membranes. I. Selective degradation of thylakoid pigment-protein complexes at the outer membrane surface. Arch Biochem Biophys 200:364–373.

Carter DP, Staehelin LA (1979b): Proteolysis of chloroplast thylakoid membranes. II. Evidence for the involvement of the light harvesting chlorophyll a/b protein complex in thylakoid stacking and for effects of magnesium ions on photosystem II-light harvesting complex aggregates in the absence of membrane stacking. Arch Biochem Biophys 200:374–386.

Clayton RK (1980): "Photosynthesis: Physical Mechanisms and Chemical Patterns." Cambridge: Cambridge University Press.

Goodenough UN, Staehelin LA (1971): Structural differentiation of stacked and unstacked membranes. J Cell Biol 48:594–619.

Gulik-Krzywicki T (1975): Structural studies of the associations between biological membrane components. Biochim Biophys Acta 415:1–28.

Horton P, Black MT (1981): Light-dependent quenching of chlorophyll fluorescence in pea chloroplasts induced by adenosine S-triphosphate. Biochim Biophys Acta 635:53–62.

Hull BE, Staehelin LA (1979): The terminal web. A reevaluation of its structure and function. J Cell Biol 81:67–82.

Izawa S, Good NE (1966): Effects of salts and electron transport on the conformation of isolated chloroplasts. II. Electron microscopy. Plant Physiol 14:544–552.

Junge W (1977): Physical aspects of light harvesting, electron transport and electrochemical potential generation in photosynthesis of green plants. In Trebst A, Aaron M (eds): "Photosynthesis I: Photosynthetic Electron Transport and Photophosphorylation." Berlin: Springer-Verlag, pp 59–93.

Kirk JTO, Tilney-Bassett RAE (1978): "The Plastids." Amsterdam: Elseiver/North-Holland.

McDonnel A, Staehelin LA (1980): Adhesion between liposomes mediated by the chlorophyll a/b light harvesting complex isolated from chloroplast membranes. J Cell Biol 84:40–56.

Miller KR, Staehelin LA (1976): Analysis of the thylakoid outer surface. Coupling factor is limited to unstacked membrane regions. J Cell Biol 68:30–47.

Mullet JE, Arntzen CJ (1980): Simulation of grana stacking in a model membrane system. Mediation by a purified light-harvesting pigment-protein complex from chloroplasts. Biochim Biophys Acta 589:100–117.

Mullet JE, Burke JJ, Arntzen CJ (1980): Chlorophyll proteins of photosystem I. Plant Physiol 65:814–822.

Mullet JE, Baldwin TO, Arntzen CJ (1981a): A mechanism for chloroplast thylakoid adhesion mediated by the chl a/b light harvesting complex. In Akoyunoglou G (ed): "Photosynthesis III. Structure and Molecular Organization of the Photosynthetic Apparatus." Philadelphia: Balban International Services, pp 577–582.

Mullet JE, Pick U, Arntzen CJ (1981b): Structural analysis of the isolated chloroplast coupling factor and the NN'-dicyclohexyl-carbodiimide binding proteolipid. Biochim Biophys Acta 642:149–157.

Myers J (1971): Enhancement studies in photosynthesis. Ann Rev Plant Physiol 22:289–312.

Ojakian GK, Satir P (1976): Particle movements in chloroplast membranes: Quantitative measurements of membrane fluidity by the freeze-fracture technique. Proc Natl Acad Sci USA 71:2052–2056.

Sculley, MJ, Duniec JT, Thorne SW, Chow WS, Boardman NK (1980): The stacking of chloroplast thyalkoids. Quantitative analysis of the balance of forces between thylakoid membranes of chloroplasts, and the role of divalent cations. Arch Biochem Biophys 201:339–346.

Sheterline P, Hopkins CR (1981): Transmembrane linkage between surface glycoproteins and components of the cytoplasm in neutrophil leukocytes. J Cell Biol 90:743–754.

Simpson DJ (1979): Freeze-fracture studies on barley plastid membranes: III. Location of the light-harvesting chlorophyll-protein. Carlsberg Res Commun 44:305–336.

Sprague SG, Morschel E, Staehelin LA (1981): Preparation of liposomes and reconstituted membranes from purified chloroplast membrane lipids. J Cell Biol 91:269a.

Staehelin LA (1968): Ultrastructural changes of the plasmalemma and cell wall during the life cycle of Cyanidium caldarium. Proc R Soc B71:249–259.

Staehelin LA (1974): Structure and function of intercellular junctions. In Bourne GH, Danielli JK (eds): "International Review of Cytology." New York: Academic Press, pp 191–283.

Staehelin LA (1976): Reversible particle movements associated with unstacking and restacking of chloroplast membranes in vitro. J Cell Biol 71:136–158.

Staehelin LA, Arntzen CJ (1979): Effects of ions and gravity forces on the supramolecular organization and excitation energy distribution in chloroplast membranes. In Ciba Foundation Symposium 61: "Chlorophyll Organization and Energy Transfer in Photosynthesis." Amsterdam: Elsevier/Exerpta Medica/North-Holland, pp 147–175.

92   Staehelin

Staehelin LA, Arntzen CJ (1982): Membrane-mediated regulation of quantum efficiency in photosynthesis (Submitted.)
Steinback KE, Burke JJ, Arntzen CJ (1979): Evidence for the role of surface-exposed segments of the light-harvesting complex in cation-mediated control of chloroplast structure and function. Arch Biochem Biophys 195:546–557.
Wollman FA, Olive J, Bennoun P, Recouvreur M (1980): Organization of the photosystem II centers and their associated antennae in the thylakoid membranes: A comparative ultrastructural, biochemical and biophysical study of Chlamydomonas wild type and mutant lacking in photosystem II reaction centers. J Cell Biol 87:728–735.

Modern Cell Biology 2:93–112
© 1983 Alan R. Liss, Inc., 150 Fifth Avenue, New York, NY 10011

Membranous Compartments and Ionic Transients in the Mitotic Apparatus

Peter K. Hepler and Stephen M. Wolniak

From the Botany Department, University of Massachusetts, Amherst, Massachusetts 01003

I. INTRODUCTION

In 1960, Porter and Machado published a classic study that delineated the contribution of the endoplasmic reticulum (ER) to the mitotic apparatus (MA) in dividing meristematic cells of the onion root tip. Membrane accumulations were noted at the spindle poles, and extensions of this system protruded into the spindle interior. These workers observed that "during mitosis in these cells a new system of ER elements is formed in orderly relation to chro-

Stephen M. Wolniak is now at Department of Botany, University of Maryland, College Park, MD 20742.

mosomes and spindle and out of this new system there seems to be fashioned a nuclear membrane, . . . (and) a phragmoplast. . . ." Without question, these observations displayed an important aspect of spindle morphology. However, they were soon eclipsed by the study of Ledbetter and Porter [1963] on microtubules. Using glutaraldehyde fixation, Ledbetter and Porter [1963] obtained a quality of cytoplasmic preservation far exceeding that available with $KMnO_4$ which had been used in the earlier study on membranes. Microtubules, ribosomes, small vesicles, and a host of general cytoplasmic details could now be resolved. Microtubules were identified in the anaphase MA, and soon numerous reports established that these long, cytoplasmic inclusions were the spindle fibers.

Given the widespread interest in the mechanism of mitosis, it is perhaps not surprising that the literature on mitotic cell morphology was and has continued to be focused on the structure and function of spindle microtubules. It has gone largely unnoticed that whereas glutaraldehyde-osmium tetroxide is good for fixing and contrasting microtubules and many other cytoplasmic details, it is much less successful in displaying elements of the ER. Ledbetter and Porter [1963] realized this shortcoming; in their original paper they stated, "Oddly, the membranes of the ER are difficult to discern and the limits of the cisternae are frequently more clearly defined by ribosomes than by anything else." In more recent years, there has been a small but growing awareness of membrane MA associations [Harris, 1975; Hepler, 1977; Paweletz, 1981]. On current evidence it seems reasonable to conclude that membranes, in one form or another, are as ubiquitous a component of the apparatus as are microtubules. It is our purpose in this chapter to explore the structure and function of mitotic membranes. We will show that elements of ER constitute a major component of the MA, and that they may be intimately associated with the mitotic cytoskeletal fibers. Subsequently, we will develop and provide support for the idea that these membranes regulate ion fluxes, especially Ca^{2+}, in the spindle.

These studies owe much to the pioneering work of Keith Porter, and it is thus appropriate that this chapter is dedicated to him on the occasion of his retirement.

II. MEMBRANE COMPONENTS OF THE MITOTIC APPARATUS
A. Membrane Organization During Mitosis

The failure of conventional glutaraldehyde-osmium tetroxide to render ER visible, especially the smooth tubular profiles, has made it difficult to study membranes in the MA or even appreciate their presence. Use of osmium tetroxide alone [Harris, 1975] and double fixations involving $KMnO_4$ plus glutaraldehyde [Moll and Paweletz, 1980; Paweletz and Finze, 1981] have achieved some success in revealing spindle membranes. In our studies we have found that glutaraldehyde followed by postfixation in a mixture of osmium tetroxide and potassium ferricyanide (OsFeCN) specifically stains the ER and nuclear envelope in different plant cells. The method is especially effective in displaying reticulate, tubular membranes while at the same time preserving microtubules, ribosomes, and other fine structure details [Hepler, 1980, 1981].

Observations of dividing cells postfixed with OsFeCN reveal an intimate association of the ER and nuclear envelope with the MA cytoskeleton, and show that these membranes undergo structural transformations and redistributions in accordance with the stages and events of mitosis. At prophase-prometaphase, elements of ER accumulate in a region that defines the spindle pole (Fig. 1). Between the membrane elements, short segments of microtubules appear, revealing the time and place of spindle formation (Fig. 2). As the nuclear envelope disperses and the spindle microtubules associate with chromosomes, profiles of smooth ER penetrate deep into the MA (Figs. 3–5). In barley leaf cells at metaphase the invaginations occur specifically along kinetochore fiber bundles; the tubular, reticulate ER intertwines among the kinetochore microtubules and extends as far as the chromosomes (Figs. 4, 5). As well as invading the interior of the MA, the ER constructs a sheath around the spindle and separates it from the rest of the cytoplasm (Fig. 3).

During mitosis, membranes progressively aggregate at the spindle pole, and by anaphase they reach their maximum degree of accumulation [Hepler, 1980]. The membranes appear largely in the form of a tubular reticulum, but lamellar elements with fenestrae are also evident. In late anaphase and telophase, as the cell plate arises, ER elements move into the interzone and intermingle with the aggregating phragmoplast vesicles. Eventually, these become closely appressed to the cell plate and give rise to the plasmodesmatal connections between neighboring cells [Hepler, 1982].

Studies of other cell types reveal membrane configurations in the MA that are similar to those reported in plants, although the specific relationship between membranes and kinetochore fibers has not been widely observed. Paweletz and Moll [1980] find that the MA in HeLa cells is surrounded by a membrane system throughout mitosis. The degree to which the MA is segregated from the rest of the cell is apparently sufficient to permit two nuclei in the same cytoplasm but in different stages of mitosis to progress through karyokinesis independently. Sea urchin spindles contain a large amount of membrane, a point repeatedly emphasized by Harris [1975, 1978]. Chains of vesicles extend along microtubules throughout this MA, and extensive accumulations of membranous elements reside in the asters.

Besides ER, there is evidence for the presence of Golgi vesicles within the MA. In certain cell types the Golgi cisternae increase in number during

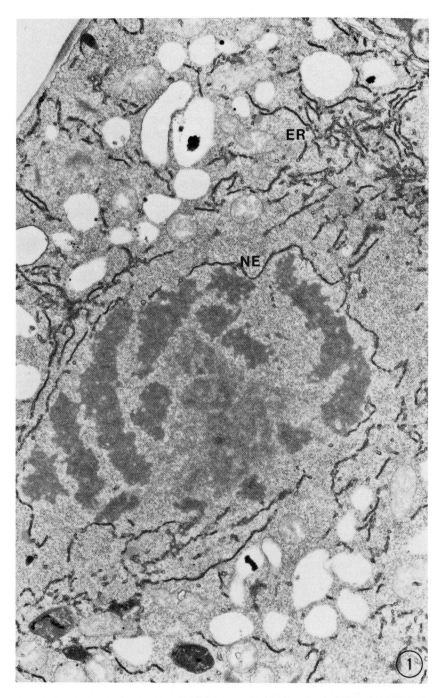

Fig. 1. Prometaphase in a lettuce root tip cell. The nuclear envelope (NE) and endoplasmic reticulum (ER) have been heavily stained through postfixation in OsFeCN. The NE has begun to break down. At this time an accumulation of ER is evident in the region of the spindle pole (upper right). × 10,000.

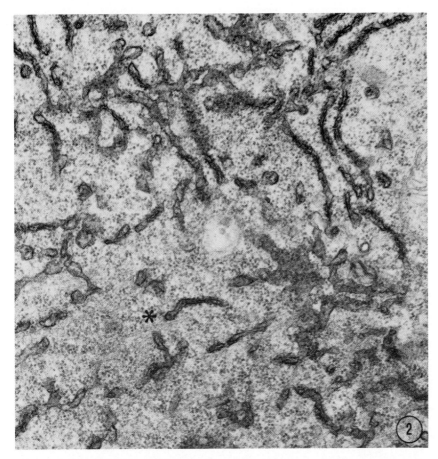

Fig. 2. The spindle pole in Figure 1 is shown here at higher magnification. Anastomosing elements of tubular ER form a loosely organized membrane network. Between the membrane elements a few segments of microtubules appear (*). × 37,000.

prophase and subsequently disintegrate into vesicles [Moskalewski et al, 1977; Paweletz, 1981]. Thus, whereas the ER is often more localized at the MA periphery, the Golgi vesicles are usually scattered throughout the spindle, with no apparent specific structural relationship to the spindle fibers. Although several reports document the breakup of the Golgi cisternae, there are some notable exceptions in which these membrane stacks remain intact and closely clustered near the MA throughout mitosis [Pickett-Heaps et al, 1980].

B. Possible Functions of Mitotic Membranes

The structural observations document the widespread occurrence of spindle membranes and provide ideas on how these elements function. Mitotic membranes might contribute to the physical integrity of the MA; they could serve

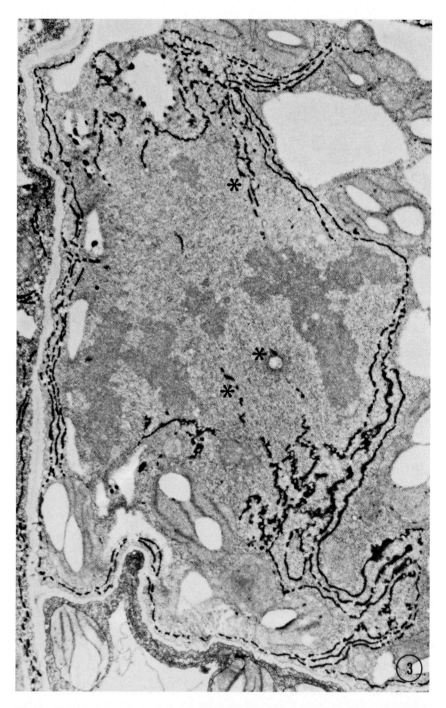

Fig. 3. Metaphase in a barley leaf cell. The ER forms a boundary between the MA and the rest of the cell. Reticulate, tubular ER extends from the spindle pole into the MA interior along kinetochore fibers (*). The upper right kinetochore region is shown at higher magnification in Figure 4. × 12,000 [from Hepler, 1980].

as key structural elements by providing anchor points to which microtubules and microfilaments might bind and against which shearing forces might be generated. With massive accumulations at the spindle poles and specific intrusions into the MA interior, membranes are well positioned to act as a scaffolding for the organization and function of force-generating macromolecules. It can be argued that membranes are fluid and not likely to provide a structural support against shearing forces. It seems plausible, however, as in red blood cells [Branton et al, 1981] that structural proteins could link and even anchor fibrous elements of the cytoskeleton to the MA membranes.

The studies on ER distribution, which show membrane changes during mitosis and the establishment of specific membrane-MA-cytoskeletal associations, provide suggestive evidence for the existence of distinct compartments or domains within the dividing cell. Thus, despite the fact that the nuclear envelope breaks down in many dividing cells, the MA remains surrounded by a membranous sheath which separates it from the rest of the cytoplasm throughout mitosis. Furthermore, the invaginations of membrane into the MA itself might create functionally localized subregions within the spindle—for example, along the kinetochore fibers (Figs. 4, 5). These separate compartments may be the structural basis for a system that is able to regulate ion concentrations locally, perhaps in the way that the sarcoplasmic reticulum of muscle works during muscle contraction. It seems noteworthy that the ER of the MA and of muscle share several properties: Both stain with OsFeCN [Forbes et al, 1977], both contain tubular smooth elements in a reticulate network, both contain fenestrated lamellae, both are intimately associated with motile force-generating structures, and finally, both contain Ca^{2+} [Hepler et al, 1981]. It therefore seems reasonable to suggest that the mitotic membranes, like the sarcoplasmic reticulum of muscle, control the concentration of free calcium in localized regions and modulate particular processes at specified times and places during mitosis.

III. MEMBRANE REGULATION OF ION FLUXES

A. Calcium Localization and Changes

Membranes of the MA contain bound deposits of calcium [Wick and Hepler, 1980; Wolniak et al, 1980]. In living endosperm cells of Haemanthus, using the fluorescent Ca^{2+}-chelate chlorotetracycline (CTC), we have observed the pattern of membrane-associated Ca^{2+} during mitosis (Fig. 6a–c) [Wolniak et al, 1980]. The spindle poles, containing numerous mitochondria and plastids, are especially fluorescent. Careful inspection of the spindle region itself in highly flattened metaphase cells reveals that Ca^{2+}-containing membranes extend from the polar regions to the MA interior in cone-shaped arrays that exactly overlie the birefringent kinetochore fibers (Fig. 6b, c) [Wolniak et al, 1981]. During anaphase the cones disappear and spindle

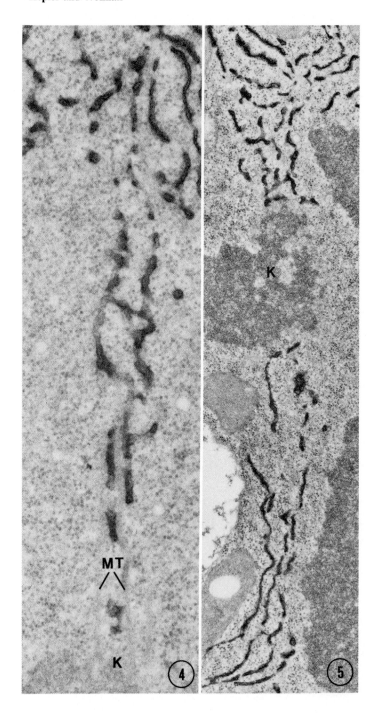

fluorescence becomes dispersed uniformly, suggesting that changes in the distribution of membrane-associated calcium have occurred.

To describe changes in CTC fluorescence more rigorously, we have quantitatively monitored light emission during mitosis with a photomultiplier tube coupled to the microscope by a light fiber. We have examined local regions within the MA, the poles, and the nonspindle cytoplasm using a 12 μm

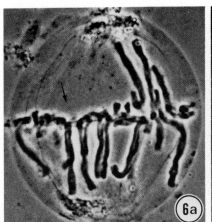

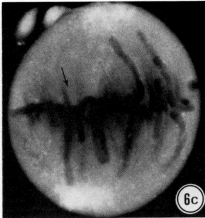

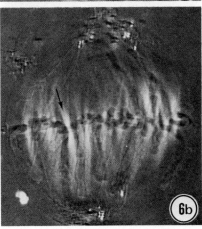

Fig. 6. Living endosperm cell of Haemanthus treated with 20 μm CTC and 2 μm taxol for 20 min and viewed by phase contrast (a) polarization (b) and fluorescence (c) microscopy. In (b) the λ/30 compensator was set 2 nm away from extinction. The arrows in all figures correspond to the same points in the mitotic apparatus and demonstrate that the distribution of fluorescent cones (c) coincides with kinetochore fiber birefringence (b) in a region of the spindle devoid of chromosome arms (a). All figures × 660 [from Wolniak et al, 1981].

Fig. 4. Tubular elements of smooth ER intrude into the spindle interior specifically along kinetochore microtubules (MT) to the point of chromosome attachment (K). Serial sections verify that the membranes are part of one continuous system. × 51,000 [from Hepler, 1980].

Fig. 5. A portion of the MA of a barley leaf cell in metaphase. Densely stained ER extends from opposing spindle poles inward to the kinetochore region (K) of sister chromatids. × 20,000.

aperture to define the area for measurement at different stages of mitosis [Wolniak et al, 1983]. We find an MA-specific CTC signal that drops during late metaphase and usually remains reduced throughout anaphase (Fig. 7), although in two of the seven cells observed the fluorescence increased again by late anaphase.

The possibility that this reduction in CTC fluorescence is simply the result of a change in the distribution of membranes has been tested by examining the fluorescence of cells treated with N-phenyl-1-naphthylamine (NPN), a hydrophobic membrane marker that does not respond to changes in electric charge [Trauble and Overath, 1973]. In contrast to CTC, the signal from NPN remains constant through metaphase and anaphase, with the exception of a transient dip at the moment of initial chromosome separation (Fig. 8) [Wolniak et al, 1983]. We conclude that the reduction in CTC fluorescence

Figs. 7–10. Fluorescence intensity changes during the metaphase/anaphase transition in living endosperm cells of Haemanthus. The onset of anaphase (time = 0) was judged visually by phase contrast microscopy. Values for fluorescence intensity were normalized to a percentage of the mean value obtained during metaphase (metaphase mean = 100%) [from Wolniak et al, 1983].

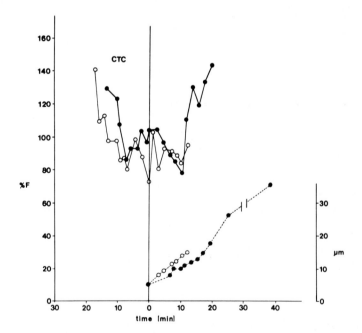

Fig. 7. Spindle fluorescence (solid line) from two endosperm cells (closed and open circles) pretreated 20 min with 25 μM CTC. Chromosome movement (dashed line) represents total separation of the centromeres.

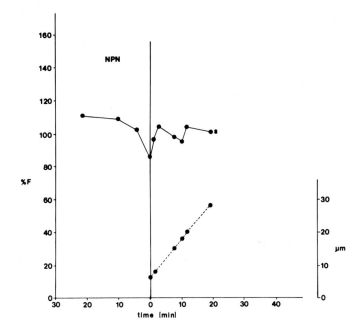

Fig. 8. Spindle fluorescence from an endosperm cell pretreated 20 min with 25 μM NPN. Chromosome movement (dashed line) represents total separation of the centromeres.

indicates an efflux of Ca^{2+} from the spindle membranes. These results suggest that there is a corresponding increase in spindle-free Ca^{2+}. Support for this conclusion is provided by in vitro studies of the sarcoplasmic reticulum that definitively correlate a diminution in CTC fluorescence with Ca^{2+} efflux [Caswell and Brandt, 1981].

B. Membrane or Surface Potential Changes

In addition to CTC and NPN we have employed two nontoxic, permeant "voltage-sensitive" dyes, 8-anilino-1-naphthalene sulfonate (ANS) and 3,3'-dipentyl dioxacarbocyanine ($diOC_5(3)$), to explore possible changes in membrane potential or surface charge during mitosis [Zimniak and Racker, 1978; Waggoner, 1979; Beeler et al, 1981]. With both probes we observe that the signal either drops or remains constant during metaphase, but that at the onset of anaphase the fluorescence rises abruptly as the chromosomes separate and move to the poles (Figs. 9, 10) [Wolniak et al, 1983]. Similar fluorescence changes are not observed at the polar regions or in the nonspindle cytoplasm, indicating that, as with CTC, the changes we observe are temporally and spatially related with the events of mitosis.

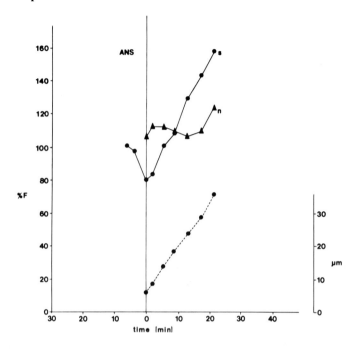

Fig. 9. Spindle fluorescence (s), and nonspindle fluorescence (n) from an endosperm cell pretreated 20 min with 15 μM ANS. Chromosome movement (dashed line) represents total separation of the centromeres.

At the moment, the specific causes for ANS and $diOC_5(3)$ fluorescence changes are not known, but some tentative conclusions and inferences can be drawn. First, the fluorescent signal probably emanates from the spindle endomembranes. If it originated from the plasma membrane, we would expect to see changes equivalent to those over the MA in all regions of the cell, which we do not. It is possible that the probes are binding to some non-membranous spindle protein such as tubulin [Bhattacharyya and Wolff, 1975]; however, we have used filter combinations more suited to fluorescence emission from membranous environments.

Second, increases in fluorescence probably indicate increases in dye binding. It is known, especially with $diOC_5(3)$, that increased dye binding is correlated with decreased fluorescence owing to dye aggregation and self-quenching [Sims et al, 1974; Waggoner, 1979]. We believe we are not witnessing changes in quenching, since the cells are treated with low dye concentrations at which the binding sites on the membranes are probably less than fully saturated.

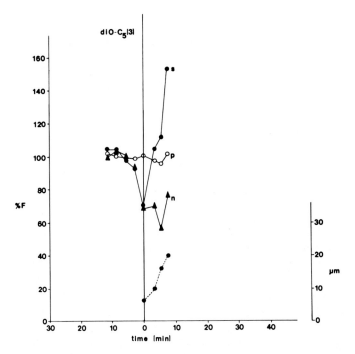

Fig. 10. Spindle (s), nonspindle (n), and pole (p) fluorescence from an endosperm cell treated with 0.2 μM diOC$_5$(3). Chromosome movement (dashed line) represents total separation of the centromeres.

Third, the fact that emission intensities with negatively charged ANS and positively charged diOC$_5$(3) change in the same direction (Figs. 9, 10) suggests that the two probes are responding to events in two different localities or compartments. For example, the two dyes may partition to opposite surfaces of a given membrane system. ANS, which is less permeant than diOC$_5$(3), may bind to the cytoplasmic (P) surface of the ER and reflect an increase in cationic charge, whereas diOC$_5$(3) may migrate to the inner membrane (E) surface owing to an increased anionic charge. Alternatively, it is possible that ANS binds to a protein-cation complex—eg, Ca^{2+} calmodulin [LaPorte et al, 1980] or tubulin [Bhattacharyya and Wolff, 1975]. Increases in diOC$_5$(3) may indicate changes in the electrical potential across the ER.

Although the underlying causes for the observed fluorescence changes cannot be specified, all of their behaviors at the metaphase-to-anaphase transition can be explained by a transmembrane flux of Ca^{2+}. As discussed earlier, the reduction in CTC fluorescence can be accounted for by an efflux of membrane-bound Ca^{2+} followed by an elevation in free Ca^{2+} in the MA.

The rise in ANS fluorescence may result from the subsequent binding of Ca^{2+} to the cytoplasm-facing surfaces of spindle membranes or to spindle proteins, whereas $diOC_5(3)$ increases may report on potential changes across spindle membranes. It is puzzling that the changes in the CTC signal are not more closely coupled in time with those from ANS and $diOC_5(3)$. Still, the fact that abrupt changes in fluorescence accompany the onset of anaphase indicates that significant ionic translocations have occurred. A complete explanation of their causes may provide new clues about the regulation of chromosome motion.

IV. CALCIUM REGULATION OF MITOSIS

It is widely thought that changes in Ca^{2+} regulate the events of mitosis [reviewed by Hepler et al, 1981], and in recent years several supporting lines of evidence have emerged: 1) Increasing Ca^{2+} through the use of the ionophore A23187 induces a variety of cell types to enter or complete mitosis or meiosis [Epel, 1978]; 2) increasing Ca^{2+} in the MA in vivo [Kiehart, 1981] and in vitro [Salmon and Segall, 1980] reduces spindle birefringence, implying a reduction in the structure or organization of the spindle microtubules; 3) increasing Ca^{2+} facilitates motion of chromosomes to poles in permealized mitotic cells [Cande, 1981]; 4) calmodulin, which promotes microtubule depolymerization when bound with Ca^{2+}, is localized in the MA, especially at the spindle poles [Andersen et al, 1978; Welsh et al, 1978] and even on the microtubules themselves [DeMey et al, 1980]; 5) a Ca^{2+}-ATPase is associated with the MA and fluctuates in activity during mitosis [Petzelt, 1979]; 6) isolated MA that contain membranes are able to sequester Ca^{2+} from the medium in an ATP-dependent manner [Silver et al, 1980]; and 7) finally, the work reported herein indicates that a Ca^{2+} efflux occurs from mitotic membranes during late metaphase. Taken together, these results support the theory that mitotic endomembranes, like the sarcoplasmic reticulum of muscle, release sequestered Ca^{2+} to the cytoplasm, perhaps in response to a signal, and thus facilitate or trigger one or more processes that participate in mitosis.

There are several processes in the MA that might be stimulated by an increase in Ca^{2+}. Foremost in our thinking is microtubule depolymerization. It is apparent that microtubule depolymerization must occur for chromosomes to move to the poles [Fuseler, 1975]; it may even be the force-generating event [Inoué and Sato, 1967]. With calmodulin localized at the spindle poles and on the microtubules themselves, the MA seems primed to accept an elevated level of Ca^{2+} that would then cause spindle fiber disassembly and permit or cause chromosomes to move to the poles.

A second process in the MA that might be controlled by Ca^{2+} could be microtubule-associated shearing forces as mediated by a dynein-like cross-bridge. Sliding between antiparallel microtubules may participate in anaphase motion [McIntosh et al, 1969]. especially in the separation of the spindle poles (anaphase B) [Cande, 1982]. It is also possible that cross-bridge-generated shearing forces occur between microtubules and membranes and that these contribute to chromosome motion. In this second possibility membranes are visualized as having a dual role: 1) as regulators of the Ca^{2+}, and 2) as structural components with which microtubule-associated dynein ATPases bind and generate force.

Increasingly, it is becoming evident that dynein participates in mitosis; proteins similar to the flagellar dynein ATPase have been isolated from sea urchin MA [Pratt et al, 1980], and inhibitors of flagellar dynein ATPase activity block anaphase B in permeabilized PtK cells [Cande, 1982; Cande and Wolniak, 1978]. That these mechanochemical ATPases may reside between microtubules and membranes in the MA is supported by the structural evidence showing close apposition of these two components, and by the previously demonstrated existence of membrane-microtubule cross bridges [Hepler et al, 1970]. It is pertinent to this discussion to note that ciliary and flagellar doublet microtubules are bridged to their plasma membrane; these associations are thought to participate in motility [Dentler, 1981]. It thus seems reasonable to conclude that mechanochemical dynein-like bridges exist between microtubules and membranes in the MA, and furthermore, that through sliding of the microtubules against the membranes, motion of chromosomes may be achieved. If dynein participates in mitotic movements it would probably be Ca^{2+}-regulated, since in flagella and cilia, beat shape and direction can be modified by changing the level of Ca^{2+} [Gibbons, 1981], and dynein ATPase activity itself is stimulated by Ca^{2+}-calmodulin [Blum et al, 1980].

The mitotic process may utilize an actomyosin microfilament system in addition to, or instead of, motile machinery mentioned above [Forer et al, 1979]. If actomyosin is the force producer, one would anticipate that increases in Ca^{2+} would stimulate its action. The specific mechanism might be a Ca^{2+}-calmodulin induction of myosin light chain phosphorylation required for the interaction of actin with myosin and the consequent stimulation of myosin ATPase activity [Adelstein and Eisenberg, 1980]. Supporting evidence for this possibility comes from the work of Guerriero et al [1981] who showed that antibody elicited against myosin light chain kinase accumulates in the chromosome-to-pole region of mitotic mammalian cells. It is even conceivable that actin microfilaments interact with microtubules and/or with membranes in a complex, multifiber array. The increased viscosity and structural apposition of microfilaments and microtubules brought about by the addition of microtubule-associated proteins of high molecular weight provide evidence in support of a close physical and possible functional relationship between these two cytoplasmic fibers [Griffith and Pollard, 1978]. Although force-generating mechanisms based on these components have not been elucidated, it seems reasonable that if they do occur, they too would be stimulated by increasing Ca^{2+}.

Finally, we draw attention to the occurrence of gel-sol transformations as possible contributors to chromosome motion. It has long been known that changes in viscosity accompany mitosis in different organisms; for example, after fertilization in Chaetopterus, the viscosity increases, then subsequently drops once the spindle is formed [Heilbrunn, 1956]. Recent studies on cell extracts of Dictyostelium, although not addressing the specific question of mitosis, provide important new information about gel-sol transformations in nonmuscle systems and their relationship to contraction [Taylor and Condeelis, 1979]. Of particular note are the observations showing that elevated Ca^{2+} converts a gel to a sol, and that conversion to a sol is a prerequisite for a subsequent contraction. In looking for an explanation of these events it seems reasonable to couple these findings with current results from studies on actin binding proteins [reviewed by Schliwa, 1981]. The identification of a class of widely occurring proteins that bind actin and are stimulated by elevated levels of Ca^{2+} to break or shorten actin microfilaments, thereby reducing viscosity, provides a plausible scheme by which Ca^{2+} could transform a gel to a sol, and thus initiate a contractile event.

The aforementioned studies on gel-sol transformation and contraction in cell extracts may find their structural and functional bases in recent studies of Porter and co-workers [Luby and Porter, 1980; Porter and Tucker, 1981] on the participation of the microtrabecular lattice in the movement of granules in erythrophores. Luby and Porter [1980] have shown that the aggregation of pigment granules involves a concomitant shortening and thickening of the microtrabeculae. Because of the insensitivity of aggregation to metabolic inhibitors, Luby and Porter [1980] suggest that the microtrabecular lattice in its extended form contains potential energy which can be released to power the inward motion of granules. Further studies show that the contraction of microtrabeculae may be stimulated by elevated levels of Ca^{2+} [Porter and Tucker, 1981].

Although highly speculative, these ideas have been extended to the MA in an effort to provide fresh thought about the mechanisms of mitosis [McIntosh, 1981]. Thus the MA may contain a structured elastic component possessing stored energy; upon stimulation (by Ca^{2+}) this component might transform the stored energy into kinetic energy that is capable of moving chromosomes [Cande, 1982; McIntosh, 1981]. It is also attractive to imagine that Ca^{2+}

release by a spatially positioned membrane system could locally alter gel structure and thus define the location of the initial contraction event.

The mechanisms mentioned above are only some of the many that can be imagined by which Ca^{2+} could regulate mitotic motion. Furthermore, they are not mutually exclusive; indeed, it is possible, perhaps likely, that two or more such mechanisms operate at once in the MA. Thus a Ca^{2+} elevation may be viewed as stimulating a variety of processes that work cooperatively or synergistically to move chromosomes.

V. CONCLUSIONS

The pioneering studies by Porter and Machado [1960] revealed that membranes are important structural elements in the MA. We now recognize that these elements possess specific and intimate structural associations with spindle microtubules. It seems increasingly evident that, like the sarcoplasmic reticulum of muscle, spindle ER regulates the ionic milieu of the MA, especially the Ca^{2+}, and thus controls one or more processes essential for the formation and function of the MA. The evidence obtained with CTC suggests that a Ca^{2+} efflux from the spindle membrane system occurs during late metaphase. These results combined with information from the voltage-sensitive probes are beginning to pinpoint the existence of spindle-specific ionic transients and redistributions that may play key roles in the regulation of mitosis.

ACKNOWLEDGMENTS

We thank Mr Dale A. Callaham for technical assistance in the preparation of this manuscript. This work has been supported by grant RO1-GM 25120 from the National Institutes of Health.

VI. REFERENCES

Adelstein R, Eisenberg E (1980): Regulation and kinetics of the actin-myosin-ATP interaction. Ann Rev Biochem 49:921–956.

Andersen B, Osborn M, Weber K (1978): Specific visualization of the distribution of the calcium-dependent regulatory protein of cyclic nucleotide phosphodiesterase (modulator protein) in tissue culture cells by immunofluorescence microscopy: mitosis and intercellular bridge. Cytobiologie 17:354–364.

Beeler TJ, Farmen RH, Martonosi AN (1981): The mechanism of voltage-sensitive dye responses on sarcoplasmic reticulum. J Membrane Biol 62:113–137.

Bhattacharyya B, Wolff J (1975): The interaction of 1-anilino-8-napthalene sulfonate with tubulin: A site independent of the colchicine binding site. Arch Biochem Biophys 167:264–269.

Blum TJ, Hayes A, Jamieson A Jr, Vanaman TC (1980); Calmodulin confers calcium sensitivity on ciliary dynein ATPase. J Cell Biol 87:386–397.

Branton D, Cohen CM, Tyler J (1981): Interaction of cytoskeletal proteins on the human erythrocyte membrane. Cell 24:24–32.

Cande WZ (1981): Physiology of chromosome movement in lysed cell models. In Schweiger HG (ed): "International Cell Biology, 1980–1981." Berlin: Springer-Verlag, pp 382–391.

Cande WZ (1982): Nucleotide requirements for anaphase chromosome movements in permeabilized mitotic cells: Anaphase B but not anaphase A requires ATP. Cell 28:15–22.

Cande WZ, Wolniak SM (1978): Chromosome movement in lysed mitotic cells is inhibited by vanadate. J Cell Biol 79:573–580.

Caswell AH, Brandt NR (1981): Ion-induced release of calcium from isolated sarcoplasmic reticulum. J Membr Biol 58:21–33.

DeMey J, Moeremans M, Geuens G, Nuydens R, VanBelle H, DeBrabander M (1980): Immunocytochemical evidence for the association of calmodulin with microtubules of the mitotic apparatus. In DeBrabander M, DeMay J (eds): "Microtubules and Microtubule Inhibitors, 1980." Amsterdam: Elsevier/North Holland, pp 227–241.

Dentler WL (1981): Microtubule-membrane interactions in cilia and flagella. Int Rev Cytol 72:1–47.

Epel D (1978): Mechanisms of activation of sperm and egg during fertilization of sea urchin gametes. Curr Top Dev Biol 12:185–246.

Forbes MS, Plantholt BA, Sperelakis N (1977): Cytochemical staining procedures selective for sarcotubular systems of muscle: Modifications and applications. J Ultrastruc Res 60:306–327.

Forer A, Jackson WT, Engberg A (1979): Actin in spindles of Haemanthus katherinae endosperm. II. Distribution of actin in chromosomal spindle fibers, determined by analysis of serial sections. J Cell Sci 37:349–371.

Fuseler J (1975): Temperature dependence of anaphase chromosome velocity and microtubule depolymerization. J Cell Biol 67:789–800.

Gibbons IR (1981): Cilia and flagella of eucaryotes. J Cell Biol 91:107s–124s.

Griffith LM, Pollard TD (1978): Evidence for actin filament-microtubule interaction mediated by microtubule-associated proteins. J Cell Biol 78:958–965.

Guerriero V, Rowley DR, Means AR (1981): Production and characterization of an antibody to myosin light chain kinase and intracellular localization of the enzyme. Cell 27:449–458.

Harris P (1975): The role of membranes in the organization of the mitotic apparatus. Exp Cell Res 94:409–425.

Harris P (1978): Triggers, trigger waves, and mitosis: A new model. In Jeter JR Jr, Cameron IL, Padilla GM, Zimmerman AM (eds): "Monographs in Cell Biology." New York: Academic Press, pp 75–104.

Heilbrunn LV (1956): "The Dynamics of Living Protoplasm." New York: Academic Press.

Hepler PK (1977): Membranes in the spindle apparatus: Their possible role in the control of microtubule assembly. In Rost TL, Gifford EM Jr (eds): "Mechanisms and Control of Cell Division." Stroudsburg, Pennsylvania: Dowden, Hutchinson and Ross, pp 212–232.

Hepler PK (1980): Membranes in the mitotic apparatus of barley cells. J Cell Biol 86:490–499.

Hepler PK (1981): The structure of the endoplasmic reticulum revealed by osmium tetroxide–potassium ferricyanide staining. Eur J Cell Biol 26:102–110.

Hepler PK (1982): Endoplasmic reticulum in the formation of the cell plate and plasmodesmata. Protoplasma 111:121–133.

Hepler PK, McIntosh JR, Cleland S (1970): Intermicrotubule bridges in mitotic spindle apparatus. J Cell Biol 45:438–444.

Hepler PK, Wick SM, Wolniak SM (1981): The structure and role of membranes in the mitotic apparatus. In Schweiger HG (ed): "International Cell Biology, 1980–1981." Berlin: Springer-Verlag, pp 673–686.

Inoué S, Sato H (1967): Cell motility by labile associations of molecules. The nature of mitotic spindle fibers and their role in chromosome movement. J Gen Physiol 50 (Suppl):259–292.

Kiehart DP (1981): Studies on the in vivo sensitivity of spindle microtubules to calcium ions and evidence for a vesicular calcium-sequestering system. J Cell Biol 88:604–617.

LaPorte DC, Wierman BM, Storm DR (1980): Calcium-induced exposure of a hydrophobic surface on calmodulin. Biochemistry 19:3814–3819.

Ledbetter, MC, Porter KR (1963): A "microtubule" in plant cell fine structure. J Cell Biol 19:239–250.

Luby KJ, Porter KR (1980): The control of pigment migration in isolated erythrophores of Holocentrus ascensionis (Osbeck). I. Energy requirements. Cell 21:13–23.

McIntosh JR (1981): Microtubule polarity and interaction in mitotic spindle function. In Schweiger HG (ed): "International Cell Biology, 1980–1981." Berlin: Springer-Verlag, pp 359–368.

McIntosh JR, Hepler PK, VanWie DG (1969): Model for mitosis. Nature 224:659–663.

Moll E, Paweletz N (1980): Membranes of the mitotic apparatus of mammalial cells. Eur J Cell Biol 21:280–287.

Moskalewski S, Thyberg J, Hinek A, Friberg U (1977): Fine structure of the Golgi complex during mitosis of cartilaginous cells in vitro. Tissue Cell 9:185–196.

Paweletz N (1981): Membranes in the mitotic apparatus. Cell Biol Int Rep 5:323–336.

Paweletz N, Finze E-M (1981): Membranes and microtubules of the mitotic apparatus of mammalian cells. J Ultrastruc Res 76:127–133.

Petzelt C (1979): Biochemistry of the mitotic spindle. Int Rev Cytol 60:53–92.

Pickett-Heaps JD, Tippit DH, Leslie (1980): Light and electron microscopic observations on cell division in two large pennate diatoms, Hantzschia and Nitzschia. II. Ultrastructure. Eur J Cell Biol 21:12–27.

Porter KR, Machado RD (1960): Studies on the endoplasmic reticulum. IV. Its form and distribution during mitosis in cells of onion root tip. J Biophys Biochem Cytol 7:167–180.

Porter KR, Tucker JB (1981): The ground substance of the living cell. Sci Am 244(3):56–67.

Pratt MM, Otter T, Salmon ED (1980): Dynein-like Mg^{2+}-ATPase in mitotic spindles isolated from sea urchin embryos (Strongylocentrotus droebachiensis). J Cell Biol 86:738–745.

Salmon ED, Segall RR (1980): Calcium-labile mitotic spindles isolated from sea urchin eggs (Lytechinus variegatus). J Cell Biol 86:355–365.

Schliwa M (1981): Proteins associated with cytoplasmic actin. Cell 25:587–590.

Silver RB, Cole RD, Cande WZ (1980): Isolation of mitotic apparatus containing vesicles with calcium sequestration activity. Cell 19:505–516.

Sims PJ, Waggoner AS, Wang CH, Hoffman JF (1974): Studies on the mechanism by which cyanine dyes measure membrane potential in red blood cells and phosphotidylcholine vesicles. Biochemistry 13:3315–3330.

Taylor DL, Condeelis JS (1979): Cytoplasmic structure and contractility in amoeboid cells. Int Rev Cytol 56:57–144.

Trauble H, Overath P (1973): The structure of Escherichia coli membranes studied by fluorescence measurements of lipid phase transitions. Biochim Biophys Acta 307:491–512.

Waggoner AS (1979): Dye indicators of membrane potential. Ann Rev Biophys Bioenerg 8:47–68.

Welsh MJ, Dedman JR, Brinkley BR, Means AR (1978): Calcium-dependent regulator protein: Localization in the mitotic apparatus of eukaryotic cells. Proc Natl Acad Sci USA 75:1867–1871.

Wick SM, Hepler PK (1980): Localization of Ca^{++}-containing antimonate precipitates during mitosis. J Cell Biol 86:500–513.

Wolniak SM, Hepler PK, Jackson WT (1980): Detection of the membrane-calcium distribution during mitosis in Haemanthus with chlorotetracycline. J Cell Biol 87:23–32.

Wolniak SM, Hepler PK, Jackson WT (1981): The coincident distribution of calcium-rich membranes and kinetochore fibers at metaphase in living endosperm cells of Haemanthus. Eur J Cell Biol 25:171–174.

Wolniak SM, Hepler PK, Jackson WT (1983): Ionic changes in the mitotic apparatus at the metaphase/anaphase transition. J Cell Biol (in press).

Zimniak P, Racker E (1978): Electrogenicity of Ca^{2+} transport catalyzed by the Ca^{2+}-ATPase from sarcoplasmic reticulum. J Biol Chem 253:4631–4637.

Organization of the Fibrous Components of Cytoplasm

Modern Cell Biology 2:115–142
© 1983 Alan R. Liss, Inc., 150 Fifth Avenue, New York, NY 10011

The Centrosome as an Organizer of the Cytoskeleton

J. Richard McIntosh

From the Department of Molecular, Cellular and Development Biology,
University of Colorado, Boulder, Colorado 80309

I. INTRODUCTION

Centrosomes are cytoplasmic structures that form the poles of the mitotic spindle in all animal cells and in many lower plants and protozoa. During interphase a centrosome is usually located near the nucleus at roughly the center of the cell. Throughout the cell division cycle, the centrosome contributes to the organization of the cytoskeleton. Its most direct action seems to be on microtubules: It works as a microtubule organizing center (MTOC), a concept first clearly stated by Porter [1966]. Indirectly, the centrosome contributes to the arrangement and coordination of other cytoplasmic fibers as well. In general, it persists from one cell generation to another, and is passed on, along with a set of chromosomes, to each successive set of daughters. This behavior led Van Beneden and Boveri to suggest during the 19th century that the centrosome was an autonomous organelle that would arise only from a preexisting body of the same kind [reviewed in Wilson,

1925]. Recent work on the polymorphism of centrosomes, on centriole loss, and on what appears to be de novo centriole formation makes it clear that although this view has merit, it is an oversimplification of reality.

In the essay to follow I will review some examples of centrosome structure and behavior in selected organisms. Essentially all currently identified functions of a centrosome can be related to its ability to act as an MTOC. The evidence concerning the mechanisms of this organizing activity will be discussed and related to the process of microtubule initiation in vitro. Finally, I will reexamine the centrosome in vivo and speculate on how the structure might work to influence cell morphology and behavior.

II. STRUCTURE AND FUNCTION OF THE INTERPHASE CENTROSOME IN VIVO

The interphase centrosome may be defined as the dominant origin of the cell's cytoskeletal microtubules. Figure 1 portrays interphase mammalian cells in which the microtubule array is visualized by indirect immunofluorescence with antitubulin. The centrosome is the juxtanuclear region that serves as the focus for most of the microtubules of the cell. Electron microscopy of this region shows the microtubules ending in an ill-defined, ellipsoidal region which usually contains two centrioles in orthogonal orientation and some granular densities called satellites distributed around the centrioles (Fig. 2) [Bernhard and de Harven, 1958; Andre, 1964; Szollosi, 1964; Murray et al, 1965; Brinkley and Stubblefield, 1970]. Some satellites are attached to the centrioles, and others are situated at a distance [Zeligs and Wollman, 1979]. The satellites usually lie at one end of a microtubule, but not all tubules end on satellites. Although the microtubule occasionally appear to make direct contact with a centriole [Brinkley and Stubblefield, 1970], they most frequently terminate in the cloud of ill-defined material that surrounds the centrioles [Porter, 1966; Roos, 1973; McGill et al, 1976; Rieder and Borisy, 1982]. Figure 3 is a diagram of the material found in a

Fig. 1. Cultured interphase cells, strain BSC, from the African green monkey, prepared for indirect immunofluorescence with a monoclonal antibubulin generously supplied by John Kilmartin. Micrograph kindness of Jonathan Izant. The centrosome lies at the center of the brightly staining region near the middle of each cell. The nucleus is just visible below the centrosome in the cell fully shown. × 1,800.

Fig. 2. A centrosome from a plasma cell of a rat. Micrograph kindness of Keith Porter. One centriole is visible; the other is in another plane of section. Several satellites are clearly visible near the centriole (arrow). Numerous microtubules invade the area, and many membrane-bound vesicles of the Golgi complex are also seen. × 65,000.

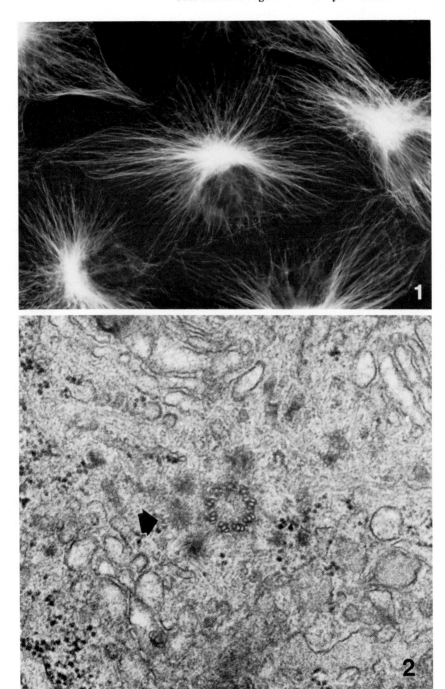

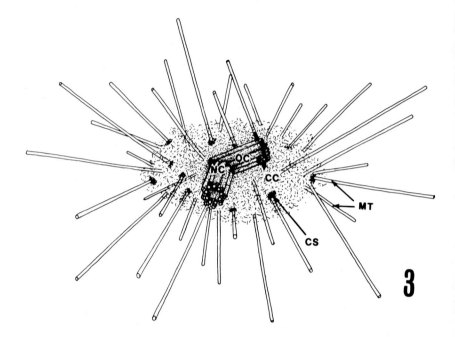

Fig. 3.   A diagram of the centrosome of an animal cell. The old centriole marked OC and the new centriole marked NC are shown, surrounded by a cloud of ill-defined material marked CC. Centriolar satellites (CS) are present at the ends of many, but not all, of the microtubules (MT). The number of satellites is not well known, but appears to be variable. The old centriole displays additional ornamentation at one end.

mammalian interphase centrosome. Microsurgical experiments using a laser microbeam to ablate the centrioles or the pericentriolar material [Berns et al, 1977; Peterson and Berns, 1978] and drug treatments that dissociate the centrioles from the pericentriolar cloud [Heidemann et al, 1980] have confirmed that in vivo it is the pericentriolar cloud material that carries out the MTOC function for the cytoskeleton, not the centrioles themselves [reviewed in Fulton, 1971; Wheatley, 1982].

Various treatments of a cell, such as addition of the drug colchicine or a trypsin-induced detachment of the cell from its substratum, cause cytoplasmic microtubules to disappear. As they reappear upon reversal of the treatment, the microtubules grow out from the centrosome, confirming the action of this structure as an MTOC in vivo [Brinkley et al, 1975; Osborn and Weber, 1976; DeBrabander et al, 1981].

All microtubules studied are structurally polar [Amos, 1979]. This polarity is essential for their participation in certain motile mechanisms [McIntosh, 1981] and contributes to the complexity of microtubule assembly properties in vitro [Margolis and Wilson, 1978]. The microtubules emanating from a centrosome are all of the same structural polarity: Their fast growing, or "plus" end is distal to the centrosome. This property has been shown both by kinetic studies of microtubule elongation rates in vitro [Bergen et al, 1980] and by decoration of centrosomal microtubules with polarity-revealing hooks assembled from tubulin dimers (Fig. 4) [Heidemann and McIntosh, 1980; Euteneuer and McIntosh, 1981]. Thus the centrosome defines not only the place where microtubules will initiate within the cell, but also the orientation of the tubules relative to that place.

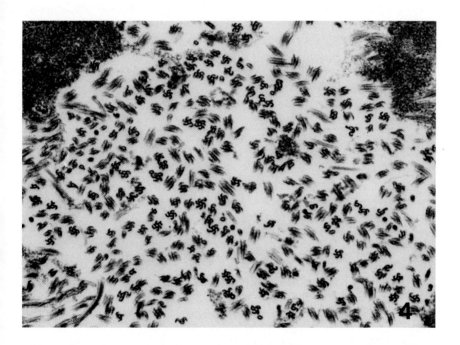

Fig. 4.   An electron micrograph of a section near the centrosome of a mitotic mammalian cell, strain PtK_1. The cell was lysed with detergent in a tubulin-containing solution that promotes formation of the hook-shaped decorations on the microtubule walls [Heidemann and McIntosh, 1980]. The direction of view is looking in toward the centrosome; since the hooks curve clockwise, most of the microtubules are oriented with their "plus" or fast-growing ends distal to the centrosome. Micrograph kindness of Ursula Euteneuer. × 50,000.

The centrosome and its associated microtubules have several functions in the cell. One is to contribute to the definition of cell polarity. The position of the centrosome relative to the nucleus is usually correlated with several other cellular vectors. For example, in a covering epithelium the centrosome and Golgi apparatus lie on the apical side of the nucleus. In some elongating epithelial cells, the centrosome migrates away from the nucleus, but still lies at the apical pole of the cell [Byers and Porter, 1964]. In a more functional example, the capacity of a cultured cell to migrate over its substratum usually requires a cytoplasmic microtubule system emanating from the centrosome. Removal of fibroblast microtubules with colcemid does not inhibit the motile activity of the cell periphery, but does block directed movement of the cell as a whole [Vasiliev and Gelfand, 1976]. Polymorphonuclear leukocytes will migrate at random in the presence of sufficient colchicine to disassemble their microtubules, but their directed movement in response to a chemotactic gradient is impaired by the drug treatment [Pryzwansky et al, 1982]. Finally, when a confluent layer of cells is scratched, causing a local disruption of the confluence, cells adjacent to the scratch migrate into the empty area. As they do so, their centrosomes move around so as to lie anterior to the nucleus in the migrating cell [Gotlieb et al, 1981]. The mechanism of this movement and its significance for cell motility remain to be discovered.

A direct effect of the centrosome and its associated microtubules on the organization of other elements in the cytoskeleton is shown by prolonged treatment of cultured cells with colchicine. Such treatment leads to a collapse of the intermediate filament array into a juxtanuclear cap [Goldman, 1971]. Reversal of the drug treatment allows both the microtubules and the intermediate filaments to reestablish their normal geometry. The reciprocal experiment of disrupting the intermediate filaments by microinjection of antibodies to their subunits into a living cell does not show a corresponding collapse of the microtubule framework [Klymkowsky, 1981]. Therefore, the centrosome appears to have some indirect but essential role in determining the cellular distribution of intermediate filaments.

The positions of the centrosomes and their associated mitotic asters define the position of the ensuing cleavage furrow. The entire surface of a marine egg is capable of forming a cleavage furrow, but the furrow always forms at the plane bisecting the line segment interconnecting two mitotic centrosomes, whether or not they reside in the same mitotic spindle [reviewed in Rappaport, 1975]. It seems, then, that the centrosome and its associated microtubules can also exert some level of regulation over the cortical actin filaments known to participate in the formation of a cleavage furrow [Schroeder, 1973].

An additional function associated with centrosomes is illustrated by movies of granule-containing cells, such as newt white blood cells shown in the published film by R. D. Allen [1972]. The granules of the cytoplasm migrate in and out toward a juxtanuclear focus along essentially straight radial paths; that is, they move along a radial family of microtubules associated with a centrosome. Similar granule movements have also been described in the mitotic asters of echinoderm eggs [Rebhun, 1972]. Pigment cells on the scales of a teleost like Holocentrus are a particularly well developed example of intracellular motility in which granules migrate radially to and from the centrosome at speeds as high as 10 μm/sec [Byers and Porter, 1977]. Exactly how the centrosome and its microtubules contribute to granule motility is a field of active current research. It is clear that in many systems the microtubules are essential for normal particle translocation, and recent evidence implicates the microtubule associated ATPase, dynein, as part of the motile mechanism [Beckerle and Porter, 1982]. However, these cells are also permeated by a network of fibers that interconnects the granules and binds them to the microtubules [Byers and Porter, 1977]. This net, the microtrabecular lattice, shows a calcium ion-dependent contraction in toward the centrosome and an ATP-cAMP-dependent expansion away from it [Luby and Porter, 1982]. The role of the centrosome in the behavior of the net is currently being explored [Porter, this volume].

In some protozoa, the centrosome serves as the organizer for an elaborate array of microtubules which is involved not only in defining cell structure but also in moving the debris of a captured food organism toward the cell body, again reflecting in a single example both a cytoskeletal and a kinetic aspect of the centrosome. The heliozoan, Heterophrys marina [Bardele, 1975] is an elegant example of these phenomena (Fig. 5). In this organism the centrosome does not contain a centriole, and the microtubule-organizing activity is not so uniformly distributed around the centrosome as in a pigment cell. The tubules that emanate from the centrosome are cross-bridged together into bundles, indicating the importance of intertubule connections in fine-tuning the arrangements of centrosome-associated microtubules [reviewed in Tucker, 1979].

In other protozoa the centrosome is not centrally located in the cell, yet its function as an organizer of cytoplasmic microtubules is still manifest. In the chrysophyte Ochromonas, the interphase centrosome is located at the anterior end of the cell. There are two flagella growing from the centrioles of this centrosome, but, in addition, there is a complex of rootlet structures that initiate an array of microtubules. These fibers run like staves beneath the plasma membrane, forming a cortical framework that helps to determine cell shape [Bouck and Brown, 1973]. In this case, the microtubule-initiating activity of the centrosome is accomplished by well-defined foci situated along the several fibrous rootlets of the centrioles. The ciliates are an interesting group to consider in the context of centrosome location because they appear

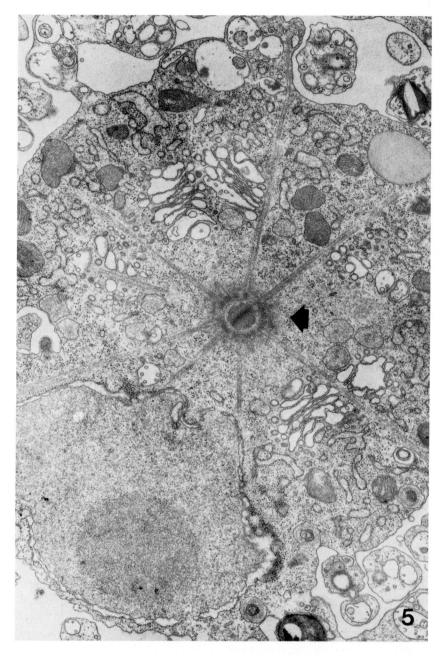

Fig. 5. The centrosome (arrow) of the centrohelidian helizoan Heterophrys marina. Micrograph kindness of Christian Bardele. The centrosome lacks centrioles but initiates the growth of many well-ordered, microtubule-containing shafts. × 20,000.

to lack a normal, juxtanuclear centrosome even at mitosis [Tucker, 1967; LaFontain and Davidson, 1980], yet their cortex is covered with centrioles serving as basal bodies for cilia and as organizers of rootlets. These rootlets in turn initiate microtubule bundles that run beneath the plasma membrane, forming a fibrous infrastructure. If the basal body-rootlet combinations of Ochromonas may legitimately be regarded as a centrosome, then the equivalent complexes in ciliates have that character as well. Indeed, the microtubule-organizing properties of the rootlets in Stentor are responsible for the formation of the cortical "Km fibers," which are motile organelles associated with changes in interphase cell shape (Fig. 6) [Huang and Pitellka, 1973].

In higher plant cells there is no obvious centrosome or centrosome analogue, during either interphase or mitosis. The mitotic centrosome function is present, but the only structural correlate to the microtubule-organizing activity of the spindle pole is an accumulation of membranous vesicles [Jackson and Doyle, 1982]. The only interphase microtubule-organizing structures known are immediately postmitotic. In the rhizoid of the water fern Azolla, there are cortical densities that appear to serve as MTOCs for the subplasmalemmal microtubules [Gunning, 1980]. Whether these structures are actually related to centrosomes in other organisms is at present unknown.

In summary, the interphase centrosome is a cytoplasmic microtubule-organizing center which initiates microtubules with a well-defined polarity from a specific cellular position. It usually lies beside the nucleus, often displacing the nucleus from the center of the cell. Many centrosomes contain centrioles, but such an association is not universal. Even in those that do, the centriole is either contained within a cloud of microtubule-initiating activity or is attached to appendages and rootlets that serve as initiators. The centrioles generally do not attach directly to the cytoplasmic tubules of the cytoskeleton. Through an association with microtubules, the centrosome is involved in multiple cellular functions: 1) establishing and maintaining cell shape and polarity; and 2) providing a framework for certain motile processes, both inside the cell and for the cell as a whole. Because the centrosome-associated microtubules can influence intermediate filament geometry and the activity of cortical actin both in cleavage and in gliding cell motility, it is tempting to attribute to the centrosome a major role in cytomorphogenesis and cell behavior. Before such a conjecture can become fact, however, it will be necessary to understand the molecular biology of the relationships between the different fiber systems of the cell. There are chemical indications of interactions between microtubules and actin microfilaments, perhaps mediated by microtubule-associated proteins [Griffith and Pollard, 1978; Sattilaro et al, 1981]. Analogous studies have begun [Pytella and Wiche, 1980] for understanding the relations between microtubules and intermediate filaments. When the chemistry of these interactions is clear, we will be better

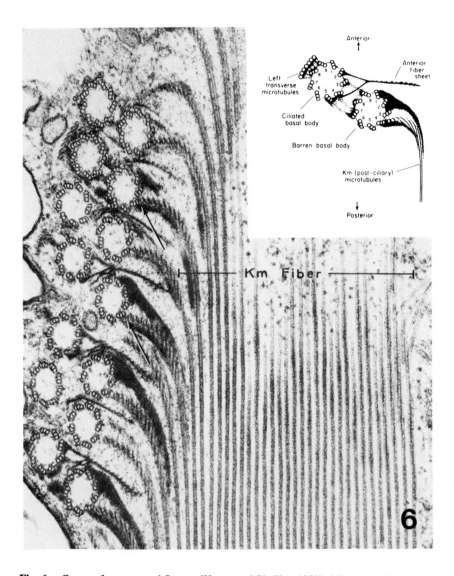

Fig. 6. Cortex of a contracted Stentor [Huang and Pitellka, 1973]. Micrograph kindness of B. Huang. The basal bodies are associated with rootlets (arrows) that initiate rows of microtubules (see diagram in insert). The rows of microtubules associate in sheets to form the Km fiber of the cortex. This fiber is involved in the shape change process of the cell, so some of the functions of the centrosome are accomplished by the ciliate basal body and its associated structures.

able to understand the impact of the centrosome on the several fiber systems of the cytoskeleton.

The microtubules emanating from the centrosome often display a geometry characteristic of a particular cell type (eg, Figs. 5, 6). Given the role of the centrosome as an MTOC, it is tempting to infer that centrosome action is characteristic for each cell type. Since centrosomes are passed from generation to generation, it has long been thought that they might contribute to the heritability of cell form [reviewed in Wilson, 1925; Solomon, 1981]. Recent work comparing the morphology of clonally related cells shows that stress fiber patterns and cell motility pathes [Albrecht-Buehler, 1977], as well as the dentritic arborizations of neuroblastoma cells [Solomon, 1981], demonstrate the tendency of cell morphology to repeat with each generation. The contribution of the centrosome to this phenomenon remains to be determined but deserves attention.

The importance of centrosome behavior for asymmetric cleavages in embryonic divisions is another subject that merits attention. It is common for differentiative divisions in embryos to occur at asymmetric cleavages preceded by an asymmetric spindle in which the two asters are of different size. Given that the whole egg cortex is capable of cleavage, the positions of the two centrosomes are significant in defining the cleavage asymmetry. It is not yet clear whether the asymmetric asters cause or are the result of asymmetric spindle placement. The questions of how sister centrosomes might become different and how this difference might be related to centrosome pedigree are interesting subjects for the future.

III. MODULATION OF CENTROSOME BEHAVIOR DURING THE CELL CYCLE

The centrosome goes through a life cycle analogous to that of the chromosomes as a cell grows and divides. During interphase the centrosome duplicates; at mitosis the two centrosomes become the spindle poles, which are then segregated to the daughter cells as the spindle elongates. The latter behavior can be followed simply by observing the microtubule distribution in cultured cells with antitubulin (Fig. 7).

The chronology of centrosome duplication has been studied by treating cells with colcemid to dissolve the bulk of the cytoplasmic microtubules, then releasing the block for a few minutes before fixing the cells and preparing them for immunofluorescence with antitubulin. The location of MTOC activity is revealed by the splash of fluorescent staining associated with the tubules formed during the brief recovery period. Brinkley and his co-workers have used this method together with synchronized populations of cells to show that the number of MTOCs per cell shifts from about one to about two

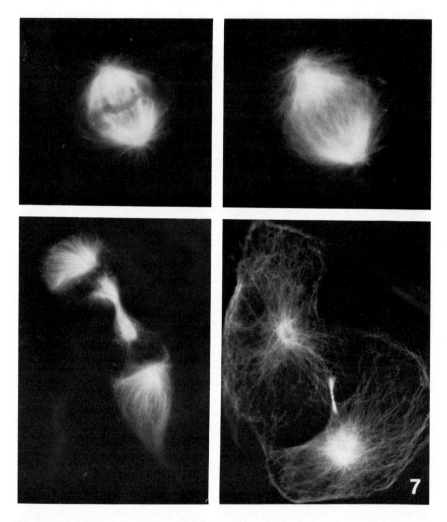

Fig. 7. BSC cells, as in Figure 1, showing indirect immunofluorescent staining of the microtubules during the mitotic cycle. The centrosomes reside at the spindle poles, and are segregated to the daughter cells. Each daughter then organizes a new interphase microtubule array, even while the two daughters are still interconnected by the midbody, seen here as a small shaft of brightly fluorescent material. × 1,800.

as the cell passes through S phase [Brinkley et al, 1981a]. The same assay shows that other cells (eg, the mouse neuroblastoma line N-115) have multiple centers, and that these can aggregate to form a single MTOC when the cell extends a neurite [Spiegelman et al, 1979]. Each of the multiple centers contains at least one centriole [Brinkley et al, 1981b; Sharp et al, 1981]. The multiple centers are all active as the cell enters mitosis; in prophase they become the vertices of a polyhedral cage of microtubules surrounding the chromosomes, and then, remarkably, they coalesce into two clusters, one forming each spindle pole [Ring et al, 1982].

The chronology of centrosome separation in cells with only one or two centrosomes is variable. In PtK cells, separation does not occur until mitosis, whereas in many other cultured mammalian cell lines it happens during the S or G2 phase [Rattner and Berns, 1976]. The machinery for moving the duplicate centrosomes apart is apparently activated by epidermal growth factor [Sherline and Mascardo, 1982] and displays a sensitivity to diazepam that merits further investigation [Anderson et al, 1981].

It has been known for many years that the centriole cycle follows a chronology similar to that described for the centrosomes [reviewed in Cleveland, 1963]. In animal cells each daughter inherits a pair of centrioles with its single centrosome. During G1 these separate slightly, and at some time near the onset of DNA synthesis, each becomes associated with a new "procentriole" that appears to grow from the side of the preexisting centriole [Robbins et al, 1968; Stubblefield, 1968; Rattner and Phillips, 1973; reviewed in Wheatley, 1982]. The procentrioles elongate during the rest of interphase so that by mitosis, each cell inherits an almost fully grown "daughter" centriole and either the two "mother" centrioles, one of which may be very old indeed. Such pedigrees are strictly analogous to those of DNA strands and are characteristic of semiconservative replication.

In the yeast Saccharomyces cerevisiae, the availability of certain temperature-sensitive cell cycle mutants and of "mating factor," a small protein pheromone which arrests cell cycle progression, has made possible a rather strict correlation between centrosome duplication and the chromosome cycle. The centrosomes of this cell are the "spindle pole bodies," specializations of the nuclear envelope that serve as initiators of the intranuclear spindle and some cytoplasmic microtubules. These must duplicate before DNA replication can begin [Byers and Goetsch, 1973, 1975]. Several genetic complementation groups have now been identified which share the phenotype of blocking spindle pole body duplication; all these arrest cell-cycle progression just before the onset of DNA replication [Reed, 1980]. In this organism the duplication of the centrosome appears to be the first event of cell duplication, and is required for all subsequent events [Hartwell, 1974, 1978; Byers, 1981]. Current evidence on this same point in other cell types is conflicting, but suggests that microtubule-related structures are active in establishing control points for DNA replication [eg, Tucker et al, 1979; Brooks et al, 1980; review by Otto, 1982].

The centrosome changes the extent of its microtubule-initiating activity during the course of the cell cycle. The well-developed interphase microtubule array disappears during prophase as the spindle begins to form. The mitotic centrosome often initiates more microtubules than are found during interphase, as indicated by a comparison of the number of interphase microtubules seen by immunofluorescence [eg, Brinkley et al, 1981a] with the number of mitotic microtubules seen by electron microscopy [eg, McIntosh et al, 1975]. Furthermore, the shape of the microtubule array changes from interphase to mitosis (compare Figs. 1 and 7). Some of this difference is certainly due to the interaction of the centrosomes and the chromosomes at mitosis, but there appear to be both qualitative and quantitative differences in centrosome activity at different stages of the cell cycle.

It is characteristic of mitotic as well as interphase centrosomes that the microtubules grow from centriole-associated material, not from the centrioles themselves. In Ochromonas the "rhizoplast" serves not only in interphase to initiate the cytoskeletal tubules but also in mitosis to initiate the spindle (Fig. 8). The hypermastigote flagellate Barbulanympha is a dramatic example in which large, centriolelike structures reside at the anteriormost portion of the cytoplasm. During interphase these grow to about 30 μm length with a bulbous structure at their distal end [Cleveland, 1954]. Each bulbous terminus serves as a mitotic centrosome to organize the spindle. It is identified by electron microscopy as the end point of a spectacular cluster of microtubules, even though there is no visible centriole at either spindle pole [Ritter et al, 1978]. The fine structure of the centriolelike organelles in two related flagellates has been studied in detail, and shows that much additional material of complex organization can be involved in connecting the true centrioles with the mitotic MTOC [Grimstone and Gibbons, 1966]. In most animal cells the spindle microtubules are initiated by the pericentriolar cloud. The amount of cloud material increases from interphase to mitosis, though the time of this change seems to preceed of the change in microtubule-initiating activity [Robbins et al, 1968]. There is some indication that the number of centriolar satellites decreases as the cloud grows [Murray et al, 1965], and it has been suggested that cloud and satellites are two forms of the same material [Rieder and Borisy, 1982].

As was first pointed out by Cleveland, the two centrioles of a centrosome are not equivalent in their association with the pericentriolar material or in their contribution to centrosome activity [reviewed by Cleveland, 1963]. In the polymastigote flagellate Saccinobacculus, there are about four normal centrioles that serve during interphase as basal bodies for flagella. These are interconnected into a complex unit that was called a centriole by Cleveland et al [1934]. More recently, the centriole-containing complexes have been shown to contain six centrioles and an interconnecting bridge composed of

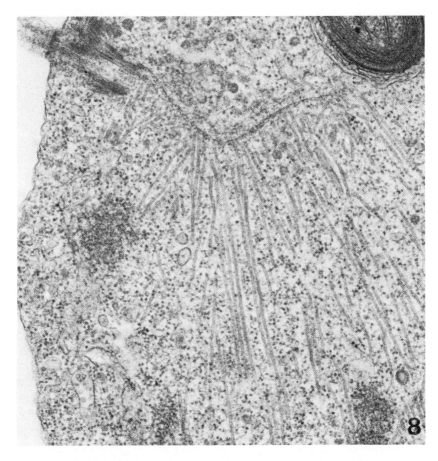

Fig. 8. A mitotic Ochromonas showing the rootlet structure serving as a spindle pole. Micrograph kindness of Benjamin Bouck. This picture displays mitotic centrosome function being accomplished by a cortical centriole (basal body) and its associated material. × 30,000.

numerous short segments of microtubules and other fibrous elements [McIntosh et al, 1973]. Each cell has two of these structures which I will call "Cleveland centrioles." During interphase, the Cleveland centrioles duplicate, so when the cell enters mitosis it has two new Cleveland centrioles and two old ones, one of each at either spindle pole. According to Cleveland [1950], the old centrioles serve to organize the spindle, and during anaphase each new Cleveland centriole initiates the growth of an axostyle, a highly ordered fiber composed of a few thousand microtubules in paracrystalline array.

Cleveland's observation on the specialized functions of new and old centrioles is now confirmed by recent work on the analogous behavior of centrioles in mammals: It is the old centriole that serves at each pole as the focus for the pericentriolar material involved in spindle formation [Vorobjev and Chentsov, 1982; Rieder and Borisy, 1982]. The old centriole is also the origin of the primary cilium in those cells that have one [Vorobjev and Chentsov, 1982]. Note, however, that one cannot say for certain whether there are "new" and "old" centrosomes associated with the new and old centrioles. As the centrioles separate and duplicate, the old pericentriolar material might be split equally between the two, and then augmented with new material, so neither centrosome would contain either simply old or new material. The work of Rieder and Borisy [1982] suggests, however, that the pericentriolar material which constitutes the MTOC activity of the centrosome does not change the centriole with which it associates; one new pericentriolar cloud appears to form at each cell cycle.

IV. ACTION OF THE CENTROSOME IN VITRO

The view that the centrosome is an MTOC [Porter, 1966] was confirmed during the mid 1970s by a large number of studies which showed that lysed cells could be used as model systems to study the initiation of microtubule growth in vitro. Weisenberg and Rosenfeld [1975] followed the development of centrosome function in homogenates of artificially activated eggs of the surf clam, Spisula. They showed that the activation process (treatment with high-potassium seawater) initiates the de novo appearance of centrioles, the pericentriolar material, and the microtubule-initiating activity of a normal centrosome. In this organism the addition of the sperm centrosome is not required for the initiation of centrosome cycles in the egg. Other investigators have studied the effects of adding neurotubulin, treated so as to be inefficient at spontaneous initiation of polymerization, to colcemid-treated mammalian cells under cell-lysing conditions. Under these conditions, the centrosomes will serve to promote efficient microtubule initiation [McGill and Brinkley, 1975; Snyder and McIntosh, 1975; Gould and Borisy, 1977; Telzer and Rosenbaum, 1979]. The majority of the tubules that form have one of their ends in the pericentriolar material, but not on the centrioles themselves, thus confirming the importance of the centriole-associated material in centrosome activity. Gould and Borisy [1977] showed that some of the in vitro tubule-initiating activity could be separated from the centrioles, yet still be identified as the pericentriolar stuff, owing to its content of particles resembling viruses. The pericentriolar material is therefore established as capable of microtubule initiation, both in vivo and in vitro.

Isolated mitotic centrosomes and their associated microtubules have been used to identify the sites of subunit addition during microtubule elongation in vitro. Under somewhat unusual polymerization conditions that promote the assembly of tubulin into decorated microtubules (Fig. 4), an aster enlarges with most of the decorations distal to the centrosome. Apparently under these conditions, subunit addition occurs in vitro at the distal ends of the centrosomal microtubules [Heidemann et al, 1980].

The kinetics of microtubule growth and disassembly in vitro are consistent with the addition and loss of subunits only at the "plus" or fast growing tubule end [Bergen et al, 1980]. These data, together with the evidence on the site of subunit addition, show that the polarity of microtubules grown from centrosomes in vitro is the same as the polarity observed in vivo as discussed above: The plus end is distal to the centrosome. They also suggest that the end proximal to the centrosome is blocked both to subunit addition and to subunit loss. There are as yet no data to confirm these conclusions for microtubules in vivo. They are, however, important points for cell physiology, because if the centrosome really blocks the proximal microtubule end, then the treadmilling of centrosome-associated microtubules—ie, subunit addition at one end and equal subunit loss at the other [Margolis and Wilson, 1978]—cannot occur in cells. Since treadmilling of centrosome-associated microtubules has been suggested as a causal factor in mitotic chromosome movement [Margolis et al, 1978] and may be relevant for centrosome function [Kirschner, 1980], the situation in vivo demands investigation.

There are several indications that a centrosome can specify the number of microtubules that grow from it. Spindle pole bodies isolated from the yeast Saccharomyces initiate about the same number of microtubules as are found in vivo [Hyams and Borisy, 1978; Byers et al, 1978]. In lysed, colcemid-treated cells, the number of microtubules formed at the centrosome increases with tubulin monomer concentration up to about 1 mg/ml tubulin and then reaches a plateau, as if there were a specific and saturable number of sites for polymer initiation [Brinkley et al, 1980]. Similarly, the number of tubules in the radial splay initiated by centrosomes isolated at a given stage of the cell cycle is fairly consistent from centrosome to centrosome [Kuriyama and Borisy, 1981]. Finally, both microtubule number and the position of microtubule initiation are strictly controlled by isolated centrosomes of the alga Polytomella [Stearns and Brown, 1981].

The number of microtubules initiated by the centrosomes in some cell types is, however, subject to modulation. Melanophores lysed with granules aggregated or granules dispersed will initiate very different numbers of microtubules from the same preparation of tubulin [Schliwa et al, 1979]. The number of tubules initiated by the centrosome in lysed cells varies with time in the cell cycle in a fashion analogous to the variation seen in vivo (discussed in II, above). The transition in the centrosome's capacity to initiate microtubules is not correlated with centrosome duplication, but occurs after the accumulation of pericentriolar material [Robbins et al, 1968; Kuriyama and Borisy, 1981], and is closely linked with the time of nuclear envelope breakdown [Snyder and McIntosh, 1975]. Recent quantitative studies show that the number of microtubules initiated by mitotic centrosomes is about five times greater than that for interphase centrosomes [Kuriyama and Borisy, 1981]. Apparently there is a fairly well defined number of microtubule initiation sites on an interphase centrosome, and as the cell enters mitosis, the number increases markedly.

Studies comparing the numbers of microtubules initiated by the interphase centrosomes of transformed and nontransformed cells show the latter to be more prolific. If, however, transformed cells are lysed, and then treated with cAMP and ATP, their microtubule-initiating capacity more closely approximates that of untransformed cells [Tash et al, 1980]. These results suggest that the number of active initiation sites may be related to some sort of protein kinase activity.

An insight into the mechanism by which centrosomes may count the microtubules they initiate derives from studies by Pepper and Brinkley [1979] on antibody treatment of centrosomes in lysed cells. When colcemid-treated, lysed PtK cells are perfused with antibodies to neurotubulin, rinsed, then challenged with purified brain tubulin, the capacity of the centrosome for in vitro microtubule initiation is abolished. The affinity-purified antibodies used in these studies were not rigorously characterized, for example by immunoelectrophoresis against solubilized PtK cells, but the results show that something in neurotubule protein can elicit antibodies that block centrosome function. The current immunochemical evidence for tubulin as a component of centrosomes is conflicting [Pepper and Brinkley, 1977; DeBrabander et al, 1979], and at least one microtubule-associated protein has been localized at the centrosome [Sherline and Mascardo, 1982]. An unambiguous interpretation of the interesting results of Pepper and Brinkley [1979] is therefore not yet possible. Clarification of the molecular mechanism for the antibody interference with centrosome activity will be of great importance.

V. THE COMPOSITION OF CENTROSOMES

Centrosome and centriole chemistry are still in their infancy. Centrioles have been isolated from several species [Gould, 1975; Anderson, 1977], and there is some description of their component proteins [Anderson and Floyd, 1980], though no structural analysis or reconstitution has yet been achieved.

An intriguing aspect of centriole composition is the now persuasive evidence that the structures contain RNA that is essential for their participation in centrosome function. Heidemann, Sander and Kirschner [1977] prepared crude basal bodies from either Chlamydomonas or Tetrahymena. When these were injected into mature, unfertilized eggs of Xenopus laevis, numerous asters formed. Treatment of the basal bodies with RNase before their injection obliterated the aster induction, while DNase treatment had no effect. Proteases had an effect only after extensive treatment which led to a structural disintegration of the basal bodies. These experiments were carefully executed and accompanied by suitable controls, so they are very convincing. Additional evidence for functionally significant centriolar RNA has been obtained by Bern and co-workers using acridine orange [Berns et al, 1977] and psoralen [Peterson and Berns, 1978] to sensitize nucleic acids to laser microirradiation. These studies confirm that the pericentriolar material is important for mitotic centrosomal activity, and add evidence that RNA is significant for the capacity of the centrosome to function. Thus far the chemistry of this suspected RNA has remained elusive, though several cytochemical methods confirm its existence [Dippell, 1976; Zackroff et al, 1979; Reider, 1979; Pepper and Brinkley, 1980; Synder, 1980].

Two kinds of immunological studies of the centrosome are in progress. Kirschner and co-workers have begun to use crude isolates of centrosomes as immmunogens for eliciting monoclonal antibodies against specific centrosomal components [Ring et al, 1980]. Several groups are capitalizing on the observation that auto antibodies against centrosomes appear at reasonable frequencies in populations of both rabbits [Connally and Kalnins, 1978] and humans [Brenner et al, 1981]. Both approaches are promising, but have yet to inform us about centrosome action.

A direct biochemical analysis of the centrosome in one system has been informative concerning not only its composition but also the mechanism of centrosome action. The alga Polytomella can be grown in quantity and lysed with comparative ease. Its centrosomes, which resemble those of Ochromonas, will function in vitro as MTOCs [Stearns et al, 1976; Stearns and Brown, 1981; reviewed in Brown and Stearns, 1982]. Chemical dissection of these centrosomes defines a set of high molecular polypeptides derived from the centriole rootlets. These proteins increase the efficiency of microtubule initiation in vitro [Stearns and Brown, 1979]. Such tubule-initiating proteins are of great potential interest, and it is hoped that an analysis of their modes of action will yield insights into the modes of action of centrosomes in general.

One can imagine that a genetic approach would help to improve our knowledge about centrosome composition and function. There are, at present, very few reports of centrosome- or centriole-defective mutants [Goodenough and St Clair, 1975; Byers, 1981]. Recently, however, a group of Chlamy-domonas genes has been found in which mutations result in a high incidence of uniflagellated cells [Huang et al, 1982]. None of these has a distinct "centrosome-defective" phenotype, but since the selection for mutant strains was for modified swimming behavior, not growth inhibition, this is hardly surprising. One can hope that these studies, and perhaps the genetic analysis of the yeast spindle pole body, will help to define the molecules of significance for centrosome function.

VI. MODELS FOR CENTROSOME FUNCTION AS A MICROTUBULE-ORGANIZING CENTER

Microtubule polymerization in vitro behaves as a two stage process: initiation of polymerization, and elongation of existing polymers [reviewed in Scheele and Borisy, 1979]. In the elongation reaction the polymerizing material—ie, dimers or oligomers of tubulin, perhaps in complex with micro-tubule-associated proteins (MAPs)—adds to one or both ends of an established microtubule. The existence of established microtubule ends defines a structure which is regenerated with every addition or loss of a subunit, so the kinetics of microtubule elongation is pseudo first order. The initiation reaction, in contrast, involves the formation of tubulin oligomers in which the association geometry may change with each new subunit added. Furthermore, local bond strains and unfilled bonding domains will in general lead to a lesser total bond free energy change per subunit added for oligomer formation than for polymer elongation [Osawa and Kasai, 1962]. Initiation of polymerization is therefore less favored energetically than elongation, and as soon as a polymer is initiated, it will tend to elongate. Polymerization from subunits in solution behaves as if it required an activation energy to begin; the addition of small amounts of polymer, for example, short pieces of preformed microtubule, can obviate this "activation energy" and permit efficient polymerization by the elongation reaction under solution conditions that will not support efficient polymer initiation. The elongation reaction will then continue until the monomer concentration drops to the "critical concentration," a level at which the probability of a subunit's falling off is equal to the probability of a subunit's going on.

The experiments described above show that centrosomes in vitro can mimic the action of pieces of microtubule by initiating assembly with initiation-incompetent microtubule protein [Snyder and McIntosh, 1975; Gould and Borisy, 1977]. It seems reasonable to conclude that the pericentriolar material reduces the "activation energy" associated with polymer initiation. There are at least three models for how this might be achieved: 1) The active principle of the centrosome might be a "seed" in which some components (perhaps including tubulin) set up a geometry sufficiently similar to the end of a microtubule that the initiation event is energetically similar to the elongation reaction. 2) There might be a "facilitated initiation" in which some factors, such as MAPs, are specifically localized at the centrosomes. As these molecules bind tubulin, the formation of oligomers is favored, either by direct cross-bridging of tubulin into oligomers or by a local increase in the tubulin concentration, thus potentiating oligomer formation. 3) The centrosome might alter the chemistry of its environment in such a way that the conditions near the centrosome are more conducive to assembly than elsewhere in the cell—eg, by lowering the Ca^{2+} concentration. In this model there would still be an activation energy barrier to initiation near the centrosome, but the whole polymerization reaction would be so favored in this region that initiation could proceed efficiently. The first two models resemble the sort of structural "catalysis" that is thought to define the pathways of assembly for the complex viruses in which a given assembly intermediate reduces the activation energy for the next assembly event [Wood, 1979]. The third model is a version of the ideas presented by DeBrabander [1982].

Antitubulin immunocytochemistry of cells treated with mitotic inhibitors has yielded conflicting results concerning the presence of tubulin in the pericentriolar cloud of the centrosome (discussed above). As I interpret the data, an active centrosome can exist without any tubulin bound, though tubulin is usually present at the centrosome, even when microtubules are absent. This means that if model 1 is correct, the ultimate initiating structure must be made of something other than tubulin. Presumably the initial event in polymer initiation is the binding of tubulin to that initiating structure. Several investigators have looked by electron microscopy for tubulelike morphology in the pericentriolar cloud, but nothing has yet been found. There is, then, no direct evidence in favor of model 1.

Model 3 is supported by one plausible interpretation of the complex behavior of centrosome activity in cells treated with taxol [DeBrabander, 1982], a drug that stabilizes polymerized microtubules and reduces the critical concentration for tubulin assembly in vitro [Schiff et al, 1979]. The capacity of isolated centrosomes to serve as assembly initiators argues, however, that any factor important for lowering the critical concentration of assembly must coisolate with the centrosome. Such factors, if they exist, must therefore be bound to the centrosome itself. Furthermore, the binding of the microtubules to the centrosome is not accounted for by this model. One must add a mechanism to form an association between the minus end of the recently initiated tubules and the pericentriolar cloud to explain existing data [DeBrabander, 1982].

Model 2 seems to me to be the most likely representation of reality. The microtubule-initiating protein from the Polytomella rootlet [Stearns and Brown,

1979] provides an example of how centrosomes in general might work. There is some immunocytochemical evidence for the localization of MAPs at centrosomes, even when no microtubules are present [Sherline and Mascardo, 1982]. MAPs could be expected to bind sufficiently strongly to the centrosome to coisolate under conditions in which the centrosome retains its activity, yet they could be extracted from the centrosome by those conditions in which the centrosome becomes inactive [Telzer and Rosenbaum, 1979]. The initiation of polymerization by a concentration of centrosomal MAPs would be expected to attach the microtubules to the centrosome and confer stability on their proximal end, accounting for the observed kinetic inactivity of this end in vitro [Bergen et al, 1980]. Finally, the capacity of a centrosome to count the tubules it initiates, and the results from Brinkley and co-workers on the effects of cAMP and antibodies on centrosome-mediated initiation in vitro are compatible with this interpretation. It will be important to probe the centrosome for specific MAPs that may concentrate there in a cell-cycle-dependent fashion, accounting for the time-dependent properties of the structure. Note however, that the same macromolecules may be located at the centrosome throughout its life, but be modified—eg, by phosphorylation—to modulate their function over time.

Centrosome action, as described above, is a temporal, spatial, and vectorial action on the assembly of microtubules. If, however, the concept of the microtrabecular lattice that is emerging from studies of erythrophores [Porter, this volume] can be generalized to all cells, then the centrosome may have a second direct effect on the fibers of the cytoplasm. In these cells it looks as if the centrosome-associated tubules may serve as a scaffold, whereas the centrosome-associated network of microtrabeculae crawls outward or contracts inward to the centrosome under the control of cytoplasmic messengers, such as the concentrations of Ca^{2+} or cAMP [Luby and Porter, 1982]. A similar situation may prevail at mitosis [McIntosh, 1981]. These features of cell behavior lead to the idea of a cytoplast, a viscoelastic meshwork in which the organelles are situated. The components of this meshwork and its relation to several aspects of cell physiology are the subject of the next few chapters of this book.

ACKNOWLEDGMENTS

This work was supported in part by grants from the ACS, CD8; NSF, PCM-8014549; and NIH, GM30213. I thank J. G. Izant for the unpublished immunofluorescent micrographs and M. Stearns and G. G. Borisy for sharing unpublished manuscripts. A. Cowan, L. Goldstein, and M. Schliwa provided invaluable assistance through reading the manuscript and critical discussion.

VII. REFERENCES

Albrecht-Buehler G (1977): Phagokinetic tracks of 3T3 cells. Cell 12:333–339.

Allen RD: "The Living Cell." (a movie available for rental) Harper & Row, N.Y., 1972.

Amos L (1979): Structure of microtubules. In Roberts K, Hyams JS (eds): "Microtubules." London: Academic Press, pp 1–64.

Andre J (1964): Le centriole et la region centrosomienne. J Microscop 3:23.

Anderson RGW (1977): Biochemical and cytochemical evidence for ATPase activity in basal bodies isolated from oviduct. J Cell Biol 74:547–560.

Anderson RGW, Floyd AK (1980): Electrophoretic analysis of basal body (centriole) proteins. Biochemistry 19:5625–5631.

Andersson LC, Lehto V-P, Stenman S, Bradley RA, Virtanen I (1981): Diazepam induces mitotic arrest at prometaphase by inhibiting centriolar separation. Nature 291:247–248.

Bardele CF (1975): The fine structure of the centrohelidian heliozoan Heterophrys marina. Cell Tissue Res 161:85–102.

Beckerle MC, Porter KR (1982): Inhibitors of dynein activity block intracellular transport in erythrophores. Nature 295:701–703.

Bergen LG, Kuriyama R, Borisy GG (1980): Polarity of microtubules nucleated by centrosomes and chromosomes of Chinese hamster ovary cells in vitro. J Cell Biol 84:151–159.

Bernhard W, de Harvan E (1958): L'ultrastructure du centriole et d'autres elements de l'appareil achromatique. IV. Int Conf Electron Microscopy Berlin, band II, pp 217–227.

Berns MW, Rattner JB, Brenner S, Meredith S (1977): The role of the centriolar region in animal cell mitosis. A laser microbeam study. J Cell Biol 72:351–367.

Bouck GB, Brown DL (1973): Microtubule biogenesis and cell shape in Ochromonas. I. The distribution of cytoplasmic and mitotic microtubules. J Cell Biol 56:340–359.

Brenner S, Pepper D, Berns MW, Tan E, Brinkley BR (1981): Kinetochore structure, duplication, and distribution in mammalian cells. J Cell Biol 91:95–102, 1981.

Brinkley BR, Stubblefield E (1970): Ultrastructure and interaction of the kinetochore and centriole in mitosis and meiosis. Adv Cell Biol 1:119–185.

Brinkley BR, Pepper DA, Cox SM, Fistel S, Brenner SL, Wible LJ, Pardue RL (1980): Characteristics of centriole- and kinetochore-associated microtubule assembly in mammalian cells. In DeBrabander M, DeMey J (eds): "Microtubules and Microtubule Inhibitors." Amsterdam: Elsevier/North-Holland, pp 281–296.

Brinkley BR, Fuller JM, Highfield DP (1975): Studies of microtubules in dividing and non-dividing mammalian cells using antibodies to 6S boving brain tubulin. In Borgers M, DeBrabander M (eds): "Microtubules and Microtubule Inhibitors." Amsterdam: Elsevier/North-Holland, pp 297–312.

Brinkley BR, Cox SM, Pepper DA, Wible L, Brenner SC, Pardue RL (1981a): Tubulin assembly sites and the organization of cytoplasmic microtubules in cultured mammalian cells. J Cell Biol 90:554–562.

Brinkley BR, Cox SM, Fistel SH (1981b): Organization centers for cell processes. Neurosci Res Prog Bull 19:107–124.

Brooks RF, Bennett DC, Smith JA (1980): Mammalian cells need two random transitions. Cell 19:493–504.

Brown DL, Stearns ME (1982): Structure and functional properties of MTOCs. In Clive T (ed): "Microtubule Systems in Plants." United Kingdom: Unilever.

Byers B (1981): Cytology of the yeast life cycle. In Strathern JN, Jones EW, Broach JR (eds): "The Molecular Biology of the Yeast Saccharomyces." New York: Cold Spring Harbor Laboratories, pp 59–96.

Byers B, Goetsch L (1973): Duplication of spindle plaques and integration of the yeast cell cycle. Cold Spring Harb Symp Quant Biol 38:123–131.

Byers B, Goetsch L (1975): Behavior of spindles and spindle plaques in the cell cycle and conjugation of Saccharomyces cerevisiae. J Bacteriol 124:511–523.

Byers B, Shriver K, Goetsch L (1978): The role of spindle pole bodies and modified microtubule ends in the initiation of microtubule assembly in Saccharomyces cerevisiae. J Cell Sci 30:331–352.

Byers B, Porter KR (1964): Oriented microtubules in elongating cells of the developing lens rudiment after induction. Proc Natl Acad Sci USA 52:1091–1099.

Byers HR, Porter KR (1977): Transformation in the structure of the cytoplasmic ground substance in erythrophores during pigment aggregation and dispersion. J Cell Biol 75:541–558.

Cleveland LR (1950): Hormone-induced sexual cycles of flagellates. III: Gametogenesis, fertilization, and one-division meiosis in Saccinobaculus. J Morphol 86:215–263.

Cleveland LR (1954): Hormone-induced sexual cycles of flagellates. XII. Meiosis in Barbulanympha. J Morphol 95:557–620.

Cleveland LR (1963): Functions of flagellate and other centrioles in cell reproduction. In Levine L (ed): "The Cell in Mitosis." New York: Academic Press, pp 3–31.

Cleveland LR, Hall SR, Saunders EP, Collier J (1934): The wood-feeding roach, Cryptocercus, its protozoa and the symbiosis between protozoa and roach. Mem Am Acad Arts Sci 17:185–336.

Connolly JA, Kalnins VI (1978): Visualization of centrioles and basal-bodies by fluorescent staining with nonimmune rabbit sera. J Cell Biol 79:526–532.

DeBrabander M (1982): A model for the microtubule organizing activity of the centrosomes and kinetochores in mammalian cells. Cell Biol Int Rep 6:901–916.

DeBrabander M, Geuens G, DeMey J, Jonaian M (1979): Light microscopic and ultrastructural distribution of immunoreactive tubulin in mitotic mammalian cells. Biol Cell 34:213–226.

DeBrabander M, Geuens G, Nuydens R, Willebrods K, DeMey J (1981): Microtubule assembly in living cells after release from nocodazole block: The effects of metabolic inhibitors, taxol and pH. Cell Biol Int Rep 5:913–920.

DeBrabander M, Geuens G, Nuydens R, Willebrods R, DeMey J (1982): Microtubule stability and assembly in living cells: The influence of metabolic inhibitors, taxol and pH. In "Organization of the Cytoskeleton." Cold Spring Harb Symp Quant Biol 46:227–240.

Dippell R (1976): Effects of nuclease and protease digestion on the ultrastructure of Paramecium basal bodies. J Cell Biol 69:622–637.

Euteneuer U, McIntosh JR (1981): Structural polarity of kinetochore microtubules in PtK$_1$ cells. J Cell Biol 89:338–345.

Fulton C (1971): Centrioles. In Reinert J, Urspring H (eds): "Origin and Continuity of Cell Organelles." Heidelberg: Springer-Verlag, pp 170–221.

Goldman RD (1971): The role of three cytoplasmic fibers in BHK-21 cell motility. 1) Microtubules and the effects of colchicine. J Cell Biol 51:752–762.

Goodenough UW, St Clair HS (1975): Bald-2: A mutation affecting the formation of doublet and triplet sets of microtubules in Chlamydomonas. J Cell Biol 66:480–491.

Gotlieb AI, May LM, Subrahmanyan L, Kalnins VI (1981): Distribution of microtubule organizing centers in migrating sheets of endothelial cells. J Cell Biol 91:589–594.

Gould RR (1975): The basal bodies of Chylamydomonas reinhardtii. Formation from probasal bodies, isolation, and partial characterization. J Cell Biol 65:65–74.

Gould RR, Borisy GG (1977): The pericentriolar material in chinese hamster ovary cells nucleates microtubule formation. J Cell Biol 73:601–615.

Griffith LM, Pollard TD (1978): Evidence for actin filament-microtubule interaction mediated by microtubule associated proteins. J Cell Biol 78:958–965.

Grimstone AV, Gibbons IR (1966): The fine structure of the centriolar apparatus and associated structures in the complex flagellates Trychonympha and Pseudotrychonympha. Phil Trans R Soc B (Lond) 250:215–242.

Gunning BES (1980): Spatial and temporal regulation of nucleating sites for arrays of cortical microtubules in root tip cells of the water fern Azolla pinnata. Eur J Cell Biol 23:53–65.

Hartwell LH (1974): Saccharomyces cerevisiae cell cycle. Bacteriol Rev 38:164–198.

Hartwell LH (1978): Cell division from a genetic perspective. J Cell Biol 77:627–637.

Heidemann SR, McIntosh JR (1980): Visualization of the structural polarity of microtubules. Nature 286:517–519.

Heidemann SR, Sander G, Kirschner MW (1977): Evidence for a functional role of RNA in centrioles. Cell 10:337–350.

Heidemann SR, Zeive GW, McIntosh JR (1980): Evidence for microtubule subunit addition to the distal end of mitotic structures in vitro. J Cell Biol 87:152–159.

Huang B, Pitellka DR (1973): The contractile process in the ciliate Stentor coeruleus. I. The role of microtubule and filaments. J Cell Biol 57:704–728.

Huang B, Ramanis Z, Dutcher SK, Luck DJL (1982): Uniflagellar mutants of Chlamydomonas: Evidence for the role of basal bodies in transmission of positional information. Cell 29:745–753.

Hyams JS, Borisy GG (1978): Nucleation of microtubule in vitro by isolated spindle pole bodies of the yeast Saccharomyces cerevisiae. J Cell Biol 78:401–414.

Jackson WT, Doyle BG (1982): Membrane distribution in dividing endosperm cells of Haemanthus. J Cell Biol 94:637–643.

Kirschner MW (1980): Implication of treadmilling for the stability and polarity of actin and tublin polymers in vivo. J Cell Biol 86:330–334.

Klymkowsky MW (1981): Intermediate filaments in 3T3 cells collapse after intracellular injection of a monoclonal anti-intermediate filament antibody. Nature 291:249–251.

Kuriyama R, Borisy GG (1981): Microtubule-nucleating activity of centrosomes in Chinese hamster ovary cells is independent of the centriole cycle, but coupled to the mitotic cycle. J Cell Biol 91:822–826.

La Fountain JR, Davidson LA (1980): An analysis of spindle ultrastructure during anaphase of micronuclear division in Tetrahymena. Cell Motil 1:41–62.

Luby KF, Porter KR (1982): The control of pigment migration in isolated erythrophores of Holocentrus ascensionis. II. The role of Ca^{++}. Cell 29:441–450.

Margolis RL, Wilson L (1978): Opposite end assembly and disassembly of microtubules at steady-state in vitro. Cell 13:1–8.

Margolis RL, Wilson L, Kiefer BJ (1978): Mitotic mechanism based on intrinsic microtubule behavior. Nature 272:450–452.

McGill M, Brinkley BR (1975): Human chromosomes and centrioles as nucleating sites for the in situ assembly of microtubules from bovine brain tubulin. J Cell Biol 67:189–199.

McGill M, Highfield DP, Monahan TM, Brinkley BR (1976): The effects of nucleic acid specific dyes on centrioles of mammalian cells. J Ultrastruc Res 57:43–53.

McIntosh JR (1981): Microtubule polarity and interaction in mitotic spindle function. In Schweiger HG (ed): "International Cell Biology 1980–1981." Berlin: Springer-Verlag, pp 359–368.

McIntosh JR, Cande WZ, Snyder JA (1975): Structure and physiology of the mammalian mitotic spindle. In Inoue S, Stephens RE (eds): "Molecules and Cell Movement." New York: Raven, pp 31–76.

McIntosh JR, Ogata ES, Landis SC (1973): The axostyle of Saccinobaculus. I. Structure of the organism and its microtubule bundle. J Cell Biol 56:304–323.

Murray RG, Murray AS, Pizzo A (1965): The fine structure of mitosis in rat thyrine lymphocytes. J Cell Biol 26:601–619.

Oasawa F, Kasai M (1962): A theory of linear and helical aggregations of macromolecules. J Mol Biol 4:10–21.

Osborn M, Weber K (1976): Cytoplasmic microtubules in tissue culture cells appear to grow from an organizing structure towards the plasma membrane. Proc Natl Acad Sci USA 73:867–871.

Otto AM (1982): Microtubules and the regulation of DNA synthesis in fibroblastic cells. Cell Biol Int Rep 6:1–18.

Pepper DA, Brinkley BR (1977): Localization of tubulin in the mitotic apparatus of mammalian cells by immunofluorescence and immunoelectron microscopy. Chromosoma 60:223–235.

Pepper DA, Brinkley BR (1979): Microtubule initiation at kinetochores and centrosomes in lysed mitotic cells: Inhibition of site specific nuclation by tubulin antibody. J Cell Biol 82:585–591.

Pepper DA, Brinkley BR (1980): Tubulin nucleation and assembly in mitotic cells: Evidence for nucleic acids in kinetochores and centrosomes. Cell Motil 1:1–15.

Peterson SP, Berns MW (1978): Evidence for centriolar region RNA functioning in spindle formation in dividing PtK₂ cells. J Cell Sci 34:289–301.

Porter KR (1966): Cytoplasmic microtubules and their functions. In Wolstenholme GEW, O'Connor M (eds): "Principles of Biomolecular Organization." London: Churchill, pp 308–357.

Pryzwansky K, Schliwa M, Porter KR: Eur. J. Cell Biol. (in press).

Pytella R, Wiche G (1980): High molecular weight polypeptides from cultured cells are related to hog brain MAPs but copurify with intermediate filaments. Proc Natl Acad Sci USA 7:4808–4812.

Rappaport R (1975): Establishment and organization of the cleavage mechanism. In Inoue S, Stephens RE (eds): "Molecules and Cell Movement." New York: Raven Press, pp 287–303.

Rattner JB, Phillips SG (1973): Independence of centriole formation and DNA synthesis. J Cell Biol 57:359–372.

Rebhun LI (1972): Polarized intracellular particle transport: Saltatory movements and cytoplasmic streaming. Int Rev Cytol 32:93–137.

Reed SI (1980): The selection of S cerevisiae mutants defective in the start event of cell division. Genetics 95:561–577.

Rieder CL (1979): Ribonucleoprotein staining of centrioles and kinetochores in newt lung cell spindles. J Cell Biol 80:1–9.

Rieder CL, Borisy GG (1982): The centrosome cycle in PtK₂ cells: Asymmetric distribution and structural changes in the pericentriolar material. Biol Cell 44:117–132.

Ring D, Hubble R, Caput D, Kirschner M (1980): Isolation of microtubule organizing centers from mouse neuroblastoma cells. In DeBrabander M, DeMey J (eds): "Microtubules and Microtubule Inhibitors." Amsterdam: Elsevier, pp 297–309.

Ring D, Hubble R, Kirschner MW (1982): Mitosis in a cell with multiple centrioles. J Cell Biol 94:549–556.

Ritter H, Inoue S, Kubai D (1978): Mitosis in Barbulamympha. I. Spindle structure, formation and kinetochore engagement. J Cell Biol 77:638–654.

Robbins E, Jentzsch G, Micali A (1968): The centriole cycle in synchronized HeLa cells. Cell Biol 36:329–339.

Ross U-P (1973): Light and electron microscopy of rat kangaroo cells in mitosis. Chomosoma 40:43–82.

Sattilaro RE, Dentler WL, LeCluyse EL (1981): Microtubule-associated proteins and the organization of actin filaments in vitro. J Cell Biol 90:467–473.

Scheele RB, Borisy GG (1979): In vitro assembly of microtubules. In Roberts K, Hyams JS (eds): "Microtubules." London: Academic Press, pp 175–254.

Schiff PB, Fant J, Horwitz SB (1979): Promotion of microtubule assembly in vitro by taxol. Nature 277:665–667.

Schliwa M, Euteneuer U, Herzog W, Weber K (1979): Evidence for rapid structural and functional changes of the melanophore microtubule organizing center upon pigment movement. J Cell Biol 83:623–632.

Schroder TE (1973): Actin in dividing cells: Contractile ring filaments bind heavy meromyosin. Proc Natl Acad Sci USA 70:1688–1692.

Sharp GA, Osborn M, Weber K (1981): Ultrastructure of multiple microtubule initiation sites in mouse neuroblastoma cells. J Cell Sci 47:1–24.

Sherline P, Mascardo RN (1982): Epidermal growth factor induces rapid centrosomal separation in HeLa and 3T3 cell. J Cell Biol 93:507–511.

Snyder JA (1980): Evidence for a ribonucleoprotein complex as a template for microtubule initiation in vivo. Cell Biol Int Rep 4:859–868.

Snyder JA, McIntosh JR (1975): Initiation and growth of microtubules from mitotic centers in lysed mammalian cells. J Cell Biol 67:744–760.

Solomon F (1981): Specification of cell morphology by endogenous determinants. J Cell Biol 90:547–553.

Spiegelman BM, Lopata MA, Kirschner MW (1979): Aggregation of microtubule initiating sites preceding neurite outgrowth in mouse neuroblastoma cells. Cell 16:253–263.

Stearns ME, Brown DL (1979): Purification of cytoplasmic tubulin and microtubule organizing center proteins functioning in microtubule initiation from the alga Polytomella. Proc Natl Acad Sci USA 76:5745–5749.

Stearns ME, Brown DL (1981): Microtubule organizing centers of the alga Polytomella exert spatial control over microtubule initiation in vivo and in vitro. J Ultrastruct Res 77:366–378.

Stearns ME, Connolly JA, Brown DL (1976): Cytoplasmic microtubule organizing centers isolated from Polytomella agilis. Science 191:188–191.

Stubblefield E (1968): Centriole replication in a mammalian cell. In "Proliferation and Spread of Neoplastic Cells," Symposium on Fundamental Cancer Research. Baltimore: Williams and Wilkins, p 175.

Szollosi D (1964): The structure and function of centrioles and their satellites in the jellyfish Phialidium gregarium. J Cell Biol 21:465–479.

Tash JS, Means AR, Brinkley BR, Dedman JR, Cox SM (1980): Cyclic nucleotide and Ca^{2+} regulation of microtubule initiation and elongation. In DeBrabander M, DeMey J (eds): "Microtubules and Microtubule Inhibitors." Amsterdam: Elsevier/North-Holland, pp 269–279.

Telzer BR, Rosenbaum JL (1979): Cell cycle–dependent, in vitro assembly of microtubules onto the pericentriolar material of HeLa cells. J Cell Biol 81:484–498.

Tucker JB (1967): Changes in nuclear structure during binary fission in the ciliate Nassula. J Cell Sci 2:481–498.

Tucker JB (1979): Spatial organization of microtubules. In Robers K, Hyams JS (eds): "Microtubules." London: Academic Press, pp 315–358.

Tucker RW, Scher CD, Stiles CD (1979): Centriole deciliation associated with the early response of 3T3 cells to growth factors but not SV40. Cell 18:1065–1072.

Vasiliev JM, Gelfand IM (1976): Effects of colcemid on morphogenetic processes and locomotion of fibroblasts. In Goldman R, Pollard T, Rosenbaum J (eds): "Cell Motility." New York: Cold Spring Harbor Laboratory, pp 279–304.

Vorobjev IA, Chentsov YS (1982): Centrioles in the cell cycle. J Cell Biol 93:938–949.

Weisenberg RC, Rosenfeld AC (1975): In vitro polymerization of microtubules into asters and spindles in homogenates of surf clam eggs. J Cell Biol 64:146–158.

Wheatley DN (1982): "The Centriole: A Central Enigma of Cell Biology." Amsterdam: Elsevier.

Wilson EB (1925): "The Cell in Development and Heredity." New York: Macmillan.

Wood WB (1979): Bacteriophage T4 assembly and the morphogenesis of subcellular structure. Harvey Lectures 73:203–224.

Zackroff RV, Rosenfeld AC, Weisenberg RC (1979): Effects of RNase and RNA or in vitro aster assembly. J Supramol Struct 5:577–589.

Zeligs JD, Wollman SH (1979): Mitosis in rat thyroid epithelial cells in vivo. II. Centrioles and pericentriolar material. J Ultrastruct Res 66:97–108.

Modern Cell Biology 2:143–162
© 1983 Alan R. Liss, Inc., 150 Fifth Avenue, New York, NY 10011

Transcytoplasmic Integration in Avian Erythrocytes and Striated Muscle: The Role of Intermediate Filaments

Elias Lazarides and Bruce L. Granger

From the Division of Biology, California Institute of Technology, Pasadena, California 91125

I. INTRODUCTION

Over the past several years the mammalian erythrocyte has provided a unique model system for the elucidation of the interaction of actin with the plasma membrane. Studies from several laboratories have revealed that the mammalian erythrocyte contains a dense, anastomosing network composed of spectrin, actin, and several associated polypeptides [reviewed by Branton et al, 1981]. This network lines the inner surface of the erythrocyte's plasma membrane and serves as its cytoskeleton, being responsible for maintaining the shape of, and confering structural integrity to, the cell. Detailed studies have shown further that spectrin interacts with the plasma membrane through its association with ankyrin, a protein that binds to a fraction of the anion transport polypeptides (band 3) [Bennett and Stenbuck, 1979a, b]. The cortical network of spectrin and actin is thus linked to the plasma membrane, and may function additionally to tether some of the transmembrane proteins and to restrict their lateral mobility in the plane of the membrane.

For a number of years electron microscopists have contested that many fibroblastic cells grown in tissue culture as well as many cells in tissues contain a network of actin filaments that underlies the plasma membrane. This cortical network of actin filaments has been thought to interact with as yet undefined membrane proteins to form the "outer boundary lamina." This lamina is operationally defined as a cortical filamentous network composed in part of actin filaments and membrane bound proteins that remains insoluble after extraction of the cell with nonionic detergents [Ben Ze'ev et al, 1979]. Until recently the scheme of molecular interactions identified for the erythrocyte membrane has been thought to be largely specific for this cell. This opinion has stemmed from the widespread belief that spectrin is a polypeptide specific to erythrocytes [Painter et al, 1975; Hiller and Weber, 1977]. Recently, however, it has been discovered that antibodies raised against the α subunit of avian erythrocyte spectrin exhibit a wide cross reactivity with a polypeptide associated with the plasma membrane in a variety of avian and mammalian nonerythroid cells [Repasky et al, 1982; Glenney et al, 1982]. This polypeptide has an electrophoretic mobility indistinguishable from its erythrocyte counterpart, suggesting that the nonerythroid protein is homologous to erythrocyte α-spectrin. These observations have led us to hypothesize that an actin-spectrin network may be a widespread constituent of the cortex of many cell types, and may be responsible for conferring structural integrity to many nonerythroid cells as well as to erythrocytes.

Is this cortical matrix integrated with the rest of the cell? Evidence for the existence of a transcytoplasmic matrix interconnecting the actin-spectrin network with cytoplasmic organelles, at least in certain differentiated cells, has come from studies on the structure and cytoplasmic distribution of a class of cytoplasmic filaments known as intermediate filaments. Two cell types that have helped to reveal the existence of such an integrating matrix are adult avian striated muscle and the adult avian erythrocyte. Below we discuss briefly the evidence which demonstrates that such a matrix, composed of intermediate filaments, exists throughout the muscle fiber, and we describe in some detail the experiments with avian erythrocytes which show that intermediate filaments may function to integrate various components of the cytoplasm by forming a crosslinked three-dimensional network. We also present evidence which indicates that interaction of this matrix with the plasma membrane is anisotropic in these two cell systems and discuss the relationship of this intermediate filament matrix to the "microtrabecular lattice" described by Porter and his colleagues in other cell types.

II. INTERMEDIATE FILAMENTS IN GENERAL

Cytoplasmic filaments with diameters of 7 to 12 nm have been described in a wide variety of cell types from diverse vertebrate sources. Known collectively as intermediate filaments, they comprise a heterogeneous class of filaments which, though similar in morphology, are composed of one or more biochemically distinct subunit proteins [for reviews see Lazarides, 1980, 1982; Anderton, 1981]. In general, we can distinguish biochemically and immunologically five major subclasses of subunits: 1) keratin (tono) filaments, expressed in epithelial cells and cells of epithelial origin; 2) desmin filaments, expressed predominantly in smooth, skeletal, and cardiac muscle cells as well as in certain nonmuscle cells; 3) vimentin filaments, expressed in many differentiating cells, cells grown in tissue culture, and certain fully differentiated cells; 4) glial filaments, in astrocytes; and 5) neurofilaments, thus far detected only in neurons. Even though biochemically and immunologically the five subunit subclasses are distinct, x-ray diffraction studies have shown that the latter four subclasses are structurally related to the keratin subclass since they contain a considerable amount of α-helix, and have a common coiled-coil structure [Steinert et al, 1980]. In addition, certain monoclonal antibodies recognize a determinant common to all classes of intermediate filament subunits [Pruss et al, 1981; Dellagi et al, 1982). These observations have lent support to the earlier postulate that all classes of intermediate filament subunits share conserved areas of amino acid sequence responsible for their conserved structure and areas of divergent amino acid sequence responsible for functional specializations in different cell types [Lazarides, 1980, 1981]. Of all intermediate filament proteins, desmin and vimentin are the most related; they share common phosphorylated peptides [Gard et al, 1979], and amino acid sequence analysis of the carboxy-terminal one third of vimentin and desmin has revealed homology in two thirds of the residues [Geisler and Weber, 1981].

A cell may possess one or more subclasses of intermediate filament. The expression of any one subclass of intermediate filament appears to be determined by the cell's type and its state of differentiation. The most extensive coexpression of intermediate filament proteins is seen in different types of muscle which express both desmin and vimentin, but in widely differing ratios depending on the muscle type and its state of differentiation [Gard and Lazarides, 1980; Granger and Lazarides, 1979; Mikawa et al, 1981; Berner et al, 1981; Osborn et al, 1981; Schmid et al, 1982].

In addition to their variable expression, these two proteins exhibit a dramatic redistribution during muscle differentiation. In immature skeletal myotubes desmin and vimentin coexist in an extensive network of cytoplasmic filaments [Bennett et al, 1979; Gard and Lazarides, 1980]. Late in myogenesis, however, both filament proteins become associated with the Z disks of myofibrils [Gard and Lazarides, 1980] and form a two-dimensional network of Z-disk collars that presumably serves to interlink Z disks of adjacent myofibrils [Granger and Lazarides, 1978, 1979].

III. TRANSCYTOPLASMIC INTEGRATION OF SKELETAL MYOFIBRILS

It has long been suspected that there is present in muscle cells a framework responsible for holding the constituent myofibrils in place. The presence of structural elements transverse to a muscle fiber's long axis is suggested by the tendency for the Z disks, M lines, and other sarcomeric units of neighboring myofibrils to be in axial register across the fiber. Numerous ultrastructural observations have suggested a linkage of the sarcomere units in cardiac and skeletal muscle. In particular, the Z disk appears connected transversely to other Z disks and to the plasma membrane [Pease and Baker, 1949; Bennett and Porter, 1953; Bergman, 1958; Garamvölgyi, 1962, 1965; Walker et al, 1968; Behrendt, 1977]. The molecular components of this integrating matrix were until recently unknown. New work on the biochemical composition of intermediate filaments, their solubility properties and cytoplasmic distribution have allowed us to approach this question from a new point of view [Granger and Lazarides, 1978, 1979]. When glycerol-extracted skeletal muscle is subjected to shearing forces, the fibers fragment along their long axes into individual myofibrils, or, more frequently, bundles of myofibrils. However, when strips of glycerinated muscle are incubated in a solution containing high concentrations of potassium iodide, a chaotropic agent capable of solubilizing actin and myosin filaments, a considerable fraction of their protein mass is solubilized. We have observed that what remains is an insoluble residue that maintains its original three-dimensional structure [Granger and Lazarides, 1978]. If these KI-extracted muscle fibers are now subjected to shearing forces, cleavage occurs preferentially along planes perpendicular to the long axis of the muscle fiber; what is generated

---

Fig. 1. A) Phase-contrast micrograph of a Z-disk sheet produced by blending a chicken skeletal muscle fiber that had been extracted with glycerol and KI. The large phase dense spots are mitochondrial remnants. Bar = 10 μm [from Granger and Lazarides, 1978]. B) Indirect immunofluorescence of an isolated Z-disk sheet showing that desmin surrounds the Z disks and thus forms a netlike lattice at right angles to the fiber axis within each Z plane. Bar = 10 μm [from Lazarides, 1980].

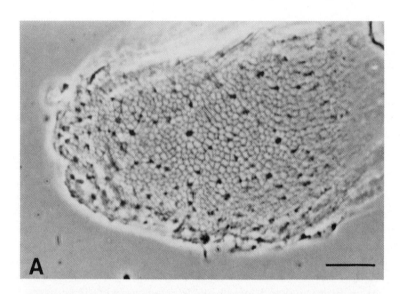

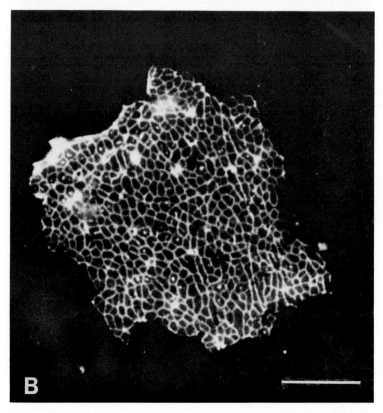

are planar arrays of Z disks with a honeycomblike appearance. Such sheets of Z disks can be clearly visualized by conventional phase contrast microscopy (Fig. 1A). This face-on view of the Z disk demonstrates that each Z disk is physically linked laterally to neighboring Z disks within a given Z-disk plane. In the isolated, intact Z disk sheets, remnants of mitochondria survive the extraction procedure and remain tightly associated with the peripheries of Z disks; these mitochondrial remnants appear as dark spots by phase contrast microscopy (Fig. 1A). Immunofluorescence microscopy of these Z-disk sheets has demonstrated that each Z disk is composed of two distinct, yet interconnected domains: a central one, which contains actin and the actin binding protein α-actinin (not shown) [see Granger and Lazarides, 1978], and a peripheral one, which contains actin and the two intermediate filament proteins desmin and vimentin (shown for desmin in Fig. 1B). Desmin and vimentin surround every Z disk, giving rise to a matrix that extends to the plasma membrane. Numerous electron microscopic observations have shown that at the peripheries of Z disks, desmin and vimentin most likely exist in the form of intermediate filaments [Kelly, 1969; Page, 1969; Sandborn et al, Eriksson and Thornell, 1979; Lindner and Schaumburg, 1968; Viragh and Challice, 1969; Ferrans and Roberts, 1973; Oliphant and Loewen, 1976; Junker and Sommer, 1977; Behrendt, 1977; Nunzi and Franzini-Armstrong, 1980; Richardson et al, 1981; Pierobon-Bormioli, 1981; Chiesi et al, 1981]. These observations are summarized in Figure 2. One important con-

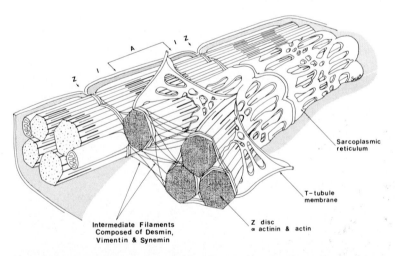

Fig. 2. Schematic representation of the distribution of intermediate filaments in skeletal muscle. The filaments extend laterally from Z disk to Z disk and ultimately terminate and anchor to the plasma membrane [modified from Lazarides, 1980].

cept to arise from such work is that this transcytoplasmic matrix is inhomogeneous, in that it is specifically located at the peripheries of Z disks and presumably interacts with the plasma membrane at specific sites that are in close apposition to Z disks.

IV. TRANSCYTOPLASMIC INTEGRATION IN THE AVIAN ERYTHROCYTE

Another cell type that has contributed evidence concerning the hypothesis that intermediate filaments form a transcytoplasmic matrix is the avian erythrocyte. Unlike mammalian erythrocytes, avian erythrocytes possess nuclei and a few mitochondria. In addition to a spectrin-based plasma membrane cytoskeleton, these cells possess a circumferential band of microtubules known as the marginal band. They also contain a system of filaments that have been identified recently as intermediate filaments [Virtanen et al, 1979; Woodcock, 1980; Granger et al, 1982] and which form a three-dimensional network in these cells, interlinking the nucleus with the plasma membrane on opposite sides of the cell [Harris and Brown, 1971; Haggis and Bond, 1979]. The details of this arrangement became evident with the development of a new means for visualizing these filaments in avian erythrocytes [Granger et al, 1982]. Cells adhering to a flat glass substrate are disrupted by sonication, and their exposed, insoluble cytoskeletons are shadowed with platinum. Examination of the platinum replicas by transmission electron microscopy allows the visualization of the three-dimensional distribution of the filaments inside the cell. Examples of such partially disrupted chicken erythrocytes are shown in Figure 3. Intermediate filaments are prominent on the lower plasma membrane and can be seen to extend and attach to the nucleus and upper membrane, forming a three-dimensional cytoplasmic network.

V. ANISOTROPIC ASSOCIATION OF INTERMEDIATE FILAMENTS WITH THE PLASMA MEMBRANE

Prolonged sonication of attached erythrocytes removes the nonadherent part of the cell, leaving attached to the coverslips the adhering part of the plasma membrane. These elliptical patches of membrane frequently contain mats of intermediate filaments, indicating that there is a stable association of a subset of the filaments with the membrane (Fig. 5A). This association is also evident upon isolation of the plasma membranes from these cells. Mechanical enucleation of erythrocyte ghosts in suspension results in a membrane fraction rich in intermediate filaments [Granger et al, 1982]; electron microscopy demonstrates a close association of the filaments with the plasma membrane (Fig. 4A).

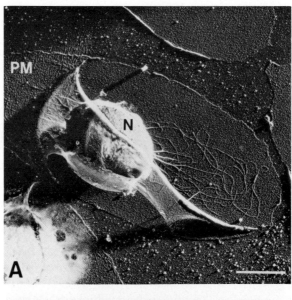

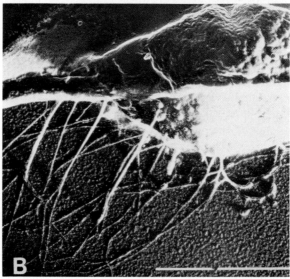

Fig. 3. Visualization of the three-dimensional disposition of intermediate filaments in soni-cated chicken erythrocytes. In (A), the nonadherent part of the plasma membrane has been partially removed revealing filaments which attach to the nucleus (N) and the plasma membrana (PM). Bar = 2 μm. The attachment of the filaments to the plasma membrane and surface of the nucleus is shown at higher magnification in Figure 3B. Bar = 2 μm. In these images cells were shadowed unidirectionally, and micrographs were printed with reversed contrast to make platinum shadows appear dark. Replicas were bleached to allow visualization of surface features of the nuclear membrane. [From Granger and Lazarides, 1982.]

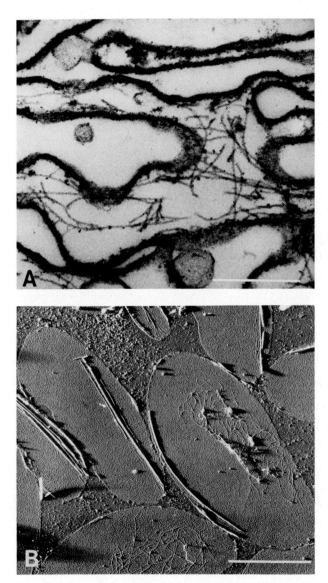

Fig. 4.   A) Thin section of isolated chicken erythrocyte membranes showing the cytoplasmic intermediate filaments and their close association with the inner surface of the plasma membrane. Bar = 500 nm [from Granger et al, 1982]. B) Association of marginal band microtubules and intermediate filaments with the plasma membrane. Chicken erythrocytes were sonicated in a microtubule stabilizing buffer. Under these conditions parts of the marginal band survive the sonication procedure and remain adherent to the membranes. The spatial segregation of the sites of membrane attachment of the two fibrous systems can be clearly visualized on the patch of membrane on the right. Bar = 5 μm.

To investigate the spatial relationships of the membrane-bound intermediate filaments and the marginal band, we carried out sonication studies in microtubule-stabilizing buffers. Under these conditions, adherent membranes can be obtained that retain both intermediate filaments and the marginal band microtubules (Fig. 4B). Such images demonstrate that both intermediate filaments and marginal band microtubules interact with the plasma membrane, perhaps indirectly via the spectrin-based plasma membrane cytoskeleton, and that they are spatially segregated from each other [Granger and Lazarides, 1982]. These observations have given rise to the important notion that the spectrin-based plasma membrane cytoskeleton is inhomogeneous with respect to these two filament systems, and that interactions between these two systems may occur through the spectrin-actin network [Granger and Lazarides, 1982].

VI. VIMENTIN AND SYNEMIN ARE COMPONENTS OF AVIAN ERYTHROCYTE INTERMEDIATE FILAMENTS

Biochemical analysis of total erythrocyte (or membrane-associated) intermediate filaments by two-dimensional gel electrophoresis and immunoautoradiography has shown that they are composed of two polypeptides: vimentin and a high molecular weight protein (MW 230,000) known as synemin [Granger et al, 1982]. Desmin is not detected in these cells. Previous biochemical and immunological studies have indicated that in adult chicken gizzard smooth muscle, intermediate filaments are composed predominantly of desmin and synemin [Granger and Lazarides, 1980]. Studies with adult chicken skeletal muscle and embryonic myotubes differentiating in tissue culture have shown that synemin is coexpressed with desmin and vimentin. However, in contrast to differentiating myotubes, unfused myoblasts express only vimentin; the synthesis of synemin and desmin is normally induced near the time of fusion [Gard and Lazarides, 1980; Granger and Lazarides, 1980]. These observations have suggested that synemin can associate with either desmin or vimentin filaments or filaments containing a mixture of both core polypeptides.

Double immunofluorescence studies with avian erythrocytes and with myotubes grown in tissue culture have shown that desmin, vimentin, and sy-

---

Fig. 5. Antibody decoration of chicken erythrocyte intermediate filaments. When the filaments are preincubated with the synemin preimmune serum, the diameter of the filaments is indistinguishable from that of controls (A). Incubation with antivimentin antibodies (B) results in uniform decoration of the filaments with no obvious periodicity. Incubation of the filaments with antisynemin antibodies (C) reveals a periodic distribution of the antigen. The average spacing of the antisynemin foci is 180 nm. For 5A–C bar = 5 μm [from Granger and Lazarides, 1982].

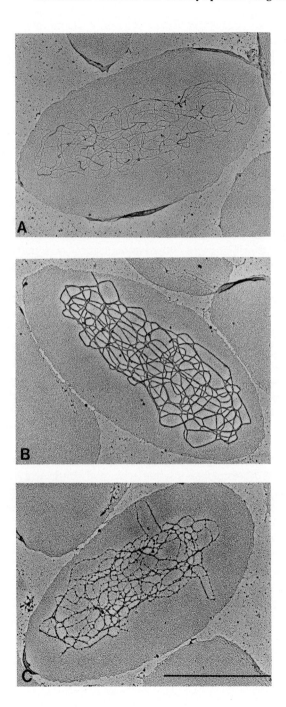

nemin colocalize, suggesting that these three proteins form heteropolymers in vivo [Gard and Lazarides, 1980; Granger and Lazarides, 1980]. To demonstrate this point more conclusively, we have examined by electron microscopy the distribution of vimentin and synemin on the filaments that remain attached to the sonicated patches of erythrocyte membranes.

VII. IMMUNOELECTRON MICROSCOPIC LOCALIZATION OF VIMENTIN AND SYNEMIN: CROSS-LINKING OF VIMENTIN FILAMENTS BY SYNEMIN

If the sonicated erythrocyte membane patches are incubated with the synemin preimmune antibodies, the distribution and diameter of the filaments are indistinguishable from those of controls (Fig. 5A). Incubation with antivimentin results in uniform decoration of all of the intermediate filaments (Fig. 5B). These results suggest that vimentin is the core polypeptide subunit of these filaments in accordance with in vitro polymerization studies [Renner et al, 1981]. These polyclonal antibodies do not reveal the presence of an axial periodicity or helicity in the filaments; the decoration pattern is similar to that seen with another polyclonal antivimentin in cultured mouse cells [Heuser and Kirschner, 1980]. In contrast, certain IgM antivimentin monoclonal antibodies reveal a 31 nm axial periodicity after thin sectioning of the decorated filaments [Blose et al, 1982]. Incubation of the erythrocyte membrane patches with antisynemin reveals periodically spaced foci along the axes of all of the filaments (Fig. 5C). Many of the segments seen in these intermediate filament networks appear to be composed of more than one filament. The axial spacing of the antisynemin decoration in filament bundles and in presumed individual filaments appears to be the same, indicating that laterally associated filaments are in axial registry. This is more evident in areas where the filaments are slightly separated. In these cases the synemin knobs appear to join end to end to form a bridge between the filaments (arrowheads in Fig. 6A, B). In addition, the filament intersection points usually coincide with decorated synemin nodes (Fig. 6A, B). These micrographs suggest that synemins of adjacent filaments can self-associate and thus mediate lateral crosslinking of the filaments; the cross-linking appears to occur only at points corresponding to synemin decoration. However, it is evident from these images that not all synemin foci are involved in filament cross-linking; perhaps the extent of filament cross-linking is altered during erythroid differentiation or when the cell is in certain physiological states [Granger and Lazarides, 1982].

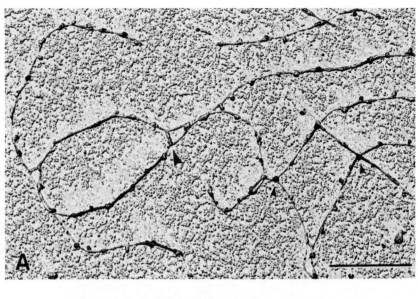

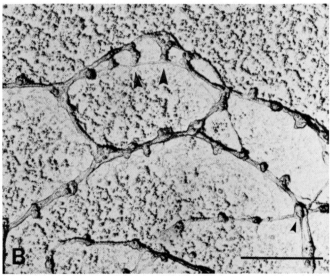

Fig. 6. Higher power views of membrane-associated intermediate filaments obtained from 10-day embryo erythrocytes and incubated with antisynemin antibodies (A and B). The large arrows point to areas where the antibodies show that synemin extends between adjacent filaments and forms a bridge. The small arrows point to areas where individual filaments or bundles of filaments converge at a single antisynemin node. For A, Bar = 1 μm; for B, bar = 500 nm [from Granger and Lazarides, 1982].

VIII. ASSOCIATION OF VIMENTIN FILAMENTS WITH MITOCHONDRIA

As we noted earlier, remnants of mitochondria remain tightly adherent to the peripheries of Z disks in isolated Z-disk sheets, suggesting some sort of interaction of the outer mitochondrial membrane with some component of intermediate filaments at the peripheries of Z disks. To investigate this point in more detail, we have noted the distribution of mitochondria with respect to intermediate filaments in sonicated avian erythrocytes. The number of mitochondria in erythroid cells declines as these cells differentiate [Weise and Ingram, 1976; Chan, 1977; Edmonds, 1966], but mitochondria are still present in mature avian erythrocytes [Harris and Brown, 1971; Beam et al, 1979; Haggis and Bond, 1979]. Organelles resembling mitochondria are often observed entangled in the intermediate filament network left after sonication of embryonic erythroid cells (Fig. 7B) and occasionally in mature erythrocytes (Fig. 7A). The possibility that intermediate filaments interact specifically with mitochondria is suggested by the micrograph in Figure 7B, which shows bridges between the filaments and a mitochondrion. One filament has four bridges with a linear spacing of approximately 60–70 nm; this spacing is clearly less than that of synemin as visualized by antibody decoration. An apparent association of mitochondria with cytoplasmic filaments in avian erythrocytes has been noted before [Harris and Brown, 1971], and the results shown here identify such filaments as vimentin-containing intermediate filaments. In addition to these observations, there is considerable electron microscopic evidence from other cell types, including rat Leydig cells and various epithelia, which suggests that mitochondria can associate with intermediate filaments [Lundgren, 1974; Bernstein and Wollman, 1975; Lee et al, 1979; David-Ferreira and David-Ferreira, 1980; Mayerson and Brumbaugh, 1981]. In our studies with chicken erythrocytes, mitochondria have not been seen in association with the marginal band microtubules; this result is in contrast with observations on cultured fibroblasts, in which mitochondria appear to associate with the microtubules and not with the intermediate filaments [Ball and Singer, 1982].

IX. CORRELATES BETWEEN MUSCLE CELLS AND AVIAN ERYTHROCYTES

The observations summarized above have led us to suggest that in the nucleated avian erythrocyte, intermediate filaments form a three-dimensional network which traverses the cytoplasm and interlinks the nucleus and mitochondria with areas of the plasma membrane on opposite sides of the cell [see Haggis and Bond, 1979]. The filaments are firmly attached to the plasma

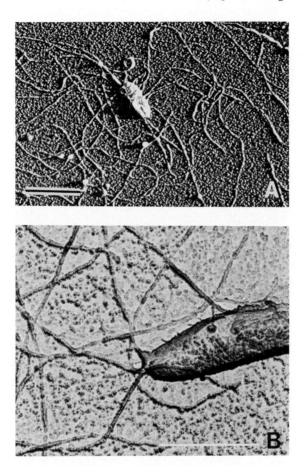

Fig. 7. Association of intermediate filaments with mitochondria in a mature chicken erythrocyte (A) and embryonic erythroid cell (B). The filaments and mitochondria are lying on adherent patches of sonicated plasma membrane. Note links between the mitochondrion and a filament in (B); these links are spaced 60–70 nm from each other. Image in (A) was printed with reversed contrast. Bars: A = 1 μm, B = 500 nm.

membrane only in a defined region on each face of the erythrocyte disk; this region is clearly separated from the area where marginal band microtubules interact with the plasma membrane. Individual filaments are linked to each other at periodic intervals by synemin, perhaps to promote rigidity in the network [Granger and Lazarides, 1982].

For the arguments presented in this section, we will assume that structural associations observed between synemin and vimentin in erythrocytes can be extended to adult skeletal muscle, even though such evidence has not yet been obtained. Synemin is expressed in adult erythrocytes as well as in adult gizzard and adult skeletal muscle. In the latter, it coexists with desmin and vimentin at the periphery of the Z disk [Granger and Lazarides, 1980]. By analogy to the adult erythrocyte, synemin may mediate the cross-linking of desmin- and vimentin-containing intermediate filaments at the periphery of the Z disk as well as desmin filaments which interconnect dense bodies in smooth muscle [Cooke, 1976].

X. RELATIONSHIP OF INTERMEDIATE FILAMENTS TO THE MICROTRABECULAR LATTICE

The studies presented here provide biochemical and morphological evidence that one of the functions of intermediate filaments in at least some differentiated cells is that of forming a transcytoplasmic integrating matrix. What is the relationship of this matrix to the "microtrabecular lattice"? Porter and his colleagues observed initially that fibroblasts grown in tissue culture contain a lattice composed of filaments with variable diameters (4–20 nm) when the cells are viewed unextracted in the high-voltage electron microscope after critical-point dying [Wolosewick and Porter, 1979]. Such a matrix is also present in neuronal axons [Ellisman et al, 1976], differentiating skeletal muscle [Porter et al, 1979], and isolated fish erythrophores [Byers and Porter, 1977]. These studies have indicated that the lattices permeate the whole cytoplasm, extending from the plasma membrane to virtually all cytoplasmic organelles, but the elements of the lattice are distinct from microtubules and the actin filaments in stress fibers [Wolosewick and Porter, 1979]. Their relationship to intermediate filaments has not been established, but it has been argued that the density of vimentin filaments seen after detergent extraction of cultured fibroblasts is sufficient to make them one of the major components of the microtrabecular lattice [Heuser and Kirschner, 1980]. However, Porter and his colleagues have shown that extensive detergent extraction of cultured fibroblasts solubilizes the microtrabecular lattice under conditions that leave vimentin filaments insoluble [Schliwa et al, 1981]. Analysis of such extracts by two-dimensional electrophoresis has revealed the presence of a complex array of polypeptides, indicating that this matrix is both detergent-labile and biochemically complex. From these observations it is difficult at present to correlate this lattice with intermediate filaments. The association of the lattice with the peripheries of Z disks during myogenesis in vitro [Porter et al, 1979; Peng et al, 1981] suggests some relationship of microtrabeculae to the desmin- and vimentin-containing intermediate filaments. Whether such a lattice exists in adult avian erythrocytes may be difficult to assess with the techniques used to demonstrate this matrix in cultured fibroblasts, owing to the high concentration of hemoglobin in these cells. Nevertheless, irrespective of the relationship of the microtrabecular lattice to the vimentin- and desmin-containing intermediate filament lattice, it is evident that the cytoplasm of cells is mechanically integrated, and that in at least some differentiated cells this matrix can be shown to have as major components, depending on the cell type and its stage in differentiation, the intermediate filament proteins desmin and vimentin. In this context it is worth noting that in the neuronal axon of the marine worm Myxicola infudibulum intermediate filaments also appear to form a cross-linked three-dimensional lattice that extends longitudinally along the whole length of the axon [Gilbert, 1975]. Similarly, in differentiated epithelial cells keratin filaments have been shown to form a cross-linked three-dimensional matrix that interconnects desmosomes [Hull and Staehelin, 1979]. From the heterogeneity in the subunit composition of intermediate filaments and from the heterogeneity in their expression during differentiation, it is evident that the development of such a matrix is tailored to the differentiated state of the cell. Elucidation of its relationship to the microtrabecular lattice may provide a deeper insight in our understanding of cytoplasmic integration.

ACKNOWLEDGMENTS

This work was supported by grants from the National Institutes of Health (PHS-GM 06965), the National Science Foundation, and the Muscular Dystrophy Association of America, and a biomedical research support grant to the Division of Biology, California Institute of Technology. B. L. Granger was also supported by a predoctoral training grant from the NIH (GM 07616). E. Lazarides is the recipient of a research career development award from the NIH.

XI. REFERENCES

Anderton BH (1981): Intermediate filaments: A family of homologous structures. J Musc Res Cell Mot 2:141–166.
Ball EH, Singer SJ (1982): Mitochondria are associated with microtubules and not with intermediate filaments in cultured fibroblasts. Proc Natl Acad Sci USA 79:123–126.
Beam KG, Alper SL, Palade GE, Greengard P (1979): Hormonally regulated phosphoprotein of turkey erythrocytes. Localization to plasma membrane. J Cell Biol 83:1–15.
Behrendt H (1977): Effects of anabolic steroids on rat heart muscle cells. I. Intermediate filaments. Cell Tissue Res 180:303–315.
Ben Ze'ev A, Duerr A, Solomon F, Penman S (1979): The outer boundary of the cytoskeleton: A lamina derived from plasma membrane proteins. Cell 17:859–865.
Bennett HS, Porter KR (1953): An electron microscope study of sectioned breast muscle of the domestic fowl. Am J Anat 93:61–105.
Bennett GS, Fellini SA, Toyama Y, Holtzer H (1979): Redistribution of intermediate filament subunits during skeletal myogenesis and maturation in vitro. J Cell Biol 82:577–584.

Bennett V, Stenbuck PJ (1979a): Identification and partial purification of ankyrin, the high affinity membrane attachment site for human erythrocyte spectrin. J Biol Chem 254:2533–2541.

Bennett V, Stenbuck PJ (1979b): The membrane attachment protein for spectrin is associated with band 3 in human erythrocyte membranes. Nature 280:468–473.

Bergman RA (1958): An experimental study of the nonfibrillar elements in frog striated muscle. Bull Johns Hopkins Hosp 103:267–280.

Berner PF, Frank E, Holtzer H, Somlyo AP (1981): The intermediate filament proteins of rabbit vascular smooth muscle: Immunofluorescent studies of desmin and vimentin. J Muscle Res Cell Mot 2:439–452.

Bernstein LH, Wollman SH (1975): Association of mitochondria with desmosomes in the rat thyroid gland. J Ultrastruc Res 53:87–92.

Blose SH, Matsumura F, Lin JJ-C (1982): Structure of vimentin 10 nm filaments probed with a monoclonal antibody that recognizes a common antigenic determinent on vimentin and tropomyosin. Cold Spring Harb Symp Quant Biol 46:455–463.

Branton D, Cohen CM, Tyler J (1981): Interaction of cytoskeletal proteins on the human erythrocyte membrane. Cell 24:24–32.

Byers HR, Porter KR (1977): Transformations in the structure of the cytoplasmic ground substance in erythrophores during pigment aggregation and dispersion. J Cell Biol 75:541–560.

Chan L-NL (1977): Changes in the composition of plasma membrane proteins during differentiation of embryonic chick erythroid cell. Proc Natl Acad Sci USA 74:1062–1066.

Chiesi M, Ho MM, Inesi G, Somlyo AV, Somlyo AP (1981): Primary role of sarcoplasmic reticulum in phasic contractile activation of cardiac myocytes with striated myolemma. J Cell Biol 91:728–742.

Cooke PH (1976): A filamentous cytoskeleton in vertebrate smooth muscle fibers. J Cell Biol 68:539–556.

David-Ferreira KL, David-Ferreira JF (1980): Association between intermediate-sized filaments and mitochondria in rat Leydig cells. Cell Biol Int Rep 4:655–662.

Dellagi K, Brouet JC, Perreau J, Paulin D (1982): Human monoclonal IgM with autoantibody activity against intermediate filaments. Proc Natl Acad Sci USA 79:446–450.

Edmonds RH (1966): Electron microscopy of erythropoiesis in the avian yolk sac. Anat Rec 154:785–805.

Ellisman MH, Rash JE, Staehelin LA, Porter KR (1976): Studies of excitable membranes. II. A comparison of specializations at neuromuscular junctions and nonjunctional sarcomas of mammalian fast and slow twitch muscle fibers. J Cell Biol 68:752–774.

Eriksson A, Thornell L-E (1979): Intermediate (skeletin) filaments in heart Purkinje fibers. A correlative morphological and biochemical identification with evidence of a cytoskeletal function. J Cell Biol 80:231–247.

Ferrans VJ, Roberts WC (1973): Intermyofibrillar and nuclear-myofibrillar connections in human and canine myocardium. An ultrastructural study. J Mol Cell Cardiol 5:247–257.

Garamvölgyi N (1962): Interfibrilläre Z. verbindungen im quergestreiften muskel. Acta Physiol Acad Sci Hung 22:235–241.

Garamvölgyi N (1965): Inter-Z bridges in the flight muscle of the bee. J Ultrastruct Res 13:435–443.

Gard DL, Lazarides E (1980): The synthesis and distribution of desmin and vimentin during myogenesis in vitro. Cell 19:263–275.

Gard DL, Bell PB, Lazarides E (1979): Coexistence of desmin and the fibroblastic intermediate filament subunit in muscle and nonmuscle cells: Identification and comparative peptide analysis. Proc Natl Acad Sci USA 76:3894–3898.

Geisler N, Weber K (1981): Comparison of the proteins of two immunologically distinct intermediate-sized filaments by amino acid sequence analysis: Desmin and vimentin. Proc Natl Acad Sci USA 78:4120–4123.

Gilbert DS (1975): Axoplasm architecture and physical properties as seen in the Myxicola giant axon. J Physiol (Lond) 253:257–301.

Glenney JR, Glenny P, Osborn M, Weber K (1982): An F-actin- and calmodulin-associated protein from isolated brush borders has a morphology related to spectrin. Cell 28:843–854.

Granger BL, Lazarides E (1978): The existence of an insoluble Z disc scaffold in chicken skeletal muscle. Cell 15:1253–1268.

Granger BL, Lazarides E (1979): Desmin and vimentin coexist at the periphery of the myofibril Z disc. Cell 18:1053–1063.

Granger BL, Lazarides E (1980): Synemin: A new high molecular weight protein associated with desmin and vimentin filaments in muscle. Cell 22:727–738.

Granger BL, Lazarides E (1982): Structural associations of synemin and vimentin filaments in avian erythrocytes revealed by immunoelectron microscopy. Cell 30:263–275.

Granger BL, Repasky EA, Lazarides E (1982): Synemin and vimentin are components of intermediate filaments in avian erythrocytes. J Cell Biol 92:299–312.

Haggis GH, Bond EF (1979): Three-dimensional view of the chromatin in freeze-fractured chicken erythrocyte nuclei. J Microsc 115:225–234.

Harris JR, Brown JN (1971): Fractionation of the avian erythrocyte: An ultrastructural study. J Ultrastruct Res 36:8–23.

Heuser JE, Kirschner MW (1980): Filament organization revealed in platinum replicas of freeze-dried cytoskeletons. J Cell Biol 86:212–234, 1980.

Hiller G, Weber K (1977): Spectrin is absent in various tissue culture cells. Nature 266:181–183.

Hull BE, Staehelin LA (1979): The terminal web: A reevaluation of its structure and function. J Cell Biol 81:67–82.

Junker J, Sommer JR (1977): Anchorfibers and the topography of the junctional SR. In Bailey GW (ed): "35th Ann Proc Electron Microsc Soc Am." Boston, pp 582–583.

Kelly DE (1969): Myofibrillogenesis and Z-band differentiation. Anat Rec 163:403–426.

Lazarides E (1980): Intermediate filaments as mechanical integrators of cellular space. Nature 283:249–256.

Lazarides E (1981): Intermediate filaments. Chemical heterogeneity in differentiation. Cell 23:649–650.

Lazarides E (1982): Intermediate filaments: A chemically heterogeneous, developmentally regulated class of proteins. Ann Rev Biochem 51:219–250.

Lee CS, Morgan G, Wooding FBP (1979): Mitochondria and mitochondria-tonofilament desmosomal associations in the mammary gland secretory epithelium of lactating cows. J Cell Sci 38:125–135.

Lindner E, Schaumburg G (1968): Zytoplasmatische filamente in den quergestreiften muskelzellen des kaudalen lymphherzens von Rana Temporaria L. Untersuchungen am lymphherzen I. Z Zellforsch 84:549–562.

Lundgren B (1974): A quantitative estimation of the association of mitochondria to septate desomsomes in the sea urchin larva. Exp Cell Res 85:429–436.

Mayerson PL, Brumbaugh JA (1981): Lavender, a chick melanocyte mutant with defective melanosome translocation: a possible role for 10 nm filaments and microfilaments but not microtubules. J Cell Sci 51:25–51.

Mikawa T, Takeda S, Shimizu T, Kitaura T (1981): Gene expression of myofibrillar proteins in single muscle fibers of adult chicken: Micro two dimensional gel electrophoretic analysis. J Biochem 89:1951–1962.

Nunzi MG, Franzini-Armstrong C (1980): Trabecular network in adult skeletal muscle. J Ultrastruc Res 73:21–26.

Oliphant LW, Leowen RD (1976): Filament systems in purkinje cells of the sheep heart: Possible alterations of myofibrillogenesis. J Mol Cell Cardiol 8:679–688.

Osborn M, Caselitz J, Weber K (1981): Heterogeneity of intermediate filament expression in vascular smooth muscle: A gradient in desmin positive cells from the rat aortic arch to the level of the arteria iliaca communis. Differentiation 20:196–202.

Page SG (1969): Structure and some contractile properties of fast and slow muscles of the chicken. J Physiol 205:131–145.

Painter RG, Sheetz M, Singer SJ (1975): Detection and ultrastructural localization of human smooth muscle myosin-like molecules in human non-muscle cells by specific antibodies. Proc Natl Acad Sci USA 72:1359–1363.

Pease DC, Baker RF (1949): The fine structure of mammalian skeletal muscle. Am J Anat 84:175–200.

Peng HB, Wolosewick J-J, Cheng P-C (1981): The development of myofibrils in cultured muscle cells: A whole-mount and thin-section electron microscopic study. Dev Biol 88:121–136.

Pierobon-Bormioli S (1981): Transverse sarcomere filamentous systems: Z- and M-cables. J Muscle Res Cell Mot 2:401–413.

Porter KR, Fleming J, Wray G (1979): The assembly of myofibrils in cultured myoblasts: An HVEM study of whole cells. J Cell Biol 83:46a.

Pruss RM, Mirsky R, Raff MC, Thorpe R, Dowding AJ, Anderton BH (1981): All classes of intermediate filaments share a common antigenic determinant defined by a monoclonal antibody. Cell 27:419–428.

Renner W, Franke WW, Schmid E, Geisler N, Weber K, Mandelkow E (1981): Reconstitution of intermediate-sized filaments from denatured monomeric vimentin. J Mol Biol 149:285–306.

Repasky EA, Granger BL, Lazarides E (1982): Widespread occurrence of avian spectrin in non-erythroid cells. Cell 29:821–833.

Richardson FL, Stromer MH, Huiatt TW, Robson RM (1981): Immunoelectron and immunofluorescence localization of desmin in mature avian muscles. Eur J Cell Biol 26:91–101.

Sandborn EB, Cote MG, Roberge J, Bois P (1967): Microtubules et filaments cytoplasmiques dans le muscle de mammiferes. J Microsc 6:169–178.

Schliwa M, Van Blerkom J, Porter KR (1981): Stabilization of the cytoplasmic ground substance in detergent-opened cells and a structural and biochemical analysis of its composition. Proc Natl Acad Sci USA 78:4329–4333.

Schmid E, Osborn M, Rungger-Brandle E, Gabbiani G, Weber K, Franke WW (1982): Distribution of vimentin and desmin filaments in smooth muscle tissue of mammalian and avian aorta. Exp Cell Res 137:329–340.

Steinert PM, Idler WW, Goldman RD (1980): Intermediate filaments of baby hamster kidney (BHK-21) cells and bovine epidermal keratinocytes have similar ultrastructures and subunit domain structures. Proc Natl Acad Sci USA 77:4534–4538.

Viragh SZ, Challice, CE (1969): Variations in filamentous and fibrillar organization, and associated sarcolemmal structures, in cells of the normal mammalian heart. J Ultrastruct Res 28:321–334.

Virtanen K, Kurkinen M, Lehto V-P (1979): Nucleus-anchoring cytoskeleton in chicken red blood cells. Cell Biol Int Rep 3:157–162.

Weise MJ, Ingram VM (1976): Proteins and glycoproteins of membranes from developing chick red cells. J Biol Chem 251:6667–6673.

Walker SM, Schrodt GR, Bingham M (1968): Electron microscope study of the sarcoplasmic reticulum at the Z line level in skeletal muscle fibers of fetal and newborn rats. J Cell Biol 39:469–475.

Wolosewick JJ, Porter KR (1979): Microtrabecular lattice of the cytoplasmic ground structure: Artifact or reality. J Cell Biol 82:114–139.

Woodcock, CLF (1980): Nucleus-associated intermediate filaments from chicken erythrocytes. J Cell Biol 85:881–889.

Modern Cell Biology 2:163–199
© 1983 Alan R. Liss, Inc., 150 Fifth Avenue, New York, NY 10011

Interactions Between Actin Filaments and Membranes Give Spatial Organization to Cells

Lewis G. Tilney

From the Department of Biology, University of Pennsylvania, Philadelphia,
Pennsylvania, 19104

I. INTRODUCTION

It is generally accepted that actin filaments play an important role in giving cells shape and asymmetry. In essence the actin filaments form a skeleton, a framework of bones, on which, by which, or with which cytoplasms can undergo their complex movements. It is apparent that this cytoskeleton must be attached to the limiting membrane or "skin" of the cell because most of the movements—eg, cytokinesis, phagocytosis, retraction or elongation of surface projections, etc are coupled reactions involving both the membrane and the internal cytoskeleton. Accordingly, a great deal of effort has been expended in trying to determine how the cytoskeleton is attached to the membrane, what proteins are involved, and whether or not these linkers are anchored to the membrane by direct coupling to intrinsic (transmembrane) proteins.

The most studied system is the red blood cell membrane, but even here there are enormous gaps in our understanding. It is known that a transmembrane protein, band 3, is bound to a cytoplasmic protein termed ankyrin which, in turn, is bound to spectrin. Spectrin is a highly asymmetric molecule which exists in situ as a tetramer that is bound, along with another protein, band 4.1, to "actin oligomers." Thus, the actin is not attached to an intrinsic membrane protein directly, but rather via a series of intermediate macromolecules [see Branton et al, 1981, for details]. What is still not known is the function of band 4.1 or whether this component can link actin to membrane proteins more directly. We also do not known how many actin monomers are associated to form an "actin oligomer" or if these monomers in situ are assembled into tiny filaments or if they are associated with one another in some novel way, perhaps being organized by band 4.1. One even wonders why actin is present in this cell type at all since spectrin displaces actin as the most abundant cytoskeleton protein in erythrocytes, comprising 75% of the cytoskeletal mass, and the actin present seems to act merely as a glue connecting adjacent spectrin tetramers. One would think that other proteins could carry out this function just as well.

It is not yet clear how applicable these interesting results on the molecular organization of the red blood cell membrane are to other eukaryotic cells. What we really want to know is how actin filaments are attached to membranes, not how "actin oligomers," present in relatively small amounts, interact with spectrin to be bound indirectly to membranes. It is worrisome that the red blood cell (in mammals) is not really a cell at all, but a specialized *nonmotile fragment* of a cell. Nevertheless, we have learned from the red blood cell that actin does not seem to be bound to intrinsic membrane proteins directly, but by a series of macromolecular intermediates. This, it seems to me, may be a legitimate hypothesis for eukaryotic cells in general.

One of the major difficulties in determining how actin filaments attach to membranes in cells other than the red blood cell is to find a system in which these attachments are common enough to allow a molecular characterization of the proteins involved. In essence this constraint means that the biological system must be highly specialized—one in which this interaction is repeated many times and from which enough material can be isolated to carry out its chemical characterization. Certainly one of the reasons why so little information is available on actin-membrane associations for cells other than the red blood cell is that the systems so far available are so specialized that they require substantial experience to handle.

In this report I will try to indicate what is known, or at least what I know, about how actin filaments are attached to membranes in cells other than the red blood cell. I will use as examples four different cell types which, because of their specialization, allow us to make some morphological and behavioral statements about how these interactions occur. These observations provide us with a direction for future experiments to locate the proteins that connect actin filaments to membranes. They also give us vague hints of how the proteins may function. I will begin by describing the morphology of these connections as seen both by classical techniques of thin sectioning and by replicas of rapidly frozen, deeply etched, and rotary shadowed material. I will then present some recent behavioral studies which demonstrate how elongation of actin filaments attached to membranes occurs. Finally, I will try to show that the distribution of actin filaments in cells is much more precise than has previously been thought, raising interesting problems for further work.

II. THE NATURE OF THE ATTACHMENT OF THE TIP OF AN ACTIN FILAMENT TO THE MEMBRANE

In order to be sure that actin filaments are associated with a membrane at one of their ends, rather than merely passing near a membrane, one requires a biological system in which this "association" occurs repeatedly and remains stable after the cell is broken open. Several systems have now been described in which such end-on attachments appear real. The best known examples are the microvilli which extend from intestinal epithelial cells [Mukerjhee and Staehelin, 1971; Mooseker and Tilney, 1975; Hull and Staehelin, 1979], the stereocilia that extend from hair cells of the cochlea [Tilney et al, 1980], and the acrosomal filament bundles in certain invertebrate sperm [Tilney et al, 1981]. In all three cell types the exact point of contact between the actin filaments and the membrane is generally obscured by some electron dense material (Fig. 1a). If the cell is broken open, this point of contact is maintained, indicating a tight coupling of the filaments to the membrane in this region [Mooseker and Tilney, 1975; Tilney et al, 1982]. Furthermore, if the cell type in question is detergent extracted, the membrane at this "contact point" resists extraction [Tilney, 1980]. Earlier we [Mooseker and Tilney, 1975] speculated that the Z-line protein, α-actinin, might be the electron-dense substance at the tips of microvilli, but this hypothesis has not been

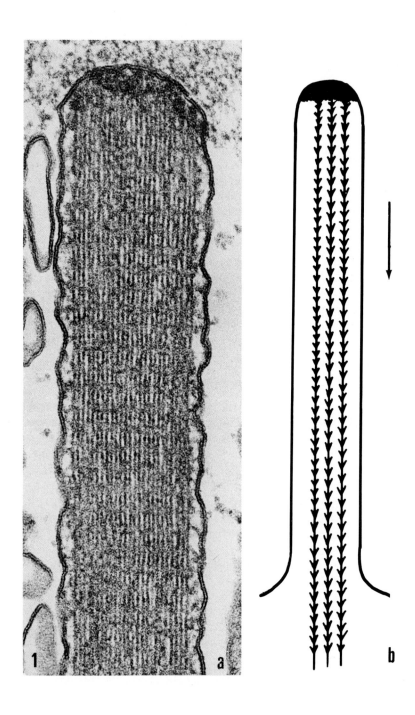

substantiated and evidence to the contrary exists [Bretscher and Weber, 1979; Craig and Pardo, 1979; Mooseker and Stephens, 1980]. Nonetheless, α-actinin appears to be part of the coupling between actin filaments and membranes in some cases because α-actinin [Lazarides, 1976] and more recently vinculin [Geiger et al, 1980] have been found by immunocytochemistry near actin attachment sites in cultured cells and at the zonula adhaerens.

III. IN GENERAL, WHEN ACTIN FILAMENTS ARE ATTACHED TO MEMBRANES AT THEIR TIPS, THEY DISPLAY UNIDIRECTIONAL POLARITY

It is well recognized that actin filaments are structurally polar. This polarity can be readily determined by "decorating" actin filaments with myosin subfragments (subfragment 1 (S1) and heavy meromyosin) and examining the arrowhead pattern. The question immediately arises, what is the polarity of the actin filaments whose tips are "attached" to membranes? This question is important in two respects. First, if the polarity is unidirectional with respect to the membrane, one must postulate that there are components at the tip of the actin filaments which not only glue the filaments to the membrane, but which also regulate the polarity of the attached actin filaments (perhaps the electron dense material previously mentioned contains this component). Second, if cytoplasmic myosin is to interact with these membrane-attached filaments, then by knowing the filament polarity we immediately know whether or not the membrane can be pulled toward the cell center by myosin.

The polarity of the actin filaments with respect to the membrane has now been determined in a variety of cell types. These include microvilli of intestinal epithelial cells [Mooseker and Tilney, 1975; Begg et al, 1978] and of sea urchin eggs [Begg et al, 1978], the stereocilia of hair cells from the inner ear [Flock and Cheung, 1977; Tilney et al, 1980], the acrosomal processes of invertebrate sperm [Tilney and Kallenbach, 1979; Tilney et al, 1981], the fertilization cone in eggs [Tilney and Jaffe, 1980], the filopodial extension of coelomocytes [Edds, 1977] and platelets [Nachmias and Asch, 1979], the surface projections of tissue culture cells [Begg et al, 1978] and Acanthamoeba [Pollard and Korn, 1973], and the attachment of the actin

---

Fig. 1.   a) Longitudinal section through a portion of a stereocilium from a hair cell of a chick cochlea. Note that the actin filaments terminate in some electron dense material located at the tip of the stereocilium. [From Tilney et al, 1982.] b) Drawing indicating the polarity of the actin filaments as revealed by decoration of the filaments with subfragment 1 of myosin. This drawing could represent a stereocilium, a microvillus, or an acrosomal process. Note that the arrowheads all point toward the cell center.

filaments in heart muscle to the intercalated disk. In all cases the arrowheads point away from the plasma membrane, toward the cell center (see Fig. 1b). Thus, if myosin were to interact with these filaments as it does in striated muscle, the only motion that could take place would be for the filament bundle, and thus the membrane, to be pulled toward the center of the cell, not pushed away from it. It is thus most unlikely that a musclelike actomyosin system is involved in the formation of microvilli and other actin-containing cytoplasmic evaginations.

At odds with the above examples is a recent report by Sugrue and Hay [1981] which shows that in motile blebs of corneal epithelial cells in culture the polarity of the actin filaments attached to the membrane is opposite that just described: The arrowheads point toward the membrane.

Why the actin filaments in these blebs have a different polarity from those in all the other systems remains obscure, but what is important to recognize is that somehow cells manage to control the polarity of the membrane-associated actin filaments. Specifically, in all the microvilli examined either by us or by others, the polarity is *always* the same. No one has yet found an exception, though exceptions have been sought [Begg et al, 1978; Tilney and Jaffe, 1980; Hirokawa et al, 1982]. One wonders how a cell can so accurately control the polarity of all the actin filaments in its microvilli or microvillar relatives, such as stereocilia. Obviously we do not know for sure, but by analogy with the red blood cell, it seems likely that filament polarity is influenced by some additional macromolecules which are either directly or indirectly attached to intrinsic membrane proteins. Two current models for these attachments have been proposed. First, already polymerized actin filaments may bind with the appropriate polarity to these membrane-attached macromolecules, and second, some of these macromolecules may regulate the assembly of the actin, perhaps by behaving as polymerization nucleators. Both of these hypotheses have the built-in advantage of not only determining the polarity of the filaments attached to the membrane, but also of regulating where actin filaments will be found relative to the membrane. This last point is important since actin filaments are not found randomly throughout cells, but rather at discrete, intracellular locations. I will discuss this point below.

IV. ELONGATION OF ACTIN FILAMENTS ATTACHED TO MEMBRANES OCCURS BY THE ADDITION OF MONOMERS TO THE END OF THE FILAMENT ATTACHED TO THE MEMBRANE

Although we have no chemical or molecular information on how actin filaments are attached to membranes, we do have some observations on how the actin filaments elongate with respect to the membrane. Before presenting this evidence, I should mention that elongation of membrane-associated actin filaments is a common occurrence in the examples cited above. For example, microvilli on the surfaces of sea urchin eggs elongate dramatically following fertilization [Begg and Rebhun, 1979; Tilney and Jaffe, 1980] and prior to the first division [Schroeder, 1978]; newly formed intestinal epithelial cells have short microvilli, but as they mature the microvilli become progressively longer [see Overton and Shoup, 1964]; and during the acrosomal reaction in sperm of echinoderms such as Thyone, there is a dramatic increase in the length of the actin filaments present [see Tilney and Inoué, 1982, for references].

As is well known, actin monomers will rapidly assemble on existing actin filaments to cause elongation. Assembly can occur on either end of the filament, but there is a strong bias (7- to 10-fold) toward the end which would appear "barbed" if the filament were decorated with S1 [Woodrun et al, 1975; see also Tilney et al, 1981]. This is the end which, in all cases of membrane-associated actin except the blebs in corneal epithelial cells, is "attached" to the limiting membrane. Here I would like to ask the following: If an actin filament attached to a membrane elongates, do monomers add to the filament at the membrane-associated end (which, if the actin filament were free, would be the preferred end for monomer addition) or to the unattached end? I will begin with evidence from the sperm of the horseshoe crab, Limulus.

In mature Limulus sperm there is an actin filament bundle approximately 60 μ in length [see Tilney, 1975]. This bundle extends from the membrane limiting the acrosomal vacuole through a canal in the center of the nucleus to the basal end of the cell where it forms five or six coils around the base of the nucleus (Fig. 2). This bundle is tapered. Near the acrosomal vacuole it is composed of only 15 filaments surrounding a dense core, whereas at the extreme basal end of the bundle there are 85 filaments (Fig. 3).

The acrosomal filament bundle can be easily isolated from the cell [Tilney, 1975]; the isolated bundle is incredibly stable, not depolymerizing or splaying in media of a variety of pH values or ionic strengths, unlike many actin bundles. Because of the taper, we always know which end of the isolated bundle would be located nearest the acrosomal vacuole in vivo. If these isolated bundles are incubated in G-actin under polymerizing conditions, the addition of monomers occurs on both ends of the bundle. However, the thin end of the bundle (that nearest the acrosomal vacuole in situ) adds monomers about 10 times more rapidly than the basal end (see Fig. 4) [Tilney et al, 1981]. This is the "barbed" end of the filament bundle if it is decorated with S1.

To determine in vivo the end of the filament bundle to which monomers add, we examined stages in the formation of the filament bundle in situ. This occurs naturally during spermatid differentiation. Just after the second meiotic

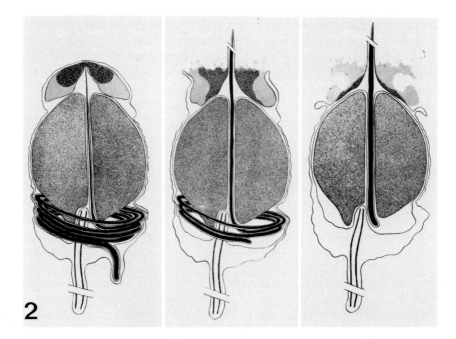

Fig. 2. Schematic view of the actin filament bundle in a Limulus sperm. The left-hand drawing illustrates the bundle in an unreacted sperm. Note that the bundle extends from the basal surface of the acrosomal vacuole through a canal in the center of the nucleus from where it is coiled around the base of the sperm head. Its most basal end extends into a short cell process called the false discharge. Important for subsequent discussion is the fact that the filament bundle tapers, being very narrow near the acrosomal vacuole, yet broad at its basal end. The drawing on the right illustrates the position of the bundle when the sperm has completed the acrosomal reaction. The middle drawing illustrates the situation when the reaction is partially complete. Note that during the acrosomal reaction the bundle gradually uncoils and at the same time pushes its way forward.

division some dense material appears on the cytoplasmic surface of the membrane limiting the acrosomal vacuole at a position from which the acrosomal filament bundle will grow (Fig. 5a). This material has been termed the "acrosomal button" by Fahrenbach [1973]. Because the bundle is tapered, there are two ways in which the bundle could grow from the acrosomal button. First, it could grow as suggested by Figure 7 in which the thin end would appear first; then as the bundle increased in length, its basal end would gradually increase in diameter. If this were the correct model, then addition of monomer should occur on the basal (thicker) end of the bundle. This is the slow-growing or "pointed" end of the decorated filament. Alternatively,

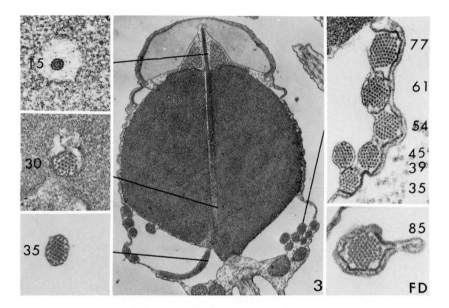

Fig. 3.   In the center of this plate is a longitudinal section through a Limulus sperm. To each side are transverse sections through the filament bundle at the position indicated on the longitudinal section. Note that the filament bundle gradually tapers, having 15 filaments near the acrosomal vacuole and 85 at the other end, which exists in the false discharge (FD). The coil makes 6–7 turns; each tends to lie apical to the last. Longitudinal sections × 41,860; transverse sections × 110,000. [From Tilney et al, 1981.]

the filament bundle could elongate as shown in Figure 8, in which the fat or basal end would form first and monomers would add at the end attached to the vacuole (see arrows, Fig. 8). The taper would develop by the sequential capping of the outermost filaments in the bundle. In this model monomers would be adding to the "preferred" or barbed end of a decorated bundle, the end on which most of the monomers add to the bundle in vitro (Fig. 4). When sections through developing spermatids are examined, it is clear that the *fat or basal end is formed first,* and as the filament elongates, the taper appears (Figs. 5b, 6). We thus conclude that the actin monomers are adding to the barbed end of the filament bundle, the end which appears to be attached to the membrane (see drawing, Fig. 9). There is additional evidence in favor of this conclusion, but it is beyond the scope of this brief review. For details see Tilney et al [1981].

We have recently strengthened the above conclusion by studying the acrosomal reaction in sperm from the sea cucumber, Thyone. Unlike Limulus

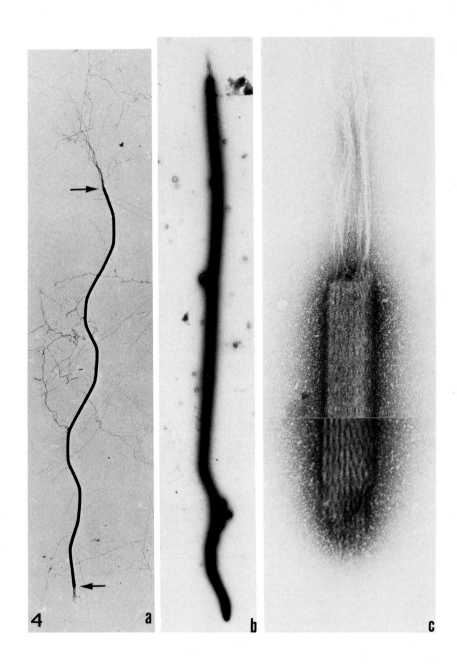

sperm, in which the actin filament bundle is fully assembled prior to the acrosomal reaction, Thyone sperm use the polymerization of actin to provide the force necessary to form the acrosomal process [see Tilney and Kallenbach, 1979]. As indicated in Figure 10, the first step in this reaction, which is triggered when the sperm comes in contact with the jelly surrounding the egg, is a fusion of the acrosomal vacuole with the plasma membrane. This is followed by the assembly of actin monomers from the "profilactin cup" onto preformed actin filaments which are located in an organelle I have called the actomere. Amazingly, an extension 90 μm long is generated in only 7 sec as a result of explosive, yet controlled, polymerization of actin. In accordance with a series of observations made in vitro and in vivo and the fact that the preferred or barbed end of the filament is at the tip of the process, we [Tilney and Kallenbach, 1979] suggested that monomers to be assembled must diffuse from their source in the sperm just anterior to the nucleus to the tip of the elongating process. This idea is diagramatically depicted in Figure 11. In considering this hypothesis, one wonders if there is sufficient time for monomers to diffuse to the tip of the growing polymers. Using video techniques, we [Tilney and Inoué, 1982] recorded the kinetics of elongation of the acrosomal process (Fig. 12). If elongation were proceeding at a rate limited by diffusion of monomers to the tip of the process, then a plot of the length of the acrosomal process as a function of (time)$^{1/2}$ should give a straight line. Our kinetic data are consistent with this prediction (Fig. 13).

We have recently obtained further evidence on this point by forming an acrosomal filament bundle in vitro. We removed the membranes of Thyone sperm with detergent at low pH (6.3), then, after washing, extracted the material comprising the profilactin cup at pH 8.0. There are two major proteins in the extract: actin in a 1:1 molar ratio with a 12,500 molecular weight protein which is profilin-like [see Tilney and Inoué, 1982]. To this extract we added actin polymerization nuclei similar to the actomere: pieces of the Limulus filament bundles, already discussed. The Thyone actin in the extract rapidly assembled on the Limulus bundles, using them as nuclei (Fig. 14). However, the results differ from those obtained when purified actin is

---

Fig. 4. a) The acrosomal filament bundle of Limulus sperm was incubated in a solution of G-actin, then salt was added. Assembly of actin occurs primarily off one end (top), but a small amount of growth occurs off the basal end. The arrows demarcate the extent of the acrosomal bundle. b) A similar experiment to that in (a), only in this case gel-filtered actin was used. × 14,000. [From Tilney et al, 1981.] c) A higher magnification of the two ends of the acrosomal filament bundle illustrated in (b). Note that the top end of the bundle is thinner in diameter than the bottom end. Because of the taper of the bundle we can see that the portion of the bundle near the top of the page would have been located in vivo nearest the acrosomal vacuole. It is this end, then, on which monomers of actin prefer to add. × 280,000. [From Tilney et al, 1981.]

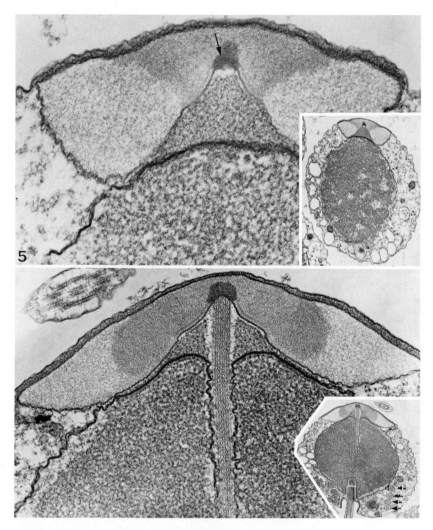

Fig. 5. a) Thin section through the apical end of a Limulus spermatid at a stage of differentiation in which the filament bundle has not yet assembled. At the position where the filament bundle will eventually contact the membrane limiting the acrosomal vacuole, we see a dense material attached to both the cytoplasmic and vacuole surfaces of this membrane (arrow). This dense material has been called the acrosomal button. × 105,000; inset, × 14,000. [From Tilney et al, 1981.] b) Thin section through the apical end of a Limulus spermatid at a stage of differentiation in which the filament bundle has been partially assembled, forming four turns of the coil (arrows in the inset). The filament bundle is attached to the acrosomal vacuole membrane at the same position as in (a). Where the filaments make contact with this membrane, there is a thin layer of dense material on the cytoplasmic side of the membrane. Of interest is that the bundle here is composed of approximately 50 filaments. × 81,000; inset, × 17,000.

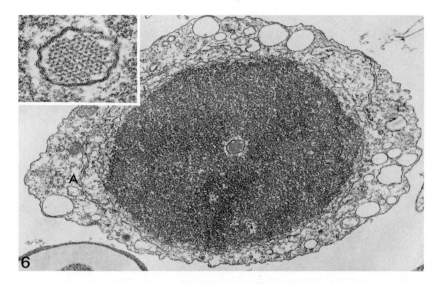

Fig. 6.   Transverse section through a developing spermatid of Limulus in which the filament bundle has just begun to form. It is of interest that the filament bundle running down the nuclear canal contains 70 filaments (see inset) organized on a hexagonal lattice. Also in this spermatid is a portion of the growing flagellar axoneme (A). × 37,000; inset, × 175,000. [From Tilney et al, 1971.]

added to Limulus bundles in vitro (see above). The Thyone actin from the extract *adds only to the preferred end; none adds to the other end* (Fig. 14). The reason for this behavior is related to the presence of the profilin-like protein. By binding to the monomer, this component regulates the amount of free or polymerizable actin so that additon can occur only at the preferred end of the filament, the end which in situ would be located at the tip of the acrosomal process. More specifically, the profilin, by binding to monomeric actin, maintains the concentration of uncomplexed actin below the critical concentration for addition of monomers to the pointed end of a decorated filament. This concentration is still high enough to allow addition on to the barbed end. Thus, in vivo, because the profilin-like protein is present, addition of the monomeric actin will effectively occur only at the tip of the elongating process, the end associated with the plasma membrane.

Also present in this extract is a protein which induces the newly polymerized actin filaments to associate into a bundle. If this component were not present, then the newly assembled filaments would splay apart as they do in vitro (compare Figs. 4 and 14). From these two examples (Limulus and Thyone sperm), the evidence is very strong that monomeric actin adds to the end of the actin filament that is associated with the membrane.

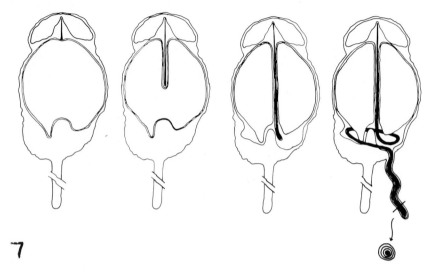

Fig. 7. Drawing illustrating the sequence of stages that one might expect to find in Limulus sperm during spermiogenesis if monomers were to add to the basal end of the filament bundle. Notice that in this model the thin end would appear first.

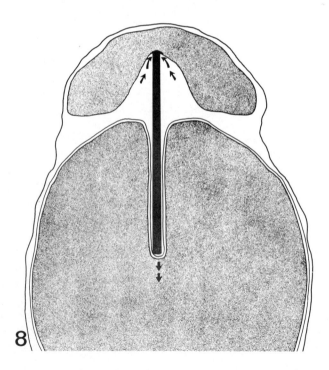

V. ACTIN FILAMENTS ARE CONNECTED TO THE MEMBRANE AT THEIR ENDS BY TINY FIBRILS REMINISCENT OF "CROWS' FEET"

Associations of actin filaments with membranes are difficult to interpret if only a single actin filament is present. However, if a bundle of filaments is connected to the membrane, the type of morphological connection present is much easier to discern. There are two reasons for this. First, if a bundle of actin filaments is present, the same type of linkage to the membrane is repeated many times, and thus the observation of a membrane-actin filament attachment seems more reliable than if only one connection is observed. Second, a cross-linked bundle of actin filaments is more stable to conventional fixation and dehydration than are free actin filaments, giving confidence to any observations.

Fig. 9. Cartoon designed to illustrate the situation in which monomers (hats) add to the end of an actin filament (head of the man) which is associated with the membrane. As each new monomer is added, the filament elongates, pushing its way basally.

Fig. 8. An alternative mechanism for the formation of the actin filament bundle. If monomers add to the end of the bundle nearest the acrosomal vacuole, as indicated by the upper arrows, the fat end of the bundle would appear first, and as it elongates the bundle would gradually extend through the nucleus as indicated by the lower arrows. Examination of thin sections of stages in spermiogenesis reveal that this is what actually happens.

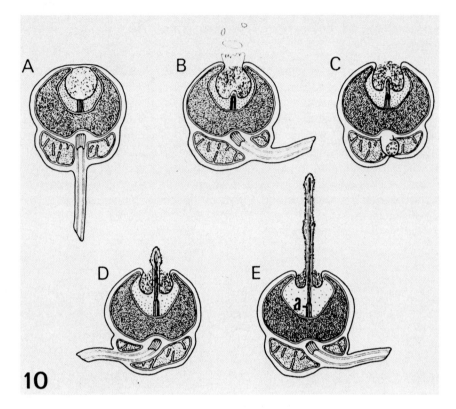

Fig. 10. Drawing depicting stages in the formation of the acrosomal process in Thyone sperm. The actomere (a), or organelle which acts as a nucleus for actin assembly, is indicated in the discharged sperm (stage E). [From Tilney and Inoué, 1982.]

There are, of course, methods that obviate some of these uncertainties. One that has proved particularly useful is to freeze a cell or portion of a cell very rapidly—eg, by throwing it against a copper block cooled to liquid helium temperatures [see Heuser and Kirschner, 1980]. This frozen tissue is subsequently placed in a freeze-fracture apparatus that allows us to deeply etch and rotary-shadow the tissue. The replica thus formed can be examined in stereo, which allows visualization in three dimensions of how the actin filaments are associated with a membrane. The advantage of this technique is that fixation and embedding of the specimen are unnecessary; thus many problems caused by these methods vanish. At the same time, because of the deep etching and rotary shadowing, we see a portion of the cell in relief; it is not the flat image that we are accustomed to in a thin section or a negative-stained preparation.

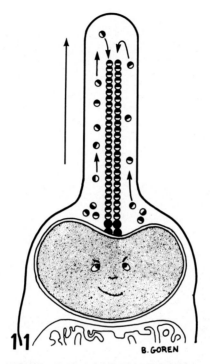

B. GOREN

Fig. 11. This cartoon is designed to illustrate the idea that actin filament elongation, and thus elongation of the acrosomal process in Thyone sperm, occurs by monomers which diffuse from the profilactin region immediately anterior to the nucleus to the tip of the process where they assemble on to the tip of existing filaments. These filaments are nucleated from the filaments present in the actomere, depicted in this drawing as solid spheres.

We [Hirokawa and Tilney, 1982] have studied the apical end of a hair cell from the vertebrate ear by this fast-freezing technique. Extending from the surface of this cell are large microvilli, termed stereocilia. Within each stereocilium, as in the microvilli of the intestine, is a bundle of actin filaments; those in the center of the bundle extend from the tip, down the length of the stereocilium, into the apical cytoplasm of the cell to serve as a rootlet (Fig. 15). Oriented perpendicular to the rootlet filaments, yet parallel to the apical end of the cell, are large numbers of actin filaments comprising what has been termed the cuticular plate [see Tilney et al, 1981]. We fast-froze the relevant portion of a chick ear, then fractured it obliquely. The fracture plane passes through the cuticular plate, thereby exposing the component filaments, and continues until it makes contact with the cytoplasmic side of the apical plasma membrane. At this point the fracture plane jumps into the bilayer

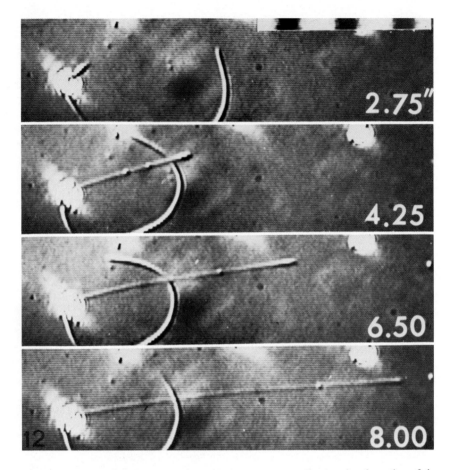

Fig. 12. Photographs of single frames of a video sequence showing the elongation of the acrosomal process of Thyone sperm. Printed to the right of each sequence is the time elapsed in seconds. [From Tilney and Inoué, 1982.]

and continues down the center of the bilayer (see Fig. 16, insert). When we etched this fracture and examined its replica, the place of particular interest was the region immediately adjacent to the membrane. It is here that we can see how the actin filaments make contact with the cytoplasmic surface of the plasma membrane. There are tiny wisps or fibrils that extend from the ends of the actin filaments to the membrane. To us these fibrils look like "crows' feet," the fibrils serving as toes. Where these fibrils contact the cytoplasmic surface of the membrane we see small lumps.

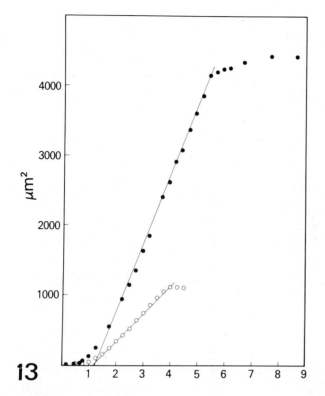

13

Fig. 13. A plot of the length^2 of the acrosomal process as a function of time (sec) for two sperm undergoing the acrosomal reaction. The curve with the black dots represents data obtained from the same sperm in the video sequence illustrated in Figure 12. Note that for both sperm, a plot of length^2 vs time is a straight line. [From Tilney and Inoué, 1982.]

VI. THERE ARE CROSS CONNECTIONS THAT BIND THE LATERAL SURFACES OF ACTIN FILAMENTS TO MEMBRANES

Extending from the lateral surfaces of the actin filaments that comprise the core filament bundle in microvilli of intestinal epithelial cells there are fine cross connections, 15–30 nm in length and 2–8 nm in width [Mukerjhee and Staehelin, 1971; Mooseker and Tilney, 1975; McNutt, 1978; Matsudaira and Burgess, 1979]. These connectives join the core filament bundle to the plasma membrane. In thin sections they appear to be spaced at 33-nm intervals [Mooseker and Tilney, 1975] (Fig. 17). Matsudaira and Burgess [1982] recently examined the distribution of these cross connections around the core filament bundle in more detail and showed that they spiral around the bundle, forming stripes like a barber pole with neighboring stripes separated by 33

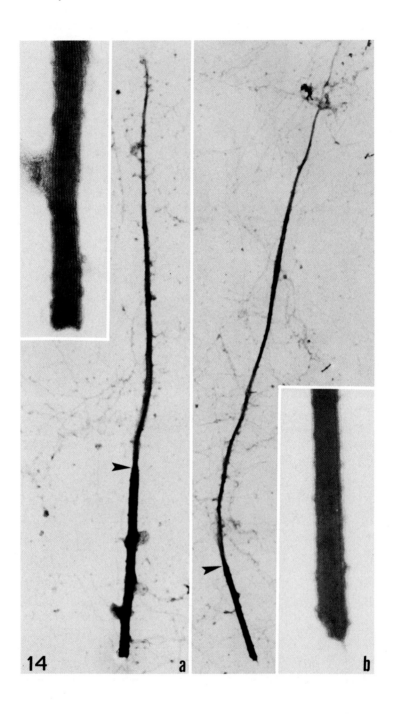

nm, as suggested by our earlier studies. They have proposed that this pattern of cross connections may be related to the geometry of the actin helix and the packing of the filaments in the core filament bundles. This idea is based on studies on the packing of actin filaments in the stereocilia of hair cells of the ear [Tilney et al, 1980; DeRosier et al, 1980; DeRosier and Tilney, 1982] in which the actin filaments are cross-bridged together to form a paracrystal such that the crossover points of the actin helices are in register. If all the filaments are in register and if the connections are attached to the filament subunits nearest the membrane, then as one moves around the filament bundle, the cross connections would move up the bundle with a helical pattern. The attractiveness of this model is that the distribution of the connections is determined by the substructure of the filament—provided, of course, the filaments are paracrystalline.

The cross connections in the brush border appear to be a protein whose molecular weight is 110,000 [Matsudaira and Burgess, 1979]. Nothing, however, is known about how these connectives are attached to the plasma membrane—ie, by what and how many intermediate macromolecules the connections are made. Nor is it known whether these connectives are attached to intrinsic membrane proteins. What is known is that neither the lateral nor the end-on attachment of actin filaments to membranes is associated with intramembraneous particles as seen in replicas of freeze fractures [Tilney and Mooseker, 1976]. This fact means only that large clusters of intrinsic membrane proteins are not involved.

Although the lateral connections between actin filaments and membranes are well established for intestinal epithelial cells, there is no information about other cell types. At least part of the failure to find these connections is due to the fact that in other systems they are easily destroyed by fixation. For example, if we examine longitudinal sections through the stereocilia of the hair cell of the ear fixed by methods that preserve the cross connections in intestinal microvilli, we fail to find any cross connections (Fig. 1a). In contrast, if we fast-freeze unfixed hair cells or cells *very briefly fixed* in glutaraldehyde, and then examine replicas of them as mentioned under the preceding subheading, we now see cross connectives clearly [Hirokawa and Tilney, 1982] (Fig. 18). Their distribution seems similar to that in microvilli.

Fig. 14. Pieces of the acrosomal filament bundle of Limulus sperm were suspended in an extract of Thyone sperm which contained actin and a profilin-like molecule. Salt was added and assembly of the Thyone actin occurs, using the Limulus filament bundles as nuclei. Of interest is that filament elongation *only* occurs on one end of the Limulus polymerization nuclei. This turns out to be the preferred end for addition of monomers, the end which in vivo is located at the tip of the acrosomal process. The arrows demarcate ends of the Limulus nuclei. Insets a and b, × 124,000.

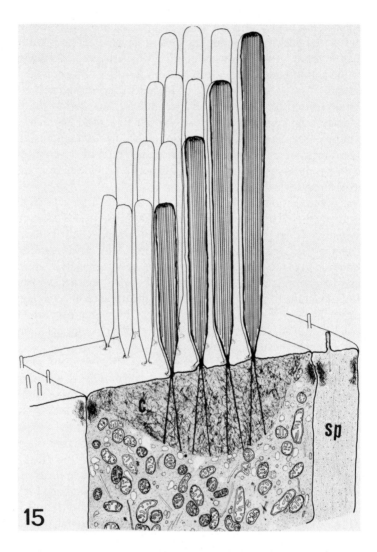

Fig. 15. Drawing illustrating the apical end of a hair cell of the cochlea. Extending from this cell are stereocilia of progressively increasing lengths. Within each stereocilium is a bundle of actin filaments; the central ones in the bundle penetrate into the apical cytoplasm as rootlets. Surrounding these rootlets is a second population of filaments which are oriented parallel to the apical surface, perpendicular to the rootlet filaments. Unfortunately, they are not obvious in this drawing. This region of the cell has been termed the cuticular plate (C). Separating adjacent hair cells are supporting cells (SP).

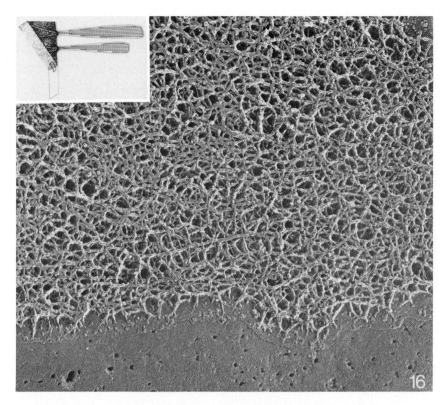

Fig. 16. Replica of a portion of the apical surface of a hair cell of the bird inner ear which had been fast-frozen and deeply etched. The inset illustrates the plane of the fracture. Of interest are the tiny fibrils that connect actin filaments to the membrane. [From Hirokawa and Tilney, 1982.]

Thus, fixation, even glutaraldehyde fixation, fails to preserve these cross connections.

Because of the extreme lability of the connectives to fixation and because of their threadlike appearance, I expect that actin filaments may be more commonly connected to membranes than has been suspected. In fact, careful scrutiny of Figure 16 shows that the actin filaments in the cuticular plate are not only connected to the membrane by crows' feet, but also by lateral connections. I thus believe that, if care is taken during tissue preparation, this type of connection will prove to be general.

VII. THE ASSOCIATION OF ACTIN FILAMENTS WITH SPECIALIZED CELL JUNCTIONS—eg, TIGHT JUNCTIONS AND INTERMEDIATE JUNCTIONS (ZONULA ADHAERENS)

Hull and Staehelin [1979] were, I believe, the first to suggest that just inside the tight and intermediate junctions is a population of fine filaments. The chemical nature of these filaments was not established in that report, nor were they preserved well enough to determine how many filaments make up this population, nor was filament polarity established. Because of the lability of actin filaments to fixation with osmium [Maupin-Szamier and Pollard, 1978] and to dehydration [Small, 1981], the best way to examine

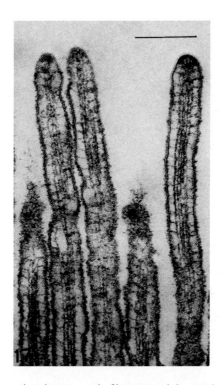

Fig. 17.   Lateral connections between actin filaments and the membrane in the microvilli of intestinal epithelial cells. [From Mooseker and Tilney, 1975.]

Fig. 18.   Lateral connections between the actin filaments and the membrane in an unfixed, yet fast-frozen and deeply etched stereocilium that extends from the hair cells of the inner ear of a bird. [From Hirokawa and Tilney, 1982.]

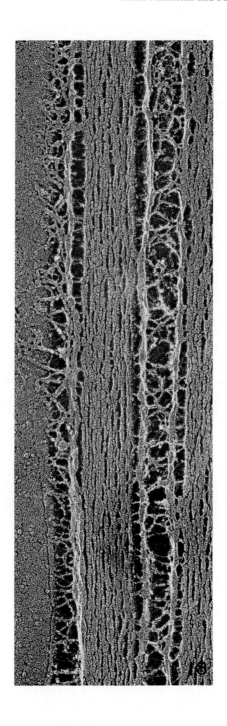

the nature of these filaments and their relation to junctions is by the fast-freezing technique.

When we [Hirokawa and Tilney, 1982] examined the junctional complexes of epithelial cells, such as the hair cells of the vertebrate ear, before and after decoration with S1, we indeed found actin filaments just inside the tight and intermediate junctions (Figs. 19, 20). Interestingly, those inside the tight junction follow closely the strings of membrane particles. They are, however, very sparse.

Beneath the intermediate junction or zonula adhaerens there is a substantial population of actin filaments that forms a ring encircling the apical margin of the cell. After decoration of these actin filaments with S1, we [Hirokawa and Tilney, 1982] were able to show that individual filaments of both polarities are present in that ring. In replicas of quick-frozen material, decorated actin filaments appear as a rope [Heuser and Kirschner, 1980]. Careful examination of this rope reveals that its gyres are angular, which allows us to determine the polarity of the actin filament (see Fig. 20). The fact that this ring is composed of filaments of mixed polarities opens the possibility that the ring might be contractile.

What is not known is how the actin filaments at the junctional complexes are attached to the membrane. We have not found any cross connections such as those seen in microvilli, stereocilia, and the cuticular plate. Vinculin, however, a 130,000-dalton protein, has been localized near, but not directly associated with, the plasma membrane at the zonula adhaerens [Geiger et al, 1980, 1981] and at the dense plaques in chicken gizzard cells. It seems possible that this protein, by means of other intermediates, attaches the filament bundle to the membrane. There is no information on how this occurs or what the chemical nature of the other intermediate macromolecules is.

VIII. THE SPATIAL ORGANIZATION OF ACTIN FILAMENTS IN CELLS—HOW PRECISE IS IT AND HOW WELL DO CELLS CONTROL THEIR SHAPE?

In a symposium on spatial organization, I would really like to be able to relate to the reader how precisely cells determine the distribution of the actin filaments within them. I would like to be able to discuss how the distributions are determined and to specify the role of the membrane in this regulation.

Fig. 20. As in Figure 19, the inner ear was detergent-extracted and the actin filaments decorated with S1 of myosin. In this figure we see a large bundle of actin filaments lying just within the zonula adhaerens. Careful examination of these decorated filaments, which appear as ropes shows that adjacent filaments exist with both polarities. The arrows indicate the polarity of some of the filaments. [From Hirokawa and Tilney, 1982.]

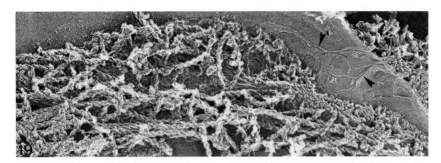

Fig. 19.   A tight junction in a replica from a rapidly frozen, deeply etched hair cell of the inner ear of a bird. Before freezing, the hair cell was briefly detergent-extracted and the actin filaments decorated with S1 of myosin. Note that actin filaments, which appear as a rope, are present within the cytoplasm immediately adjacent to the tight junctions. [From Hirokawa and Tilney, 1982.]

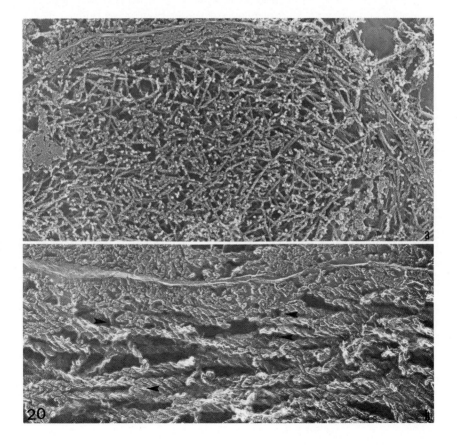

Clearly certain cells exert rather precise controls over the number, polarity, and distribution of actin filaments within their cytoplasms. For example, in Limulus sperm the only actin filaments in the cell are in the acrosomal filament bundle which is 60 ± 5 μm in length and is composed of approximately the same number of filaments in every cell examined. Other examples include the brush border of intestinal epithelial cells where again the number, polarity, and length of the core bundle of actin filaments in the microvilli are controlled; thus in the chick microvilli we find 18–25 actin filaments, each the same length for a given cell and each oriented with the barbed end at the plasma membrane. The filament bundles in the stereocilia of the hair cells from the ear are similarly precisely controlled [Tilney et al, 1980; Tilney and Saunders, 1982]. What is fascinating about the latter is that the lengths and widths of the stereocilia, and thus the lengths and numbers of actin filaments within these cell extensions, change in a predictable way depending on where the hair cell is located on the cochlea. Thus, individual hair cells are different from each other; each hair cell, depending on its position, "knows" how long to grow its actin filaments, how many actin filaments to cross-bridge together into a bundle, and how many bundles to assemble [Tilney and Saunders, 1982]. Finally, we know that the actin filaments in skeletal and heart muscle cells have precise lengths, polarities, and distributions.

There is a tendency to regard the above mentioned systems as not representative of the "more typical" situation in cells since these systems are highly specialized. Instead, most investigators want to know how organized are the actin filaments in the "cortex" of a "typical cell" such as a macrophage or a tissue culture cell. One expects that they are less organized than the above cell types and probably contain a tangle of disorganized actin filaments. Although it would be foolish and probably incorrect to state that the actin filaments in these cell types are also extremely ordered, it is my impression that there is much more order here than one would intuitively predict. My reasons for this suspicion are that in three examples of what superficially look like disordered actin tangles, close examination reveals remarkable order.

First, let us consider the cuticular plate of hair cells. When we initially studied thin sections through this region, it appeared as a tangle of filaments such as one might imagine would be the image if one were to cut a section through an actin gel. However, when the filaments making up the cuticular plate are decorated with S1, we find filaments of only two orientations. One group comprises the rootlet filaments, a group that extends from the basal end of the stereocilium towards the center of the cell; the other group lies perpendicular to the rootlet filaments, or parallel to the apical surface of the cell [see Tilney et al, 1980; Hirokawa and Tilney, 1982] (Fig. 21). The filaments of the second group are oriented parallel to the apical surface, but

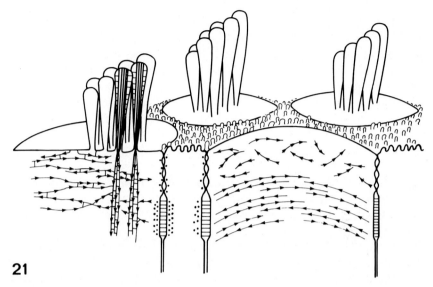

21

Fig. 21. Drawing illustrating the distribution of actin filaments in the apical end of the hair cell. The arrowheads indicate the polarity of the actin filaments. [From Hirokawa and Tilney, 1982.]

lie in all directions within that plane, as if one had dropped straws on the floor. It is striking that there are almost no filaments that run from the apical surface toward the basal surface at an angle oblique to the cell's axis, nor are there any filaments extending basally that are not rootlet filaments, as defined by their termination in the stereocilium proper. Furthermore, the lateral margins of the cuticular plate do not extend all the way to the intercellular junctions; in essence, they are not just filling the apical cytoplasm of the cell with what might correspond to glass wool (see Fig. 15). Instead, the confines of the cuticular plate are demarked with precision. It was largely two facts that initially led us to think of the undecorated cuticular plate as a tangle of filaments: 1) fixation and dehydration, prerequisites for thin sectioning, distort this structure dramatically; and 2) interconnecting the rootlet filaments and the population of actin filaments that lie perpendicular to them are numerous, 3 nm, cross-connecting filaments [see Tilney et al, 1980; Hirokawa and Tilney, 1982]. These connections, which are not actin, extend at all angles to the existing actin filaments. Thus, the apparent tangle is actually not a tangle at all, but an ordered arrangement of two populations of filaments. How the cell specifies the distribution of the actin filaments parallel to the apical surface is still a real puzzle.

Second, let us examine the cytoplasmic region directly beneath the microvilli of an intestinal epithelial cell, the so-called terminal web region. Again, in thin sections this region looks like a tangle of filaments running between the bases of the core filament bundles which extend into the microvilli. Recently we [Hirokawa et al, 1982] examined replicas of this region following fast-freezing under a variety of conditions. We found that the only actin filaments in this region, except for the bundles of actin filaments that are associated with the tight and intermediate junctions, are the core filament bundles which extend from the microvilli into the terminal web. These rootlet bundles do not splay out as originally thought [Mooseker and Tilney, 1975; Rodewald et al, 1976; Drenckhahn and Groschel-Stewart, 1980; Hull and Staehelin, 1980] (Figs. 22–24). Instead, there are two other kinds of filaments in the terminal web which can be distinguished from actin filaments in fast frozen tissue: intermediate filaments (Fig. 22) and fine filaments that interconnect adjacent actin filament bundles. Some of the latter are bipolar myosin filaments composed of 2–4 myosin molecules. The bulk of the intermediate filaments lie beneath the core filament rootlet bundles, but course up and around these bundles, connecting to them by threadlike cross connectives. The second population, fine filaments composed in part of bipolar myosin molecules, connects adjacent rootlet bundles. We have thus found, by using the fast-freezing method, that the actin filaments in the terminal web are, in fact, highly organized; they are not randomly oriented as expected from fixed, dehydrated, and thin-sectioned material [Hirokawa et al, 1982]. The apparent randomness seen in thin sections is due to the failure of that method to distinguish actin filaments from intermediate and myosin filaments. Since the "cortex" of many cells—eg, an egg—is really similar to the terminal web of intestinal epithelial cells, we suspect that what appears in many cells to be an actin tangle may, in fact, not be a tangle at all, but an ordered arrangement of actin filaments and their cross connections.

Third, Albrecht-Buehler [1977] was the first to bring to the attention of the scientific community the fact that when tissue culture cells divide, the two daughter cells have strikingly similar shapes, as if there were a "cell lineage" that is passed from cell to cell. This concept has been more recently confirmed and extended by Solomon [1981]. Both microtubules and microfilaments [Yamada et al, 1970] seem to be important in this shape determination. Although details about the length, number, distribution, and polarity of the actin filaments in these cells are not known, the data suggest that mitotic sisters are indeed very similar, reinforcing the notion that cells exercise a rather precise control over their actin filaments. Of course these patterns can be modulated in a moving cell such as a tissue culture cell—eg, if the cell bumps into something [see Solomon, 1981]—but one still suspects that the organization of the actin filaments is not random. This last

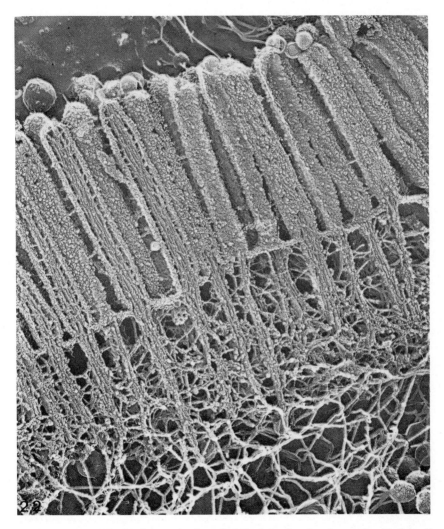

Fig. 22.   Replica of a fast-frozen, deeply etched brush border of an intestinal epithelial cell [From Hirokawa et al, 1982.]

example is clearly the weakest since so very little is known, but the phenomenon is intriguing enough to mention here.

From these three examples (the cuticular plate, terminal web, and mitotic sisters), and from the other examples mentioned earlier (skeletal muscle, microvilli, stereocilia, and sperm), we see that the distribution of actin filaments in cells is much more ordered than one might have predicted. In the

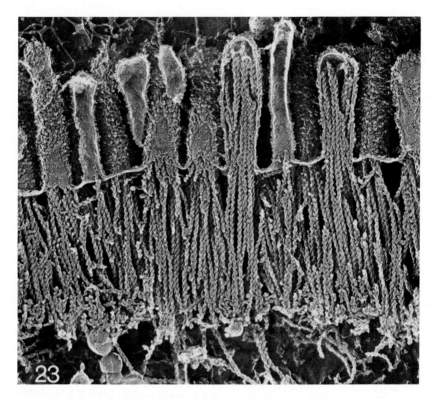

Fig. 23. Replica of a fast-frozen, deeply etched brush border of an intestinal epithelium that was incubated in S1. Note that there is no splaying of the filament bundles that extend down from the microvilli into the terminal web, nor are there any decorated filaments that run perpendicular to these core filament bundles. [From Hirokawa et al, 1982.]

case of Thyone sperm, a cytoplasmic organelle, the actomere, acts to order the filaments, but in other cases, with the exception of skeletal muscle, the actin filaments seem somehow to be organized with respect to the cell surface. Perhaps there is some positional information associated with the membrane that acts through some unknown number of intermediate components and specifies the polarity of the filaments associated with it, by controlling either their polymerization or their attachments. This magical material or materials, then, must be nonrandomly distributed on membranes because individual cells often have fixed numbers of cell extensions (eg, the stereocilia of the hair cells of the cochlea [see Tilney and Saunders, 1982]). Since the width of these extensions can also be fixed, (eg, stereocilia, microvilli, etc) and thus the number of actin filaments within them somehow counted, the size

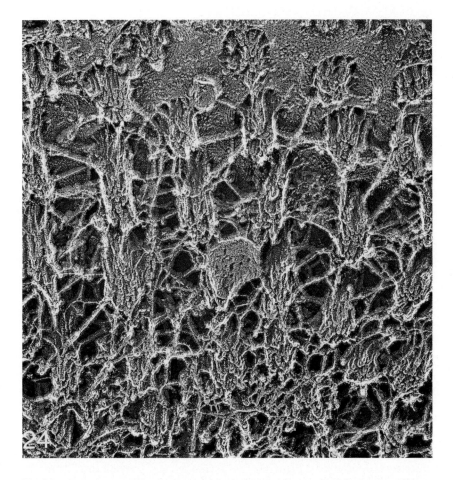

Fig. 24. Cross fracture through the terminal web of a fast-frozen, deeply etched brush border of an intestinal epithelium. [From Hirokawa et al, 1982.]

of these magical spots must also be determined. Yet even this fanciful idea cannot be the whole story, since in the cuticular plate many of the actin filaments (ignoring the rootlet filaments) lie parallel to the apical cell surface and could never make an end-on connection with a membrane. In short, we do not know how cells control the distribution of their actin filaments, but it is clear that the distribution, polarity, length, and number are controlled and in some way this control, in most cases, occurs relative to the plasma membrane.

IX. SUMMARY AND CONCLUSIONS

In this review I have attempted to cover what is known about how actin filaments interact with membranes to provide a cell with a cytoskeleton. Not only is this cytoskeleton important in helping to give the cell its shape, but it provides "bones" on which movements can be elicited and makes asymmetries useful for sensory transduction. Unfortunately, our knowledge of how the actin filaments are attached to the membrane, with its complement of intrinsic and extrinsic proteins, is fragmentary at best. The most studied case is the erythrocyte, but the information here is probably not directly applicable to other cell types, since actin filaments are not present in this cell type, nor are there any obvious movements or asymmetries in the cell. In other eukaryotic cells there is no chemical or molecular characterization of how actin is associated with membranes. We do know, however, that there is a variety of morphological connections. These include 1) end-on connections in some electron-dense material of unknown composition—eg, in microvilli, stereocilia, and the acrosomal process of Limulus sperm; 2) end-on connections by tiny fibrils or "crows' feet"—eg, in the cuticular plate of hair cells; 3) lateral connections between the surfaces of the actin filaments and membranes—eg, in microvilli, stereocilia, and the cuticular plate; and 4) some type of undefined connection involving α-actinin and vinculin between actin filaments and both the zonula adhaerens and the tight junction. I have argued that the lateral connections may be general in occurrence in spite of the infrequency with which they have been reported, since they are extremely labile to fixation. Furthermore, we know from studies both in vivo and in vitro, that the actin filaments that are connected to membranes at one end elongate by the addition of monomers to the membrane-associated end, the end that corresponds to the barbed end of a decorated actin filament.

Although our ignorance of the intermediary macromolecules involved in attaching actin filaments to membranes is somewhat overwhelming, the amazingly rapid progress in biochemical characterization of the proteins that bind to actin in vitro encourages the view that even the factors governing the distribution, length, number, and polarity of the actin filaments may soon be understood. Admittedly, there is a battery of macromolecules now described which cap actin filaments, fragment actin filaments, inhibit assembly of actin monomers, cross-link actin filaments, etc, but an understanding of the way these components lead to the magnificent organization of actin filaments in cells is not yet forthcoming. We must have some way to relate test tube chemistry to the precision of the spatial organization of cells.

This paper is dedicated to Keith Porter, who, more than any other individual, has stimulated people like myself to address these exciting questions.

ACKNOWLEDGMENTS

It is a particular pleasure for me to contribute to a symposium on "Porterplasm" because I was reared by Keith on the mysteries of the cytoplasm. It is amazing that many of the ideas expressed here have not changed or improved in the 15 years since I was a happy postdoc in his lab. Supported by NIH grant HD-144-74.

X. REFERENCES

Albrecht-Buehler G (1977): Daughter 3T3 cells. Are they mirror images of each other? J Cell Biol 75:595–603.

Begg DA, Rebhun LI (1979): pH regulates the polymerization of actin in the sea urchin egg cortex. J Cell Biol 83:241–248.

Begg DA, Rodewald R, Rebhun LI (1978): The visualization of actin filament polarity in thin sections. Evidence for uniform polarity of membrane associated filaments. J Cell Biol 79:846–852.

Branton D, Cohen LM, Tyler J (1981): Interaction of cytoskeletal proteins on the human erythrocyte membrane. Cell 24:24–32.

Bretscher A, Weber K (1979): Localization of actin and microfilament-associated proteins in the microvilli and terminal web of the intestinal brush border by immunofluorescence microscopy. J Cell Biol 79:839–845.

Burgess DR, Schroeder TE (1977): Polarized bundles of actin filaments within microvilli of fertilized sea urchin eggs. J Cell Biol 74:1032–1037.

Craig SW, Pardo JV (1979): Alpha-actinin localization in the junctional complex of intestinal epithelial cells. J Cell Biol 80:203–210.

DeRosier DJ, Tilney LG, Egelman E (1980): Actin in the inner ear: The remarkable structure of the stereocilium. Nature (Lond) 287:291–296.

DeRosier DJ, Tilney LG (1982): How actin filaments pack into bundles. Cold Spring Harb Symp Quant Biol 46:525–540.

Drenckhahn D, Groschel-Stewart U (1980): Localization of myosin, actin, and tropomyosin in rat intestinal epithelium: Immunohistochemical studies at the light and electron microscope levels. J Cell Biol 86:475–482.

Edds KT (1977): Microfilament bundles. I. Formation with uniform polarity. Exp Cell Res 108:452–456.

Fahrenbach WH (1973): Spermiogenesis in the horseshoe crab, Limulus polyphemus. J Morphol 140:31–52.

Flock AH, Cheung C (1977): Actin filaments in sensory hairs of inner ear receptor cells. J Cell Biol 75:339–343.

Geiger B, Dutton, AH, Tokuyasu KT, Singer SJ (1981): Immunoelectron microscope studies on membrane-microfilament interactions: Distribution of α actinin, tropomyosin, and vinculin in intestinal epithelial brush border and in chicken gizzard smooth muscle cells. J Cell Biol 91:614–628.

Geiger B, Tokuyasu KT, Dutton AH, Singer SJ (1980): Vinculin, and intracellular protein localized at specialized sites where microfilament bundles terminate at cell membranes. Proc Natl Acad Sci USA 77:4127–4131.

Heuser JE, Kirschner MW (1980): Filament organization revealed in platinum replicas of freeze-dried cytoskeletons. J Cell Biol 86:212–234.

Hirokawa N, Tilney LG, Fujiwara K, Heuser JE (1982): The organization of actin, myosin, and intermediate filaments in the brush border of intestinal epithelial cells. J Cell Biol 94:425–443.

Hirokawa N, Tilney LG (1982): Interactions between actin filaments and membranes in quick frozen and deeply etched hair cells of the chick ear. J Cell Biol 95:249–261.

Hull BE, Staehelin LA (1979): The terminal web: A reevaluation of its structure and function. J Cell Biol 81:67–82.

Lazarides E (1976): Actin, α actinin, and tropomyosin interaction in the structural organization of actin filaments in non-muscle cells. J Cell Biol 68:202–219.

Matsudaira PT, Burgess DR (1979): Identification and organization of the components in the isolated microvillus cytoskeleton. J Cell Biol 83:667–673.

Matsudaira PT, Burgess DR (1982): Organization of the cross-filaments in intestinal microvilli. J Cell Biol 92:657–664.

Maupin-Szamier P, Pollard TD (1978): Actin filament destruction by osmiun tetroxide. J Cell Biol 77:837–852.

McNutt NS (1978): A thin-section and freeze-fracture study of microfilament-membrane attachments in choroid plexus and intestinal microvilli. J Cell Biol 79:774–787.

Mooseker MS, Stephens RE (1980): Brush border alpha-actinin? Comparison of two proteins of the microvillus core with alpha-actinin by two dimensional peptide mapping. J Cell Biol 86:466–474.

Mooseker MS, Tilney LG (1975): The organization of an actin filament–membrane complex: Filament polarity and membrane attachment in the microvilli of intestinal epithelial cells. J Cell Biol 67:725–743.

Mukerjhee TM, Staehelin LA (1971): The fine structural organization of the brush border or intestinal epithelial cells. J Cell Sci 8:573–599.

Nachmias VT, Asch A (1979): Regulation and polarity: Results with mixomycete plasmodium and with human platelets. In Goldman R, Pollard T, Rosenbaum J (eds): "Cell Motility," Vol 3. Cold Spring Harbor New York: Cold Spring Harbor Laboratories, pp 771–783.

Overton J, Shoup J (1964): Fine structure of cell surface specializations in the maturing duodenal mucosa of the chick. J Cell Biol 21:75–85.

Pollard TD, Korn ED (1973): Electron microscopic identification of actin associated with isolated amoeba plasma membranes. J Biol Chem 248:448–450.

Rodewald R, Newman SB, Karnovsky MJ (1976): Contraction of isolated brush borders from intestinal epithelium. J Cell Biol 70:541–554.

Schroeder TE (1978): Microvilli on sea urchin eggs: A second burst of elongation. Dev Biol 64:346–342.

Small JV (1981): Organization of actin in the leading edge of cultured cells: Influence of osmium tetroxide and dehydration on the ultrastructure of actin meshworks. J Cell Biol 91:695–705.

Solomon F (1981): Specification of cell morphology of endogenous determinants. J Cell Biol 90:547–553.

Sugrue SP, Hay ED (1981): Response of basal epithelial cell surfaces and cytoskeleton to solubilized extracellular matrix molecules. J Cell Biol 91:45–54.

Tilney LG (1975): Actin filaments in the acrosomal reaction of Limulus sperm: Motion generated by alterations in the packing of the filaments. J Cell Biol 64:289–301.

Tilney LG (1980): Membrane events in the acrosomal reaction of Limulus and Mytilus sperm. In Gilula B (ed): "Membrane-Membrane Interactions." New York: Raven Press, pp 59–80.

Tilney, LG, Bonder, EM, DeRosier DJ (1981): Actin filaments elongate from their membrane associated ends. J Cell Biol 90:485–494.

Tilney LG, DeRosier DJ, Mulroy MJ (1980): The organization of actin filaments in the stereocilia of cochlea hair cells. J Cell Biol 86:244–259.

Tilney LG, Egelman E, DeRosier DJ, Saunders J (1982): The organization of the actin filaments in the stereocilia of the hair cells of the bird cochlea II. J Cell Biol (accepted).

Tilney LG, Inoué S (1982): The acrosomal reaction of *Thyone* sperm. II. The kinetics and possible mechanism of acrosomal process elongation. J Cell Biol 93:820–827.

Tilney LG, Jaffe LA (1980): Actin, microvilli, and the fertilization cone of sea urchin eggs. J Cell Biol 87:771–782.

Tilney LG, Kallenbach N (1979): The polymerization of actin. VI. The polarity of the actin filaments in the acrosomal process and how it might be determined. J Cell Biol 81:608–623.

Tilney LG, Mooseker MS (1976): Actin filament–membrane attachment: Are membrane particles involved? J Cell Biol 71:402–416.

Tilney LG, Saunders J (1982): The organization of the actin filaments in the stereocilia of the hair cells of the bird cochlea I. J Cell Biol (submitted).

Woodrun DT, Rich S, Pollard TD (1975): Evidence for biased unidirectinal polymerization of actin filaments using heavy meromyosin by an improved method. J Cell Biol 67:231–237.

Yamada KM, Spooner BS, Wessells NK (1970): Axon growth: Roles of microfilaments and microtubules. Proc Natl Acad Sci USA 66:1206–1212.

Cytoskeletal Action in Cytoplasmic Organization

Modern Cell Biology 2:203–223
© 1983 Alan R. Liss, Inc., 150 Fifth Avenue, New York, NY 10011

The Spatial Organization of Cortical Cytoplasm in Macrophages

Thomas P. Stossel

From the Hematology-Oncology Unit, Massachusetts General Hospital, and the Department of Medicine, Harvard Medical School, Boston, Massachusetts 02114

I. INTRODUCTION

This review concerns the spatial organization of a special domain of cytoplasm in a particular type of cell. The cell is the mammalian macrophage, a phagocytic leukocyte derived from the blood monocyte. The macrophage is essential for host defense against microbial infection and for the clearance of other foreign matter; and it is a mediator of inflammation. The region of cytoplasm in focus is the organelle-excluding hyaline cortex of the cell periphery. This area of the cell, which can be considered the "motor" of the macrophage, is mechanically very active. It forms the pseudopodia, filopodia, and lammellae that actively extend from the cell during spreading, locomotion, and phagocytosis, the activities by means of which the macrophage performs its immunological role.

II. THE STRUCTURE OF CORTICAL CYTOPLASM

As the cell moves, the cortical cytoplasm changes its shape by what appear to be propulsive and retractive movements, redistributing itself about the cell periphery. An early theory as to the basis of these movements held that the cortical cytoplasm of macrophages is a colloidal system that exists as a gel capable of exerting varying degrees of "contractile tension" on the cell [Lewis, 1939]. As will become evident, this view is generally correct, even in the light of recent findings.

Interest in gelation as it relates to cytoplasmic structure revived about 7 years ago when a number of investigators began to study the solidification of extracts prepared from diverse cells [reviewed by Taylor and Condeelis, 1979]. Such studies had been undertaken previously [eg, Bensley, 1938], but the discovery that actin fibers are a major component of both cytoplasm and the solidified cytoplasmic extracts represented a major advance over this earlier work.

Abundant numbers of 4 to 6 nm diameter "microfilaments" are the principal structures in electron micrographs of thin sections of macrophage cortical cytoplasm [Reaven and Axline, 1973]. Filaments with these dimensions in macrophage cytoplasm bind heavy meromyosin, establishing their identity as actin filaments [Allison et al, 1971]. One might therefore refer to the cell periphery as the "actoplasm."

From studies of macrophage cytoplasm it was concluded that the solidification of extracts from this cell resulted from the assembly of actin into fibers and the cross-linking of these fibers by a high molecular weight actin-binding protein which had previously been purified [Hartwig and Stossel, 1975; Stossel and Hartwig, 1975, 1976]. This conclusion was based in part on a comparison of the efficiency of actin-binding protein in actin gelation relative to that of other proteins, as determined by the ability of these proteins to link actin polymers into large aggregates. A sedimentation assay was employed to measure aggregation: A solution of purified muscle actin filaments was subjected to centrifugal forces at which the actin filaments sedimented very slowly. When small samples of macrophage extracts were added to the actin filaments, the actin sedimentation rate increased. By use of this assay, actin sedimenting activity was purified from the macrophage extracts, and 75% of the total actin-sedimenting activity copurified with actin-binding protein [Brotschi et al, 1978].

The concentrations at which three "actin-binding proteins"—macrophage actin-binding protein, chicken gizzard smooth muscle filamin (a protein structurally very similar to macrophage actin-binding protein [Wang, 1977; Wallach et al, 1978]), and rabbit skeletal muscle myosin—converted purified skeletal muscle actin filaments from a sol to a gel were also determined. This transition between sol and gel states of polymers characteristically occurs with abruptness when a critical density of cross linker within the polymer is attained. The potency of a polymer cross-linking agent in causing gelation is inversely related to the concentration required for it to produce incipient gelation of the polymer. The consistency of actin solutions in these studies was measured by means of a hydrostatic device which determined the force required to produce flow in an actin filament solution containing varying concentrations of the added actin-binding proteins. This parameter, called the yield force, is described below.* On a molar basis, macrophage actin-binding protein was fourfold more potent than filamin and tenfold more potent than myosin in gelling actin as defined by this criterion. In addition, actin-binding protein was more efficient than filamin or myosin in increasing the sedimentation rate of actin filaments [Brotschi et al, 1978].

In subsequent studies [Hartwig and Stossel, 1979; Yin et al, 1980] it was shown that the molar concentration of actin-binding protein required for incipient gelation of actin filaments is very close to that theoretically predicted for the gelation of flexible polymers oriented as random coils [Flory, 1953]. An assumption of the theory is that each added cross link recruits a new polymer, and that this recruitment leads to the formation of a "giant" molecule which is the gel. The model for gelation on which the theory is based requires that the orientation of polymers undergoing cross linking is such that the chains can become cross-linked wherever they overlap. Should the polymers tend to line up in parallel, it is likely that cross-linking would occur between adjacent chains more than once. With such redundant cross linking, each added cross linker would not tend to recruit a new chain into the three-dimensional network.

---

*Although the glutinous consistency of actin filament solutions has been appreciated since the discovery of the protein, the basis of this property is not well understood. Oosawa and co-workers concluded that actin filaments formed an entangled network, but also inferred that an actin filament solution had gel-like properties, because the extrapolated value of its viscosity at zero shear was infinite [Kasai et al, 1960].

Recent studies on the mechanical properties of actin filament solutions have been facilitated by the availability of instrumentation capable of exerting very low shear forces [Zaner et al, 1981]. The results of these investigations show that actin filaments in solution are entangled, and that these entanglements are responsible for the high apparent viscosity of actin solutions, for the dependence of this viscosity on the rate of sheer, and for the elastic properties of actin filaments. The flow of actin filaments cross-linked by actin-binding protein in response to shear is limited by the existence of true cross links [Stossel et al, 1981; Zaner and Stossel, 1982]. Therefore, at the molecular level, the properties of viscous actin filament solutions and of true actin gels cross-linked by actin-binding protein are important in different ways, although these differences are not intuitively obvious when one observes what appear to be very gooey actin sols and gels. In the case of the yield point measurements mentioned above, the beginning of flow of the actin in response to a given applied force could represent either rupture of an actin gel or disentanglement of a viscoelastic actin sol.

The efficiency with which actin-binding protein cross-links actin filaments suggests of the formation of a network in which the parallel alignment of actin filaments is minimal. The results of the sedimentation assay described above are also consonant with this conclusion, because of the efficiency with which small quantities of actin-binding protein link individual actin fibers into large aggregates which sediment rapidly.

Unlike many synthetic polymers for which the theory of gelation was derived, actin filaments are relatively straight and rigid [Nagashima and Asakura, 1980]. These properties are expected to cause filaments to line up into parallel bundles [Flory, 1956], although this stiffness is in part offset by the long lengths of actin filaments, since semirigid chains can approximate random coils if they are long enough. However, the efficiency of actin-binding protein in the gelation of stiff actin chains appears to reside in its large size and flexibility, and in its capacity to promote the *perpendicular* branching of actin fibers.

The large size and double helical structure of the actin filament require that the cross-linking molecule have substantial size and flexibility in order to be able to link these large polymers in a configuration other than as bundles. Macrophage actin-binding protein is a dimer composed of large flexible subunits, each having a contour length of 80 nm. The subunits are linked "head"-to-"head" leaving long flexible segments free to bind actin filaments (Fig. 1a). The large size and flexibility of the actin-binding protein dimers permit them to bind to specific domains of the actin filaments more or less wherever such filaments overlap at random. It is possible that this domain in the actin filaments is in the groove of the double helix [Hartwig and Stossel, 1981]. Since the binding affinities of actin-binding protein, filamin, and myosin for actin filaments are similar, and because all of these molecules are large, it has been theorized that the differences in actin cross-linking efficiency of these proteins might arise from variations in their flexibility [Hartwig and Stossel, 1981].

A major mechanism by which actin-binding protein produces a bundle-free actin network is the promotion of perpendicular branching of actin filaments. This idea was initially suggested to explain how actin-binding protein was so efficient in producing a rapidly sedimenting actin gel "particle." It was reasoned that this particle grew as a sphere and that spherical growth of the gel could arise from perpendicular branching [Stossel et al, 1979]. This idea was subsequently supported by direct observations with the electron microscope. When actin assembles in the presence of actin-binding protein, actin fibers branching with striking perpendicularity are observed in the electron microscope after negative staining or rotary shadowing of the specimens [Hartwig et al, 1980]. When actin is assembled in the presence of actin-binding protein, the actin-binding protein molecules nucleate the

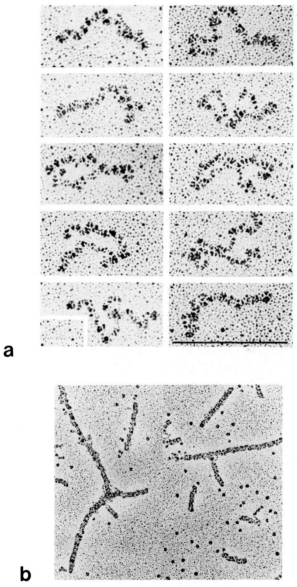

Fig. 1. Electron micrographs of actin-binding protein molecules alone (a) and of molecules cross-linking actin filaments at right angles (b). The specimens were rotary-shadowed with platinum. The bar in (a) represents 0.1 micron.

assembly of the actin filaments which then grow in a branching fashion off of the "tails" of the actin-binding protein dimers [Hartwig et al, 1980] (Fig. 1b). The mechanism by means of which actin-binding protein limits the growth of actin filaments to a perpendicular orientation is not currently known. However, this property of actin-binding protein indicates that the molecule has some restrictions on its flexibility. A minor variation in this flexibility parameter might explain why chicken gizzard filamin, which is structurally very similar to macrophage actin-binding protein [Wang, 1977; Kotelianski et al, 1982], is less efficient than macrophage actin-binding protein as an actin cross-linking protein [Brotschi et al, 1978] and does not promote the perpendicular branching of actin (Fig. 2) (Hartwig, unpublished results).

On the basis of evidence summarized above it can be concluded that the solidification of macrophage extracts is explained by the assembly of actin and the cross linking of the actin filaments by actin-binding protein. The efficiency with which actin-binding protein cross-links actin resides in its large size and flexibility and in its capacity to promote the perpendicular branching of actin.

The initial evidence that the interaction between actin and actin-binding protein was relevant for the structure of macrophage cortical cytoplasm was the finding that actin-binding protein resides in the cell periphery. Fragments of cortical cytoplasm bounded by plasma membrane can be isolated in high purity. Fifteen to 20 percent of the total protein of these fragments by weight is actin and about 1.5% of the protein is actin-binding protein [Davies and Stossel, 1977; Hartwig et al, 1977]. Subsequently actin-binding protein was localized by means of indirect immunofluorescence in the cortical cytoplasm

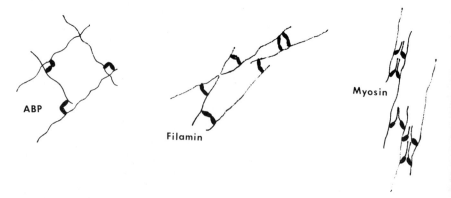

Fig. 2.   Schematic illustration of how differences in the shape of high molecular weight actin cross-linking proteins, which affect the angle at which they cross-link actin filaments, might account for variations in their actin cross-linking efficiency.

of spreading macrophages. Pseudopodia which the cells had extended to engulf yeast particles also contained actin-binding protein [Stendahl et al, 1980].

A preliminary analysis of the ultrastructure of spreading macrophages has revealed that the spread lamella of the cells contains a three-dimensional network of actin filaments which branch at right angles. Actin-binding protein molecules reside at the vertices of the branches.

The acquisition of this information was based on the detergent permeabilization technique originally applied to sperm by Gibbons and Gibbons [1972] and extended to other eukaryotic cells by others [eg, Clarke et al, 1975; Heuser and Kirschner, 1980; Schliwa et al, 1981]. Rabbit lung macrophages were permitted to spread on carbon- and Formvar-coated copper electron microscope grids for 45 minutes. The cells were then extracted with the nonionic detergent Triton X-100 in a buffered salt solution containing 2 mM magnesium chloride and the calcium chelating agent EGTA at a concentration of 5 mM. Under these conditions the morphology of the spread macrophages did not change when viewed in the light microscope. The detergent-permeabilized cells were then fixed, critical-point dried, and rotary-shadowed with platinum. The morphology of the spread lamella was analyzed in stereo pairs of electron micrographs.

The images obtained revealed that the lamella contains a three-dimensional array of actin filaments. The stereo views permitted us to ascertain that the filaments make contact with one another primarily by means of end-to-side interactions. It was possible to verify the impression that the cortical lamella contains an actin network that branches at close to perpendicular angles by means of a computer-assisted analysis of the branching angles in the stereo electron micrographs. Three fourths of the actin intersections are within 15° of a right angle. The average distance between junctions of filaments is 100 nm.

Also evident from these images was the fact that the principal axes of the actin filaments are parallel to and perpendicular to the plane of the substrate. This finding suggests that the initial assembly of actin within the cortex occurs along the substrate, and that this initial growth defines the axes of the branching network.

An additional observation was that the actin network is not uniformly distributed throughout the spread lamella. The perpendicular branching network is restricted to the two thirds of the volume of the lamella that is closest to the substrate. On top of the plane of the network nearer to the free surface of the lamella are small vesicles associated with sheets of material which had not been extracted by the detergent. Individual actin fibers project above this plane, and give the impression that they had supported the free surface of the plasma membrane which was removed by the detergent. The morphology of the lamella is summarized schematically in Figure 5.

As mentioned above, actin-binding protein nucleates actin assembly and promotes the formation of a perpendicular branching actin network in vitro. That such an interaction between actin and actin-binding protein may occur in the cortical cytoplasm of macrophages was suggested by the high resolution localization of actin-binding protein molecules in the cortical lamella. The detergent-extracted cells were reacted with anti-actin-binding protein IgG and subsequently with gold microbeads coated with staphylococcus protein A. The specimens were then prepared for electron microscopy as described above. The gold beads localized at the T-shaped intersections of actin filaments, establishing the existence of actin-binding protein molecules at these junctions.

Three fourths of the total macrophage actin and all of the actin-binding protein remained associated with the cells during extraction with detergent. The actin that was solubilized was either monomeric or complexed to some other small molecule such as profilin [Lindberg et al, 1979], as determined by gel filtration and ultracentrifugation of the extract. From these observations it can be concluded that the morphological appearance of actin in the detergent-extracted cell is representative of the bulk of actin in the cell. The findings also suggest that a sizable fraction of the total cell actin may be sequestered in a nonpolymerizing monomeric form.

In summary, the interaction between actin-binding protein and actin, initially discovered in extracts of macrophages, is responsible for the dominant structure of the cortex of spreading macrophages. The observations establish that a gel exists in cortical cytoplasm. The perpendicular branching configuration permits a maximum of extension of the network with a minimum of polymer mass. Furthermore, the interlocking of short, stiff, perpendicularly branched actin chains can generate considerable rigidity within the network beyond that generated by the cross linking of the filaments per se. The distances between actin filament junctions defines a pore size that is small enough to explain the exclusion of large organelles which characterizes the hyaline cortex.

To what extent can one generalize about the structure of cortical cytoplasm from what we have learned in studying the proteins of rabbit lung macrophages? The presence in cortical cytoplasm of abundant numbers of actin filaments has been known for over a decade [Ishikawa et al, 1969]. Proteins that are similar in structure and function to macrophage actin-binding protein have been purified from various nonmuscle cells including human polymorphonuclear leukocytes [Boxer and Stossel, 1976], human blood platelets [Schollmeyer et al, 1978; Rosenberg et al, 1981], cultured epithelial cells [Schloss and Goldman, 1979], and toad eggs (Corwin and Hartwig, unpublished results). In the case of the last example, the high molecular weight actin-binding protein accounted for the bulk of the actin-sedimenting activity of toad egg extracts. Schliwa and co-workers have detected the presence of actin filaments joined by end-to-side junctions in detergent-extracted cultured epithelial cells. Arrowheads formed by the binding of myosin subfragment one to the actin filaments invariably pointed to the vertexes of these T-shaped intersections [Schliwa and Van Blerkom, 1981; Schliwa, 1982]. This is the same orientation of the arrowheads as that detected on the branches of actin filaments assembled in the presence of purified macrophage actin-binding protein [Hartwig et al, 1980]. From this information it is likely that at least the portion of Porter's "microtrabecular network" [Porter and Tucker, 1981] which resides in the cell periphery is the perpendicular branching network of actin cross-linked by actin-binding protein.

III. THE REGULATION OF CORTICAL CYTOPLASMIC STRUCTURE

Of the many substances affecting the gelation of cytoplasmic extracts, two are of particular experimental and theoretical interest: the fungal metabolite, cytochalasin B, and the divalent cation, calcium. Both agents inhibited the gelation of cytoplasmic extracts from a variety of cells [reviewed by Taylor and Condeelis, 1979]. Cytochalasin B was interesting because this drug altered the shape of many cells and inhibited a large number of cell movements, including phagocytosis and the ruffling movements of macrophages [Allison et al, 1971]. Calcium attracted attention because of its role in excitation-contraction coupling in muscle cells and stimulus-secretion coupling in nonmuscle cells.

In the case of the macrophage, it was found that both compounds were acting on the gelation of actin-containing extracts by means of a similar mechanism. Cytochalasin B bound directly to actin filaments and caused the length of the filaments to shorten without reducing the total mass of filaments. This reaction was therefore equivalent to fragmentation of the actin [Hartwig and Stossel, 1979; Maruyama et al, 1980]. Calcium in submicromolar concentrations bound to gelsolin, a protein present in macrophage extracts which in turn bound to actin filaments and shortened them without significant depolymerization—ie, it fragmented them [Yin and Stossel, 1979, 1980; Yin et al, 1980]. Gelsolin is present in macrophage cytoplasm at a molar ratio to actin of 1:50. Moreover, as determined by immunofluorescence microscopy, gelsolin resides in the cortical cytoplasm [Yin et al, 1981b]. The fragmentation of actin filaments by cytochalasin B and by calcium in the presence of gelsolin is a reversible reaction. Lowering of the calcium concentration allows filaments shortened by the calcium-gelsolin complex to anneal. Removal of cytochalasin B from cytoplasmic extracts permits these extracts to solidify [Hartwig and Stossel, 1976], implying that the shortened filaments become capable of lengthening.

The power of fragmentation as a mechanism for regulating actin network structure is illustrated in Figure 3. For the purposes of the analysis, the regulation of network structure is depicted as the crossing of the gel point between sol and gel states. Alternatives for making this transition are shown in the figure. The reversible addition of cross links connecting the filaments would suffice, but this mechanism would operate at the cost of losing the perpendicular architecture of the filaments. Furthermore, there is no evidence that actin-binding protein alters its binding to actin filaments under physiological conditions tested thus far. Reversible fragmentation of actin is much faster than reversible polymerization of actin for changing filament length. Furthermore, reversible polymerization does not cause a transition across the gel point, although assembly is required to form polymers that can be cross-linked in the first place. The relationship between filament length and the gel point derived by Flory [1953] and shown in the figure illustrates this point. The molar concentration of cross linker required for incipient gelation (the gel point) is equal to the total concentration of monomer units in the polymers divided by the weight-average molecular weight of the filaments. If the filament length is changed by polymerization or depolymerization, the equation shows that the critical cross-linker concentration does not change. However, if the filament length is changed by reversible fragmentation, it is evident that a large alteration in length takes place with no change in the total numerator of the equation, and the critical cross-linker concentration varies accordingly.

The mechanisms by which cytochalasin B and the calcium-gelsolin complex fragment actin filaments is not clear. These compounds might actively fragment filaments, prevent the annealing of spontaneously breaking filaments, or act by both mechanisms. In any case, it has been shown that cytochalasin B and gelsolin-calcium bind to the "barbed" end of the shortened actin filaments when they are decorated with heavy meromyosin arrowheads [MacLean-Fletcher and Pollard, 1980; Yin et al, 1981a].

Another protein of macrophages that affects actin filament length was first isolated from human polymorphonuclear leukocytes. This protein was shown to account for the lower than expected viscosity of actin in leukocyte extracts containing salts which ordinarily promote the maximal assembly of purified actins [Southwick and Stossel, 1981]. This protein was subsequently purified from rabbit lung macrophages and found to lower the actin viscosity, because it bound to the "pointed" end of actin filaments decorated with heavy meromyosin arrowheads [Southwick and Hartwig, 1982]. Therefore, this protein was named acumentin (from the Latin for "on the point"). Acumentin is

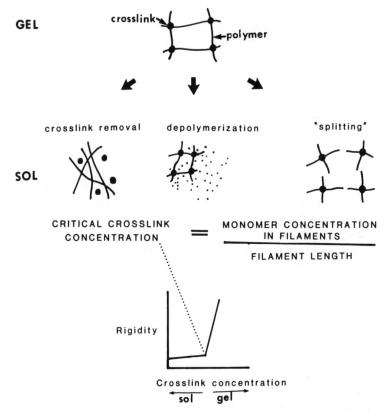

Fig. 3.   Regulation of actin gel structure in the macrophage. A gel is defined as linear polymers cross-linked into a three-dimensional network. In the case of actin gels cross-linked by actin-binding protein, the actin polymers branch at right angles. Solation of the gel occurs when the cross-link density falls below the critical point. This critical cross-link concentration is inversely related to the weight-average degree of polymerization as shown. Solation can be accomplished by removal of cross links, but at the expense of losing the perpendicular branching configuration of the polymers. Solation does not occur rapidly with depolymerization, as explained in the text. Fragmentation is the most efficient mechanism for solation.

present in high concentrations in macrophages, one mole of acumentin being present for every five actin monomers. However, the affinity of acumentin for the "pointed" end of actin filaments is much lower than that of the gelsolin-calcium complex for the "barbed" end of actin filaments [Yin et al, 1981a; Southwick and Hartwig, 1982; Southwick et al, 1982].

The differences between acumentin and gelsolin permit the proteins to act cooperatively and to confer a high degree of control over actin filament length. This control is shown schematically in Figure 4. Irrespective of whether these proteins are added to monomeric actin undergoing polymerization or to filamentous actin, the final filament length distribution is shortened, and the gelsolin-calcium complex and acumentin are bound respectively to the "barbed" and "pointed" ends of the shortened filaments. The covering of both ends of the filaments prevents the dissociation of monomers from the ends of filaments. It also inhibits the annealing of filaments. Following a fall in the free calcium concentration, gelsolin dissociates from the "barbed" ends of the actin filaments.

The free "barbed" ends have a higher affinity than acumentin for "pointed" filament ends [Southwick and Hartwig, 1982]. Therefore filaments anneal, and a longer filament length distribution emerges. The free acumentin concentration is expected to increase somewhat when filaments anneal. This rise would in effect buffer the extent of the filament length increase. Therefore, the filaments do not become as long as expected for pure F-actin. In the total control of actin filament length, gelsolin can be viewed as a regulator and acumentin as a buffer.

Whether macrophages also contain molecules that sequester actin monomers, thereby adding an additional dimension of control over the length

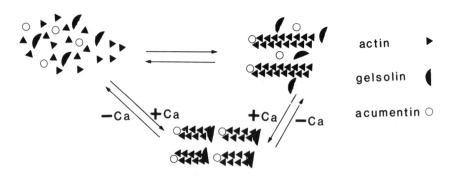

Fig. 4. Regulation of actin filament length in macrophage cytoplasm. Gelsolin-calcium and acumentin bind to the "barbed" and "pointed" ends of actin filaments respectively, either during or after assembly. The polarity of actin monomers in filaments is indicated by the arrowhead symbols depicting actin. Acumentin binds constitutively, but with relatively low affinity, to actin, resulting in a shortened length distribution. In the presence of submicromolar free calcium concentrations, gelsolin further shortens the length distribution. When the free calcium concentration is below this value, gelsolin is inactive.

and number of actin filaments in the cortical cytoplasm, remains to be determined. The reported existence of such molecules in other cells [eg, Lindberg et al, 1979] and the small fraction of nonsedimentable actin in detergent extracts of macrophages suggest that this may be the case.

IV. THE SPATIAL ORGANIZATION OF MACROPHAGE CORTICAL CYTOPLASM

The information summarized in the preceding sections of this discussion can be integrated to create a working hypothesis as to how the cortical actin network might form and move the peripheral cytoplasm of macrophages. The basic premise of the hypothesis is that the free calcium concentration regulates the state of assembly of the cortical cytoplasmic actin network and, secondarily, its structure and response to stress. The structural basis of the hypothesis is the reversible growth of the actin network under the influence of actin-binding protein, acumentin, and gelsolin. This activity in turn regulates the actin filament length as outlined in Figure 4.

The ultimate regulation of the spatial organization of the peripheral cytoplasm must reside in the control of free calcium levels in different regions of the actoplasm. There is relatively little information about this control at the present time. Macrophage plasma membranes possess a pump which transports calcium against a concentration gradient from the cytoplasm to the extracellular medium. The pump requires ATP and magnesium, and its activity is increased by calmodulin [Lew and Stossel, 1980]. Changes in the activity of localization of such a pump could be one factor controlling the free calcium in various domains of the actoplasm.

The only assumption required to concoct a mechanism for the assembly of the actin lattice from its components as a distinct and discrete structure in the macrophage cortex is that the free calcium concentration be high where the components are synthesized (Fig. 5). If actin-binding protein, myosin, gelsolin, and acumentin are made in the proximity of monomeric actin, the actin is exposed to four nucleating agents besides itself. This is especially true if the free calcium concentration is high enough for gelsolin to bind tightly to actin. The actin filaments formed in the presence of these nucleating proteins would be branched by actin-binding protein and be very short because of the full activation of gelsolin by calcium.

As the branched actin fragments complexed with acumentin and gelsolin diffuse into the cell periphery where membrane calcium pumps reduce the free calcium concentration, gelsolin becomes inactive, and the actin oligomers anneal. In addition, monomeric actin or actin complexed to profilin in this region can assemble onto barbed ends of filaments no longer tightly capped by gelsolin.

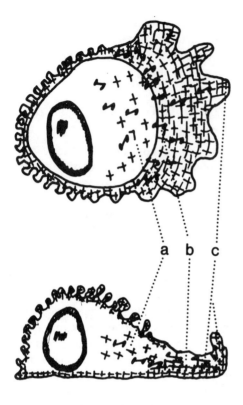

↬ myosin filament

+ lattice subunit

Fig. 5. Schematic drawing depicting how the cortical cytoplasm could arise from subunits and how a lamella of a macrophage might spread. Actin, and/or actin complexed with profilin, actin-binding protein, gelsolin, and acumentin are synthesized in domain (a) which is presumed to have a sufficiently high free calcium concentration to activate gelsolin and the myosin light-chain kinase-calmodulin complex. This activation leads to the formation of units composed of short actin filaments branched by actin-binding protein, capped by gelsolin and acumentin and bound to myosin filaments. After diffusion of these units to region (b), presumed to have a lower free calcium concentration, actin fragments anneal and/or monomers add to filaments, because "pointed" actin filament ends can bind to "barbed" ends previously capped by gelsolin. Myosin filaments may remain assembled in this "border" zone, because of kinetic considerations or a different calcium concentration dependence on assembly. In domain (c), presumed to have the lowest free calcium concentration, actin lattice assembly is maximal, and myosin, if present, is disaggregated into inactive monomers. Myosin filaments in region (b) tend to draw actin lattice subunits from the central (c) region and cortical actin fragments including fibers attached to the plasma membrane from the (b) region toward the (a) region. Myosin filaments attached to immobilized actin filaments with their "barbed" ends facing the leading edge may also move themselves centripetally. Note how the actin lattice-poor "channel" near the top surface of the leading lamella can serve as a conduit for the outward advance of lattice subunits.

The assembly of an actin network on a pedestal of preexisting actin network (or some other cellular structure) could generate a propulsive force. Propulsive movements produced by network assembly in the cortex require no special filament-membrane connections, because the growing actin network could simply push on the lipid membrane bilayer. The prediction follows from this idea that protrusions of cytoplasm will occur in regions where the free calcium concentration is relatively low. Furthermore, the calcium concentration is expected to be lowest at the tip of an advancing protrusion. One requirement for the propulsive mechanism proposed is that actin monomers or fragments must be available to build onto the preexisting scaffolding. These subunits could diffuse or be actively propelled from regions of the cytoplasm with higher free calcium concentrations.

Does assembly occur at the proximal or distal portion of the cortical lattice? Although proximal assembly apparently simplifies the job of transporting lattice subunits to the site of growth, the evidence concerning the protrusion of actin bundles in other cell types favors distal assembly [Tilney, this volume]. If the filaments grow distally off of uncapped "barbed" ends, filaments at the leading edge of an expanding cortex will have their "barbed" ends facing outward. This orientation of actin filaments is what has been found in cortices of various cells [Tilney, this volume].

The existence of myosin molecules in the cortical cytoplasm of macrophages suggests an additional "contractile" element in the movements of the cell periphery [Davies and Stossel, 1977; Hartwig et al, 1977; Stendahl et al, 1980; Painter et al, 1981]. The myosin molecules of these cells are hexameric proteins each containing two large and four small subunits. They are similar in structure to other vertebrate muscle and nonmuscle cell myosins, and are capable of generating contractile forces energized by ATP hydrolysis [Hartwig and Stossel, 1975; Stossel and Hartwig, 1976; Trotter and Adelstein, 1979]. The force-generating mechanism involves cyclic cross-bridge formation between globular actin binding domains in the two large myosin subunits and actin monomers in filaments. Reversible phosphorylation of two of the small myosin subunits by a kinase-phosphatase enzyme system probably regulates the rate of cross-bridge formation. The activity of the kinase enzyme may in turn be regulated in a positive fashion by the free calcium concentration and by the calcium-binding protein, calmodulin. In the presence of micromolar free calcium concentrations, a calmodulin-calcium complex binds to and activates the myosin light-chain kinase enzyme. This chain of events has been well documented in the case of vertebrate smooth muscle and platelets [Adelstein and Eisenberg, 1980]. The magnesium-ATPase activity of macrophage myosin in the presence of actin filaments, an activity correlated with cross-bridge movement, correlates with the extent of phosphorylation of the myosin light chains [Trotter and Adelstein, 1979], and calmodulin is a component of macrophage cytoplasm [Lew and Stossel, 1981]. Therefore, there is evidence for this activation mechanism in the case of macrophages.

Like most muscle and nonmuscle cell myosins, macrophage myosin forms bipolar filaments, structures in which helical "tail" domains of the large myosin subunits aggregate, leaving the globular actin-binding "heads" protruding from the ends of the aggregates. Scholey et al [1980] have reported that phosphorylation of some of the light chains of thymocyte myosin leads not only to increased actin-activated magnesium-ATPase activity of the myosin but also is a requirement for bipolar filament formation.

Synthesizing the available facts and adding some speculation, we might imagine macrophage myosin molecules to be in a dispersed state in cytoplasmic domains with low free calcium concentrations; they are thus inactive in cross-bridge formation with actin. This disaggregated and inactive disposition would permit myosin molecules to diffuse into regions of cytoplasm where the free calcium concentration is higher. Approaching these regions, the myosin molecules might be phosphorylated by calcium-dependent kinase, assemble into bipolar filaments, and begin to generate movements of actin filaments in that part of the lattice.

It is interesting that the control of myosin filament assembly by calcium is "positive" (ie, calcium promotes myosin filament assembly) and involves a calmodulin-dependent enzyme, myosin light-chain kinase, whereas the control of actin filament assembly by calcium is "negative" (calcium disaggregates actin filaments), and is conferred by gelsolin independently of calmodulin. A possible explanation for this different approach to control of assembly of actin and myosin is that it may be advantageous if the kinetics and extent of actin and myosin assembly in response to changes in calcium concentrations differ somewhat in order to position the filaments relative to one another in the proper configurations. For example, during the initial assembly of cortex (Fig. 5), a delay in myosin disaggregation, as actin filaments assemble centripetally, could preserve a region of "working" myosin filaments for some distance into the actin lattice. This domain of assembled myosin filaments could be involved in drawing lattice components toward the more tightly cross-linked regions of the lattice.

The predicted overall direction of actin filament movement is from regions of high to regions of low cytoplasmic free calcium concentrations. This prediction has been verified experimentally with a model system in which actin filaments were assembled in horizontal capillaries in the presence of skeletal muscle myosin filaments, macrophage gelsolin, and macrophage actin-binding protein. The ends of the capillaries were inserted into wells containing buffer solutions with either low or high free calcium concentrations. In the presence of ATP some of the protein mixtures contracted. When contraction occurred in the presence of a calcium gradient from one end of the capillary to the other, the direction of contraction was from high to low calcium concentrations [Stendahl and Stossel, 1980].

The effect of cross linking of the actin filaments by macrophage actin-binding protein on the contraction of actin by myosin was also shown in this system. Cross linking of actin filaments by actin-binding protein reduced the quantity of myosin required to produce a gross contraction. The results indicate that the amount of stress generated by the interaction of actin and myosin filaments under these conditions is sufficient to disentangle the actin filaments. Therefore, cross linking of the actin filaments by actin-binding protein is required in order for the contraction to propagate throughout a given volume. Myosin filaments may contribute to this cross linking, but they are probably less potent than actin-binding protein molecules in this activity [Brotschi et al, 1978]. The relative inefficiency of myosin molecules as actin cross linkers explains why high myosin:actin ratios are required to promote a contraction ("superprecipitation") of actomyosin solutions.

In the intact cell, the cortical actin network can be pictured as engaged in a "tug-of-war" around the cell periphery. The cortical actin network builds in areas of relatively low free calcium concentrations at the expense of actin network from regions of higher calcium concentration. This formulation can account for the "conservation of mass" of the intact cortex during cell movements which has been documented frequently [eg, Trinkaus, 1980]. This phenomenon, in which the appearance of new protrusions of cortex is precisely coordinated with the retraction of preexistent pseudopodia, has previously been likened to a tug of war [Weiss and Garber, 1952].

The dissolution of the cortical actin network in regions where the calcium concentration rises can lead to the collapse of overlying plasma membrane which is no longer supported by the actin network. In this case, the function of myosin is simply to transport the cortical cytosol. Because the membrane is a continuous structure, there is no need to pull it actively. It can move passively from places where the actin network is collapsing into enlarging pseudopodia where the network is building. On the other hand, segments of plasma membrane may also be pulled actively from local attachments of actin filaments which are being retracted by myosin toward regions where the actin network is growing. The assembly of the actin lattice close to the substrate-associated surface of the membrane would tend to immobilize that surface. Studies of lamellae of tissue culture cells in the interference contrast microscope demonstrate close contacts between the substrate and the plasma membrane in this region [Kolega et al, 1982]. In contrast, the top surface of the lamellae may be relatively unencumbered with actin, and is certainly not adherent to the substrate. This free surface of membrane could well be the mobile source of membrane for distal expansion of the lamellae being propelled by the assembly of actin network.

Selective exclusion is a property of gels, and variable organelle exclusion from the cell periphery is one of the first phenomena invoked as evidence for gel-sol transformations in cortical cytoplasm. Changes in the stability of the cortical actin network occurring in response to fluctuations in the free calcium concentration could regulate the size of the gel pores. Together with corresponding effects of these network variations on myosin-based contractions, the cell could generate movements of organelles. This mechanism of organelle movement has the simplicity of requiring no specific attachments between the organelle membranes and actin filaments. The lengths of actin filaments between junctions described above are sufficiently small to define pore sizes that would exclude organelles not ordinarily found in the cell periphery. The pore size of the actin network near the base of a lamella is also small enough to retard the movement of all but the smallest actin filaments, especially if they are branched by actin-binding protein. These pore aperture sizes would also restrict the movement of myosin filaments. Therefore, if actin lattice grows near the base of the lamella, actin monomers or the profilin-actin complex would have to serve as the principal source of material for lattice enlargement.

In summary, calcium gradients may regulate the direction of growth and the regional stability of a perpendicular branching actin network in the cortex of macrophages. Retractive forces exerted on the actin filaments by myosin draw actin filaments from regions of lower network stability to areas with greater strength. In this way cortical actin flows to domains where it can generate new pseudopodia. Plasma membrane can be pushed by propulsive forces generated by the growth of the lattice and collapse upon dissolution of the network. In some cases, actin filaments may be attached to membrane-binding sites and actively pull the membrane. Changes in the pore size and flow of the cortical actin network may move the internal organelles. Porter's "microtrabecular network" in the cell cortex is, as postulated many years ago, a contractile gel. We have now defined many of its components and can begin to determine how it works in the living cell.

ACKNOWLEDGMENTS

The work summarized in this review was supported by grants from the National Institutes of Health, the Council for Tobacco Research, the American Heart Association, the Muscular Dystrophy Association, the E. S. Webster Foundation, and E. W. Hiam.

Comments by J. Scholey concerning this paper are gratefully acknowledged.

V. REFERENCES

Adelstein RS, Eisenberg E (1980): Regulation and kinetics of the actin-myosin-ATP interaction. Ann Rev Biochem 49:921–956.

Allison AC, Davies P, DePetris S (1971): Role of contractile microfilaments in macrophage movement and endocytosis. Nature New Biol 232:153–155.

Bensley RR (1938): Plasmosin. The gel- and fiber-forming constituent of the protoplasm of the hepatic cell. Anat Rec 72:351–369.

Boxer LA, Stossel TP (1976): Interactions of actin, myosin and an actin-binding protein of chronic myelogenous leukemia granulocytes. J Clin Invest 57:964–976.

Brotschi EA, Hartwig JH, Stossel TP (1978): The gelation of actin by actin-binding protein. J Biol Chem 253:8988–8993.

Clarke M, Schatten G, Mazia D, Spudich JA (1975): Visualization of actin fibers associated with the cell membrane in amoeba of Dictyostelium discoideum. Proc Natl Acad Sci USA 72:1758–1762.

Davies WA, Stossel TP (1977): Peripheral hyaline blebs (podosomes) of macrophages. J Cell Biol 75:941–955.

Flory PJ (1953): "Principles of Polymer Chemistry." Ithaca, New York: Cornell University Press.

Flory PJ (1956): Statistical thermodynamics of semi-flexible chain molecules. Proc R Soc (Lond) Ser A 234:60–72.

Gibbons BH, Gibbons IR (1972): Flagellar movement and adenosine triphosphatase activity in sea urchin sperm extracted with Triton X-100. J Cell Biol 54:75–97.

Hartwig JH, Davies WA, Stossel TP (1977): Evidence for contractile protein translocation in macrophage spreading, phagocytosis and phagolysosome formation. J Cell Biol 75:956–967.

Hartwig JH, Stossel TP (1979): Cytochalasin B and the structure of actin gels. J Mol Biol 134:539–554.

Hartwig JH, Stossel TP (1976): Interactions of actin, myosin and an actin-binding protein of rabbit pulmonary macrophages. III. Effects of cytochalasin B. J Cell Biol 71:295–303.

Hartwig JH, Stossel TP (1975): Isolation and properties of actin, myosin and a new actin-binding protein in rabbit alveolar macrophages. J Biol Chem 250:5699–5707.

Hartwig JH, Stossel TP (1981): Structure of macrophage actin-binding protein molecules in solution and interacting with actin filaments. J Mol Biol 145:563–581.

Hartwig JH, Tyler J, Stossel TP (1980): Macrophage actin-binding protein promotes the bipolar and branching polymerization of actin. J Cell Biol 87:841–848.

Heuser JE, Kirschner MW (1980): Filament organization revealed in platinum replicas of freeze-dried cytoskeletons. J Cell Biol 86:212–234.

Ishikawa H, Bischoff R, Holtzer H (1969): Formation of arrowhead complexes with heavy meromyosin in a variety of cell types. J Cell Biol 43:312–328.

Kasai M, Kawashima H, Oosawa F (1960): Structure of F-actin solutions. J Polymer Sci 44:51–69.

Kolega J, Shure MS, Chen W-T, Young ND (1982): Rapid cellular translocation is related to close contacts formed between various cultured cells and their substrata. J Cell Sci 54:23–34.

Kotelianski VE, Glukhova MA, Shirinsky VP, Smirnov VN, Bushueva TL, Filimonov VV, Venyaminov SY (1982): A structural study of filamin, a high-molecular weight actin-binding protein from chicken gizzard. Eur J Biochem 121:553–559.

Lew PD, Stossel TP (1980): Calcium transport by macrophage plasma membranes. J Biol Chem 255:5841–5846.

Lew PD, Stossel TP (1981): Effect of calcium on superoxide production by phagocytic vesicles from rabbit alveolar macrophages. J Clin Invest 67:1–9.

Lewis WH (1939): The role of a superficial plasmagel layer in changes of form, locomotion and division of cells in tissue cultures. Arch Exp Zellforsch 23:1–7.

Lindberg U, Carlsson L, Markey F, Nystroem LE (1979): The unpolymerized form of actin in nonmuscle cells. Method Achiev Exp Pathol 8:143–170.

MacLean-Fletcher S, Pollard TD (1980): Mechanism of action of cytochalasin B on actin. Cell 20:329–341.

Maruyama K, Hartwig JH, Stossel TP (1980): Cytochalasin B and the structure of actin gels. II. Further evidence for the splitting of F-actin by cytochalasin B. Biochim Biophys Acta 626:494–500.

Nagashima H, Asakura S (1980): Dark-field light microscopic study of the flexibility of F-actin complexes. J Mol Biol 136:169–182.

Painter RG, Whisenand J, McIntosh AT (1981): Effects of cytochalsin B on actin and myosin associated with particle binding sites in mouse macrophages: Implications with regard to the mechanism of action of the cytochalasins. J Cell Biol 91:373–384.

Porter KR, Tucker JB (1981): The ground substance of the living cell. Sci Am 57–67.

Reaven EP, Axline SG (1973): Subplasmalemmal microfilaments and microtubules in resting and phagocytizing cultivated macrophages. J Cell Biol 59:12–27.

Rosenberg S, Stracher A, Lucas RC (1981): Isolation and characterization of actin and actin-binding protein from human platelets. J Cell Biol 91:201–211.

Schliwa M (1982): Action of cytochalasin D on cytoskeletal networks. J Cell Biol 92:79–91.

Schliwa M, Van Blerkom J (1981): Structural interaction of cytoskeletal components. J Cell Biol 90:222–235.

Schliwa M, Van Blerkom J, Porter KR (1981): Stabilization of the cytoplasmic ground substance in detergent-opened cells and a structural and biochemical analysis of its composition. Proc Natl Acad Sci USA 78:4329–4333.

Schloss JA, Goldman RD (1979): Isolation of a high molecular weight actin-binding protein from baby hamster kidney (BHK-21) cells. Proc Natl Acad Sci USA 76:4484–4488.

Scholey JM, Taylor RA, Kendrick-Jones J (1980): Regulation of nonmuscle myosin assembly by calmodulin-dependent light chain kinase. Nature 287:233–234.

Schollmeyer JV, Rao GHR, White JG (1978): An actin-binding protein in human platelets. Interactions with alpha-actinin on gelation of actin and the influence of cytochalasin B. Am J Pathol 93:433–446.

Southwick FS, Hartwig JH (1982): Acumentin, a protein in macrophages which caps the "pointed" end of actin filaments. Nature 297:303–307.

Southwick FS, Stossel TP (1981): Isolation of an inhibitor of actin polymerization from human polymorphonuclear leukocytes. J Biol Chem 256:3030–3036.

Southwick FS, Tatsumi N, Stossel TP (1982): Acumentin, an actin modulating protein of rabbit pulmonary macrophages. Biochemistry 21:6321–6326.

Stendahl OI, Hartwig JH, Brotschi EA, Stossel TP (1980): Distribution of actin-binding protein and myosin in macrophages during spreading and phagocytosis. J Cell Biol 84:215–224.

Stendahl OI, Stossel TP (1980): Actin-binding protein amplifies actomyosin contraction, and gelsolin confers calcium control on the direction of contraction. Biochem Biophys Res Commun 92:675–681.

Stossel TP, Hartwig JH (1975): Interactions of actin, myosin and an actin-binding protein of rabbit alveolar macrophages. Macrophage myosin Mg-adenosine triphosphatase requires a cofactor for activation by actin. J Biol Chem 250:5706–5712.

Stossel TP, Hartwig JH (1976): Interactions of actin, myosin and a new actin-binding protein of rabbit pulmonary macrophages. II. Role in cytoplasmic movement and phagocytosis. J Cell Biol 68:602–619.

Stossel TP, Hartwig JH, Yin HL, Davies WA (1979): Actin-binding protein. In Hatano S, Ishikawa H, Sato H (eds): "Cell Motility: Molecules and Organization." Tokyo: University of Tokyo Press, pp 189–209.

Stossel TP, Hartwig JH, Yin HL, Zaner KS, Stendahl OI (1982): Actin gelation and the structure of cortical cytoplasm. Cold Spring Harb Symp Quant Biol 46:569–578.

Taylor DL, Condeelis JS (1979): Cytoplasmic structure and coutractility in ameboid cells. Int Rev Cytol 56:57–144.

Trinkaus JP (1980): Formation of protrusions of the cell surface during tissue cell movement. In "Tumor Cell Surfaces and Malignancy." New York: Alan R Liss, pp 887–906.

Trotter JA, Adelstein RS (1979): Macrophage myosin. Regulation of actin-activated ATPase activity by phosphorylation of the 20,000-dalton light chain. J Biol Chem 254:8781–8785.

Wallach D, Davies PAJ, Pastan I (1978): Purification of mammalian filamin. Similarity to high molecular weight actin-binding protein in macrophages, platelets, fibroblasts and other tissues. J Biol Chem 253:3328–3335.

Wang K (1977): Filamin, a new high-molecular-weight protein found in smooth muscle and non-muscle cells. Biochemistry 16:1857–1865.

Weiss P, Garber B (1952): Shape and movement of mesenchyme cells as functions of the physical structure of the medium. Contributions to a quantitative morphology. Proc Natl Acad Sci USA 38:264–280.

Yin HL, Albrecht JH, Fattoum A (1981a): Identification of gelsolin, a Ca-dependent regulatory protein of actin gel-sol transformation and its intracellular distribution in a variety of cells and tissues. J Cell Biol 91:901–906.

Yin HL, Hartwig JH, Maruyama K, Stossel TP (1981b): Ca^{++} control of actin filament length. Effects of macrophage gelsolin on actin polymerization. J Biol Chem 256:9693–9697.

Yin HL, Stossel TP (1979): Control of cytoplasmic actin gel-sol transformation by gelsolin, a calcium-dependent regulatory protein. Nature 281:583–586.

Yin HL, Stossel TP (1980): Purification and structural properties of gelsolin, a Ca^{++}-activated regulatory protein of macrophages. J Biol Chem 255:9490–9493.

Yin HL, Zaner KS, Stossel TP (1980): Ca control of actin gelation. Interaction of gelsolin with actin filaments and regulation of actin gelation. J Biol Chem 255:9494–9500.

Zaner KS, Fotland R, Stossel TP (1981): A low-shear, small volume viscoelastometer. Rev Sci Instrum 52:85–87.

Zaner KS, Stossel TP (1982): Some perspective on the viscosity of actin filaments. J Cell Biol 93:987–991.

Modern Cell Biology 2:225–240
© 1983 Alan R. Liss, Inc., 150 Fifth Avenue, New York, NY 10011

Rheological Properties of Cytoplasm:
Significance for the Organization of Spatial Information and Movement

John Condeelis

From the Department of Anatomy, Albert Einstein College of Medicine, Bronx, New York 10461

I. INTRODUCTION

During the past decade our concept of cytoplasmic structure has changed enormously. It is now apparent that the so-called cytosol is not a sol but rather a semisolid scaffold on which many events occur, mediated by the "free" components of the cytoplasm. This scaffold is such a ubiquitous component of the ground substance of cells that Porter and his colleagues have given it a name, the microtrabecullar lattice, and attribute to it the status of an organelle [Porter, 1976; Wolosewick and Porter, 1979]. The implications of the scaffold concept of cytoplasmic organization become apparent when we realize that much of this structure is composed of various polymeric forms of the cytoskeletal proteins actin and tubulin. Since these proteins are involved in the determination of cell shape, organelle location, and motility, the scaffold concept suggests a mechanism by which the cell can integrate the spatial distribution of membrane bounded organelles and cell motility with the metabolic capabilities of "free" cytoplasm.

It is my job here to consider one of these relationships: that between cytoplasmic structure and cell motility. This is fortunate since the most dramatic examples of correlations between cytoplasmic structure and cell function come from experiments that were designed to unravel the mechanisms of cell motility.

II. CYTOPLASMIC STRUCTURE AND MOVEMENT

Some of the most remarkable examples of this relationship can be found in the work of Chambers [Chambers and Chambers, 1961]. Using micromanipulation he was able to demonstrate that during cytokinesis, the solidity of cortical cytoplasm in the region where the cleavage furrow is about to form undergoes first a decrease and then an increase—ie, solation followed by gelation. He showed that this gelled region of cytoplasm persists until cleavage is complete, even if the daughter cells are mechanically lysed, releasing the remainder of the cytoplasm in the form of a sol. These experiments demonstrated that cytoplasm has structural variation, and that the variation can be correlated with discrete cell functions. Subsequent generations of investigators have used combinations of micromanipulation, centrifuge microscopy, and polarized light microscopy to define regions of structural variation within cytoplasm with more precision and to correlate these with a variety of cell movements, as reviewed elsewhere [Taylor and Condeelis, 1979]. However, an understanding of the interrelations between movement and structure, particularly during cytoplasmic streaming and amoeboid movement, awaited the discovery of nonmuscle actin and its associated proteins.

Since actin comprises such a large percentage of the cell's protein, it was thought initially that polymerization of actin alone might be sufficient to explain the changes in cytoplasmic viscosity that accompany cell movement [Pollard and Ito, 1970]. It became apparent however, that the type of interaction between actin and myosin filaments found during rigor mortis could also account for the rather pronounced changes in cytoplasmic structure and, in particular, the development of structures that are capable of subsequent contraction [Taylor et al, 1973]. Subsequently, numerous laboratories began to discover actin binding proteins (ABPs) that are capable of causing dramatic changes in the organization of actin, and hence cytoplasmic viscosity, by

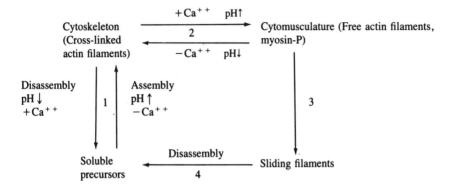

Fig. 1.   Possible steps in the cycle of contraction of a non-muscle cell. The events at each step and proteins involved are detailed in the text.

promoting filament growth, cross-linking, severing, and depolymerization. The number of such proteins has now reached epidemic proportions, and the rate of discovery of new ones shows no sign of abating [Craig and Pollard, 1982].

It is not yet clear how to classify these proteins in terms of their functions during cell motility, in part because of ignorance of where they reside in vivo and in part because of difficulty in reconciling the multiple activities that many of these proteins exhibit in vitro with discrete motile events in vivo. In this chapter I would like to attempt a simple classification based on how these proteins could be used to achieve a cycle of contraction in a nonmuscle cell and in so doing identify some of the most likely events in which these proteins might function to generate cell movement.

III. CYCLICAL CONTRACTION

The event under discussion might be any of several. Consider, for example, cytokinesis, which is thought to involve the assembly of an actin containing structure, conversion of this structure into a form capable of generating force, shortening via a sliding of filaments employing myosin and ATP, and, finally, disassembly (Fig. 1). Such an event could also be cyclic, involving all of the above steps repeated many times to achieve an on-going motion like amoeboid movement.

IV. STEP 1—ASSEMBLY

Assembly of actin-containing structures within the cytoplasm of nonmuscle cells may involve two steps: nucleation and growth of filaments from definite nucleation centers in the cell (nucleation), and subsequent organization of these filaments into structures which, because of their geometry, are capable of accomplishing the intended function (organization). Assembly may be a discrete event in both space and time, as exemplified by the assembly of the contractile ring in the cortex of mitotic cells [Schroeder, 1975], assembly of the acrosomal process in Thyone sperm during fertilization [Tilney, 1976], and extension of the mt$^+$ mating structure in Chlamydomonas during cell fusion to form the zygote [Detmers, 1982]. Assembly may also occur with much less precision in both space and time—for example, in advancing regions of the cortex of motile cells or in the entire cell cortex during ligand-induced redistribution of cell surface components.

A. Nucleation

How cells regulate the precise location and directionality of actin polymerization is unknown, but several actin binding proteins (ABPs) have been isolated that will nucleate assembly in vitro. Macrophage ABP [Hartwig et al, 1980] and Dictyostelium 120K protein [Condeelis, 1981, Condeelis et al, 1982] have been shown to nucleate the polymerization of actin, resulting in the formation of branched filaments where the structural polarity of the filaments, as seen by decoration with heavy meromyosin (HMM), points toward the intersection of the two filaments. Electron microscopy and viscometry suggest that the ABP resides at the branch point and cross-links the two filaments together. It is not known if the ABP caps the pointed end of the actin filament at the branch point to inhibit monomer addition there. However, nucleation of filament growth by these ABPs results in a population of cross-linked filaments which are elongating by addition of subunits to their free ends—ie, the ends that appear "barbed" after decoration with HMM. The concentration of macrophage ABP and Dictyostelium 120K protein in advancing regions of the cell cortex, as demonstrated by immunofluorescence [Stendahl et al, 1980; Condeelis et al, 1981], suggests that nucleation of new filament growth may occur in these regions of the cell. The availability of numerous filaments with free barbed ends could be used to advantage by the cell to achieve the barbed end filament-membrane interaction seen in many cells. Unfortunately, the role that these ABPs play in nucleating growth of actin filaments in vivo remains uncertain, owing to a lack of information concerning their location in situ at high resolution and the mechanism by which their interaction with actin is regulated.

Additional ABPs that may bind to the pointed end of actin filaments and cause nucleation are the β-actinin's from skeletal muscle [Maruyama et al,

1977] and rat kidney [Maruyama and Sakai, 1981]. Both immunofluorescence in situ and viscometric data in solution are consistent with the interpretation that muscle β-actinin binds at or near the pointed end of the actin filament. In addition, a protein called acumentin isolated from macrophages [Stossel, this volume] has been suggested to bind at or near the end of actin filaments. However, these proteins neither cross-link nor branch actin filaments, so they are functionally distinct from macrophage ABP and Dictyostelium 120K protein.

Structures that are visible in the electron microscope, that contain proteins like macrophage ABP or Dictyostelium 120K protein, and that act as nucleating sites for filament growth have not yet been found in situ. However, the nucleation of new filament growth from discrete, cytoplasmic structures has been observed in two cases.

In Thyone sperm, actin filaments in the acrosomal process grow from the actomere [Tilney, 1978]. In Chlamydomonas actin filaments in the mt^+ mating structure grow from the doublet zone located in the cell cortex [Detmers, 1982]. In both cases nucleation results in a population of actin filaments that grow at their "barbed" ends only. These free, growing ends extend out toward and presumably push against the cell membrane whereas the pointed end of the filament remains anchored in the nucleating structure. The complete chemical identity of these nucleating structures remains unknown.

There is an additional group of proteins that also nucleates the assembly of actin filaments in vitro. Acanthamoeba capping protein [Isenberg et al, 1980], villin [Bretscher and Weber, 1980; Mooseker et al, 1980], gelsolin [Yin et al, 1980], and fragmin [Hasegawa et al, 1980] appear to bind to the barbed end of actin dimers or trimers, stabilizing them as nuclei for actin polymerization. This results in filament growth from the pointed end—ie, opposite to the direction discussed above. Villin, gelsolin, and fragmin require micromolar Ca^{++} to induce nucleation; at submicromolar Ca^{++} they are ineffective. Villin is unique among these proteins in that it aggregates actin filaments into bundles when the Ca^{++} concentration falls below micromolar. In fact, villin appears to be restricted in occurrence to the microvilli of epithelial cells [Bretscher and Weber, 1978] where it acts to cross-link filaments and form the microvillar core bundles. Nucleation by Acanthamoeba capping protein is not regulated by Ca^{++} under any known conditions. Hence, these proteins are functionally heterodisperse; their common characteristic is the ability to bind to the barbed end of actin filaments.

It is not clear if these proteins that nucleate microfilament growth at the pointed end in vitro, while blocking the barbed end to monomer addition, actually function as nucleation sites in vivo. Since the preferred end of filament growth is the barbed end, more rapid filament growth would tend to occur from nucleation sites containing components like macrophage ABP or the Dictyostelium 120K protein, even at actin monomer concentrations exceeding those in equilibrium with the pointed end of the filament. Furthermore, at monomer concentrations intermediate between those in equilibrium with the pointed and the barbed ends of the filament, net filament assembly would only occur on filaments initiated from sites containing proteins that either lower the critical concentration at the pointed end or cap the pointed end—eg, macrophage ABP and Dictyostelium 120K protein. In this case, growth would only be at the barbed end. Hence, the ability of barbed-end capping proteins to act as nucleation sites in vivo would depend on the local concentration of free monomer over a narrower range than if nucleation were from the barbed end. It seems more likely that these barbed-end capping proteins function to sever filaments during disassembly of actin structures and/or to link the barbed ends of filaments to other components in the cytoplasm as discussed below.

B. Organization

Once actin filaments have been nucleated from ABPs at specific sites within the cell, additional information must be superimposed upon the filaments to ensure that the resulting actin-containing machinery has the correct geometry and linkage relative to the structures that are to be immobilized or moved.

1. Geometry. The particular function exhibited by an actin-containing structure depends in large part on the geometry of the filaments within the structure. The contractile ring is a good example. Filaments are assembled and then cross-linked into a geometry (antiparallel, circumferential) that determines the function of the ring during filament sliding. ABPs that cross-link actin filaments into antiparallel arrays, but can be released from actin to permit sliding under conditions which induce contraction, are likely to be crucial in organizing actin filaments to generate force via a sliding filament mechanism with myosin. ABPs that cross-link filaments to form branches or meshworks such as macrophage ABP, filamin, and Dictyostelium 120K protein result in geometries that are likely to impede force generation by a sliding filament mechanism. This idea is consistent with the observations that both filamin [Davies et al, 1977] and Dictyostelium 120K protein [Condeelis et al, 1982] inhibit actin-activated HMM-ATPase, and that 120K protein inhibits superprecipitation of actomyosin [Vahey and Condeelis, 1983].

Nonmuscle α-actinin [Burridge and Feramisco, 1981], actinogelin [Mimura and Asano, 1979], Acanthamoeba 90K protein [Pollard, 1981], and Dictyostelium 95K protein [Condeelis and Vahey, 1982] are all molecules with the ability to cross-link actin filaments at submicromolar Ca^{++} concentrations. Cross-linking is suppressed as the Ca^{++} concentration is raised to micromolar levels, a condition that has been shown to dissociate Dictyostelium 95K protein from actin. Like muscle α-actinin, Dictyostelium 95K protein increases the rate of superprecipitation of actomyosin in vitro, but unlike muscle α-actinin this phenomenon is much more pronounced if actomyosin is first incubated with 95K protein in the absence of Ca^{++}, and then superprecipitation is allowed to proceed in the presence of micromolar Ca^{++}. The mechanism by which 95K protein accomplishes this enhancement is unknown, but if 95K protein cross-links actin filaments into antiparallel arrays in the absence of Ca^{++}, the efficiency of sliding between actin and myosin filaments might be increased. Subsequent addition of Ca^{++} to inhibit the interaction between 95K protein and actin would break the cross links between actin filaments that would otherwise resist sliding. Another possibility is that 95K protein might act to space actin filaments at a convenient distance, which would then permit myosin to penetrate the lateral array of filaments.

Clearly, more work is needed to determine the mechanism by which these α-actinin-like proteins organize actin into structures that can interact with myosin to produce force.

2. Linkage. Sliding filament interactions between actin and myosin can be used to generate meaningful movement only if the filaments are linked to specific structures that are to be moved. These structures include the cell and cytoplasmic membranes, microtubules, and intermediate filaments. The majority of current evidence (which is primarily morphological) indicates that actin forms the primary link with these structures whereas myosin is usually associated with them indirectly through its interaction with actin. Since actin-containing structures and their links with other cellular components are transient in nonmuscle cells, the links may play an important role in determining the site and polarity of actin filament assembly. Actin can also form a rigid gel that resists deformation and contraction via its interaction with various ABPs. Therefore, links between actin filaments and various cellular structures could be used to lock the position of these structures within the cytoskeleton, thereby assuring spatial specificity of organelle location, cell shape, and the topography of integral membrane components.

The most common link seen between actin filaments and various cell structures is that between microfilaments and membranes. The polarity of filaments relative to the membrane as revealed by HMM decoration is usually away from the membrane, suggesting that the barbed end of the filament is attached to the membrane. In several well-documented cases, however, actin filaments are found in lateral association with the cell membrane. In microvilli of intestinal epithelial cells, lateral links between actin filaments in the core bundle and the cell membrane have been identified [Mooseker and Tilney, 1975]. A protein measuring 110,000 daltons on SDS-PAGE has been implicated in this association [Matsudaira and Burgess, 1979]. In human erythrocytes actin oligomers containing as many as 30 subunits [Atkinson et al, 1982] are linked to the plasma membrane via the spectrin-ankyrin–band 3 complex [Branton et al, 1981]. Spectrin binds to the side of the actin oligomer, not to the pointed or barbed end. It is not known how general this mechanism for attachment of polymeric actin to membranes might be, but the recent discovery of spectrinlike molecules in epithelia [Glenney et al, 1982] and brain [Levine and Willard, 1981] and the wide distribution of immunoreactive analogues to spectrin [Lazarides, this volume] suggest that a spectrin-mediated actin-membrane type of interaction is general.

The evidence for attachment of the barbed end of the actin filament to membranes is currently morphological. Hence, it is possible that many of these barbed end attachment sites are poorly resolved lateral attachments, similar to those occurring in microvilli and erythrocytes. Nonetheless, the discovery of ABPs that bind to the barbed end of actin filaments in vitro has made it tempting to suggest that some of these proteins may mediate membrane binding to filament ends. Gelsolin [Yin et al, 1980], fragmin [Hasegawa et al, 1980], and Dictyostelium 40K protein [Brown et al, 1982] are all thought to be capable of binding to the barbed end of actin filaments in vitro in the presence but not the absence of Ca^{++}. The cytoplasmic distribution of gelsolin is not consistent with the idea that this protein is located exclusively on the cytoplasmic surface of cell and cytoplasmic membranes [Yin et al, 1981], but the distribution of the other proteins is not yet known. A more serious objection to the idea that these proteins mediate the interaction between actin and membranes is the finding that their binding to the barbed end occurs only in the presence of micromolar Ca^{++}, a situation likely to occur only in regions of the cell where solation of actin gels and cytoplasmic contraction are in progress. There are additional proteins such as Acanthamoeba capping protein that bind to the barbed end of actin filaments regardless of the Ca^{++} concentration [Isenberg et al, 1980]. This sort of protein is a more likely candidate for mediating the interaction between actin filament ends and membranes which might occur in situ at low Ca^{++} concentrations.

Vinculin, a protein first isolated from smooth muscle, has been proposed as a mediator for the attachment of actin filaments to the plasma membrane based on immunolocalization at the electron microscope level [Geiger et al, 1981]. The ability of this protein to bind to the barbed end of actin filaments has been inferred from viscometric and binding studies. It has been suggested that vinculin might mediate barbed end binding to membranes [Wilkins and Lin, 1982], but its ability to bundle actin filaments in vitro suggests that its function in the cell cortex may be much more complex [Jockusch and Isenberg, 1981].

In summary, there is no compelling evidence at the molecular level that actin filaments are bound to membranes at their barbed ends. Further work on various ABPs that bind to this region of the actin filament, particularly regarding the location of these proteins in situ and their ability to bind to membrane-associated components in vitro, will be required to resolve this question.

Morphological evidence for an interaction between actin filaments and microtubules has been reported [Griffith and Pollard, 1978]. Biochemical evidence suggests that this interaction may be mediated by microtubule-associated proteins. The finding that ankyrin, the spectrin and band 3 binding protein of erythrocytes, has immunoreactive analogues that are localized on microtubules in a variety of cells and that ankyrin competes with microtubule-associated proteins for binding to microtubules [Bennett and Davis, 1981] suggests mechanisms by which microtubules could bind to both membranes and microfilaments. Since ankyrin binds to spectrin (an ABP) in human erythrocytes, its counterpart in nucleated cells might mediate the association of spectrinlike ABPs with microtubules which would then be capable of binding to actin filaments.

V. STEPS 2 AND 3—CONVERSION OF CYTOSKELETON TO CYTOMUSCULATURE

Some types of cytoskeletal structures are unsuitable for use in the generation of contractile force via a sliding filament mechanism involving myosin, owing to the geometry of their actin filaments, the extent of cross-linking, and/or the length of the actin filaments. Two extreme examples are the actin filament bundle of the Limulus acrosomal process and the spectrin-actin complex on the erythrocyte membrane. Other actin-containing structures, such as the cortical gel of amoeboid cells and stress fibers in cultured vertebrate cells, are capable of acting either as cytoskeletal elements to constrict random movement of organelles and determine cell shape, or as cytomuscular elements to generate contractile force to move the cell and its contents. Hints as to how actin is reversibly shuttled between these two roles has come from experiments with cell extracts and more recently from work with isolated ABPs. Work with cell extracts that are enriched for actin and various ABPs demonstrated that 1) the formation of actin gels requires the presence of ABPs; 2) such gels inhibit superprecipitation of actomyosin; and 3) that solation of the gel, produced either by micromolar Ca^{++}, pH above 7.2, pressure, or even cytochalasin B, induces superprecipitation of actomyosin [Condeelis and Taylor, 1977]. A mechanism was proposed in which cross-linking of actin filaments by ABPs to form the gel inhibited their sliding interaction with myosin and hence the contraction of the gel. This is called the solation-contraction-coupling hypothesis, and is described in detail elsewhere [Condeelis, 1981a]. Inhibition of cross-linking of actin filaments by micromolar levels of Ca^{++}, a physiological trigger, would then produce both solation and contraction. Recent purification of the α-actinin-like ABPs that cross-link actin filaments in the absence of Ca^{++} but not at micromolar concentrations of Ca^{++}, as described above, suggests a molecular mechanism by which solation and contraction might be coupled.

Additional steps in a conversion of the cytoskeleton to cytomusculature are probably required. One of these is certainly phosphorylation of myosin light chains by light chain kinase in response to an increase in Ca^{++} concentration to micromolar levels. Light chain phosphorylation results in the stimulation of actin-activated myosin ATPase and in thick filament formation [Adelstein and Eisenberg, 1980; Scholey et al, 1980]. (However, the regulation of myosin activity by phosphorylation is more controversial than indicated here: Witness the varied results regarding the effect of phosphorylation of myosin heavy chain on the activity of myosin isolated from Acanthamoeba [Collins and Korn, 1980], Dictyostelium [Kuczmarski and Spudich, 1980], and Physarum [Ogihara et al, 1983]. A discussion of this complexity is beyond the scope of this chapter.)

VI. STEP 4—DISASSEMBLY

Contraction of cytoplasm is not reversed by the pull of cytoplasmic structures analogous to an antagonistic muscle. Reversal of contraction in nonmuscle cells appears to be accomplished by a disassembly of the actin-containing structure that was used to produce the force for movement. The clearest example of such a situation is the behavior of the contractile ring during and after cytokinesis. This structure is assembled, shortens circumferentially, and, during shortening, disassembles [Schroeder, 1975]. Studies on myosin light chain kinase, on solation in cell extracts, and on contraction of whole cytoplasm in vitro agree that micromolar Ca^{++} is required to induce contraction in nonmuscle cells. Therefore, a single signal, the rise in free Ca^{++} to micromolar levels, may cause three events which are separable on the molecular level: first, conversion of cytoskeleton to cytomusculature, involving the release from actin of ABPs such as Dictyostelium 95K protein that organize filaments into a form capable of generating force via a sliding interaction with myosin; second, activation of myosin involving Ca^{++}-induced myosin light chain kinase; third, activation of ABPs such as gelsolin that are capable of severing actin filaments during or immediately following contraction. Proteins in this latter category would also be used to disassemble actin-containing cytoskeletal structures that may have been formed during step 1 of Figure 1. That is, they would be responsible for running step 1 in reverse.

ABPs with the ability to sever actin filaments have been isolated from a variety of cell types. Villin, gelsolin, and fragmin are ABPs that sever actin filaments at micromolar Ca^{++} concentrations and then probably remain bound

**TABLE I. Classification of ABP According to
Potential Function**

Assembly

  A. Nucleation

    Macrophage ABP
    Dictyostelium 120 K
    β-actinin

  B. Organization

    1. Geometry

      Actinogelin
      α-actinin
      Dictyostelium 95K
      Villin
      Acanthamoeba 90K

    2. Linkage

      Acanthamoeba capping protein
      Vinculin
      TW 260/240
      Fodrin
      Spectrin
      MAPs

Disassembly

  A. Filament severing proteins

    Villin
    Gelsolin
    Fragmin
    Dictyostelium 40K

  B. Proteins that regulate actin monomer pool size

    Brain depolymerizing protein
    Profilin

to the barbed end of the severed filament. This activity has been correlated with solation of a variety of actin-containing structures in vitro. Dictyostelium 40K protein also probably falls into this group of proteins.

As discussed above, the ability of these proteins to bind to dimers and trimers of actin accounts for the nucleation activity they exhibit in vitro in the presence of micromolar Ca^{++}. Their location on the barbed end of the actin filament suggests that they could be involved in the association of the barbed ends with membranes. Hence, these proteins could be classed as either nucleation, linkage, or disassembly proteins. However, this imprecision is more likely due to our current level of ignorance regarding the location and types of events in which these proteins participate in vivo, rather than any true flexibility of function. Since there are reservations about classifying these proteins as nucleation and linkage ABPs as discussed above, I have tentatively classified them as disassembly proteins according to the scheme in Table I.

Once an actin-containing structure has been fragmented into short actin oligomers, these oligomers could be then shuttled to new locations within the cell by a cytoplasmic flow resulting from local contraction. After they have passed into a region of cytoplasm with a submicromolar Ca^{++} concentration, they could act as building blocks for assembly of new actin structures. Alternatively, this newly formed population of oligomers with free-pointed ends could be depolymerized by proteins like the "depolymerizing factor" isolated from brain [Bamberg et al, 1980]. The resultant pool of monomer might then be buffered from spontaneous polymerization by proteins like profilin that bind to monomeric actin and inhibit its polymerization [Carlsson et al, 1977]. Unfortunately, the mechanisms for regulating interactions between actin, depolymerizing proteins, and profilinlike proteins remain unknown, so it is currently difficult to hypothesize how these proteins might function to control the location and pool size of soluble, monomeric actin in nonmuscle cells.

Once soluble precursors of actin-containing structures are available, lowering the Ca^{++} concentration and unmasking nucleation sites could lead to another round of assembly, starting the cycle over again at step 1.

VII. ALTERNATIVE MECHANISMS OF FORCE TRANSDUCTION

The discussion so far has been restricted to the generation of force by sliding interactions between actin and myosin filaments. This, however, is not the only mechanism by which force is produced by actin in nonmuscle cells. For example, polymerization of actin as it occurs in the acrosomal reaction of Thyone [Tilney, 1976], or changes in the packing of bundled actin filaments as occurs in the acrosomal process of Limulus [Tilney, 1975], can generate force for the extension of cell processes. These examples point out the existence of additional, important mechanisms involving actin but not requiring the intervention of myosin. ABPs that are involved in cyclical contraction described above in Figure 1 and Table I may also be important in these alternative mechanisms of force transduction. However, the sequence of events and key regulatory steps would differ from those outlined in Figure 1.

One example of a cellular event that might result from the interaction between actin and various ABPs is anaphase chromosome movement. The use in recent years of microinjection techniques for studying the effect of injected molecules on cell function in vivo [Mabuchi and Okuno, 1977] and lysed cells in which anaphase movement persists [Cande et al, 1981] has led to a more detailed appreciation of the diversity of possible force mechanisms that may function during mitosis. Anaphase A (movement of chromosomes to the poles), anaphase B (separation of the poles), and cytokinesis are now distinguished. Anaphase B is inhibited by vanadate, erythro 9(3-(2-hydroxynonyl)) adenine and nonhydrolyzable ATP analogues. These are all drugs that inhibit dynein-mediated microtubule movement as it occurs in cilia. Cytokinesis is inhibited by cytochalasin B, n-ethylmaleimid-poisoned subfragment one of HMM, and by the injection of antibodies directed against myosin [Cande et al, 1981; Mabuchi and Okuno, 1977]. Anaphase A, however, is not inhibited by any of these agents, suggesting that neither an actomyosin nor a dynein-microtubule based mechanism functions in chromosome-to-pole movement [Cande et al, 1981; Cande, 1982]. An alternative mechanism that may explain chromosome-to-pole movement is ABP-induced gel syntheses—ie, the isotropic contraction of the gel, accompanied by the extrusion of water. Filamin, an ABP from smooth muscle that closely resembles macrophage ABP, will promote extensive syneresis of actin gels in vitro [Wang and Singer, 1977]. This phenomenon does not require ATP and myosin; it is not inhibited by cytochalasin B. The molecular mechanism by which this occurs is unknown, but may involve changes in packing density of actin filaments within the gel, as more filamin-actin binding sites become progressively satisfied.

It is not clear whether proteins that exhibit filaminlike activity in their binding to actin, such as macrophage ABP and Dictyostelium 120K protein, also promote syneresis of actin gels. However, Dictyostelium 120K protein gels actin and the packing density of filaments within the gel increases as the 120K protein-to-actin molar ratio is increased [Condeelis, 1981b]. Therefore, by increasing the concentration of 120K in a region of a cytoplasmic actin gel, it might be possible to decrease the volume occupied by the gel—ie, to achieve syneresis. Syneresis could be used to generate directed movements if an ordered linear structure, such as the microtubule component of the spindle, were superimposed on the gel as its volume decreased.

The above is only rank speculation. Clearly much work needs to be done to investigate this possibility further. The question of whether actin is present within or near the spindle in quantities sufficient to carry out such a function remains controversial. In addition, it would be necessary to establish whether filamin or 120K-like ABPs are concentrated within or near the spindle during Anaphase A. Finally, such a model for anaphase A movement would predict that this movement could be inhibited by proteins that depolymerize actin filaments such as brain depolymerizing protein.

ACKNOWLEDGMENTS

Supported by PHS GM 25813 and the Rita Allen Foundation.

VIII. REFERENCES

Adelstein R, Eisenberg E (1980): Regulation and kinetics of the actin-myosin-ATP interaction. Ann Rev Biochem 49:921–956.

Atkinson M, Morrow J, Marchesi J (1982): The polymeric state of actin in the human erythrocyte cytoskeleton. J Cell Biochem 18:493–505.

Bamberg J, Harris H, Weeds A (1980): Partial purification and characterization of actin depolymerizing factor from brain. FEBS Lett 121:178–182.

Bennett V, Davis J (1981): Erythrocyte ankyrin: Immunoreactive analogues are associated with mitotic structures in cultured cells and with microtubules in brain. PNAS 78:7550–7554.

Branton D, Cohen C, Tyler J (1981): Interaction of cytoskeletal proteins on the human erythrocyte membrane. Cell 24:24–32.

Bretscher A, Weber K (1980): Villin is a major protein of the microvillus cytoskeleton which binds both G and F actin. Cell 20:839–847.

Bretscher A, Weber K (1978): Localization of actin and microfilament associated proteins in the microvilli and terminal web of the intestinal brush border by immunofluorescence microscopy. J Cell Biol 79:839–845.

Brown S, Yamamoto K, Spudich JA (1982): A 40,000 dalton protein from Dictyostelium discoideum affects assembly properties of actin in a Ca^{2+} dependent manner. J Cell Biol 93:205–210.

Burridge K, Feramisco J (1981): Non-muscle α-actinins are calcium sensitive actin binding proteins. Nature 294:565–567.

Cande Z, McDonald K, Meeusen R (1981): A permeabilized cell model for studying cell division: A comparison of anaphase chromosome movement and cleavage furrow constriction in lysed PTK cells. J Cell Biol 88:618–629.

Cande Z (1982): Inhibition of spindle elongation in permeabolized mitotic cells by erythro-9 (3-(2 hydroxynonyl)) adenine. Nature 295:700–701.

Carlsson L, Nystrom L, Sundkvist I, Markey F, Lindberg U (1977): Actin polymerizability is influenced by profilin, a low molecular weight protein in nonmuscle cells. J Mol Biol 115:465–483.

Chambers R, Chambers E (1961): "Explorations Into the Nature of Living Cell." Cambridge, Massachusetts: Harvard University Press.

Collins J, Korn E (1980): Actin activation of Ca^{2+}-sensitive Mg^{2+}-ATPase activity of Acanthamoeba myosin II is enhanced by dephosphorylation of its heavy chains. J Biol Chem 255:8011–8014.

Condeelis J, Taylor D (1977): The control of solation, gelation and contraction in extracts from Dictyostelium discoideum. J Cell Biol 74:901–927.

Condeelis J (1981): Reciprocal interactions between the actin lattice and cell membrane. Neurosci Res Prog Bull 19:83–101.

Condeelis J, Salisbury J, Fujiwara K (1981): An actin binding progein from Dictyostelium discoideum that gels F actin is concentrated in the cell cortex. Nature 292:161–163.

Condeelis J (1981a): Reciprocal interactions between the actin lattice and cell membrane. Neurosci Res Prog Bull 19:83–101.

Condeelis J (1981b): Microfilament-membrane interactions in cell shape and surface architecture. In Schweiger H (ed): "International Cell Biology, 1980–1981." Berlin: Springer-Verlag, pp 306–320.

Condeelis J, Salisbury J, Fujiwara K (1981): An actin binding protein from Dictyostelium discoideum that gels F actin is concentrated in the cell cortex. Nature 292:161–163.

Craig S, Pollard T (1982): Actin-binding proteins. Trends Biochem Sci March, 1982.

Davies P, et al (1977): Filamin inhibits actin activation of heavy meromyosin ATPase. FEBS Lett 77:228–232.

Detmers P (1982): Transient polymerization of actin occurs from an organizing structure in the fertilization tubule of Chlamydomonas. (Submitted.)

Geiger B, Dutton A, Tokuyasu K, Singer S (1981): Immunoelectron microscope studies of membrane-filament interactions: Distribution of α-actinin, tropomyosin and vinculin in intestinal epithelial brush border and chicken gizzard smooth muscle cells. J Cell Biol 91:614–628.

Glenney J, Glenney P, Osborn M, Weber K (1982): An F-actin and calmodulin binding protein from isolated intestinal brush borders has a morphology related to spectrin. Cell 28:843–854.

Griffith L, Pollard T (1978): Evidence for actin filament-microtubule interaction mediated by microtubule-associated proteins. J Cell Biol 78:958–965.

Hartwig J, Tyler J, Stossel T (1980): Actin binding protein promotes the bipolar and perpendicular branching of actin filaments. J Cell Biol 87:841–848.

Hasegawa T, Takabashi S, Hayashi H, Hatano S (1980): Fragmin: A calcium insensitive regulatory factor on the formation of actin filaments. Biochemistry 19:2677–2683.

Isenberg G, Aebi U, Pollard T (1980): An actin-binding protein from Acanthamoeba regulates actin filament polymerization and interactions. Nature 288:455–459.

Jockusch B, Isenberg G (1981): Interaction of α-actinin and vinculin with actin: Opposite effects on filament network formation. PNAS 78:3005–3009.

Kuczmarski E, Spudich J (1980): Regulation of myosin self assembly: Phosphorylation of Dictyostelium myosin heavy chain inhibits formation of thick filaments. PNAS 77:7292–7296.

Levine J, Willard M (1981): Fodrin: Axonally transported polypeptides associated with the internal periphery of many cells. J Cell Biol 90:631–643.

Mabuchi I, Okuno M (1977): The effect of myosin antibody on the division of starfish blastomers. J Cell Biol 74:251–263.

Maruyama K, Kimura S, Ishii T, Kuroda M, Ohashi K, Muramatsu S (1977): Beta-actinin, a regulatory protein of muscle. J Biochem 81:215–232.

Maruyama K, Sakai H (1981): Cell beta-actinin, an accelerator of actin polymerization isolated from rat kidney cytosol. J Biochem 89:1337–1340.

Matsudaira P, Burgess D (1979): Identification and organization of the components in the isolated microvillus cytoskeleton. J Cell Biol 83:667–673.

Mimura N, Asano A (1979): Ca^{2+}-sensitive gelation of actin filaments by a new protein factor. Nature 282:44–48.

Mooseker M, Tilney L (1975): Filament polarity and membrane attachment in the microvilli of intestinal epithelial cells. J Cell Biol 67:725–743.

Mooseker M, Graves T, Wharton K, Falco N, Howe C (1980): Regulation of microvillus structure: Calcium-dependent solation and cross-linking of actin filaments in the microvilli of intestinal epithelial cells. J Cell Biol 87:809–822.

Ogihara S, Ikebe M, Takahashi K, Tonomura Y (1983): Requirement of phosphorylation of Physarum heavy chain for thick filament formation, actin activation of Mg^{2+} ATPase activity and Ca^{2+} inhibitory superprecipitation. J Biochem (in press).

Pollard T, Ito S (1970): Cytoplasmic filaments of Amoeba proteus. I. The role of filaments in consistency changes and movement. J Cell Biol 46:267–289.

Pollard T (1981): Purification of a calcium-sensitive actin gelation protein from Acanthamoeba. J Biol Chem 256:7666–7670.

Porter KR (1976): Introduction: Motility in Cells. In Goldman R, Pollard T and Rosenbaum J (eds): "Cell Motility" New York: Cold Spring Harbor, p 1.

Scholey J, Taylor K, Kendrick-Jones J (1980): Regulation of nonmuscle myosin assembly by calmodulin-dependent light chain kinase. Nature 287:233–236.

Schroeder T (1975): Actin in dividing cells: Evidence for its role in cleavage but not mitosis. In Inoue S, Stephens R (eds): "Molecules and Cell Movement." New York: Raven Press, p 305.

Stendahl O, Hartwig J, Brotschi E, Stossel T (1980): Distribution of actin-binding protein and myosin in macrophages during spreading and phagocytosis. J Cell Biol 84:215–224.

Taylor D, Condeelis J, Moore P, Allen R (1973): The chemical control of motility in isolated cytoplasm. J Cell Biol 59:378–394.

Taylor D, Condeelis J (1979): Cytoplasmic structure and contractility in amoeboid cells. Int Rev Cytol 56:57–144.

Tilney L (1975): Actin filaments in the acrosomal reaction of Limulus sperm. Motion generated by alterations in the packing of the filaments. J Cell Biol 64:289–310.

Tilney L (1976): Nonfilamentous aggregates of actin and their association with membranes. In Goldman R, Pollard T, Rosenbaum J (eds): "Cell Motility." New York: Cold Spring Harbor, p 513.

Tilney L (1978): A new organelle, the actomere, that initiates the assembly of actin filaments in Thyone sperm. J Cell Biol 77:551–564.

Vahey M, Condeelis J (1982): Properties of 120K and 95K actin binding proteins for Dictyostelium and solation-contraction coupling. J Cell Biol 95:299a.

Wang K, Singer S (1977): Interaction of filamin with F actin in solution. PNAS 74:2021–2025.

Wilkins J, Lin S (1982): High-affinity interaction of vinculin with actin filaments in vitro. Cell 28:83–90.

Wolosewick J, Porter K (1979): Microtrabecular lattice of the cytoplasmic ground substance. Artifact or reality. J Cell Biol 82:114–139.

Yin H, Zaner K, Stossel T (1980): Ca^{2+} control of actin gelation. Interaction of gelsolin with actin filaments and regulation of gelation. J Biol Chem 255:9494–9500.

Yin H, Albrecht J, Fattoum A (1981): Identification of gelsolin, a Ca^{2+}-dependent regulatory protein of actin gel-sol transformation, and its intracellular distribution in a variety of cells and tissues. J Cell Biol 91:901–906.

Modern Cell Biology 2:241–258
© 1983 Alan R. Liss, Inc., 150 Fifth Avenue, New York, NY 10011

Morphogenesis in Desmids: Our Present State of Ignorance

Jeremy D. Pickett-Heaps

From the Department of Molecular, Cellular and Developmental Biology,
University of Colorado, Boulder, Colorado 80309

I. INTRODUCTION

Although modern cell biology has placed great emphasis on defining the structure of the cell and its components, many modern techniques are limited in that this analysis (for example, cell fractionation and electronmicroscopy) necessarily result in the destruction of the cell's organization and/or the irrevocable cessation of cell activities. This limitation is of great significance in subsequent attempts to understand cell division, differentiation, morphogenesis, and senescence, for example, in terms of the functioning and interaction of the various cytoplasmic components. As a result, biologists have to be very careful in interpreting results obtained under these limitations, a risky necessity. This essay will demonstrate how current discussion on cellular differentiation provides little insight into one phenomenon that may be fundamental in cell morphogenesis in general.

The control exercised by cells over their size and shape has long fascinated biologists. There are some indications that cellular form and internal organization may be controlled by certain of its components, particularly by two organelles, the microtubules (MTs) and microfilaments (MFs) found in most eukaryotic cells. Their roles in certain types of cell organization and morphogenesis are so well reviewed that they need not be discussed in detail here. In summary, MTs are often implicated in relatively slow, precisely controlled changes in cell organization, as well as constituting a cytoskeletal system that often influences certain cytoplasmic activities. MFs, conversely, are often associated with directed forms of active movement (eg, particularly streaming and ruffling of cell margins) or contractile movements such as those involved in cell cleavage and certain specific shape changes (eg, during embryogenesis). In other cells, the outer cell membrane appears important in morphogenesis, but its role, organizing capabilities, and the means by which it exerts its control are far from clear. In the case of the desmids, this latter possibility appears the only reasonable choice for the location of a mysterious system that controls a morphogenesis of unparalleled elegance.

II. THE DESMIDS

Desmids range from small to large; they are usually single-celled, green algae of the order Zygnematales (or Conjugales) [Fritsch, 1935]. The placoderm desmids, the main subject of this paper, are far more ornate than their primitive relatives, the saccoderm desmids. Most placoderm desmids have a pronounced central constriction which divides each cell into two "semicells." In normal conditions, these semicells are similar or identical, and are notable for the complexity and diversity of their ornamentative features, characteristic for each species (Figs. 1, 2). A mature desmid is thus a remarkably beautiful object under the microscope. The nucleus at interphase lies in the "isthmus" defined by the central constriction separating the semicells, each of which is largely filled by an elaborate chloroplast containing a variable number of pyrenoids. Many cells are normally flattened or disciform and possess three main axes of symmetry. In these cells, each semicell can be thought of as consisting of a polar lobe flanked by two wings which are often deeply cleft in a precise, species-specific pattern. The surface of the cell often displays a variety of warts, ridges, and/or spines. These species, in which the two wings flank the central lobe, are termed "biradiate." Other species, normally triradiate or tetraradiate, have three or four wings symmetrically arrayed around the polar lobe (Fig. 4), or if this is not distinguishable, an axis of symmetry that passes through both semicells and the

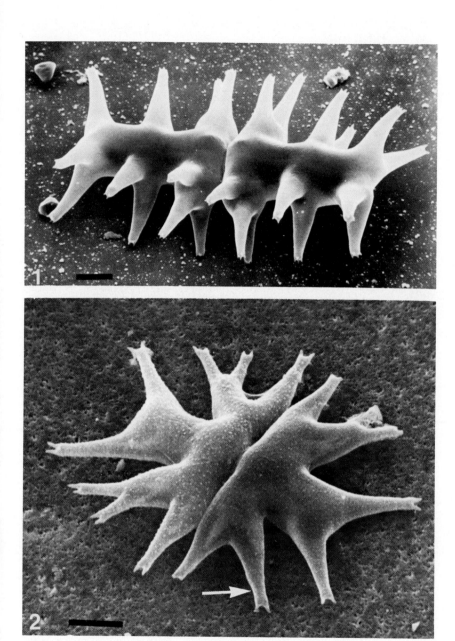

Fig. 1. Micrasterias muricata. Note that the two semicells defined by the central constriction are symmetrical, even though the pattern of spikes is complex. × 500; bar = 20 μ.

Fig. 2. Micrasterias radiata; a more typical flattened desmid. In this species, there is frequently some variation in the semicell's morphology; one of the lobes (arrow) here is not bifurcated like the others. × 750; bar = 20 μ.

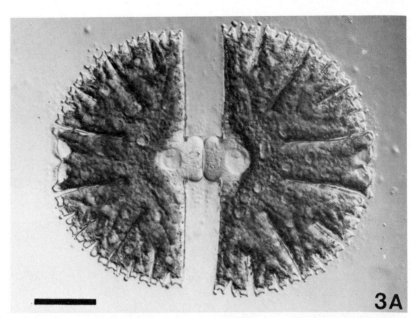

3A

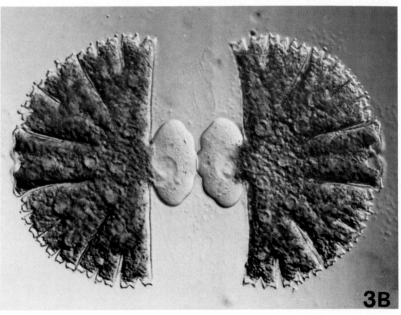

3B

Fig. 3.   Micrasterias denticulata; successive stages in semicell morphogensis after cytokinesis. A) Early bulge stage; B) marginal differentiation leading to the five-lobe stage; C) expansion of the five lobes; D) two further steps in marginal differentiation, with the polar lobe showing no further differentiation. A–C: × 350; bar = 50 μ; D: × 290; bar = 50 μ.

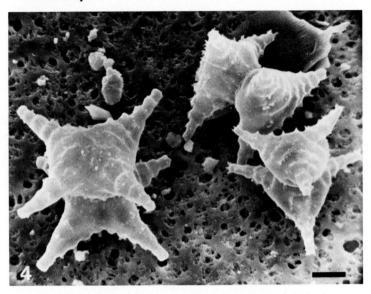

Fig. 4. Staurastrum pingue. Normally, these cells are triradiate, but tetraradiate individuals (one shown) are sometimes encountered. × 1,000; bar = 10 μ.

isthmus [Pickett-Heaps, 1975]. Pentaradiate and higher degrees of radiateness are known, but are not common.

Following cell division, these organisms reveal their potential for the study of the cytoplasmic control over cell form. The nucleus divides mitotically as usual, and then a cross wall divides the parent cell at the isthmus. Each daughter cell now consists of one semicell containing one chloroplast and a nucleus. Over the next few hours, each daughter cell regenerates a new semicell into which, as it expands, the single chloroplast moves, partly filling it (Fig. 3A–D). Finally, as the new semicell attains the size of the older one, the chloroplast is divided at the isthmus where the nucleus has come to rest. The result is two daughter cells identical to the parent cell in form and organization. However, since the volume of each daughter cell has doubled, its cytoplasm now appears pale and vacuolated as it enters interphase. Note that each daughter cell inherits one semicell wall and synthesizes a new one; these are attached tightly to each other at the isthmus, but are clearly separable entities, one always older than the other.

Formation of the new semicell results from several closely coordinated events. The force driving cell expansion is turgor pressure [Kiermayer, 1964, 1970]. The wall enclosing the expanding semicell is a typical extensible primary cell wall; originally derived from the cross wall formed at cytokinesis, new wall material is continuously added to it as the semicell expands. Once the new semicell has reached full size, a tough, thick secondary wall is laid down inside the primary wall, which is subsequently shed in many species. No significant cell expansion is undergone by the secondary wall, which therefore assumes the shape of the primary wall (and any abnormalities that may have arisen or been induced experimentally in it). Thus, the secondary wall can never alter its shape. Deviations in its form are often due to nongenetic changes in the cell. The form of the new semicell is therefore dictated during primary wall expansion, although the secondary wall does generate some less obvious morphogenetic features (for example, patterns of mucilage pores [Pickett-Heaps, 1972]).

A. Variations in Cell Form

Under normal growth conditions, the size, symmetry, and ornamentation of a given species of desmid is maintained within fairly strict limits. However, variation can be encountered in both normal and experimentally manipulated cells. In this regard, we will now distinguish between semicell *symmetry* (ie, the degree of radiateness) and *ornamentation* (degree of elaboration of the semicell margin and surface). However, such a distinction should not be regarded as absolute, since the cellular entities involved in controlling symmetry may not be different from those controlling ornamentation [Pickett-Heaps, 1975]. The volume of the cell of given ploidy appears constant [Brandham, 1965]; when morphological variants lose one or more wings, the polar lobe (and remaining wing) compensates by being appreciable larger.

B. Cell Symmetry

While a given species of desmid maintains a characteristic symmetry, variations can occur. For example, Staurastrum paradoxum is consistently found in the wild in two forms, either biradiate or triradiate [Reynolds, 1940; see also Tews, 1969]. The form predominating in a given ecological situation seems to depend on the season, and is probably influenced by external conditions such as temperature [Brandham and Godward, 1965], day length, etc. When desmids were first cultured, similar variants were soon encountered (Fig. 4). Waris and Kallio [1964] in particular, have characterized several such variants, some quite bizarre. For example, some Micrasterias (Fig. 6) will, rarely, generate semicells which are uniradiate (Fig. 7) or even aradiate (ie, consisting solely of a polar lobe). These changes in symmetry are brought about during semicell expansion, and do not usually require intermediate transitions between degrees of symmetry [eg, Brandham and Godward, 1965]. The immediate product of such divisions is a "Janus" cell, whose semicells are of different symmetry (Fig. 5). However, Lacalli (personal communication) suggests that sometimes intermediates can be formed. When isolated and cultured, these morphological variants appear usually to breed true in

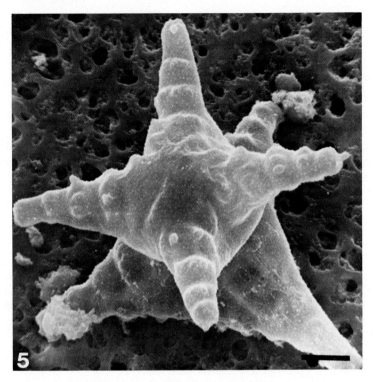

Fig. 5. S pingue, as in Figure 4. This is a "Janus" cell whose two semicells have different symmetry. × 1,500; bar = 10 μ.

the sense that they perpetuate the changed symmetry they have acquired. However, strains display a considerable variation in behavior. Some aradiate strains are stable and have never been seen to revert back to the normal morphology. However, many uniradiate strains consistently tend to revert back to the biradiate condition; in fact, what often happens is that one individual will spontaneously develop a semicell with biradiate symmetry. When this Janus cell divides, the biradiate half usually will produce a complementary biradiate semicell. The resultant biradiate daughter cell grows faster than its uniradiant variants, and consequently the culture dish soon becomes dominated by the biradiate forms. These changes in symmetry do not seem to involve genetic alterations, since the cells show the ability, however irregular, to revert back to normal morphogenesis. Other variants, particularly aradiates, may be true structural mutants (these results are summarized in Pickett-Heaps [1975]).

In some desmids, an increase in radiateness is often associated with sexual reproduction, but again, this does not seem to be the result of mutation [eg, Tews, 1969]. Starr [1958] found that tri- and even tetraradiate variants of Cosmarium often appeared when zygotes germinated following sexual reproduction. However, the tetraradiates appeared very unstable and soon reverted to forming triradiate semicells. Triradiates themselves were more stable, but generally tended to generate biradiate daughter semicells. Thus, a culture arising from germinating zygotes rapidly became dominated by the normal biradiate form. The symmetry of a given desmid is therefore probably somewhat plastic, being most stable in a given form. If biradiate desmids are experimentally made diploid (vegetative desmids are haploid), they may now become more stable as triradiates. Thus, as ploidy of the cell increases, forms with higher radiateness appear to become more stable [eg, Brandham, 1965].

Waris and Kallio [1964] have shown that when daughter cells are experimentally enucleated, the symmetry arising in the short-lived, enucleate daughter cell is that of the parent. They conclude that symmetry is cytoplasmically determined. However, I believe it is equally likely that pertinent genetic information could have been expressed before mitosis via entities (eg, proteins) that become operative immediately after the cell wall starts expanding [Pickett-Heaps, 1975].

Genetic analysis of cell symmetry is difficult, mainly because the conditions that induce sexuality in many desmids are poorly understood, and it is difficult to get the product of sexual reproduction, the diploid zygote, to germinate at will. Bradham and Godward [1964], using complementary mating strains of Cosmarium, investigated the genetics of symmetry by crossing biradiate cells, triradiate cells, and bi- and triradiate cells. All the zygotes from these crosses germinated to produce about the same proportion of triradiate cells versus normal biradiate cells (around 12%) regardless of the parentage of the zygote. Thus, there is no compelling evidence that the interconversion of bi- and triradiates is controlled genetically. As expected, cultures of these cells rapidly became dominated by the most stable form, the biradiate.

In summary, there is no useful indication as to how symmetry is controlled. Most experiments suggest that although the genetic information of the organism is ultimately induced, the expression of symmetry in forming semicells is determined by entities that are totally mysterious.

C. Ornamentation

The degree of ornamentation and its particular type in a semicell of a given species is determined largely by the genetic information in the cell. Again, however, variation can be induced in ornamentation under certain conditions. If cells are grown in poor conditions, semicell expansion becomes increasingly stunted and may ultimately result in cells that have little resemblance to healthy individuals [Pickett-Heaps, 1975]. This stunting of ornamentation is not genetically derived, since stunted cells, upon return to good conditions, immediately divide to form semicells of normal morphology. Waris and Kallio [1964] in a variety of experiments (including enucleation and UV irradiation) presented evidence that ornamentation was under direct nuclear control, and that this control is expressed after cytokinesis.

One can conceptualize ornamentation in terms of a given amount of morphogenetic information (from the nucleus) exerting its effects upon the expanding cell wall. This concept becomes useful with uniradiate variants in which the single wing formed at each division is larger than those in the biradiate cell (Figs. 6, 7—remember that the cell volume for a given ploidy remains constant) and where the margin is considerably more ornamented.

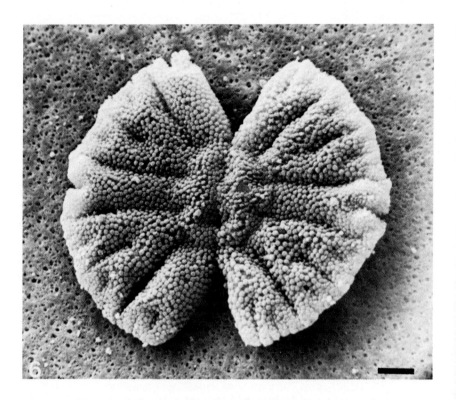

Fig. 6. Micrasterias thomasiana; the normal biradiate, haploid cell. The covering of spherical globules represents small mucilage secretions retained after processing for scanning microscopy. × 400; bar = 25 μ.

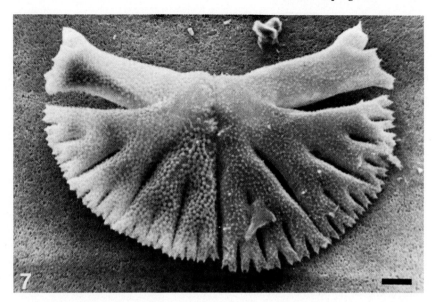

Fig. 7. M thomasiana; uniradiate variant of the cell in Figure 6. Note how the margin of the semicells is appreciably more differentiated than that in Figure 6. × 350; bar = 25 μ.

A given amount of morphogenetic information is apparently being expressed in one semicell instead of two. Similarly, a biradiate diploid cell has a margin and surface more highly ornamented than that of its haploid biradiate cousin. If, however, the diploid is triradiate, the cell margin now appears less highly differentiated and more like the haploid biradiate.

Kiermayer [1964, 1967] demonstrated that the expansion of a forming semicell ceases when cells are put into a medium of high osmotic pressure that balances the internal turgor pressure driving expansion. Deposition of primary wall material continues, but this deposition is uneven, its pattern being strictly controlled and related to the way the semicell would have continued expanding [see also Kiermayer, 1967, 1981]. Thus, control over the pattern of wall deposition could be involved with control over developing cell shape.

Both protein and RNA synthesis are intimately involved in normal morphogenesis. When expanding cells of Micrasterias are placed in cycloheximide, after a short delay, control over the spatial differentiation of the wall appears lost, even though the semicell continues expanding until approximately full size (Fig. 8). Actinomycin D gives similar results [Selman, 1966; Tippit and Pickett-Heaps, 1974], except that the cells usually end up by bursting. More significantly, the cycloheximide effects are reversible. Thus,

8

Fig. 8. Micrasterias denticulata, treated with cycloheximide (25 g/ml) continuously from just before the five-lobe stage (Fig. 3B). Differentiation of the margin ceased shortly afterward, although semicell expansion continued. × 250; bar = 50 μ.

when a cell is treated early in morphogenesis with cycloheximide, and then the cycloheximide is later removed, cleft formation recommences in the margin of the still expanding semicell, which by now has become fairly abnormal (Fig. 9). I believe that these experiments offer a promising way of investigating morphogenesis in Micrasterias, since they indicate that isolation and characterization of new proteins (ie, ones not previously present in the parent cell) synthesized specifically for morphogenesis, would be possible.

Other ways of affecting morphogenesis have been described. For example, laser microbeam irradiation of expanding lobes locally stunts their further expansion [Lacalli, 1975a, b], but the destruction exerted on cytoplasmic organization has not been characterized. When electric fields are applied to growing Micrasterias [Brower and McIntosh, 1980], expanding lobes show a galvinotropism toward the cathode, and if cell expansion is concurrently also inhibited, wall material accumulates along the cathode-facing side of the lobes. These results suggest that the vesicles which may be involved in primary wall deposition (see below), could be influenced by electric fields.

These defects in ornamentation are not permanent in the sense that cells possessing them can, upon subsequent divisions, generate normal semicells if not otherwise interfered with. Tourte's [1972] report to the contrary has not, I believe, been repeated. I emphasize again that once the shape of a semicell has been rendered permanent by deposition of a secondary wall, the shape of that particular semicell cannot thereafter be changed.

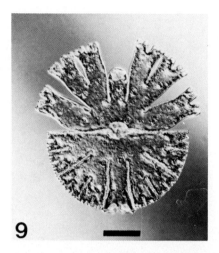

Fig. 9. As for Figure 8, except that the cycloheximide was removed after about 60 minutes. Differentiation of the margin had subsequently recommenced although the semicell is abnormal because of the earlier interruption in protein synthesis. × 200; bar = 50 μ.

D. Cell Form Mutants

There have been few attempts to generate true mutants of cell form in desmids. Korn [1970] induced some in Cosmarium by UV irradiation. Waris and Kallio [1964] had earlier shown that UV irradiation of expanding sem-icells could severely influence morphogenesis, an effect attributable to an interference with protein and RNA synthesis (see above); however, there was no indication that these alterations were permanent. Korn's mutants developed some alterations in morphology after the irradiation (ie, in subsequent generations of daughter cells), and these changes were likely the result of nuclear mutations. In our laboratory, two interesting variants have been induced in Micrasterias by mutagens; these have subsequently bred true over several years. One lacks the polar lobe (Fig. 10), and the other is consistently stunted [Tippit et al, 1982].

From such results, it is reasonable to assume that desmid morphogenesis is basically under nuclear control, and that mutations can profoundly affect cell form.

E. Ultrastructure of Expanding Semicells

At this level in the analysis of morphogenesis, we are confronted by its mystery—there is absolutely no indication of what it is that controls the shape of the expanding cell wall. First, MTs are demonstrably not involved in morphogenesis, unless they constitute a class specifically destroyed during fixation, an unlikely possibility. MTs are found in characteristic locations in

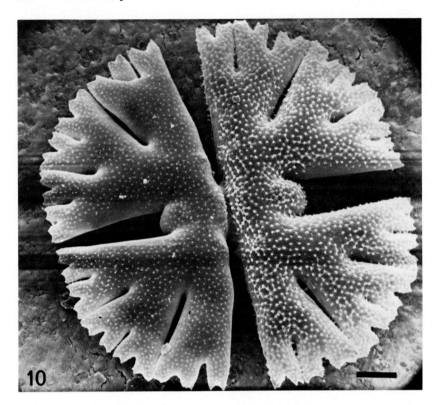

Fig. 10. M denticulata; an abnormal strain lacking a polar lobe. This strain was induced by treatment of normal cells with the mutagen EMS (ethyl-methane sulphonate), and it bred true for several years (ie, hundreds of generations) before the culture was lost. × 430; bar = 25 μ.

desmids both at interphase and during the cellular reorganization accompanying morphogenesis [Kiermayer, 1968; Pickett-Heaps, 1972]. In particular, a pronounced band of MTs encircles the isthmus of interphase cells. However, MTs are not associated in any way with the expanding wall of new semicells. This noninvolvement of MTs is confirmed by treating cells with a variety of MT-disrupting agents; none significantly affect the shape of an expanding semicell [Kiermayer, 1981].

Neither do MFs appear to be involved in pattern formation. Waris and Kallio [1964] invoked hypothetical cytoplasmic axes, presumably fibrous, to explain control over cell symmetry, but their existence has never been demonstrated. Neither are filamentous systems apparent at the ultrastructural level, although the preservation presently achieved by most workers may not be adequate to reveal them. The cytoplasm of expanding semicells streams actively. When studied using time-lapse cinematography, we have not been able to determine any pattern in the streaming (which appears random) that could be related to the shape of the forming semicell. However, one intriguing experiment merits further investigation. We [Tippit and Pickett-Heaps, 1974] found that brief treatment of expanding semicells with cytochalasin B results in the complete cessation of semicell expansion and consequently, morphogenesis. Surprisingly, these effects are irreversible, and viable cells arise which consist of one normal and one very stunted semicell. Furthermore, these cells can divide subsequently and the semicell arising from the original stunted semicell, though smaller than normal, was relatively well ornamented [see Tippit and Pickett-Heaps, 1974, for details].

There are numerous other papers on the ultrastructure of expanding semicells [reviewed by Kiermayer, 1981]. However, none has useful structural information concerning spatial control over wall morphogenesis. Golgi bodies probably contribute to primary wall deposition [reviewed by Lacalli, 1974; Kiermayer, 1981], and Kiermayer and Dobberstein [1973] describe an unusual, flattened type of Golgi vesicle which appears to be incorporated into the plasma membrane and which may be involved in secondary wall deposition. The plasmalemma displays one structural differentiation during secondary wall deposition [Pickett-Heaps, 1972]: where each mucilage pore is being formed, the plasmalemma is decorated by tiny projections. Thus, one minor component in the morphogenesis (the pattern of mucilage pores) can be traceable to a differentiation of the cell membrane. There are also interesting particle arrays in the plasma membrane [Giddings et al, 1980] which may be involved in formation of the massive secondary wall fibrils [Kiermayer and Staehelin, 1972], but these particles display no predetermined pattern that reflects wall morphogenesis. Even if these particles turn out to be important in affecting the shape of an expanding wall, how their own disposition and spatial rearrangements are accomplished remains totally obscure.

In summary, therefore, the information currently available on the ultrastructure of expanding semicells gives no indication as to how spatial control over morphogenesis is brought about by the cytoplasm. Although Kiermayer and others suggest that the "cell cortex" is involved, this statement is too vague to be really useful, since no differentiation has been detected here that anticipates or reflects the processes of morphogenesis.

F. Morphogenesis in Closterium

The desmid Closterium is unusual, in that it is basically cylindrical with no central constriction. In fact, the cell is usually lunate or symmetrically curved to varying degrees, depending on the species. In this desmid, there *is* clear indication of an involvement of MTs in morphogenesis. The cross wall is laid down after mitosis in the middle of the cell as normal, and at first there are no MTs near it. As it thickens, increasing numbers of MTs rapidly accumulate close to it [Pickett-Heaps and Fowke, 1970], and as semicell expansion gets under way, these MTs come to adopt the transverse orientation reminiscent of that found near the cortex of expanding higher plant cell walls [eg, Ledbetter and Porter, 1963]. These MTs are localized adjacent to the cell membrane *only* along the primary, expanding wall, never near the older secondary wall. This precise organization of MTs accompanies semicell expansion to completion. Colchicine treatment of expanding sem-icells produces the anticipated affect—namely, bulbous swelling due to loss of control over cylindrical expansion [Hogetsu and Shibaoka, 1978]; how-ever, these experiments are complicated by colchicine usually causing ex-pansion to cease, perhaps by a more general poisoning effect.

III. CONCLUSIONS

The work briefly summarized here suggests several conclusions. First, there is no cytological basis yet for understanding how most desmids exert their remarkable control over semicell morphogenesis. This control could in principle be exerted several ways—for example, by localization of enzyme complexes that "harden" the wall (cross-link wall polysaccharides) in specific regions where it is thereby prevented from expanding. Lobe differentiation can thus be visualized as classical "tip growth," interrupted at regular sites [eg, Lacalli, 1975a, b]. However, such concepts are of little value in grappling with the problem of how the overall pattern of such spatial determinants is set up in the first place. I therefore conclude that desmids exemplify a phenomenon that may be widespread: the expression of precisely controlled morphogenesis determined by unknown cytoplasmic entities or organizations. Thus, though it has been useful to study the role of MTs and MFs in mor-phogenesis generally, there are clearly other undescribed entities that may be equally or even more important.

In this context, the example of Closterium is significant, since it apparently has come, during its evolution, to use MTs during morphogenesis of the tubular semicell in a manner reminiscent of many higher plant cells. These MTs may represent an evolutionary innovation that reinforces the more prim-itive morphogenetic system exemplified by other desmids. This evolutionary modification has parallels elsewhere, for example, during cytokinesis in a variety of cells. Many primitive cells (bacteria, fungi, small algae, etc) simply furrow their membrane to achieve cell partitioning. No cytoplasmic feature besides the membrane has been associated with this primitive cleavage. In many larger animal cells and some protozoa (and notably the diatoms in the algae), a prominent system of MFs, shown in some cells to contain actin, line the ingrowing cleavage furrow. I deduce that the involvement of actin here represents an innovation in cells as they become much larger and need structural and functional reinforcement of their cleavage membrane. In conclusion, I suggest that in many cases where MFs and MTs are involved in morphogenesis, they may be ancillary (even if vitally important) organelles involved in a process that is basically more mysterious and primitive. If so, then we will never attain a useful understanding of morphogenesis until we have characterized this primitive and elusive system—and desmids seem as good a place as any to start looking at it.

ACKNOWLEDGMENTS

The author gratefully acknowledges the continuing support from the Systematics Section, N.S.F. (grant DEB 79-22200) which enabled this work to be carried out.

IV. REFERENCES

Brandham PE (1965): Polyploidy in desmids. Can J Bot 43:405–417.

Brandham PE, Godward MBE (1964): The production and inheritance of the haploid triradiate form in Cosmarium botrytis. Phycologia 4:75–83.

Brandham PE, Godward MBE (1965): The effect of temperature on the radial symmetry of Staurastrum polymorphum. J Phycol 1:55–57.

Brower DL, McIntosh JR (1980): The effects of applied electric fields on Micrasterias. I. Morphogenesis and the pattern of cell wall deposition. J Cell Sci 42:261–277.

Fritsch FE (1935): "The Structure and Reproduction of the Algae," Vol 1. Cambridge: Cambridge University Press.

Giddings TH, Brower DL, Staehelin LA (1980): Visualization of particle complexes in the plasma membrane of Micrasterias denticulata associated with the formation of cellulose fibrils in primary and secondary cell walls. J Cell Biol 84:327–339.

Hogetsu T, Shibaoka H (1978): Effects of colchicine on cell shape and on microfibril arrangement in the cell wall of Closterium acerosum. Planta 140:15–18.

Kiermayer O (1964): Untersuchengen uber die Morphogenese and Zellwandbildung bei Micasterias denticulata Breb. Protoplasma 59:76–132.

Kiermayer, O (1967): Das Septum-Initialmuster von Micrasterias denticulata und seine Bildung. Protoplasma 64:481–484.

Kiermayer O (1968): The distribution of microtubules in differentiating cells of Micrasterias denticulata Breb. Planta (Berl) 83:223–236.

Kiermayer O (1970): Causal aspects of cytomorphogenesis in Micrasterias. Ann NY Acad Sci 175:686–701.

Kiermayer O (1981): Cytoplasmic basis of morphogenesis in Micrasterias. In "Cytomorphogenesis in Plants" (Cell Biol Monographs, Vol 8). Vienna, New York: Springer-Verlag, pp 147–189.

Kiermayer O, Dobberstein B (1973): Membrankomplexe dictyosomaler Herkunft als "Matrizen" fur die extraplasmatische Synthese und Orientierung von Microfibrillen. Protoplasma 77:437–451.

Kiermayer O, Staehelin LA (1972): Feinstruktur von Zellwand und Plasmamembran bei Micrasterias denticulata Breb. nach Gefrieratzung. Protoplasma 74:227–237.

Korn RW (1970): Induction and inheritance of morphological mutations in Cosmarium turpinii Breb. Genetics 65:41–49.

Lacalli T (1974): Golgi function and cell wall formation in the Conjugales. Rev Can Biol 33:125–133.

Lacalli T (1975a): Morphogenesis in Micrasterias. I. Tip growth. J Embryol Exp Morph 33:95–115.

Lacalli T (1975b): Morphogenesis in Micrasterias. II. Patterns of morphogenesis. J Embryol Exp Morphol 33:117–126.

Ledbetter MC, Porter KR (1963): A "microtubule" in plant cell fine structure. J Cell Biol 19:239–250.

Pickett-Heaps J (1972): Cell division in Cosmarium. J Phycol 8:343–360.

Pickett-Heaps JD (1975): "Green Algae; Structure, Reproduction and Evolution in Selected Genera." Stamford, Connecticut: Sinauer.

Pickett-Heaps JD, Fowke LC (1970): Mitosis, cytokinesis and cell elongation in the desmid Closterium littorale. J Phycol 6:189–215.

Reynolds N (1940): Seasonal variations in Staurastrum paradoxum Meyen. New Phytol 39:86–89.

Selman GG (1966): Experimental evidence for the nuclear control of differentiation in Micrasterias. J Embryol Exp Morphol 16:469–485.

Starr RC (1958): The production and inheritance of the triradiate form in Cosmarium turpinii. Am J Bot 45:243–248.

Tews LL (1969): Dimorphism in Cosmarium botrytis var. depressum. J Phycol 5:270–271.

Tippit DH, Pickett-Heaps JD (1974): Experimental investigations into morphogenesis in Micrasterias. Protoplasma 81:271–296.

Tippit DH, Smith H, Pickett-Heaps JD (1982): Cell form mutants in Micrasterias. Protoplasma (in press).

Tourte M (1972): Modifications morphogenetiques induites par la puromycin et la cyclohex-imide sue la Micrasterias fimbriata au cours du bourgeonnment. C R Acad Sci (Paris) 274:2295–2298.

Waris H, Kallio P (1964): Morphogenesis in Micrasterias. Adv Morphog 4:45–82.

Modern Cell Biology 2:259–302
© 1983 Alan R. Liss, Inc., 150 Fifth Avenue, New York, NY 10011

The Cytoplasmic Matrix

K. R. Porter, M. Beckerle, and M. McNiven

From the Department of Molecular, Cellular and Developmental Biology,
University of Colorado, Boulder, Colorado 80309

I. INTRODUCTION

In 1942 the American Society of Plant Physiologists published a monograph entitled "The Structure of Protoplasm." It comprised a collection of papers presented at a symposium in Philadelphia 18 months earlier. In large part the papers deal with the structural features of the cytoplasm, and to a lesser extent with function. Though the time is only 40 years ago, the vocabulary seems very old indeed. There are references to the ectoplast, the plasmagel, and plasmasol, and to such esoteric parts of the cytoplasm as kinoplasm and trophoplasm. Some of the contributors favored the "brush-heap" concept for the structure of the cytoplasmic matrix—a loose bonding of chains of molecules or micellae to account for the gel-like properties of the matrix [Seifriz, 1942]. Equilibria between sol and gel states were identified as existing between the cortical and central regions of Arbacia eggs. In support of this idea it was reported that only small centrifugal forces were needed to displace the yolk in the central cytoplasm, whereas much greater forces were required to sediment the pigment in the cortical zone. Thus the viscosity in this and numerous other cell types was described as being much greater in the cytoplasmic cortex. (This "cortex," by the way, would seem to coincide with the actin and spectrin-rich regions in the current picture of the cell cortex [see Branton et al, 1981] or the cytoplasmic lamina [Penman, this volume].) The gels of cells were tested exhaustively by Marsland [1942] and co-workers for their sensitivity to high hydrostatic pressures and, without notable exception, they were found to solate under pressures of 4,000–10,000 pounds per square inch; with release of pressure, the viscosity and form of the cell remarkably returned to normal [Tilney et al, 1966]. The literature of this period is replete with references to such physical properties of protoplasm as elasticity, thixotropy, and contractility. Enormous centrifugal forces (100,000–400,000g) were used to demonstrate the capacity of the cell both to survive the stratification of its contents and to return the visible particulates to their original location after the force was relaxed [Beams, 1943]. It was observed by Scarth [1927] that the nucleus of Spirogyra could be displaced with a needle and yet return promptly to its original position. A more ingenious approach was used by several investigators who introduced iron or nickel particles into cells and followed their displacement with externally applied magnetic fields [Heilbronn, 1922; Seifriz, 1942]. Uniformly, the displaced particles returned toward their original position when the magnetic field was turned off. These and similar experiments obviously indicate a significant level of elasticity in the cytomatrix, and this kind of evidence inspired Frey-Wyssling [1948] to write imaginatively on the "structural scheme of the cytoplasm" as being basically filamentous. He predicted, quite accurately we think, that when observed microscopically the cytomatrix would appear as "an extremely fine network," essentially a "molecular framework." "The meshes of this framework contain the interstitial substances: a solution of salts in water, and lipids, including phosphatides." In the same period, the continuous phase of protoplasm was described as aqueous by Moyer [1942].

The most analytical study of cytoplasmic elasticity was published by Crick and Hughes in 1950. They succeeded in getting cultured chick embryo cells to take up small particles of oxidized iron, and then in moving these particles through the cytoplasm with magnetic fields applied in various ways. The movement of the particles through the viscous cytoplasmic milieu in response to the applied magnetic field was recorded by time-lapse cinematography. In each case, when a particle was moved there was a small, rapid recoil when the field was turned off. The recoil never returned the magnetic particle to its starting point; usually recoil was about one third of the distance originally traveled. Repeated, magnetically induced excursions of particles seemed to generate a larger space for unimpeded motion; ie, eight reiterations of the on-and-off action of the field lowered the viscosity in the space through which the particles had moved. They concluded from this that the cytoplasm is a thixotropic gel. The limited recoil one might interpret as reflecting the elasticity of a zone of compression that develops in a filamentous meshwork immediately in advance of the moving particle. The distance traversed minus the recoil could represent a part of the matrix in which the structure was damaged and the viscosity markedly altered (a thixotropic effect). Quite obviously, these experiments and others describe the presence of a viscous and structured matrix in the cytoplasm, a matrix with enough elasticity (or structural information) to return it to its undistorted form after some kinds of perturbations.

This period of indirect observations on the physical properties of the cytomatrix, which essentially came to an end in the 1940s, was succeeded with some overlap by a period in which centrifugal fractionation of tissue extracts [Bensley and Hoerr, 1934; Claude, 1940] and electron microscopy became the productive approaches. Little if any thought was given to the destructiveness of tissue homogenization by most practitioners of the technique. Particles of various kinds (mitochondria, microsomes, and ribosomes) emerged in resuspensions of the pellet, and these were viewed as floating freely in the cytosol much as they do in the centrifuge tube. It was further concluded that a large fraction of the chemical reactions of intermediary metabolism occur in the cytosol as a product of random association of enzyme and substrate. That these reactions might occur in relation to structure and that randomness might not be involved has not until recently been given much thought. Only now, as images of a structured cytomatrix (and the cytoskeleton) begin to appear in the literature, do we find any substantial number of biochemists discussing the possible existence of organization in the disposition of biologically active molecules [for review see Clegg, 1982]. The trend away from the cytosol concept (a bag of enzymes) may by now have advanced to the point where the idea is largely an interest of history.

II. WHOLE-CELL PHENOMENA AND THE CYTOMATRIX

Quite apart from observations on the physical properties of protoplasm, there are whole-cell phenomena which argue for the existence of a structured unity and organization. For example, it is common knowledge that cells have polarity. This is evident in egg cells through the distribution of yolk, pigment, and developmental potential, and it is characteristic of all derivatives of the fertilized egg. Epithelial cells have apical and basal poles; nerve cells have axonal and dendritic poles; and wandering cells such as lymphocytes and neutrophils have advancing and trailing poles. These all bear some relation to the cell center and known populations of microtubules.

Then also, descriptive of an organization not ordinarily displayed by sols are the uniform size and shape of cells of a single type. The size varies with ploidy but not remarkably otherwise. An exception to this rule is exemplified by tumor cells which exist in many sizes and shapes, especially in the highly malignant transformants. It would seem that in the normal cell there exists an information frame that supervises the development and maintenance of size and form under ordinary environmental conditions.

The same conclusion is indicated by the nonrandom distribution of the more obvious cytoplasmic systems and organelles such as ER, Golgi, stress fibers, microtubules, and the so-called free ribosomes. A structured matrix with unity and an extraordinary capacity to maintain its design is seemingly required to contain and control the disposition of these otherwise unconnected components.

The capacity of cells of many kinds to recognize the loss of a part also implies a sense of form and wholeness—a sense which probably influences the subsequent regeneration of the lost part. It is difficult to conceive of this as a property of anything but a structured unit.

A picture of a highly differentiated structure attends as well the capacity of cells to provide rapid and directed transport for such physiologically active components as steroid hormones from the cell surface to nucleus. In many instances, cells seem not to rely on random diffusion and its wastefulness to get physiologically active agents from point of entrance to the specific targets.

From among these various properties of cells, the translocation of cytoplasm—as, for example, during a change in cell shape—seemed to us the most amenable to experimental manipulation. In a study of cell elongation that takes place during lens development in vertebrate morphogenesis, microtubules were found to be coextensive with the event [Byers and Porter, 1964]. It was postulated on the basis of this observation that microtubules function as an essential frame for achieving a nonrandom distribution of the cytomatrix. Not all form changes appear to be microtubule-dependent; for example, the development of lamellopodia and microvilli where microtubules are not found [Porter, 1978]. These surface excrescences are obviously movements on a small scale compared with whole-cell shape changes or the morphogenesis of cilia or axopods where microtubules certainly are involved.

For a closer look at vectorial motion and its relation to microtubules, the chromatophore seemed a good choice [Bikle et al, 1966]. Thus began a series of studies that are still in progress and which have brought us closer to a definitive understanding of the mechanism for pigment motion and possibly all directed motion in cells.

A. Chromatophores

The ability of animals, especially invertebrates and lower vertebrates, to change their color was noted by the earliest Greek naturalists. This extraordinary capacity resides in cells known as chromatophores, which are found in the dermal layers of the animal's integument. The cells are able to move pigment granules through their cytoplasms from dispersed to centrally aggregated positions with concomitant changes in the color of the whole animal. Such cells, derivatives of the neural crest, are differentiated for this purpose. Some can aggregate their pigment in a few seconds; others take minutes or even hours. In most instances, the purpose of the motion is to achieve protective coloration.

The pigment is usually in the form of granules or small crystals. The most common is the black pigment melanin, but carotenoids are also used, as well as crystals of proteins and purines [Rohrlich and Porter, 1972]. A chromatophore will usually contain only one type of pigment, but instances do exist in which the cell contains as many as four differently colored granules [Robison and Charlton, 1973]. These can be moved separately or in different combinations to achieve different hues. In other instances, the same or similar effects are achieved by different chromatophores adopting organized associations and superimpositions [Bagnara et al, 1968].

Chromatophores are heavily innervated, a fact that has been known since the observations of Ballowitz [1893]. Nerve endings (synapses) are all over the cell, but are especially numerous over the central part. Since the chromatophore aggregates its pigment in response to epinephrine, it is not surprising that these endings show the dense secretory granules of adrenergic neurons. Depolarization of the pigment cell with high extracellular concentrations of K^+ will also initiate aggregation; cAMP or caffeine will induce dispersion. In some animals (decapods), the control is totally by circulating hormones derived from the eye stalk [see Parker, 1948, for review].

Of the many interesting questions one may investigate with chromatophores, none is more significant than that which asks how the cells move their pigment granules. What structure provides the motive force? How is the "ghost in the machine" controlled? Is there a single macromolecular complex, like the dynein of cilia and flagella, which through some configurational change pushes or pulls the pigment? It is reasonable to infer that the mechanism resides in the cytoplasmic matrix and is beyond the resolving powers of the light microscope.

These questions have also motivated many of the investigations of our laboratory over the past 10 to 15 years. Our studies have spread into investigations of the broader issues that can be asked of cells in general and which seem to require a closer look at the cytoplasmic matrix or ground substance. How, as mentioned earlier, does the cell establish and maintain its organization? How does it recognize the loss of a part and regenerate it? How does it control its size and its shape (see below)? Though the solutions to these larger issues are not immediately apparent, we do think that we have achieved some understanding of the mechanisms involved in pigment motion and presumably in the motion of the several other components of cells that, like chromosomes, move in a microtubule-dependent fashion.

We have thought that on this occasion it would be appropriate to review our observations on chromatophores and speculate more generously than is usually permitted on the roles of the structured ground substance in chromatophores and in other cells less obviously specialized than chromatophores. It does seem to be time (or even past time) for cell biologists and biochemists to probe the properties of the larger organizational entities, the polymeric meshworks, so to say, that make up a large part of the living cell [Fulton, 1982].

B. Erythrophores

The initial work with chromatophores focused on the melanophores of Fundulus and asked essentially whether these cells have microtubules, and, if so, how they might be involved in the motion of the pigment [Bikle et al, 1966]. The presence of microtubules was easily demonstrated, but experimental studies then attempted did not characterize the contributions of microtubules to the granule motions. Since then it has become amply clear that microtubules function as a frame (or guide) for pigment motion (see below) [Green, 1968; Murphy and Tilney, 1974; Schliwa and Euteneuer, 1978].

For several reasons we then chose to use the erythrophore, first described by Smith and Smith [1935] and later by Porter et al [1970]. It moves its pigment more rapidly than melanophores, and has pigment granules that are semitransparent to high-energy electrons, permitting one to view the morphology of pigment-associated structures even though these are below or above the granule.

Erythrophores are disk-shaped cells about 40 µm in diameter, which give a red color to the squirrel fish, Holocentrus ascensionis (or rufus) Osbeck. They reside just under the epidermis of the fishes scales [Junqueira et al, 1970]. They contain a large number of small spherical granules of red pigment (carotenoid) which are about 0.5 µm in diameter *and lack a membrane*. It is apparent, even at light-microscopic resolutions, that the pigment granules are arranged in linear files oriented radially with respect to the cell center (Figs. 1, 10). Thus, the cell, with its pigment dispersed, reminds one of the radial organization shown strikingly by the asters which appear in many animal cells at the poles of the mitotic spindle. There is an equivalence also in the radial saltatory motion shown by various particulates in these two structures as they occur in living cells [Rebhun, 1972].

At high resolutions provided by high-voltage microscopy of whole cells, certain important structural features appear. First, it is obvious that they contain a large number of microtubules. The number in the erythrophore just exceeds 1,000; in melanophores of the angelfish it is greater [Schliwa, 1978]. These are all arranged radially and appear to represent the tracks along which the pigment moves in aggregation and dispersion (Fig. 2). It seems probable, though it has not been proved, that all the tubules insert in the central apparatus and have their plus (distal) ends free within the cell margin. They all appear to have the same polarity [Euteneuer and McIntosh, 1981]. The cell center is a complex structure that includes two centrioles and a number (20) of "dense bodies" (MTOCs) which are continuous with the minus ends of the tubules [McNiven and Porter, 1981]. The MTOCs are peculiar; they look like a much finer and denser version of the three-dimensional meshwork that exists elsewhere in the cytoplasmic matrix. These MTOCs become more intimately associated with one another when the cell aggregates its pigment, and separate again during granule dispersion.

Each erythrophore has a kidney-shaped nucleus which remains in a peripheral location whether the pigment is aggregated or dispersed, though it does move a little closer to the cell center when the pigment aggregates (compare Figs. 1 and 2). The mitochondria are few in number (20–30) and not obviously exceptional in any other respect. They also are moved during pigment aggregation and dispersion but independently of the pigment. Their motion, like that of the nucleus, seems relatively minor.

There are two other structural components in the cytoplasm of these cells which have current importance because of their probable involvement in pigment motion. One is a meshwork of slender strands that connect the dispersed pigment granules to one another and also to the microtubules and to the adjacent cytoplasmic cortices. The strands in dispersed erythrophores are roughly uniform in diameter at 5–10 nm and extremely variable in length. In its dimensions this irregular lattice appears more fragile than its putative equivalent, the microtrabecular lattice (MTL) in growing cells from other sources [Wolosewick and Porter, 1979; Porter and Tucker, 1981; Porter and Anderson, 1982]. The other component is evident in thin sections of erythrophores as well as in images of whole cells; it appears as a complex system of membrane-limited channels and vesicles, evidently a form of the smooth endoplasmic reticulum (SER). The system extends from the cell margins to the cell center but does not invade the center itself [Porter and McNiven, 1982]. Its limiting membrane has the same dimension (thickness) as the membranes comprising the nuclear envelope. There are reasons to suspect

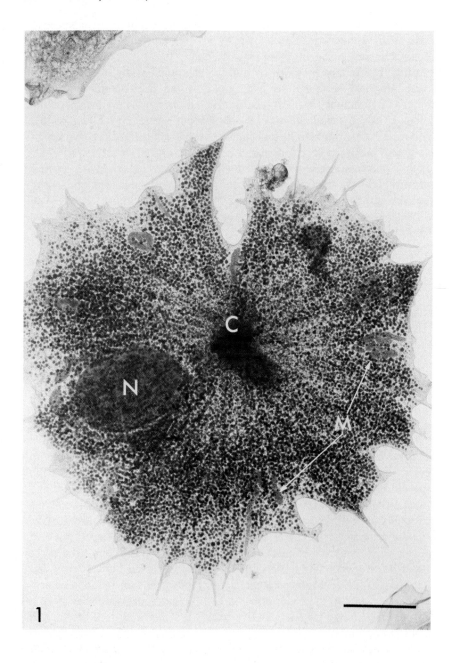

that this SER should be equated with the sarcoplasmic reticulum of striated muscle in both its structural and functional properties (see below).

III. THE MECHANISM FOR PIGMENT MOTION

The motion of the pigment granules in aggregation and dispersion is distinctly different. During aggregation the granules move at a uniform velocity of about 10 μm per second and aggregation may be completed in 2–3 seconds. Dispersion, in contrast, takes twice as long, and the individual granules shuttle back and forth (in and out) along radial tracks. This saltatory (shuttling) motion persists for some granules during dispersion and is especially noticeable in the middle zone, where the pigment is frequently less abundant (Fig. 1).

Several theories have been proposed to explain pigment motion, and all have been shown for one reason or another to be inadequate [see Green, 1968]. One has survived, and, for reasons that will become apparent, it deserves special attention. On the basis of light microscope observations, Lorna Green suggested 15 years ago that the pigment (melanosomes) in melanophores of Fundulus heteroclitus are contained in a "structured continuum" and that the forces required to move the pigment develop in this matrix. This interpretation was contained in a remarkably prescient paper which has not received the attention it deserves. She proposed further that the continuum "contracts" in aggregation and "expands" during dispersion. Had Green been able to observe the continuum in electron micrographs of thin sections, she would doubtless have described it in some detail and would have avoided assigning to the ER or microtubules a primary role in propelling the pigment. Though she did not clearly state it, she probably thought of the continuum as a fairly homogeneous matrix containing ER and microtubules. Our ability to observe the matrix structure in whole cells by high-voltage microscopy and without the impediments of embedding resins has permitted us to avoid this error.

The idea of a continuum which contains the pigment suggests that it may constitute a unit structure. We know that the radially confined motion of the pigment is enforced by the radially oriented microtubules. In their absence, following treatments with colchicine or nocodazole or low temperatures, the

---

Fig. 1.   Low-magnification image of a Holocentrus erythrophore with pigment dispersed. For culturing, this cell was first dissociated from the scale with enzymes according to methods in Luby and Porter [1980], then, when suitably spread, it was fixed with glutaraldehyde and OsO_4, critical-point dried, and viewed with a 1 MeV electron microscope. The pigment granules are arranged in radial files along microtubules that extend from the centrosphere (C) to the cell cortex. M, mitochondria; N, nucleus. Magnification 4,000 ×. Bar = 5 μm.

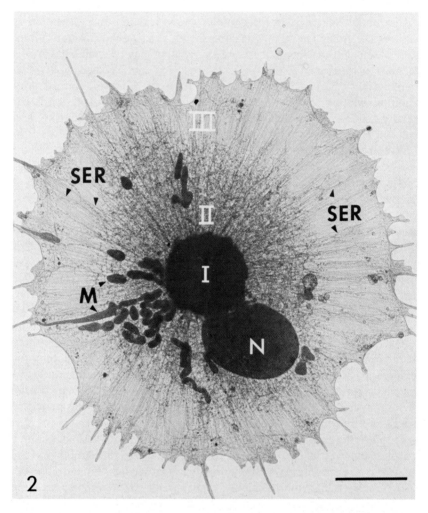

Fig. 2. Low-magnification image of a cultured erythrophore, first aggregated with epinephrine, then prepared for microscopy as cell in Figure 1. Hundreds of microtubules, extending out of the pigment mass (I), are visible as is also the smooth endoplasmic reticulum (SER). Three zones are recognized in pigment-aggregated cells. I, The central pigment mass; II, a middle zone, exclusive of pigment but including MTL, SER, mitochondria and nucleus, and a peripheral zone; III, consisting of cytoplasmic cortices, microtubules and marginal extensions of the SER. M, mitochondria; N, nucleus. Magnification 4,000 ×. Bar = 5 μm.

pigment clumps randomly when stimulated to aggregate [Porter et al, 1970; Schliwa and Euteneuer, 1978; Beckerle and Porter, 1983]. The fact that in dispersion the pigment does not pile up at the cell's periphery suggests that the granules are constrained by some structural property of the matrix. The pigment in dispersion is obliged by the matrix to scatter over the entire cytoplast, thus to achieve the functional purpose of the erythrophore.

The logic of this argument is supported by recent observations on the movement of individual and identifiable granules [Porter and McNiven, 1982]. In this study, we observed that such granules do in fact occupy fixed positions in a unit structure, which we have called the cytoplast. In repeated pigment dispersions (25 repetitions), the same granule returned to the same position ($\pm$ 1–2 μm) relative to the whole cell (Fig. 10). Those identified at the very edge of the cytoplast are the easiest to follow, but the fixed position applies as well to granules that normally stop their centrifugal motion at some site more central to the pigment mass.

These observations and reasonings support seemingly only one conclusion: that this pigment cell has a centrally organized system (Green's continuum) which has the capacity to change its shape from disk to sphere and in so doing to transport contained pigment. Little is left to chance or the possibility of pigment granules changing their positions; they are "contained" in a lattice of stable design. This lattice includes the proximal ends of many microtubules and the centrally localized MTOCs into which the tubules insert. But the system that moves the pigment *does not move the peripheral microtubules, the SER, the mitochondria, or the nucleus.* In other words, there appears to be a separate three-dimensional lattice for moving the pigment.

A. Structural Changes Associated With Pigment Aggregation

The onset of aggregation is marked by a sudden freezing of all motion, including the saltations of individual granules. In a fraction of a second, this is followed by the general migration toward the cell center (at the same velocity) of almost every (99% est.) pigment granule in the cell. The motion is described as resolute. Electron micrographs show that the pigment-containing lattice moves with the pigment and changes its morphology (compare Figs. 3, 4, 5, and 7). Trabeculae that were smooth and relatively straight in the dispersed cell now become beaded and variously distorted (Figs. 3, 4). These structural changes are presumed to be significant in the "contraction" of the unit structure and the motion of the associated pigment. A more detailed analysis will be part of a separate publication [McNiven and Porter, 1981]. As part of these events, the pigment-associated lattice separates from the bulk of the microtubules and the SER which, in their distal extensions, remain integral to the cell's cortices (Figs. 2, 7). The microtubules, in preserving

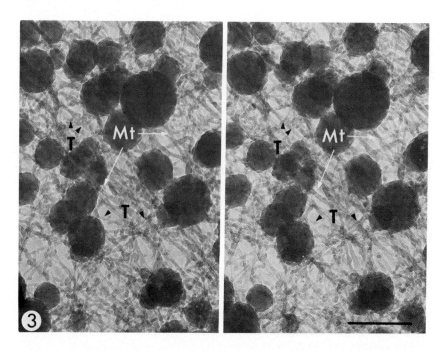

Fig. 3. High-magnification stereo* image of a small part of a dispersed erythrophore. Pigment granules of varying sizes, without membranes, are suspended (and interconnected) in a meshwork of slender (15 nm) trabeculae (T). Microtubules (Mt) pass above and below the pigment in a direction from lower right to upper left. In this and other images, trabeculae insert on these tubules, and presumably by these connections the pigment is given direction in its motion in both aggregation and dispersion. Magnification 80,000 ×. Bar = 0.25 μm.

their preaggregation position and associating with the cortices, presumably function as a "memory" for the radial organization; and the SER, whatever its function (possibly a control of local Ca^{++} concentration), preserves the distribution of that function for the redispersion of the pigment.

Not being content with the adequacy of the relatively slow glutaraldehyde fixation, we have recently made generous use of freeze substitution in which cell structure is frozen rapidly by propane cooled to $-185°C$ with liquid

*The stereo pairs were prepared to be observed through a simple type (desk model) viewer such as can be purchased from the Adams Instrument Corporation, Lansing, MI 48915, or from Gordon Enterprises, Inc, 5362 Cahuenga Blvd, North Hollywood, CA 91601. It is possible with a little practice to see stereo without any mechanical aids. The procedure to follow during training is to focus on an object about 10 meters distant and then insert the stereo pair to be observed into the line of vision with eyes still focused on the distant point. The observer should see three images instead of two and the central one, now focused on, will appear in 3-D.

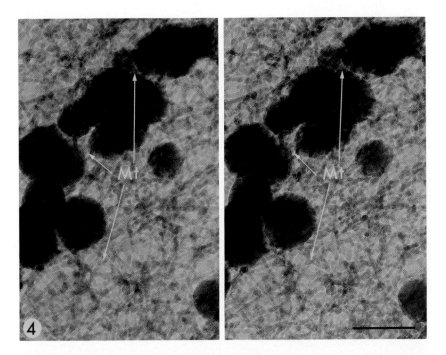

Fig. 4. High-magnification stereo images of a small area of an erythrophore fixed during aggregation. Pigment granules have been drawn together. Microtubules (Mt), which run across the area from lower left to upper right, are difficult to discern in the confusion of microtrabeculae. These latter are individually shorter and thicker than the equivalent in Figure 3. They are also beaded (20 nm). This is the form of the microtrabecular lattice consistently associated with the aggregation process. Magnification 80,000 ×. Bar = 0.25 μm.

nitrogen. This approach has the added advantage that the solvent as a vehicle for OsO₄ (2-methoxyethanol) removes the pigment but leaves the MTL and microtubules for better viewing (Figs. 5–7). Micrographs of these preparations (aggregated cells) make it quite clear that there is a separate MTL for the pigment and a residual MTL associated with the mitochondria, the SER, and the nucleus (Figs. 2, 6, 7). The trabeculae of the pigment-associated lattice are uniformly beaded during aggregation, whereas those of the dispersed and residual lattices are for the most part unbeaded. (Figs. 8, 9). The morphology of aggregation includes also wavelike domains of beaded trabeculae (Fig. 6). These apparently exist behind the moving pigment as part of the mechanism that rapidly achieves pigment aggregation.

These observations seem to say that the erythrophore has two MTL systems—one to move pigment, and one possibly to maintain the cell's integrity

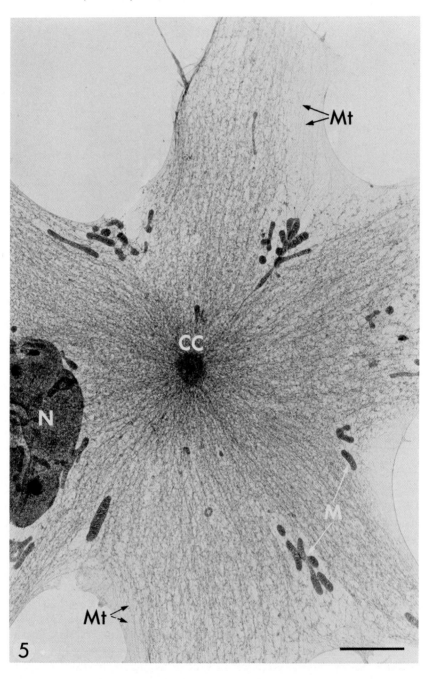

and the disposition of other components essential to this and other cell functions. That these two systems can separate when the pigment aggregates and commingle when the pigment disperses seems most surprising.

B. The Modes of Pigment Dispersion

In pigment dispersion, the events are equally dramatic. The pigment mass seems to explode initially as all the granules begin simultaneously to move from the cell center. Early within the 4- or 5-second period occupied by dispersion, the granules begin to shuttle (display saltatory motion), but the net motion is centrifugal. This motion is linear (proximally and distally) and parallel to the radial orientation of the adjacent microtubules. It seems probable that shuttling is a local expression of the same mechanisms that cause aggregation and dispersion. Why both appear during dispersion and not during aggregation is unclear; the expansion of the pigment-containing lattice is thus irresolute compared with contraction.

We have sought an explanation for these differences in studies of the involved fine structure. Though we remain uncertain in terms of a finite explanation, we can offer the view, supported by published evidence [Junqueira et al, 1974; Luby and Porter, 1980], that the restructuring and shape change involved in dispersion are structurally complex and require ATP. The basic event is the elongation of individual trabeculae associated with the pigment; in the absence of ATP, elongation fails. It is as though normally the basic structure expands. At the same time, a continuity with microtubules is reestablished presumably so that the linearity of motion and radial distribution of pigment will be maintained. The beading of the individual trabeculae disappears except where fixation preserves the local changes in trabeculae dimensions associated with saltation (Fig. 8). There is no obvious distinction during dispersion between the pigment-associated lattice and that which remains outside involvement in pigment aggregation; the two appear to blend.

Fig. 5. Low-magnification image of a dispersed erythrophore fixed by freeze substitution. This cell was quick-frozen in liquid propane at $-185°C$, and maintained in a 2-methoxyethanol-osmium tetroxide mixture for 48 hours at $-90°C$. It was then brought to room temperature, transferred to liquid CO_2, and critical-point dried. The pigment dissolved away during this procedure, leaving an intact cytoplast. This is represented by a finely divided complex of microtubules and microtrabeculae. The numerous small circular areas in which there is less structure mark former locations of pigment granules now dissolved. There remains only a faint basket (not evident at this magnification) which contained the granule. CC marks the cell center which doubles as the MTOC. N, nucleus; M, mitochondria. Microtubules (Mt) can be seen only in lattice free areas at the cell margins (arrows). Magnification 6,000 ×. Bar = 3 μm.

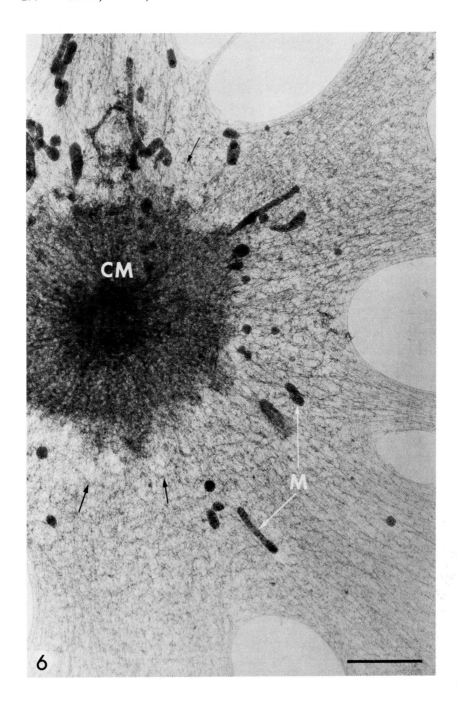

6

Summary. High-voltage electron microscopy of erythrophores prepared for observation as whole cells indicates that the pigment granules are contained by a fine meshwork of strands, the microtrabecular lattice. This pigment-bearing lattice appears to be one of at least two separate lattices within this cell.

The fundamental event in pigment aggregation is the change in the dimensions and configuration of the pigment-associated lattice. In shortening, the trabeculae take on a beaded character which may reflect some manner of coiling of a component central to the axis of the trabeculae. These beads (or gobbets, as we have also called them), are about 200 Å in diameter as observed in the whole-cell preparations. They are commonly evident in other cells, especially in cells exposed to low temperatures [Porter and Anderson, 1980; Porter and Tucker, 1981] or to metabolic inhibitors [Luby and Porter, 1980]. We have come to regard them as representing a low-energy state within the system. They have been observed in fibroblasts by Buckley and Raju [1976] and called "globoid bodies"; and in melanophores by Schliwa [1979]. What they actually mean in terms of biochemistry has not been revealed, and it seems too early to speculate further on their significance.

C. Inhibitors of Dynein ATPase Activity Block Pigment Motion

Because the microtubule framework of the erythrophores is essential for organized intracellular transport of pigment granules [Beckerle et al, 1979; Beckerle and Porter, 1983], we were interested in the possibility that a dynein-microtubule interaction might be responsible for generating the motive force in these cells, as it appears to be in cilia and flagella. There are two probes that have been used to assess the level of dynein involvement in a cellular process. These are vanadate (in the +5 oxidation state) and erythro-9-3-(2-hydroxynonyl)adenine (EHNA). Vanadate is a potent inhibitor of dynein ATPase activity in vitro [Kobayashi et al, 1978; Shimizu, 1981], and it also blocks reactivated flagellar beat [Gibbons et al, 1978]. In pioneer studies Cande has used vanadate with detergent-permeabilized cells to study the

---

Fig. 6. Low magnification image of an erythrophore quick-frozen during aggregation (at −185°C) and prepared by freeze-substitution as described in the legend to Figure 5. The peripheral constituents of the cytoplast (MTL), including pigment, have withdrawn in part toward the cell center to form a circular mass of contracted lattice material (CM). The residual pigment and pigment-associated lattice were in transit when fixed. Small, wavelike condensations of trabeculae (arrows) are characteristic of the cytoplast just peripheral to the dense, aggregated zone CM. The mitochondria (M) have also moved centrally. Microtubules, now partly free of the lattice elements, are most apparent in the margins of the micrograph. Magnification 7,000 ×. Bar = 3 μm.

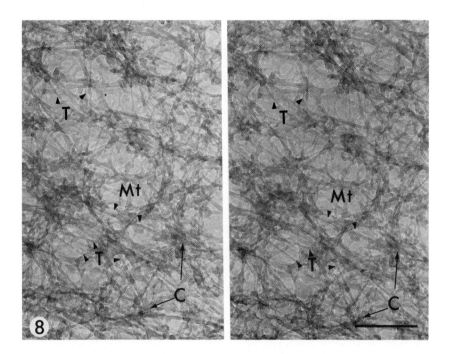

Fig. 8. High-magnification stereo images of a small area from a freeze-substituted, dispersed erythrophore from which the pigment has been dissolved. Numerous thin elongate trabeculae (T) interconnect with each other and with adjacent microtubules (Mt). Condensations of lattice material (C) are evident without order in their disposition or structure. They include beaded trabeculae and are interpreted as associated locally with saltatory motion which goes on all the time in dispersed cells. Magnification 80,000 ×. Bar = 0.25 μm.

Fig. 7. Low-magnification image of a fully aggregated quick-frozen (at −185°C) erythrophore prepared for microscopy by freeze-substitution. The pigment-moving cytoplast has fully contracted to the cell center forming a dense and greatly constricted central lattice mass (zone I). This is obviously more condensed than in the half aggregated cell in Figure 6. A second form of lattice (zone II) extends outward from the central mass and appears to provide structural support for mitochondria (M) and maintenance of cell shape during pigment migration. It seems never to become part of the central zone I. This micrograph shows further that the bulk of the pigment-bearing lattice has been swept out of the cell margins (zone III), leaving only the microtubules and a few slender extensions of the SER. Magnification 6,000 ×. Bar = 3 μm.

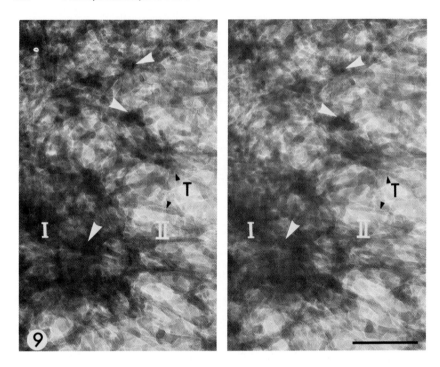

Fig. 9. High-magnification stereo images of the interface between the aggregated zone I and the more peripheral zone II from the aggregating freeze-substituted cell in Figure 6. In zone I, individual trabeculae have become shorter, thicker, and beaded, while also forming numerous small condensations (arrowheads). Zone II is occupied in part by MTL and nonheaded trabeculae (T), apparently not involved in pigment motion. Magnification 80,000 ×.

involvement of dynein in mitotis and has shown that anaphase B in particular is inhibited by vanadate [Cande and Wolniak, 1978; Cande, 1982]. Fibroblast intracellular transport, which is thought to be microtubule-mediated, is also perturbed by vanadate [Forman, 1982]. EHNA has more recently been described as an inhibitor of dynein ATPase activity [Bouchard et al, 1981], and the evidence thus far indicates that it is a fairly specific inhibitor of dyneins [Penningroth et al, 1982]. It has the advantage that it can permeate intact biological membranes.

Microinjection of erythrophores with vanadate rapidly inhibits pigment motility [Beckerle and Porter, 1982]. Such inhibition is strictly dependent on the inclusion of vanadate in the injection buffer. Even if the erythrophores are in the process of aggregating their pigment during the injection, the motility ceases as soon as injection is accomplished, and the pigment granules seem to drift toward the dispersed state. The word "drift" is used to describe

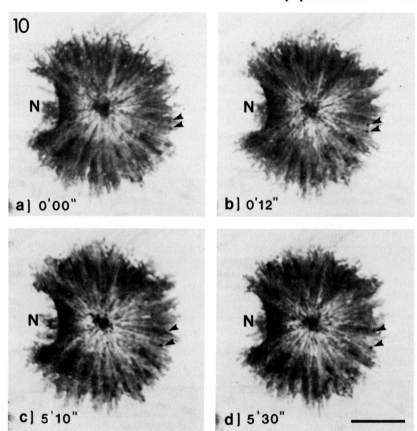

Fig. 10. Snapshots of the same erythrophore taken at the end of four different dispersions. Arrowheads point to two larger, denser granules which return to approximately the same location after each pulsation. Pictures (a) and (b) are of successive dispersions separated by 12 seconds; (b) and (c) are separated by 5 min during which the cell went through 30–40 pulsations; (c) and (d) are separated by 20 seconds. The two identifiable granules retain their relative positions during this time span and for longer. Small changes in location are more apparent than real because the constant and rapid saltatory motions make recording difficult. Magnification 1500 ×. Bar = 10 μm.

this motion because this dispersion is much slower than the normal and is not saltatory in nature as it is in control cells. The injection of about one tenth the cell's volume of a vanadate solution more concentrated than 250 μM (in the injection needle) causes aggregation of the pigment to be completely inhibited. Motility is perturbed even at somewhat lower concentrations (50–100 μM), but at these lower concentrations aggregative motility can still sometimes be stimulated though it is slower than in control cells. The radially organized shuttling of the pigment granules is rapidly inhibited when vanadate is injected.

Similar findings have been made in our laboratory using erythrophores permeabilized with low concentrations (0.001%) of digitonin [Stearns and Ochs, 1982]. Despite the fact that their membranes were full of small holes resolved by scanning microscopy (SEM), these cells continued to move their pigment in a saltatory fashion for extended periods (1 h +) of time. Vanadate in concentrations between 0.4 and 10 μM inhibits this transport totally in 3 min, and this effect is reversed in a few minutes by treatment with 1 mM catechol. Vanadate also blocks pigment motion in detergent-permeabilized melanophores [Clark and Rosenbaum, 1982]. Inhibitors of oxidative phosphorylation (azide and DNP) are also active in stopping granule motion in these permeabilized cells. This block is reversed by Mg^{++}/ATP. The digitonin "open" model for studying transport is sensitive to variations in the concentration of Ca^{++} in the cell's environment at the 0.1 μM level. The usefulness of this model for studying the functional aspects of the pigment moving mechanism in chromatophores is obvious; it should provide an increasingly rich source of information in the future.

EHNA applied to erythrophores likewise blocks saltatory particle motion [Beckerle and Porter, 1982], but the cells (in contrast with vanadate-injected cells) remain aggregation-competent even after extended treatment. Although EHNA is known to inhibit carboxymethylase activity [Pike et al, 1978], this is not the cause of the EHNA-induced inhibition of pigment transport [Beckerle and Porter, 1982]. Since both of these dynein inhibitors block saltatory motion, there exists the possibility that a dyneinlike complex participates in force generation in the erythrophore. However, the inhibition of aggregative motility by vanadate, but not EHNA, is somewhat baffling. It is difficult to imagine how a dyneinlike molecule might be involved in the uniform velocity pigment aggregation, particularly since the aggregated state has long been known to be a low energy state [Luby and Porter, 1980; Green, 1968; Junqueira et al, 1974; Saidel, 1977]. Moreover, as mentioned earlier, the microtubules in chromatophores are of a single polarity, with their assembly-preferred (+) end distal to the cell center [Euteneuer and McIntosh, 1981]. Since the oriented binding of dynein to a microtubule is defined by the polarity of the microtubule, one might expect dynein to mediate pigment motion in only a single direction.

D. Is Actin Involved in Pigment Motion?

Despite the fact that microtubules are required for the radially directed pigment motion, both erythrophores [Beckerle and Porter, 1983] and melanophores [Obika et al, 1978; Schliwa and Euteneuer, 1978] are capable of some pigment movement when all microtubules are disassembled. The ability to induce motility in microtubule-depleted cells suggests that some nonmicrotubular component is generating at least some of the motive force in these cells. Although actin and actomyosin are ubiquitous proteins and are known to be involved in many instances of nonmuscle motility, they do not seem to participate in pigment translocation. It is true that actin filaments have been identified in chromatophores [Schliwa et al, 1981b], but they are present only in loose cortical arrays and in microspikes extending from the cell surface; very few, if any, filaments are associated with the pigment granules. Moreover, microinjection into erythrophores of agents that perturb actin (cytochalasin B, DNase I, and phalloidin) has no effect on the ability of erythrophores to move their pigment granules, although they do produce cell shape changes that might be anticipated, given actin's cortical distribution in the erythrophore [Beckerle and Porter, 1983]. Injection of buffer alone has no effect on cell shape. Control experiments with amoebae indicated that these agents were indeed biologically active when injected, despite their lack of effect on erythrophore intracellular motility. Moreover, the fixed position of the granules in aggregation and dispersion argues against any sol-gel phenomena involving actin [Porter and McNiven, 1982].

E. Another Possible Source of the Motive Force

If actin is not involved in force generation in erythrophores and dynein-microtubule interactions are not sufficient to explain pigment motion, what molecules are responsible for intracellular translocation in these cells?

Available evidence warrants a consideration of spasmin. This is a protein (MW 20,000) which is a prominent component of the spasmoneme, a rapidly contracting organelle of the vorticellid ciliates [Amos, 1972; Routledge, 1978]. It avidly binds Ca^{++}, and 10^{-6} M free Ca^{++} is sufficient to induce spasmoneme contraction. Electron microscopy shows the spasmoneme to consist in large part of 2–4 nm filaments, contractile elements that are beaded, and membrane-limited vesicles which appear to sequester Ca^{++} like the SER in muscle cells. Mitochondria are few in number. In the presence of Mg^{++} and ATP, the intact spasmoneme undergoes cyclic contractions and elongations, an activity that is stopped by ATPase inhibitors [Routledge, 1978]. Isolated spasmonemes respond simply to shifts in Ca^{++} concentration between 10^{-6} and 10^{-8} M. Relaxation of the spasmoneme takes 1–2 seconds and is slow relative to contraction which takes only 4–10 msec.

Parallels between the behavior and control of the spasmoneme and the pigment-associated MTL of the erythrophore are obvious, though in need of verification in some details. In the first place, it is known that an increase in intracellular calcium above a threshold concentration of 5×10^{-6} M triggers the typical aggregation response in erythrophores, whereas concentrations below this threshold are essential for dispersion [Luby and Porter,

1982]. It is also evident from electron micrographs that a form of the SER is distributed in the cytoplast in concentrations that roughly parallel the concentration of pigment-associated lattice [Porter and McNiven, 1982]. Equated, as the SER frequently is, with Ca^{++} sequestration and release, it is appropriately disposed in erythrophores to control Ca^{++} concentration, both locally and generally in the cytoplast. The dispersive phase of pigment motion is blocked by inhibitors of oxidative phosphorylation. And, finally, there are remarkably few mitochondria in erythrophores and certainly not so many as one would expect in a cell that moves, by any of the more common motility mechanisms, a massive amount of pigment. Like the pigment-moving lattice in erythrophores, the spasmoneme is an intracellular unit structure in the stalk(s) of the vorticellid (colonial) protozoa, populated by innumerable filaments in the size range of microtrabeculae.

It seems to us, then, on first consideration, that several of these properties of spasmin could comfortably fit the contractile component of the pigment-associated lattice and that the presence of the SER in close association with lattice could provide for Ca^{++} control using only a minor shift in Ca^{++} concentration to regulate the mechanochemical events of pigment aggregation and dispersion.

IV. THE MICROTRABECULAR LATTICE IN OTHER CELLS

In parallel with studies on the erythrophore and its MTL, our laboratory has maintained an active program of investigations into (1) the occurrence and characteristics of the lattice (or a closely analogous structure) in other kinds of cells; (2) the possibility that the MTL is an artifact; and (3) the responses of the structured matrix (the MTL) to such common and readily varied conditions of the environment as temperature, osmolarity, and divalent cations. Since most of these studies have been published in abstract, if not in full papers, we hope to be excused from giving a detailed and fully documented account for our findings.

In its occurrence, the so-called lattice has been found to be ubiquitous. Its structure and relation to other components of the eukaryotic cell are similar wherever observed. Differences in preparation procedures notwithstanding, whether one uses glutaraldehyde plus OsO_4 or freeze substitution with OsO_4 in 2-methoxyethanol as the fixative, the structure that emerges is basically the same qualitatively and quantitatively [Wolosewick and Porter, 1979; Porter and Anderson, 1982]. It is true that the form of individual trabeculae varies slightly depending on whether one uses slowly penetrating glutaraldehyde or rapid fixation by freezing at $-185°C$ in propane followed by drying at $-95°C$ (Fig. 14). These differences are readily explained by the failure of glutaraldehyde to act rapidly enough to stop all motion in an obviously dynamic system without concomitant distortions (see above). Also, we can agree with Hans Ris [1980] that the CO_2 used in critical-point drying must be dry, and precautions toward this end are routine in our laboratory. With superdry CO_2 (10 parts H_2O/million CO_2), the lattice and its characteristic trabecular form are reproducible.

The same structure, with some minor variations, has been reported in several types of cultured cells and by a number of investigators [Buckley, 1975; Wolosewick and Porter, 1975, 1979; Buckley and Porter, 1975; Buckley and Raju, 1976; Fulton et al, 1980; Pudney and Singer, 1980; Hama, 1980; Porter and Anderson, 1980]. A variety of procedures, including freeze-fracture followed by deep etch have been used to visualize it in nerve cells [Metuzals, 1969; Ellisman and Porter, 1980]. Attempts have recently been made to explore plant protoplasts using matrix-free sections cut from polyethylene glycol embeddings. The resulting images depict the MTL in typical form [Hawes et al, 1982; Wardrop, 1982].

The major features of the MTL as they appear in our preparations are shown in Figures 11, 12, and 13. It comprises a very large number of slender strands which connect all the better-known components of the cytoplasm to form a space-filling lattice (Figs. 12, 13). Thus one can identify filaments running from ER to microtubules, from microtubules to the cytoplasmic cortex and in between, where the trabeculae intersect it is common to find polysomes (especially in growing cells). The thickness dimension of the trabeculae varies (5–10 nm or more) depending on whether the trabeculae are close to the cortex or more central and whether the preparation procedures used glutaraldehyde and critical-point drying or rapid freezing and subsequent drying [Porter and Anderson, 1982]. Their lengths are equally variable (50–100 nm). These variations seem to be more pronounced in the growing, more highly hydrated cell than in such fully differentiated cells as erythrophores (see above) or lymphocytes [Guatelli et al, 1982]. In the vicinity of the centrioles and satellites, the meshwork is condensed and more finely structured than at the cell's margins. Since most of our observations have been made on the thinner margins of whole growing cells that are in various phases of the cell cycle, we may in some instances be observing differences related to these facts. As noted previously, it is regrettable that more studies do not use normal, nonproliferating diffentiated cells. For these and other reasons, it is encouraging that techniques are being developed for looking at thick sections of fully differentiated tissue cells, cut from tissue blocks embedded in polyethylene glycol (PEG) [Wolosewick, 1980; Guatelli et al, 1982]. The PEG is subsequently dissolved out of the section which is then dehydrated and dried by the critical-point method. Such preparations are consistent in showing the MTL and microtubules. What is also important is that when such thick sections are reembedded in epon and sectioned for comparison with matrix-free sections, it is evident that the "wispy strands" which have

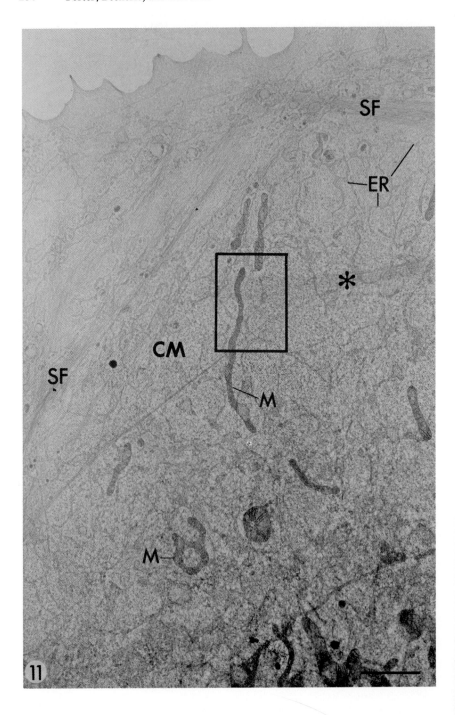

been repeatedly noted in micrographs of stained epoxy sections are the equivalent of the trabeculae in the PEG preparations [Guatelli et al, 1982].
Thus we are encouraged to note that many earlier studies have described
similar 3-D lattices of strands except for those parts lost to view because
their electron scattering properties closely match those of the epoxy resins.*
It follows as well that (1) those "wispy" strands comprising "filamentous
networks" [Yamada et al, 1971; Wessels et al, 1973]; (2) those bridges
between microtubules in mitotic spindles and elsewhere [McIntosh, 1974]
and in axostyles [Mooseker and Tilney, 1973; McIntosh, 1973]; (3) the spokes
in cilia axonemes, which, as well as outer and inner dynein arms, are probably
equivalents of trabeculae, or differentiated variations of the same. Of these,
the outer dynein arms are larger and more substantial when viewed in freeze-
etch preparations of Chlamydomonas flagella [Goodenough and Heuser, 1982]
than they are in images of thin sections. It is important to note also in this
connection that in freeze-etch preparations of axonemes there are structure-
free spaces (water-rich?) between the dynein arms and the microtubules, an
observation noted as well by Wolosewick [1980] in PEG preparations of
sperm flagella. One would hope that when these generalizations have been
evaluated and possibly accepted, the cytomatrix in differentiated cells may
become more popular for investigation (see below).

A. Fact or Artifact?

It is not easy to discuss this issue objectively and without equivocation.
Electron microscopists have been describing artifacts ever since 1940, and
there is no end in sight. We are, after all, attempting to preserve without
major distortion a very complex object with chemicals whose actions, broadly
speaking, are not understood. Fortunately, in several instances, the element
of fine structure discovered has been identified in the living cell by the use
of light microscopy and appropriate optics. However, when we begin to
study structures that are not obviously ordered and are not more than a few

---

*Tannic acid in conjunction with glutaraldehyde and OsO$_4$ does much to enhance the visibility
of the cytomatrix in electron micrographs, presumably because it increases the scattering
properties of the matrix components.

Fig. 11.   This micrograph depicts the fine structure of part of a BSC cell grown in culture
and prepared for viewing by conventional techniques of glutaraldehyde plus OsO$_4$ fixation and
critical-point drying. It includes images of mitochondria (M), the endoplasmic reticulum (ER),
stress fibers (SF), and a ribosome-rich cytomatrix (CM). The area outlined is shown in Figure
12. One stress fiber (*) is or particular interest because it is included in Figures 12 and 13
where, at higher magnifications, it appears as a differentiation of the lower cytoplasmic cortex.
Magnification 8100 ×. Bar = 2 μm.

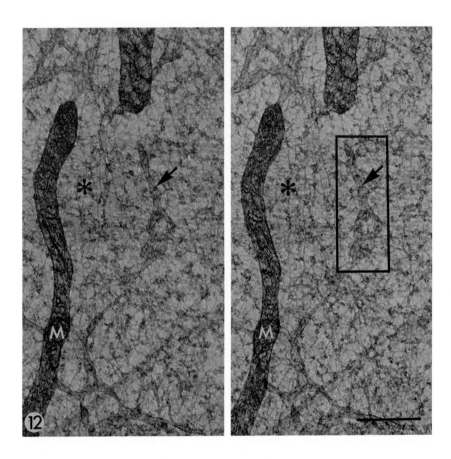

Fig. 12. Stereo pair at higher magnification of region outlined in Figure 11. Here it is possible more readily to view the smaller elements of the cytoskeleton as well as mitochondria and ER (arrow). Microtubules are numerous and run mostly from top to bottom of micrograph. When studied with the aid of a 2× stereo viewer, the micrograph is observed to show numerous fine strands (microtrabeculae) linking microtubules to one another and to the upper or lower cortices. Similar strands join polysomes and serve apparently to suspend them in space. The immature stress fiber (*) comprises a complex of broad trabeculae essentially in the plane of the lower cortex. In a detergent-extracted cell, this structure would probably be revealed as rich in actin filaments. Intermediate filaments are not easily identified because, like the microfilaments, they are coated by other matrix components. Some of the longer and more slender trabeculae probably contain segments of 100 Å filaments. The spaces between the lattice elements are structure-free and are referred to as water-rich. Magnification 40,500 ×. Bar = 0.5 μm.

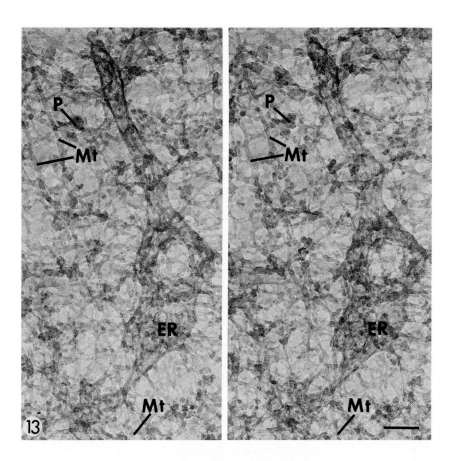

Fig. 13.   Stereo images of area outlined on Figure 12. Polysomes (P) as suspended in lattice are shown as well as ER and microtubules (Mt). The latter seem to be pulled aside from their strictly linear form by connections with the lattice. In addition to the dense spherical particles representing ribosomes, there are spherical bodies of lower density which we have called "gobbets" or "varicosities" on trabeculae in other cells and reports. These are prominent in the pigment-bearing lattice in chromatophores where the trabeculae have been characterized as "beaded." The image of the MTL includes a number of broadly flattened trabeculae that appear to be flattened against the lower cortex of the cytoplast. These are part of the ribbon-like stress fiber that crosses the field at (*) in Figures 11 and 12. Magnification 108,000 ×. Bar = 0.1 μm.

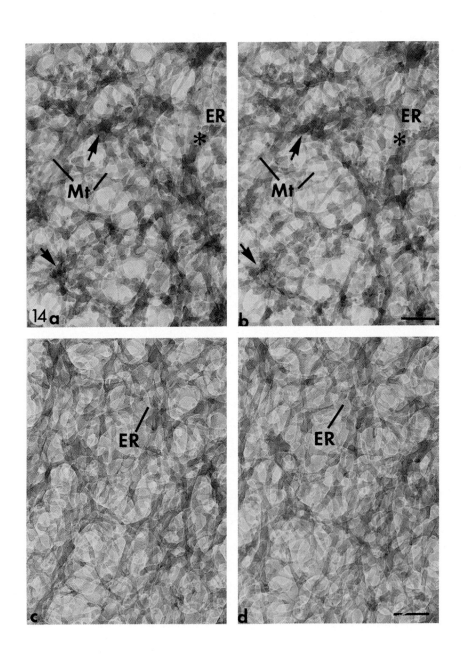

nanometers in size, we have problems. One might argue that if all the various procedures of fixation used yield a structure that is basically the same, we are safe in assuming that we have identified essentially the correct morphology. In this instance such a conclusion is complicated by the fact that many cells have a "cytoskeleton" consisting of the well-known threesome: microfilaments, intermediate filaments, and microtubules. In evaluating the relation of these to the MTL, it can be reasonably suggested that the MTL is an artifactual condensation of soluble proteins on the framework of the more traditional cytoskeleton (see below for a discussion of MTL-cytoskeleton relationship). It is not totally convincing to present available evidence that there is probably little if any protein in solution in cells [Kempner and Miller, 1968; Clegg, 1982] or that the cytoskeleton, as evidenced in detergent-extracted cells, does not in number of strands and complexity match the MTL.

To avoid a more extended discussion than space limitations will allow, we choose to mention two or three observations that we find most pertinent to the artifact question. First, the image of the erythrophore offers information on an MTL that changes its form with cell function and which appears to have a fairly stable structure. This is built around and is continuous with the central complex of centrioles and dense bodies or microtubule organizing centers. No microfilaments or intermediate filaments that might serve for condensation have been seen or otherwise identified in these cells. Microtubules are present in abundance but microtrabeculae are all around the tubules and show no evidence of condensation on them. The trabeculae have, in fact, a remarkably uniform size which changes slightly when they participate in pigment aggregation. All of these features support the belief that

---

Fig. 14. a, b   These are stereo images of a small area from a cultured NRK cell. In preparation for microscopy, it was first fixed in glutaraldehyde and OsO_4, then quick-frozen in propane chilled to $-185°C$ and finally dried for 36 h at $-95°C$. The experiment had two major purposes: (1) to observe the effects (destructive or otherwise) of glutaraldehyde, and (2) to avoid critical-point drying identified by H. Ris as a hazardous step in the preparation procedure. Comparisons are made with the images 14c and d of NRK matrix. Here the cell was simply frozen and dried in the same way and at the same time as the cell for Figure 14a and b. The similarities that are close are more important than the differences. To us, it is evident that fixation with glutaraldehyde has induced some clumping of the lattice (arrows), probably because fixation is slow. This leads to the production of larger intertrabecular spaces and some stretching and breaking of trabeculae. Mt, microtubule; ER, endoplasmic reticulum. Magnification 108,000 ×. Bar = 0.1 μm. **c, d)** Stereo images of NRK matrix frozen at $-185°C$ and dried during 36 h at $-95°C$. The dimensions of the microtrabeculae and the intertrabecular spaces vary within a relatively narrow range. There is no evidence of clumping as in Figure 14a and b. It should be stressed that this cell had never seen glutaraldehyde or critical-point drying. The endoplasmic reticulum is represented by a network of small cisternae (ER). Magnification 108,000 ×. Bar = 0.1 μm.

they are parts of a system that exists independently of the so-called cyto-skeleton. If the procedures we use are trustworthy in the case of the ery-throphore, why not when applied to other cell types?

The relatively simple fact that glutaraldehyde fixation yields a morphology in NRK cells that is similar to but not identical with that obtained with freeze-drying describes a limited distortion by the chemical fixative and also implies a preparative distortion of a living lattice structure. Furthermore, if conden-sation takes place during freeze-drying to produce the lattice, would soluble proteins be equally mobile after glutaraldehyde fixation, and if not, as seems likely, where is the residue in the images (see stereo pairs in Fig. 14).

Finally, in limited studies of the effects of environmental factors such as low temperatures (4°C), osmolarity, concentration of Ca^{++} and Mg^{++}, and cytochalasin on the morphology of the MTL [Porter and Tucker, 1981; Porter et al, 1982], it has been shown that the lattice changes uniquely and dra-matically in easily resolvable features. To account for these as expressions of glutaraldehyde and/or OsO_4 fixation would require assumptions greatly in excess of those which explain the changes in terms of hydrophobic bonds, osmotic pressure and ATPases dependent on Mg^{++} or Ca^{++} ions.

These few examples may suffice to relieve a few anxieties as to the equivalence of the MTL to the in vivo structure of the cytomatrix or hyalin cytoplasm.

B. Speculations on the Form and Function of the MTL

The interpretative diagram in Figure 15 provides a basis for a general discussion of the microtrabecular lattice. It is valuable for its depiction of the structure as we currently view it and so constitutes a useful communication device. The hazards in such devices reside chiefly in their being accepted as entirely factual and used to illustrate text books. In this instance, we propose to use the diagram for an explicit and convenient speculation on the principal putative functions of the structure. The various parts represented in the draw-ing are identified in the legend.

It is probable that the microtrabeculae are composed largely of protein. We have chosen, therefore, to refer to them collectively as the *protein-rich* phase of the cytomatrix. This phase is complementary with the more volu-minous, structure-free space that exists between the trabeculae and which we refer to as the *water-rich* phase. Both phases are continuous. The protein-rich comprises a space-filling lattice as noted earlier. It has been shown to be soluble in a number of detergents, including Triton X-100 and Brij 58 [Schliwa et al, 1981] and is presumably a metastable polymer. The water-rich phase has been assumed to include such solutes as K^+, Na^+, Ca^{++}, PO_4^{++}, and metabolites O_2, CO_2, glucose, and amino acids. We have not thus far recognized any precipitate or condensation that might indicate the

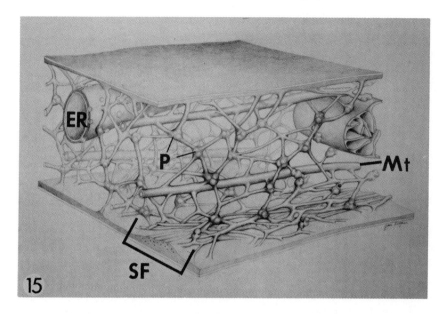

Fig. 15. Interpretative diagram of the MTL and associated organelles and systems of the cytoplast. Representations of the ER, mitochondrion, microtubules (Mt), polysomes (P), and stress fibers (SF) are included. The upper and lower cortices are shown as layers inside the plasma membranes that are continuous with the MTL. The latter, as the name implies, is an irregular three-dimensional lattice of protein-rich strands. This structure plus the water-rich phase between gives the cytoplast its relatively high viscosity (see text). Magnification 100,000 ×

presence of a protein solution in the water-rich phase. This is true no matter what the method of preparation employed.

The drawing (Fig. 15) illustrates that trabeculae of the lattice are continuous with the cytoplasmic cortex, the surfaces of the ER, stress fibers, and polysomes. Expressed another way, the lattice material covers these structures so that the lattice as a unit *contains* all the better-known organelles, etc of the cytoplasm. The so-called zone of exclusion frequently seen around microtubules in thin sections may represent a layer of lattice material (with electron-scattering properties similar to epoxy), and some of the trabeculae associated with the microtubules may be recognized eventually as a form of MAPs. The important concept is containment in the same sense that in the erythrophore a contractile lattice contains the pigment [Luby and Porter, 1980]. We thus have reason to ask whether nonrandom dispositions of polysomes and mRNA may not be integral to the structure of the lattice, especially in differentiating cells. In cultured chick myoblasts, for example, large polysomes are clustered in the vicinity of the forming myofibrils [Porter

and Anderson, 1980]; in epidermal cells, they are numerous around kera-
tohyaline granules [Bonneville, 1968]. There are doubtless other instances
of the same organization that have been or remain to be identified. In this
connection, the observations of Penman and co-workers are very significant
[Fulton and Penman, 1980; Lenk et al, 1977]. They report that mRNA is
attached to elements of the cytoskeleton and that the protein synthesized is
retained at these sites. In our view, it makes reasonable sense for the differentiating cell to synthesize the involved proteins in the places (or within
domains) where they are to be used in assembly and not, alternatively, have
them diffuse away willy-nilly to all parts of the cytoplasm. This reasoning
would seem to apply as well to cellular components that repeatedly assemble
and disassemble, such as microfilaments and microtubules. Thus tubulin
(monomeric or dimeric) released during tubule disassembly is probably retained in the adjacent lattice and kept available for the rapid reassembly of
the tubule [De Mey et al, 1980]. Reorganization and division of the MTL,
as in the formation of daughter cytoplasts and the preprophase asters, would
presumably include transport of the tubulin with the MTL to sites where it
is assembled into astral rays. Such speculation presumes a knowledge of the
MTL greatly in excess of what we have, and is excusable only because it
mentions interesting questions.

Returning again to Figure 15, we find it instructive to view the relationship
of the water-rich phase to the sites of protein synthesis. Assuming that the
amino acids required exist as solutes in this phase, it follows that they are
available within very short diffusion distances (< 10 Å) for combination
with tRNA and the sites of mRNA translation. This should contribute to the
efficiency of the system.

The biphasic structure of the cytoplasmic matrix offers much more for
contemplation. The matrix should have and apparently does have the structure
of a gel. Where the intertrabecular space is large in volume relative to that
of the protein-rich phase, the viscosity of the cytomatrix is presumed to be
low; where small, the viscosity is higher, as in most fully differentiated,
nonmotile cells where the total protein content is 25% by weight. This gel
in cells differs from synthetic noncellular gels in being dynamic. It can change
its shape with the assembly or disassembly of microtubules and actin microfilaments, which seem to provide varying degrees of anisometry to the
whole cell or to lamellopodia and filopodia (or microvilli). The water-rich
phase of this gel structure gives the cell a medium for rapid diffusion of
metabolites and for homeostasis of its internal environment.

In most respects the disposition of cell water is not well understood. In
the grossest terms, it makes up 70–80% of the cell's contents by weight.
For those students of the subject who believe as we do that it is compartmentalized, about 60% of the water is available as a solvent and is commonly

called bulk water [Clegg, 1982; Horowitz and Paine, 1979]. Besides this, there is good evidence for another form that may be ordered with respect to the trabecular surfaces. This is called "vicinal water" or "water of hydration," and constitutes somewhere in the range of 40% of total H_2O. Obviously, the MTL as imaged here is surface-rich, and according to some recent computer-assisted evaluations [Gershon et al, 1982], it provides about 50,000 μm^2 of surface in one cultured NRK cell. Just how this vicinal water fits into the physiology of the cell is essentially unknown. The existence of a bulk water phase in oocytes [Horowitz and Paine, 1979] and Artemia (brine shrimp) [Clegg, 1982] and Euglena [Kempner and Miller, 1968] is now well established. In the latter case, the investigators centrifuged the cells suspended at a 5–25% Dextran interface and stratified their contents into six layers. One of these was water-clear and when appropriately examined was found to be without enzymes or other proteins. It follows from this and other evidence that the bulk water phase in the cytomatrix may contain little if any protein and conversely that the biologically active molecules are structure bound [Kempner, 1980].

By its very nature as a gel, the cytomatrix depends on the presence of water to retain its normal dimensions. In other words, if the water is removed, the gel collapses; if augmented, the gel swells. In contrast, the disposition of cell water and its retention rely on the existence of the lattice and the plasma membrane. These thoughts have prompted us to perform some simple experiments that involve little more than changing the osmolarity of the environment with sucrose and observing the resulting changes in the morphology of the cytomatrix [Porter et al, 1982]. We decided as well to take note of conditions that effected saltatory motion (directed) as opposed to Brownian motion (random) in the responding cell.

The results were essentially as expected. Increasing the osmolarity of the external environment to 600 mOsm (hypertonic) reduced the mean diameters of the intertrabecular spaces (water-rich) by almost 50%. It stopped both saltatory and Brownian motion. The substrate-attached cells collapsed to about half their normal depth, but did not change their dimensions in other respects. The exposure time before fixation or observation was 1 minute. Reducing the osmolarity to 150 mOsm (hypotonic) had the opposite effect. Water was taken into the spaces by osmotic pressure and caused their expansion to diameters that exceeded those of the control by 70%. A recognizable trabecular lattice persisted but included trabeculae with free ends and evidence of fragmentation. Many trabeculae seemed to collapse into the adjacent cortices, and appeared as shadows in these locations. The thickness of the cell increased. Brownian motion became prominent and saltatory motion ceased. Apparently, the integrity of the lattice was partly destroyed, and particles released into the enlarged intertrabecular spaces were freed for

Brownian motion. The cells, however, were not destroyed by these effects. When returned to the normal environment (300 mOsm) for as little as 1 min, they recovered both normal morphology and normal activity.

These observations reinforce our contention that the MTL is a unit of structure within the cell capable of reacting to tonicity variations in the cell's environment without total loss of integrity. It appears to have a preferred morphology to which it quickly returns when the environment has normal osmolarity. We presume it gives the cell its viscoelastic properties. The volume changes of the intertrabecular spaces in response to these experiments indicate that these spaces are the water-rich compartment of the cell.

Low temperatures (4°C), if applied over extended periods of 2–3 hours, are also effective in altering the MTL morphology. In this instance, the cells (NRK) round up and leave only a trace of their original dimensions in the form of slender filamentous attachments to the substrate. First to disassemble in response to chilling are the microtubules, followed within the first half hour by the stress fibers. Thereafter, the continuity and reticular form of the MTL are affected, and the bulk of the lattice either aggregates into clumps or associates closely with the cytoplasmic cortices. Brownian motion becomes obvious. When the chilled cells are returned to physiological temperatures, they rapidly recover their normal morphology [Porter and Tucker, 1981]. Within 5 min, the clumps disperse and the lattice is restructured. After this initial response, the MTL spreads rapidly to fill the available space, and by 30 min the cell and lattice are back where they were before the experiment started. Brownian motion ceases. Again, it appears that the lattice returns to a preferred form when the normal metabolism and supply of ATP (presumed) are restored [Luby and Porter, 1980].

It has already been demonstrated in our study of the erythrophore that the MTL, or some derivative of it, is dynamic. Some observations indicate that the trabeculae can shorten and become beaded (the appearance of varicosities) and then elongate again if the supply of ATP is not cut off and the Ca^{++} buffering system (SER) is intact and functioning. Apparently, saltatory motion is a product of local cycles of these events [Ellisman and Porter, 1980]. The beading phenomenon is not limited to erythrophores and the associated changes with lattice contraction; it appears as well in the clumps encountered in chilled cells, and disappears as the trabeculae elongate in recovery during incubation. The origin of the beads (200 Å in diameter) is not understood, but may reflect some sort of coiling. The basic molecular unit involved is the force-generating mechanism for granule movement, the ghost in the machine. Whether it is some form of dynein or spasmin, as discussed earlier, is not settled at this time. The elongation of the trabeculae is more ATP-dependent than the contraction [Luby and Porter, 1980]. An analogous motion and ATP independence are shown by the mechanisms responsible for moving chromosomes in mitosis [Cande, 1982].

The answer to a number of questions may become available when we know the composition of the lattice and the distribution of its components. Currently, we know that actin is a prominent electrophoretic component in all extracts of the MTL (eg, Triton extracts). There is reason to suppose that both MAPs and actin binding proteins are also present. We find tubulin, of course, and there is evidence of myosin and calmodulin [Hartwig et al, 1977]. Besides these commonly occurring proteins, which appear in two-dimensional gels of Triton extracts, it is possible to count about 150 separate polypeptides [Schliwa et al, 1981a]. Just how many of these come from the MTL and how many from membranes, ribosomes, and nucleoplasm is not clear at this time. In an attempt to make a more representative analysis of the lattice, Kuroda [1982] has recently isolated what seem to be preparations of cytoplasts (the MTL). When these are solubilized and run on 2-D gels, the number of spots is fewer than from extracts of whole cells, but the identity of most of them is unknown.

C. The Cytoskeleton and the MTL

It was evident to us early on that the MTL and the cytoskeleton were not the same. Rather, it seemed that the MTL, in addition to coating or containing, elements such as the microfilaments (as in stress fibers), also contributed many cross links and elongated strands that did not survive extraction of the unfixed cell with Triton X-100 or other detergents. Among these, Brig 58 was relatively mild in its action; it could remove the plasma membrane and ER cisternae without removing all of the MTL [Schliwa et al, 1981a]. Triton, on the other hand, washed away practically everything except the three filament types usually taken to represent the cytoskeleton. Even with this detergent, however, there is no end point to the extraction, making its reproducible use a bit tricky. About 50% of the cell's protein is removed before the residue is fixed for observation. As far as we can tell, all microfilaments are reduced in diameter by this extraction, which means to us that they were coated by MTL proteins before extraction.

Various conclusions may follow from this observation. One is that the filaments of actin, of desmin, and microtubules are contained by the lattice. Another is that the lattice may control the disposition of these cytoskeletal elements, at least during morphogenesis [Porter and Anderson, 1980]. A third is that if they disassemble in the physiological life of the cell, the monomeric actin or tubulin may become part of the lattice [Schliwa, 1979].

This last conclusion should in time be tested, for until now only the slightest of evidence in support of it is available [De Mey et al, 1980].

It is easier to establish, as indicated earlier, that there is much more to the MTL than is represented in the cytoskeleton. Any comparison of images of extracted and nonextracted cultured cells makes that clear [see, eg, Schliwa et al, 1981a]. The similarity is closest at the cell surface (cortex) where extracted cells usually show a wealth of actin filaments. Deeper in the cytoplast, the cytoskeleton is usually represented by 100 Å filaments and microtubules and here in the presence of ER cisternae, mitochondria, polysomes, and secretory or pigment granules the MTL appears to be the component of the gel in which all organelles are suspended.

This story would not be complete if we failed now to relate the present image, revealed by high-voltage microscopy, with that available in the 1940s before electron microscopy destroyed the privilege of one's being a cell detective and putting together small bits of indirect evidence to achieve the whole picture. Frey-Wyssling, in the preface to his 1948 edition, laments the passing of this latter approach. Crick and Hughes are more sanguine about the future as they provide a realistic view of the limitations evident in the technics of moving magnetic particles and the interpretation of their behavior. These viewpoints notwithstanding, it is valuable to equate the current image with that of the 1950s before electron microscopy had penetrated the mysteries of the cytomatrix. They have much in common, we think, for it is easy now to read and accept the speculations of Scarth and Frey-Wyssling and to interpret the motion of magnetic particles through the matrix. Such motions obviously are inhibited by the "polymeric meshwork," in which the particles catch and from which they are then released; they create in their backward and forward motions regions of solation in which they then move more freely.

The water-rich phase did not receive much thought in those earlier papers. Since then, this part of the cell has been separated from the rest by substantial centrifugal forces and has been found to lack proteins and enzyme activity [Kempner and Miller, 1968; Clegg, 1982]. Equivalent data have come from Horowitz and colleagues [1979], who by introducing blobs of gelatin as a reference phase into amphibian eggs have obtained samples of solvents and solutes. They have been able to verify the existence of a water-rich phase, explore its content, and test its capacity to accept a solute such as sucrose. It can be said that the pieces of information are coming together and that in a few years, "with a little bit o' luck," we should have a fairly complete and accurate image of the cytomatrix.

ACKNOWLEDGMENTS

We are pleased to acknowledge the help of Mary Ulrich with the illustrations, and Joyce Albersheim with the typing and preparation of the manuscript. We are grateful also to Richard McIntosh for suggested improvements in the manuscript.

The work reported here was supported by grants from the NIGMS (5R01 GM 27324) and from the Muscular Dystrophy Association of America. The HVEM we used is featured in a shared facility supported by the Biotechnology Resources Program of the Division of Research Resources, NIH (grant 5P41 RR00592). Support was also provided by Ernest Fullam.

Micrographs for the accompanying figures were taken at various magnifications by Karen Anderson using the 1 MeV electron microscope installed by the Division of Research Resources, NIH, as a shared facility in the Department of Molecular, Cellular and Developmental Biology, University of Colorado, Boulder.

V. REFERENCES

Amos WB (1972): Structure and coiling of the stalk in the peritrich ciliates Vorticella and Carchesium. J Cell Sci 10:95–122.

Bagnara JT, Taylor JD, Hadley, ME (1968): The dermal chromatophore unit. J Cell Biol 38:67–79.

Ballowitz E (1893): Die nervenendigungen der pigmentzellen, ein beitrag zur kenntnis des zusammenhanges der endverzweigungen der nerven mit dem protoplasma der zellen. Z Wiss Zool 56:673–706.

Beams HW (1943): Ultracentrifugal studies on cytoplasmic components and inclusions. In Cattell J (ed): "Biological Symposia: Frontiers in cytochemistry," Vol 10. Lancaster, PA: Jaques Cattell Press, pp. 71–90.

Beckerle MC, Porter KR (1982): Inhibitors of dynein activity block intracellular transport in erythrophores. Nature 295:701–703.

Beckerle MC, Porter KR (1983): Analysis of the role of microtubules and actin in erythrophore intracellular motility. J Cell Biol (in press).

Beckerle MC, Byers HR, Fujiwara K, Porter KR (1979): Indirect immunofluorescence and stereo high voltage electron microscopic evidence for microtubule associated migration of pigment granules in erythrophores. J Cell Biol 83(2):352a.

Behnke O, Kristensen BI, Engdahl Nielsen L (1971): Electron microscopical observations on actinoid and myosinoid filaments in blood platelets. J Ultrastr Res 37:351–369.

Bensley RR, Hoerr NJ (1934): Studies on cell structure by the freeze-drying method. V. The chemical basis of the organization of the cell. Anat Rec 60:251–266.

Bikle D, Tilney LG, Porter KR (1966): Microtubules and pigment migration in the melanophores of Fundulus heteroclitus. Protoplasma 61:322–345.

Bonneville MA (1968): Observations on epidermal differentiation in the fetal rat. Am J Anat 123:147–164.

Bouchard P, Penningroth SM, Cheung A, Gagnon C, Bardin CW (1981): Erythro-9-(3-(2-hydroxynonly))adenine is an inhibitor of sperm motility that blocks dynein ATPase and carboxylmethylase activities. Proc Natl Acad Sci USA 78:1033–1036.

Branton D, Cohen LM, Tyler J (1981): Interaction of cytoskeletal proteins on the human erythrocyte membrane. Cell 24:24–32.

Buckley IK (1974): Subcellular motility: A correlated light and electron microscopic study using cultured cells. Tissue Cell 6:1–20.

Buckley IK (1975): Three-dimensional fine structure of cultured cells: Possible implications for subcellular motility. Tissue Cell 7:51–72.

Buckley IK, Porter KR (1975): Electron microscopy of critical-point dried whole cultured cells. J Micros 104:107–120.

Buckley IK, Raju TR (1976): Form and distribution of actin and myosin in non-muscle cells: A study using cultured chick embryo fibroblasts. J Micros 107:129–149.

Burton PR, Fernandez HL (1973): Delineation by lanthanum staining of filamentous elements associated with the surfaces of axonal microtubules. J Cell Sci 12:567–583.

Byers B, Porter KR (1964): Oriented microtubules in elongating cells of the developing lens rudiment after induction. Proc Natl Acad Sci USA 52:1091–1099.

Byers HR, Porter KR (1977): Transformations in the structure of the cytoplasmic ground substance in erythrophores during pigment aggregation and dispersion. J Cell Biol 75:541–558.

Byers HR, Fujiwara K, Porter KR (1980): Visualization of microtubules of cells *in situ* by indirect immunofluorescence. Proc Natl Acad Sci USA 77:6657–6661.

Cande WZ (1982): Nucleotide requirements for anaphase chromosome movements in permeabilized mitotic cells: Anaphase B but not anaphase A requires ATP. Cell 28:15–22.

Cande WZ, Wolniak SM (1978): Chromosome movement in lysed mitotic cells is inhibited by vanadate. J Cell Biol 79:573–580.

Clark TG, Rosenbaum JL (1982): Pigment particle translocation in detergent-permeabilized melanophores of Fundulus heteroclitus. Proc Natl Acad Sci USA 79:4655–4659.

Claude A (1940): Particulate components of normal and tumor cells. Science 91:77–78.

Clegg JS (1982): Interrelationships between water and cell metabolism in *Artemia* cysts. IX. Evidence for organization of soluble cytoplasmic enzymes. In Cold Spring Harbor Symposium on Quantitative Biology, Vol. XLVI, Cold Spring Harbor Laboratory, pp 23–37.

Crick FHC, Hughes AFW (1950): The physical properties of cytoplasm. A study by means of the magnetic particle method. Part I: Experimental. Exp Cell Res 1:37–80.

De Mey J, Wolosewick JJ, De Brabander M, Geuens G, Joniau M, Porter KR (1980): Tubulin localization in whole, glutaraldehyde fixed cells, viewed with stereo high-voltage electron microscopy. In De Brabander J (ed): "Cell Movement and Neoplasia." Oxford: Pergamon Press, pp 21–28.

Ellisman MH, Porter KR (1980): Microtrabecular structure of the axoplasmic matrix: Visualization of cross-linking structures and their distribution. J Cell Biol 87:464–479.

Euteneuer U, McIntosh JR (1981): Polarity of some motility-related microtubules. Proc Natl Acad Sci USA 78:372–376.

Forman DS (1982): Vanadate inhibits saltatory organelle movement in a permeabilized cell model. Exp Cell Res 141:139–148.

Frey-Wyssling A (1948): "Submicroscopic Morphology of Protoplasm and Its Derivates." Amsterdam: Elsevier.

Fulton AB (1982): How crowded is the cytoplasm. Cell 30:345–347.

Fulton AB, Wan KM, Penman S (1980): The spatial distribution of polyribosomes in 3T3 cells and the associated assembly of proteins into the skeletal framework. Cell 20:849–857.

Gershon N, Porter K. Trus B (1982): The microtrabecular lattice and the cytoskeleton. Their volume, surface area and the diffusion of molecules through it. Proc of the Aharon Katzir-Katchalsky Memorial Symposium (in press).

Gibbons IR, Cosson MP, Evans JA, Gibbons BH, Houck B, Martinson KH, Sale WS, Tang WY (1978): Potent inhibition of dynein adenosine triphosphatase and of motility of cilia and sperm flagella by vanadate. Proc Natl Acad Sci USA 75:2220–2224.

Green L (1968): Mechanism of movements of granules in melanocytes of Fundulus heteroclitus. Proc Natl Acad Sci USA 59:1179–1186.

Guatelli JC, Porter KR, Anderson KL, Boggs DP (1982): Ultrastructure of the cytoplasmic and nuclear matrices of human lymphocytes observed using high voltage electron microscopy of embedment-free sections. Biol Cell 43:69–80.

Hama K (1980): Whole mount observations of cultured cells. In Bailey GW (ed): "Proc 38th Ann EMSA Meeting." Baton Rouge: Claitor's Publishing, pp 802–805.

Hartwig JH, Davies WA, Stossel TP (1977): Evidence for contractile protein translocations in macrophage spreading, phagocytosis, and phagolysosome formation. J Cell Biol 75:956–967.

Hawes CR, Juniper BE, Horne JC (1982): Electron microscopy of resin-free sections of plant cells. Protoplasma (in press).

Heilbronn A (1922): Jahrb Wiss Bot 61:284.

Horowitz SB, Paine PL (1979): Reference phase analysis of free and bound intracellular solutes. II. Isothermal and isotopic studies of cytoplasmic sodium, potassium and water. Biophys J 25:45–62.

Horowitz SB, Paine PL, Tluczek L, Reynhout JK (1979): Reference phase analysis of free and bound intracellular solutes. I. Sodium and potassium in amphibian oocytes. Biophys J 25:33–44.

Junqueira LCU, Toledo AMS, Porter KR (1970): Observations on the structure of the skin of the teleost Fundulus heteroclitus (L). Arch Hist Jpn 32:1–15.

Junqueira LC, Raker E, Porter KR (1974): Studies on pigment migration in the melanophores of the teleost Fundulus heteroclitus (L). Arch Hist Jpn 36:339–366.

Kempner ES (1980): Metabolic compartments and their interactions. In Nover L, Lynen F, Mothes K (eds): "Cell Compartmentation and Metabolic Channeling." Jena (GDR): VEB Gustav Fischer Verlag; Amsterdam: Elsevier/North-Holland, pp 217–224.

Kempner ES, Miller JH (1968): The molecular biology of Euglena gracilis. V. Enzyme localization. Exp Cell Res 51:150–156.

Kobayashi T, Martensen T, Nath J, Flavin M (1978): Inhibition of dynein ATPase by vanadate, and its possible use as a probe for the role of dynein in cytoplasmic motility. Biochem Biophys Res Comm 81:1313–1318.

Kuroda M, Olson G, Porter KR (1982): The structure and protein composition of isolated cytoskeletons with and without the microtrabecular lattice. (In Preparation.)

Lenk R, Ransom L, Kaufman Y, Penman S (1977): A cytoskeletal structure with associated polyribosomes obtained from HeLa cells. Cell 10:67–78.

Luby KJ, Porter KR (1980): The control of pigment migration in isolated erythrophores of Holocentrus ascensionis (Osbeck). I. Energy requirements. Cell 21:13–23.

Luby-Phelps K, Porter KR (1982): The control of pigment migration in isolated erythrophores of *Holocentrus ascensionis* (Osbeck). II. The role of calcium. Cell 29:441–450.

Marsland DA (1942): Protoplasmic streaming in relation to gel structure in the cytoplasm. In Seifritz W (ed): "The Structure of Protoplasm: A Monograph of the American Society of Plant Physiologists." Ames, Iowa: Iowa State College Press, pp 127–161.

McIntosh JR, Ogata ES, Landis S (1973): The axostyle of Saccinobaculus. I. Structure of the organism and its microtubule bundle. J Cell Biol 56:304–323.

McIntosh JR (1974): Bridges between microtubules. J Cell Biol 61:166–187.

McNiven MA, Porter KR (1981): The microtubule organizing center in erythrophores; its 3-dimensional structure and behavior during pigment motion. J Cell Biol 91(2):334a.

Metuzals J (1969): Configuration of a filamentous network in the axoplasm of the squid (Loligopaclii L) giant nerve fiber. J Cell Biol 43:480–505.

Mooseker MS, Tilney LG (1973): Isolation and reactivation of the axostyle. Evidence for a dynein-like ATPase in the axostyle. J Cell Biol 56:13–26.

Moyer LS (1942): Proteins and protoplasmic structure. In Seifritz W (ed): "The Structure of Protoplasm: A Monograph of The American Society of Plant Physiologists." Ames, Iowa: Iowa State College Press, pp 23–40.

Murphy DB, Tilney LG (1974): Role of microtubules in the movement of pigment granules in teleost melanophores. J Cell Biol 61:757–779.

Obika M, Turner WA, Negishi S, Menter DG, Tchen TT, Taylor JD (1978): The effects of lumicolchicine, colchicine and vinblastin on pigment migration in fish chromatophores. J Exp Zool 205:95–110.

Parker GH (1948): "Animal Colour Changes and Their Neurohumours: A Survey of Investigations, 1910–1943." Cambridge, England: Cambridge University Press.

Penningroth SM, Cheung A (1982): Pharmacology and structure of flagellar and cytoplasmic dyneins. In Proc of International Conference on Development and Function in Cilia and Sperm Flagella, Siena, Italy, p 46.

Penningroth SM, Cheung A, Bouchard P, Gagnon C, Bardin CW (1982): Dynein ATPase is inhibited selectively in vitro by erythro-9-(3-2-(hydroxynonyl)) adenine. Biochem Biophys Res Comm 104:234–240.

Pike MC, Kredich NM, Snyderman R (1978): Requirement of S-adenosyl-L-methionine-mediated methylation for human monocyte chemotaxis. Proc Natl Acad Sci USA 75:3928–3932.

Porter KR (1973): Microtubules in intracellular locomotion. In "Locomotion of Tissue Cells," Ciba Foundation Symposium 14. Amsterdam: Elsevier/North-Holland, pp 149–169.

Porter KR (1978): Organization in the structure of the cytoplasmic ground substance. In Sturgess JM (ed): "Organization of Intracellular Filament Systems." Mississauga, Ontario: Imperial Press, pp 627–639.

Porter KR, Anderson KL (1980): The morphogenesis of myofibrils from the microtrabecular lattice as observed in cultured myoblasts. In Ebashi S (ed): "Muscle Contraction: Its Regulatory Mechanisms." Berlin: Springer-Verlag, pp 527–540.

Porter KR, Anderson KL (1982): The structure of the cytoplasmic matrix preserved by freeze-drying and freeze-substitution. Eur J Cell Biol (in press).

Porter KR, Beckerle MC, McNiven MA (1982): A mechanism for intracellular transport. In Guerri C (ed): Symposium "Cajal: Horizons in Neuroscience," Valencia, Spain.

Porter KR, Bennett GS, Junqueira LC (1970): Microtubules in intracellular pigment migration. In Proc 7th International Congress on Electron Microscopy, Grenoble, France, pp 945–946.

Porter KR, Boggs DP, Anderson KL (1982): The distribution of water in the cytoplasm. Proc 40th Ann EMSA Meeting, Washington, DC.

Porter KR, McNiven MA (1982): The cytoplast: A unit structure in chromatophores. Cell 29:23–32.

Porter KR, Tucker JB (1981): The ground substance of the living cell. Sci Am 244(3):57–67.

Pryzwansky KB, Schliwa M, Porter KR (1982): Comparison of the three-dimensional organization of unextracted and Triton-extracted human neutrophilic polymorphonuclear leukocytes. Eur J Cell Biol (in press).

Pudney J, Singer RH (1980): Intracellular filament bundles in whole mounts of chick and human myoblasts extracted with Triton X-100. Tissue Cell 12:595–612.

Rebhun LI (1972): Polarized intracellular particle transport: Saltatory movements and cytoplasmic streaming. In Bourne GH, Danielli JF (eds): "International Review of Cytology." New York: Academic Press, pp 93–137.

Ris H (1980): The cytoplasmic "microtrabecular lattice"—Reality or artifact. In Bailey GW (ed): "Proceedings of 38th Annual EMSA Meeting." Baton Rouge: Claitor's Publishing, pp 812–813.

Robison WG Jr, Charlton JS (1973): Microtubules, microfilaments and pigment movement in the chromatophores of Palaemonetes vulgaris. J Exp Zool 186:279–304.

Rohrlich ST, Porter KR (1972): Fine structural observations relating to the production of color by the iridophores of a lizard, Anolis carolinensis. J Cell Biol 53:38–52.

Routledge LM (1978): Calcium-binding proteins in the Vorticellid spasmoneme. J Cell Biol 77:358–370.

Saidel WM (1977): Metabolic energy requirements during teleost melanophore adaptations. Experientia 33:1573–1574.

Scarth GW (1927): The structural organization of plant protoplasm in the light of micrurgy. Protoplasma 2:189–205.

Scarth GW (1942): Structural differentiation of cytoplasm. In Seifritz W (ed): "The Structure of Protoplasm: A Monograph of the American Society of Plant Physiologists." Ames, Iowa: Iowa State College Press, pp 99–107.

Schliwa M (1978): Microtubular apparatus of melanophores. J Cell Biol 76:605–614.

Schliwa M (1979): Stereo high voltage electron microscopy of melanophores. Exp Cell Res 118:323–340.

Schliwa M (1981): Microtubule-dependent intracellular transport in chromatophores. In Schweiger HG (ed): "International Cell Biology 1980–1981." Berlin: Springer-Verlag, pp 275–285.

Schliwa M, Euteneuer U (1978): A microtubule-independent component may be involved in granule transport in pigment cells. Nature 273:556–558.

Schliwa M, Van Blerkom J, Porter KR (1981a): Stabilization of the cytoplasmic ground substance in detergent-opened cells and a structural and biochemical analysis of its composition. Proc Natl Acad Sci USA 78:4329–4333.

Schliwa M, Weber K, Porter KR (1981b): Localization and organization of actin in melanophores. J Cell Biol 89:267–275.

Seifritz W (1942): Some physical properties of protoplasm and their bearing on structure. In Seifritz W (ed): "The Structure of Protoplasm: A Monograph of the American Society of Plant Physiologists." Ames, Iowa: Iowa State College Press, pp 245–264.

Shimizu T (1981): Steady-state kinetic study of vanadate-induced inhibition of ciliary dynein adenosine triphosphate activity in Tetrahymena. Biochem 20:4347–4354.

Silver RB, Cole RD, Cande WZ (1980): Isolation of mitotic apparatus containing vesicles with calcium sequestration activity. Cell 19:505–516.

Smith DC, Smith MT (1935): Observations on the color changes and isolated scale erythrophores of the squirrel fish, Holocentrus ascensionis (Osbeck). Biol Bull 68:131–139.

Stearns RE, Ochs RL (1982): A functional *in vitro* model for studies of intracellular motility

in digitonin permeabilized erythrophores. J Cell Biol (in press).
Tilney LG, Hiramoto Y, Marsland D (1966): Studies on the microtubules in heliozoa. III. A pressure analysis of the role of these structures in the formation and maintenance of the axopodia of Actinosphaerium nucleofilum (Barret). J Cell Biol 29:77–95.
Wardrop A (1982): Evidence for the possible presence of a microtrabecular lattice in plant cells. Protoplasma (in press).
Wessells NK, Spooner BS, Ludueña MA (1973): Surface movements, microfilaments and cell locomotion. In Abercrombie (ed): "Locomotion in Tissue Cells." A Ciba Foundation Symposium.
Wolniak SM, Hepler PK, Jackson WT (1980): Detection of the membrane-calcium distribution during mitosis in Haemanthus endosperm with chlorotetracycline. J Cell Biol 87:23–32.
Wolosewick JJ (1980): The application of polyethylene glycol (PEG) to electron microscopy. J Cell Biol 86:675–681.
Wolosewick JJ, Porter KR (1975): High voltage electron microscopy of WI-38 cells. Anat Rec 181:511–512.
Wolosewick JJ, Porter KR (1979): The microtrabecular lattice of the cytoplasmic ground substance: Artifact or reality. J Cell Biol 82:114–139.
Yamada KM, Spooner BS, Wessells NK (1971): Ultrastructure and function of growth cones and axons of cultured nerve cells. J Cell Biol 49:614–635.

Organization of Nuclear Structure and the Control of Gene Action

Modern Cell Biology 2:305–328
© 1983 Alan R. Liss, Inc., 150 Fifth Avenue, New York, NY 10011

Strategies for Gene Organization and Information Expression

Leroy E. Hood, Tim Hunkapiller, and Ellen Kraig

From the Division of Biology, California Institute of Technology, Pasadena, California 91125

I. INTRODUCTION

During the last 10 years, recombinant DNA technology has revolutionized our view of eukaryotic gene structure and chromosomal organization. We would like to discuss three concepts that have recently emerged from this work: 1) the existence of split genes with their exons, or coding regions, and introns, or intervening DNA sequences; 2) the organization of related genes into tandemly linked arrays, the multigene families; and 3) the dynamic nature of eukaryotic chromosomes, which are capable of changing by DNA rearrangements during evolution and even during the development of a single organism.

These ideas are important because they allow one to recognize facets of the complexity of genome structure that have profound implications, both for the mechanisms of control of gene expression and for the information content of the genes themselves. It will be seen that individual exons can encode discrete structural domains of the resulting gene products, so the duplication of an exon can lead to the modular construction of very complex genes. Similarly, an individual gene can undergo repeated duplications to generate a multigene family, each member of which can evolve in subtly different ways. The resulting diversity can in turn be organized into batteries of genes, which together may constitute a program for building a part of a complex organism, such as a hand. Although many aspects of this structural complexity are not yet understood, the information available permits insights into the problem of how a limited number of genes can code for the structural complexity which is manifest in a multicellular, eukaryotic organism.

II. SPLIT GENES—INTRONS AND EXONS

Until recombinant DNA techniques were applied to the study of eukaryotic genomes, it was generally assumed that eukaryotic genes would be arrayed and transcribed much like those of prokaryotes, in a linear and contiguous manner. However, we now know that many eukaryotic genes are split (Fig. 1), containing as few as two exons in certain insulin genes or as many as 50 or more exons present in the collagen genes [Abelson, 1979]. Exons generally encode the polypeptide sequence, as well as the 3' and 5' untranslated portions of the mRNA. The function of introns has been less clear and hence is much discussed. An intron of one gene, yeast mitochondrial cytochrome b, encodes a protein or "maturase" that is involved in processing the message of the larger gene [Lazowska et al, 1980]. However, it appears certain that this is not a common phenomenon. Most intervening sequences do not code for regulatory peptides. In the discussion that follows, it will become obvious that introns can play a fundamental role in strategies of gene expression by dividing genes into discrete units which can then be rearranged

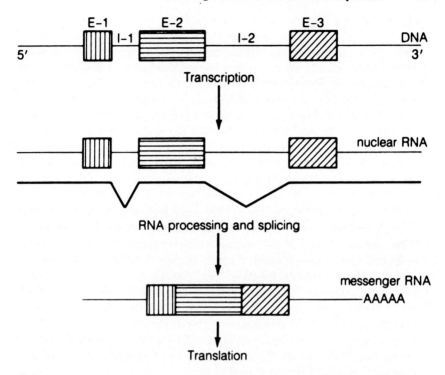

Fig. 1. A model of RNA splicing. The original RNA transcript of a split gene includes both coding (exons or E) and noncoding (introns or I) sequences. RNA processing, which includes the splicing together of all exons and the subsequent loss of introns, produces a mature, translatable messenger RNA. Non-protein-coding RNAs (transfer RNA, ribosomal RNAs) undergo a similar RNA splicing process. [From Hunkapiller et al, 1982].

and joined in alternative assemblies either through DNA rearrangements or alternative patterns of RNA splicing.

A. Exons, Domains, Function, and Evolution

In many genes the exons encode separate structural, functional, and evolutionary units. For example, antibody molecules are divided into discrete structural domains that correspond precisely to the separate exons encoding them [Sakano et al, 1979; Calame et al, 1980]. A similar correlation between exon and functional domain has been noted for two other families of cell-surface recognition proteins, transplantation antigens and Ia molecules encoded in the major histocompatibility complex [Steinmetz et al, 1981; McNicholas et al, 1982].

To demonstrate how the exon-intron structure may be employed to amplify information during development and in the process of evolution, we use as an example the antibody gene. Antibodies are composed of two polypeptides, light and heavy chains, which in turn are divided into two distinct regions, an amino-terminal portion of variable (V) amino-acid sequence, and a carboxy-terminal region which is constant (C) in sequence from one immunoglobulin to another. These polypeptides fold into four or five discrete domains which carry out distinct functions—pattern recognition by the variable domains and effector functions by the constant domains (Fig. 2). Individual exons encode each of the structural domains which perform individual functions, so exons encode functional as well as structural units of information (Fig. 2A).

The exons of the antibody gene exhibit sequence homology to one another, indicating that the various individual antibody gene exons diverged from a common ancestral exon. In this manner, complex genes can evolve to encode multidomain proteins composed of many discrete functions, joined together as modules. The multiple functions (eg, pattern recognition and effector functions for antibodies) can then operate synergistically. The exons become units of evolution as well as units of structure and function. Through rearrangement or duplications and deletions of such elements or "exon shuffling," successful structural and functional information can be employed within new contexts [Doolittle, 1978] (Fig. 2B). Hence the exon-intron organization provides the potential for rapid evolution.

B. Informational Expansion via Alternative Patterns of RNA Splicing

In order to express the polypeptide encoded by a split gene, the introns must be deleted. The gene is transcribed in its entirety, and the RNA is then spliced or processed such that the introns are deleted and the corresponding exons are joined in a contiguous fashion to produce a mature mRNA (Fig. 1). RNA splicing can be employed by a cell to generate two or more mRNAs from a single gene. For example, the antibody heavy chain genes encode polypeptides that must exist in two physiological states—membrane-bound (m) and secreted (s). An analysis of the messenger RNAs encoding these two forms of the mu (μ) heavy chain demonstrate that they are identical throughout their coding regions but differ at their 3' ends. The μ_m chain has a hydrophobic tail and the μ_s chain a hydrophilic tail (Fig. 3). The μ_s and μ_m RNAs are generated from a single heavy chain μ gene by alternative patterns of RNA splicing (Fig. 3). The hydrophilic tail of the secreted μ chain is 20 residues in length and is encoded within the last exon of the μ gene proper, whereas the hydrophobic membrane tail is 41 codons in length and is encoded by two exons located approximately 1,850 nucleotides from the 3' end of the μ gene [Early et al, 1980; Kehry et al, 1980; Rogers et al, 1980].

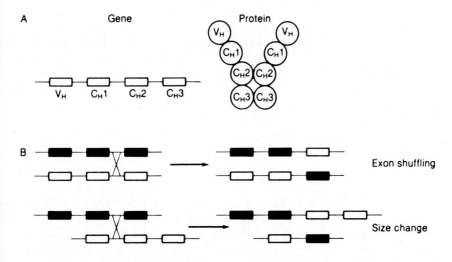

Fig. 2. A) The correlation of immunoglobulin exons and globular domains. The DNA organization of an immunoglobulin heavy chain gene and domain structure of the corresponding heavy chain polypeptide are represented. The boxes represent exons, and the lines represent flanking sequences or introns for the gene. The circles represent discrete globular domains for the protein (the light chains are omitted for clarity). There is perfect correlation between the exons and the corresponding globular domains. B) Exon shuffling. Classical genetic phenomena can result in the rearrangement or "shuffling" of exon information. Shown are possible rearrangement events involving immunoglobulin heavy chain constant region genes that could result in new exon arrangements and, subsequently, new synergistic associations of functionally and/or structurally discrete domains. [From Hunkapiller et al, 1982.]

For the antibody μ heavy chain gene, the alternative patterns of RNA splicing are developmentally regulated. B cells at an early developmental stage contain approximately equal amounts of each of the mRNA species. However, in mature plasma cells secreting large quantities of antibody molecules, 95–98% of the messenger RNA is of the μ_s type. Therefore, differential RNA splicing is one way the cell can increase, or "amplify," the potential information encoded within a single gene. It is likely that each of the eight heavy chain genes in the mouse immune system employs alternative patterns of RNA splicing to generate both secreted and membrane-associated forms of the polypeptide chains. In addition, during B-cell development, two or more classes of immunoglobulins may be expressed simultaneously in individual B cells, again regulated at least in part by differential RNA splicing [Moore et al, 1981]. Several additional examples of eukaryotic genes that employ a similar strategy of alternative patterns of RNA splicing have been described, including the α-amylase gene [Hagenbüchle et al, 1980]. It is

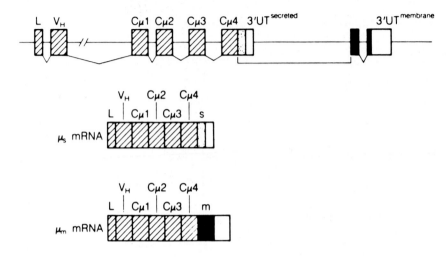

Fig. 3. Antibody molecules are expressed in either the secreted (s) or the membrane-bound (m) form, depending on the RNA splicing and processing pathways employed after transcription of the genes. The exons (boxes) and introns (lines) of the μ heavy chain gene (encoding antibodies of the IgM class) are depicted. Alternative splicing patterns result in two mRNAs, $\mu_{membrane}$ and $\mu_{secreted}$, that differ only at the 3' (C-terminal) coding end. Open boxes denote the 3' untranslated regions. Black boxes represent the membrane tail coding region and stippled boxes the secreted tail-coding region. L, V_H, and C_μ denote the leader, variable gene, and constant domain exons, respectively. [From Hunkapiller et al, 1982].

likely that many eukaryotic genes employ this strategy to amplify genomic information so that the resulting gene products can function in disparate environments (membrane and blood) or acquire new functions (different immunoglobulin effector functions).

III. MULTIGENE FAMILIES

Just as exons may be duplicated to create more complex genes, so individual genes may be duplicated to generate the multigene families characteristic of all eukaryotic DNA. Such families are composed of informational units that 1) are homologous in structure, 2) overlap in function, and 3) are often linked in tanden arrays (Fig. 4). Multigene families range in size from a few copies, like the globin families, to the thousands of gene copies of ribosomal genes found in some organisms. The gene copies within a family can be virtually identical or quite divergent. Multigene families are dynamic entities over the course of evolution. They provide special opportunities for processing, expressing, expanding, and evolving genetic information. The

Chromosome

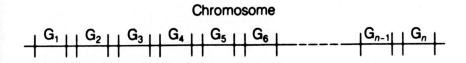

Fig. 4.   Model of a multigene family. [From Hood et al, 1975].

general properties of multigene families have fascinating implications for the generation and developmental regulation of information in eukaryotic organisms.

A. Variation in Gene Copy Number

Multigene families can vary greatly in size from one species to another. For example, the clawed toad Xenopus laevis has 24,000 copies of 5S ribosomal genes, whereas its cousin Xenopus borealis, has only 6000 copies [Brown et al, 1971]. Expansion and contraction events may occur in all multigene families. For example, the λ immunoglobulin variable gene family in mouse has only two gene members, whereas its human counterpart has many more. Presumably the expansion-contraction of multigene families is promoted through homologous, unequal crossing over (Fig. 5). Therefore, on an evolutionary scale, the genetic potential of multigene families can change rapidly within an appropriate selective context.

B. Heterogeneity

A single gene generally can encode only a single chemically homogeneous gene product. In contrast, a multigene family can encode a heterogeneous collection of closely related protein species. For example, the ribosomal and histone gene families have members whose sequences are virtually identical to one another within a given species. These gene families are termed "identical" or "multiplicational" multigene families. In contrast, the antibody V genes may differ from one another by as much as 50% of their nucleotide sequence. Such heterogeneous multigene families are termed "informational." Thus, identical multigene families may express protein or RNA products of a single type that are needed in very large quantities, such as histone or ribosomal genes. In contrast, the heterogeneous multigene families, like the one encoding the antibodies, generate many discretely different gene products which, in the case of antibodies, provide the basis for the enormous range of specificities in the immune system.

The organization of genes within a multigene family may have a major role in determining the extent of heterogeneity within that family. For example, identical multigene families tend to be composed of closely spaced

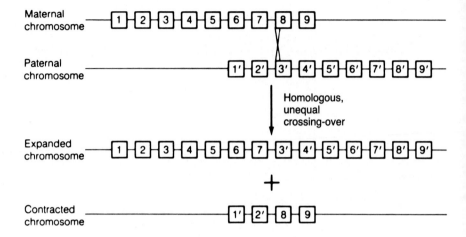

Fig. 5. A model of homologous but unequal crossing over. Homologous crossing-over events can be promoted by the mispairing of homologous genes between chromosomes carrying closely linked, homologous multigene families. This process yields one chromosome with an increased number of genes and a second with a decreased number of genes. [From Hood et al, 1975].

units with virtually identical flanking sequences. In contrast, the members of the various antibody V-gene families, numbering up to 250 or more are widely dispersed and separated by long stretches of DNA, ranging from 5,000 to more than 40,000 nucleotides. Presumably those gene families whose individual members and flanking sequences are more homologous can undergo unequal crossing over more easily. In this way, the family will maintain multiple, very similar copies which undergo extensive expansion and contraction [Smith, 1976]. In contrast, families like the antibody V genes, with a great deal of interspersed nonhomologous DNA, may be more restricted in the frequency of unequal crossing over, resulting in greater diversity. In the absence of this corrective mechanism, the individual members of the multigene family are free to evolve independently, resulting in greater structural diversity [Hood et al, 1980].

C. Multigene Families Shield Their Gene Copies From Natural Selection

Gene copies that overlap in function are shielded from selection in the sense that deleterious mutations of one gene copy will be buffered if the other gene copies can assume, at least in part, its function. For example, hundreds of virtually identical histone genes are present in many higher organisms. The effect on the organism from the loss of one functional histone gene copy through mutation would be negligible. Thus, in the absence of correction, redundant gene copies would be expected to diverge spontaneously over evolutionary time. Such shielding is exemplified by the presence of pseudogenes. A pseudogene occurs when a gene sequence has been altered so that it no longer is functionally expressed. Pseudogenes result from nonsense mutations, changes in the RNA processing signals, or sequence interruptions that lead to translational frameshifts. Indeed, 25% of the known globin-like genes are pseudogenes. The frequency with which pseudogenes occur underscores the fact that individual members of the gene family may be under reduced selective constraints. Presumably pseudogene sequences will eventually randomize through mutation unless they have selectable functions within the gene family [Li et al, 1981].

D. Developmental Control of Gene Expression

The hemoglobin synthesized by humans is encoded by two multigene families, the α and β globins, whose individual members are expressed in a developmentally regulated fashion (Fig. 6) [Proudfoot et al, 1980]. Different globins have distinct oxygen-binding properties and are regulated by various effector molecules in quite different fashions. Distinct globins are expressed at the embryonic ($\zeta_2 \, \epsilon_2$, $\alpha_2 \, \epsilon_2$), the fetal ($\alpha_2 \, \gamma_2$), and the adult ($\alpha_2 \, \delta_2$, $\alpha_2\beta_2$) stages of development (Fig. 6B). The order of expression of the β-like genes is correlated directly to the β gene genomic order (5' to 3'). This observation raises the intriguing possibility that other more complex multigene systems, like the antibodies, will be capable of developmental regulation of the expression of individual gene expression.

E. Multigene Families Are a New Unit of Evolution

The emergence of informationally complex multigene families may have occurred first during the emergence of metazoan organisms. Indeed, the appearance of early multigene families may have reflected the enormous increase of information required at the metazoan cell surface for cellular interactions, transmembrane signaling, and cellular migrations. Once generated, these gene families presented the organism with a new evolutionary unit—the multigene family itself. A multigene family or any part of it may be duplicated and could then diverge just as do duplicate genes. Some families could come to encode new phenotypic attributes. Since the duplications of an entire family would copy not only the structural genes, but all cis-controlling elements as well, complex multigene families could be generated rapidly, without reenacting the evolutionary process that initially shaped the parent gene family. Thus, in time, a superfamily, composed of many multigene families, each devoted to distinct aspects of the eukaryotic phenotype, could have emerged (Fig. 7) [Hood et al, 1975]. This attractive hypothesis has recently been supported by the intriguing observation that the mammalian genes encoding the transplantation antigens, the Ia (Ir) polypeptides, and antibodies all have similar general structures and encode polypeptides with

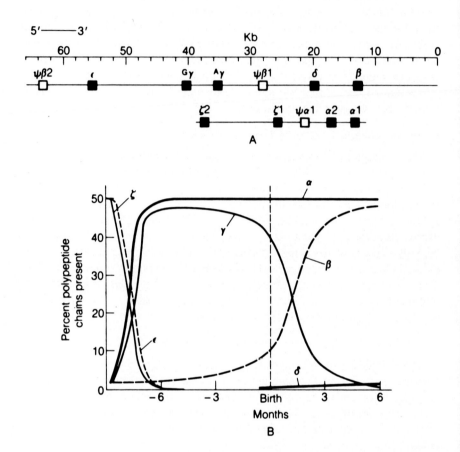

Fig. 6. A) Models of the beta (β) and alpha (α) globin gene families in man. Individual genes (boxes) denote separate genes and do not indicate exon-intron configurations. Open boxes represent pseudogenes (ψ). The ε and ζ genes are embryonic; the γ genes are fetal; the α, β, and δ genes are adult. The size of these families is indicated in kilobases (kb). B) Time course of concentrations of the respective polypeptides. (Courtesy of T. Maniatis.)

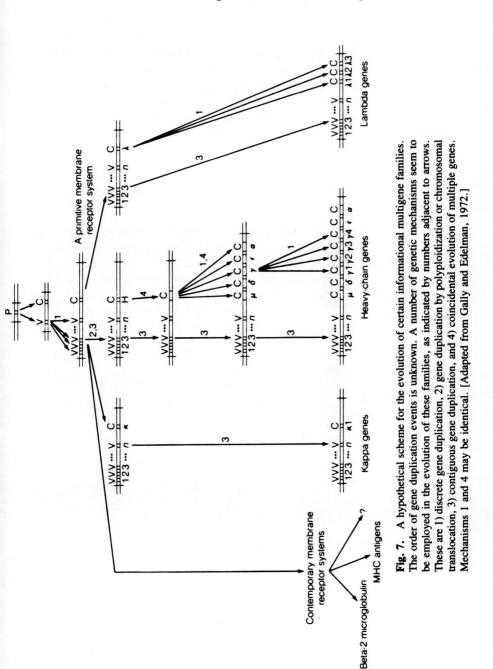

Fig. 7. A hypothetical scheme for the evolution of certain informational multigene families. The order of gene duplication events is unknown. A number of genetic mechanisms seem to be employed in the evolution of these families, as indicated by numbers adjacent to arrows. These are 1) discrete gene duplication, 2) gene duplication by polyploidization or chromosomal translocation, 3) contiguous gene duplication, and 4) coincidental evolution of multiple genes. Mechanisms 1 and 4 may be identical. [Adapted from Gally and Edelman, 1972.]

similar domains (Fig. 8). Each of these types of genes contain one or more exons that exhibit striking sequence homology to its counterparts in the other gene families. Thus, three multigenic families encoding various aspects of immune responsiveness appear to share a common evolutionary heritage (Fig. 7). Functional interactions and informational feedback schemes also can be maintained between disparate families by the evolution of trans-acting regulatory mechanisms. Thus very sophisticated gene families could have evolved through a limited number of genetic events and possibly within a very short period of evolutionary time.

IV. DNA REARRANGEMENTS AND THE DYNAMIC NATURE OF EUKARYOTIC CHROMOSOMES

Within prokaryotic and eukaryotic genomes there are numerous quasistable genetic elements, able to alter rapidly the genetic structure and potential of an organism primarily through rearrangement and/or expansion of the genetic information. Often these changes are a physiologic response to the organism's environment. Particularly interesting are mobile genetic forms that are able to transfer information between organisms and even between species. Many of these elements are capable of autonomous replication. Such elements have

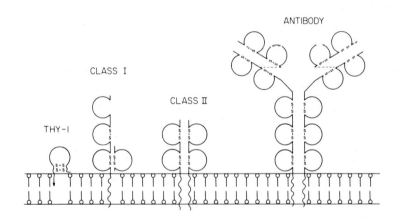

Fig. 8. Molecules that share homology with antibodies. A schematic representation of cell-surface molecules that share structural homology with antibodies. Included are Thy-1, an antigen found on T cells and brain, class I and class II molecules encoded by the major histocompatability complex, and an antibody molecule. Disulfide bonds are indicated (S-S). [Adapted from Williams, 1982.]

been implicated theoretically in the evolution of eukaryotic genes, like the antibody genes. Another class of DNA rearrangements has also been identified as necessary for the function of certain genes, including antibodies, mating type in yeast, and the trypanosome coat antigens.

A. Quasistable Genetic Elements in Prokaryotes

The mobile quasistable systems range in size and complexity from the tiny viroids of plants and insertion (IS) sequences of bacteria to sophisticated viruses. Their effects upon the host organism range from benign parasitism to the subversion of an entire genome to their own purposes. Although most of the well-studied examples of quasistable elements are found in prokaryotes, analogous systems have been identified in eukaryotes.

The best characterized group of mobile genetic elements are the transposable elements of prokaryotes [for review see Kleckner, 1977; Bukhari et al, 1977]. These are relatively short segments of DNA with direct or inverted repeats at either end. The repeats are usually, though not always, identical. Transposable elements are characterized by their participation in "illegitimate recombination" events (independent of the cell's normal recombinational machinery), which may involve the duplication and translocation of their genetic information. Some are highly site-specific in their transposition, whereas others insert randomly throughout the host DNA. Their deletion from the genome occurs independently of transposition. Many even encode the enzymes necessary for their own transposition. Similar elements that are near one another can transpose as one unit and thereby duplicate the genetic information in the DNA sequence between them, thus generating larger and more complex assemblies of mobile elements. Transposable elements are therefore present in multiple forms which may be arranged in a hierarchy of complexity (Fig. 9). Beginning with insertion elements, transposons can be generated which contain within their boundary sequences genes that are potentially useful to the survival of the transposon, or indeed to its deletion and rearrangement from a particular chromosomal region (Fig. 9). Transposons in turn may become more complex and generate autonomously replicating plasmid particles. An even more sophisticated form of these elements is represented by the bacteriophage which in some cases has alternative developmental programs either for integrating into host chromosomal DNA or for replication independent of host DNA.

B. Quasistable Genetic Elements in Eukaryotes

McClintock and Rhoades in the 1940s and 1950s observed heritable genetic instabilities in maize and proposed that transposition of discrete genetic elements was responsible [for review see Nevers and Saedler, 1977]. Though this was an unpopular idea at the time, it is consistent with the findings of

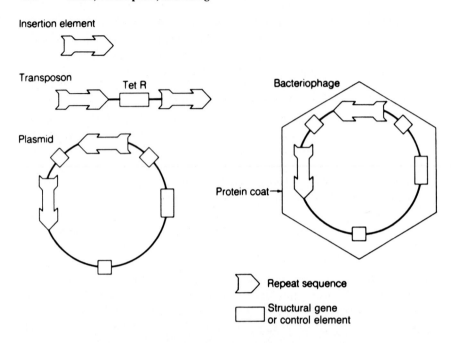

Fig. 9. Four classes of prokaryote mobile genetic elements. Note the overlapping structural organizations. [From Hunkapiller et al, 1982.].

recent molecular investigations of these phenomena. In fact, it seems likely that transposable elements of one kind or another are resident in the genomes of all eukaryotes. Examples include the copia elements in Drosophila [Finnegan et al, 1977], Ty in yeast [Cameron et al, 1979], and Alu and retroviruses in man [Houck et al, 1979; Martin et al, 1981; Reitz et al, 1981; Calabretta et al, 1982]. These mobile genetic elements are generally dispersed throughout the eukaryotic genome and are present in a large number of copies. Since the multiple elements of a given transposable gene such as copia in Drosophila are remarkably similar to one another, it is an intriguing question how this sequence similarity is maintained in spite of the dispersion of the individual members of the multigene family. It is possible that a novel type of gene correction mechanism might function. Alternatively, such elements might be under selective constraints as to the kinds of sequences that can readily transpose and hence expand in number. Certain transposable elements in Drosophila also have been associated with mutations of the host genome [Rasmuson et al, 1974]. Thus, the movement of these quasistable genetic elements can affect the structure of the "host" genome and may be a factor in the control of gene expression in developmental systems.

In mammalian cells the retroviruses are infectious agents capable of integration into the host cell genome. Like prokaryotic transposons, retroviral sequences not only can expand within a genome through transposition, but also can carry with them extraneous DNA. Thus they transfer information from one organism to another. Also like transposons, retroviral sequences are bounded by direct or inverted terminal repeats. This repeat structure is also seen in several very large dispersed sequence families in vertebrates such as the 300,000-member Alu family of man. These sequences are generally very simple and transposable, analogous to the IS elements of prokaryotes. It is unclear whether, like IS elements, Alu sequences represent the lower end of a spectrum of increasingly complex elements.

One of the most interesting of quasistable elements is the Ti plasmid of Agrobacterium tunefaciens, a gram-negative bacteria and agent of crown gall neoplasia in many dicotyledonous plants. Upon infection, information from the plasmid integrates into the genome of the host cell and directs the synthesis of bacterial products responsible for the cancerous growths [Zambryski et al, 1980]. Ti's ability to function across such a phylogentic gulf raises intriguing questions as to its role in the horizontal transmission of genetic information between widely disparate organisms.

C. Antibody Gene Families and DNA Rearrangements

1. Antibody gene families. Transposition and deletion of quasistable elements can have profound effects on the genetic structure and expression of their host genomes. Most such events as with any other class of mutation have deleterious effects. However, there are several examples which suggest that eukaryotic genomes have been able to utilize quasistable genetic elements and their mobile nature to the organism's own advantage. One of the most intriguing of these suggestions concerns the DNA rearrangements involved in the expression of antibody genes.

Antibody molecules are encoded by three multigene families, two for the light chains λ and κ and a third for heavy chains (Fig. 10). Light chains are encoded by three separate gene elements, variable (V_L), joining (J_L), and constant (C_L). Heavy-chain gene families include these three elements and a fourth gene segment, diversity (D) (Fig. 10). Table I gives our current estimates as to the number of germline gene segments within each of the three antibody gene families of mice.

The initial expression of antibody genes requires DNA rearrangement events which bring into contiguous apposition the elements of V and C genes with the subsequent loss of intervening DNA sequences (Fig. 11) [Brack et al, 1978; Seidman et al, 1978; Davis et al, 1980a]. In the light chains, the V_L and J_L gene segments are joined, whereas in the heavy chains the V_H, D, and J_H elements are joined together. The DNA rearrangements leading to V-gene formation occur during the development of a particular antibody-

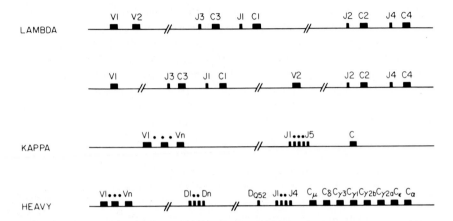

Fig. 10. Organization of the three antibody gene families of the mouse. Exon-intron organization of heavy chain constant region genes is not shown. V, D, J, and C denote variable, diversity, joining, and constant genes, respectively. [From Early and Hood, 1981.].

TABLE I. Mouse Antibody Elements and Combinatorial Mechanisms for Generating Diversity

Germline	
Kappa	$\sim$250 V_κ
	4 J_κ
Heavy	$\sim$250 V_H
	$\sim$10 D
	4 J_H
Lambda	2 V_λ
	4 J_λ
Combinatorial joining	
Kappa	250 V_κ × 4 J_κ = 1,000 κ genes
Heavy	250 V_H × 10 D × 4 J_H = 10,000 H genes
Combinatorial asociation	
1,000 κ × 10,000 H = 10^7 antibody molecules	

producing or B cell. In the mouse heavy-chain family, the rearranged V_H gene can be juxtaposed with any one of eight different C_H genes through a second type of developmentally regulated DNA rearrangement called class switching (Fig. 11) [Davis et al, 1980b]. In this way, identical antigen specificities encoded by the V domains can be associated with various effector functions encoded by the different C_H sequences.

   2. Recognition sequences for V-gene formation. Two types of recognition sequences, denoted R and R', have been found associated with antibody gene elements that are joined in V-gene formation (Fig. 12A). These two recognition elements have a striking symmetry, reminiscent of the repeated

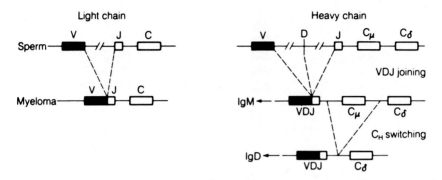

Fig. 11. Two types of DNA rearrangements that occur during the differentiation of antibody-producing cells. In light- and heavy-chain genes V-J or V-D-J joining juxtaposes the gene segments encoding the V_L and V_H genes. Subsequently in heavy-chain genes class, or C_H, switching may occur. The class switch leads to the expression of different immunoglobulin classes. Sperm indicates DNA undifferentiated with regard to antibody function, whereas myeloma denotes DNA in which one or more DNA rearrangements have occurred. [From Hunkapiller et al, 1982].

sequences seen at the boundaries of transposable genetic elements (Fig. 9). Each recognition sequence, from the many V, J, and D gene segments sequenced, has at one end a highly conserved heptamer sequence and at the other end a highly conserved nonamer sequence (Fig. 12B). The heptamer and nonamer are separated by more variable spacer sequences of approximately 11 (R sequence) or 22 (R' sequence) nucleotides. The R sequence of the V_κ and the R' sequence of the J_κ gene segments are distinguished by their different spacer lengths (Fig. 12B) and are in inverted orientation with respect to one another. In the λ genes, the R' sequence is associated with the V_λ and the R sequence with the J_λ sequences. These observations suggest that the recognition of R and R' sequences is necessary for the DNA rearrangements of V-gene formation. The generalization applies to the heavy chain gene segments as well (Fig. 11). Since the R and R' recognition sequences have been conserved in the three antibody gene families that probably diverged from one another more than 500 million years ago, they probably play a fundamental role in the DNA rearrangement mechanism of V-gene formation. It is likely that the recognition elements arose from a common precursor. We suggest that these recognition sequences were derived from a quasimobile genetic element that was captured by a primordial antibody gene and that a single such genetic event could explain the nature of all the recognition elements seen in the antibody gene families.

3. The evolution of a mechanism for DNA rearrangements in antibody genes. The primordial antibody gene was probably composed of three exons—one encoding a leader peptide, the second a V domain, and the third

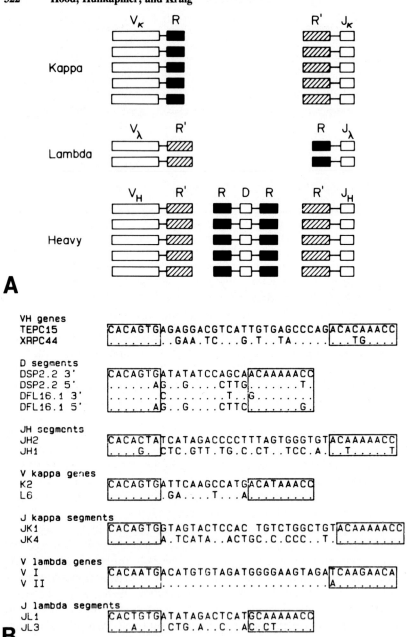

Fig. 12. The recognition sequences for V-gene formation. A) Schematic representation of the conserved heptamers and nonamers and the nonconserved spacer regions comprising the recognition sequences for V-gene formation. B) Examples of actual recognition sequences. The heptamer and nonamer sequences are indicated.

a C domain. A transposable element carrying a primordial D gene segment with recognition sequences R and R' (indicated by short and long arrows in Fig. 13) could have been inserted into the primordial V gene, thus generating a primordial heavy chain gene with V, D, and J gene segments and their attendant recognition sequences. Appropriate deletions of one or the other terminal repeat of the transposable element after duplication of the entire primordial heavy chain gene could lead to κ- or λ-like antibody genes (Fig. 13). Thus, from a single insertion of a quasistable genetic element one can generate the mechanism for V-gene formation in the three antibody gene families. In this manner one complex developmental system may have acquired the capacity for programmed DNA rearrangements during development. The capacity for DNA rearrangement of the V-gene elements enormously enhances the potential for generating antibody diversity, as will be discussed below.

   4. Strategies for information amplification. There are several distinct mechanisms for the generation and amplification of information in the antibody gene families. Numerous germline V, D, and J gene segments exist for most of the antibody gene families (Table I). Since each antibody-producing cell synthesizes just a single type of antibody molecule, the germline multiplicity ensures that different forms of antibody molecules can be ex-

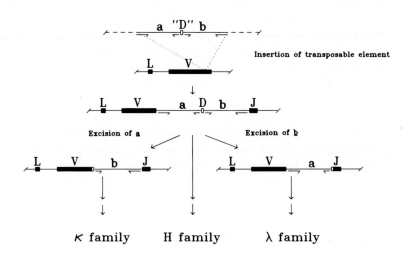

Fig. 13.   Generation of separate V-gene segments with associated recognition elements. Model proposed for insertion of transposable element containing D-like element into the primordial V gene. The "D" is flanked by inverted repeats, designated a and b. Subsequent duplication of the gene, followed by excision of either a or b generate the λ and κ families. A copy which deletes neither a nor b generates the H chain V family. Conserved heptamer and nanomer sequences are indicated, respectively, by short and long arrows. [From Hood et al, 1982.]

pressed in any individual organism. Also, the appropriate gene segments may be joined together in all combinations, for example, any V_L with any J_L gene segment. Thus through the processes of combinatorial joining of gene elements and combinatorial association of light and heavy chains, the diversity of antibody molecules can be increased enormously (Table I). Indeed, from approximately 518 germline κ and H gene elements, the mouse may generate 10^7 different antibody molecules. The remaining mechanisms for generating V region diversity occur somatically. 1) The mechanism that joins together the V, D, and J gene elements itself generates variability. The site of the joining—for example, V_L and J_L gene segments—is not fixed and may occur at many different points in the junctional sequences, thereby leading to hybrid codons and junctional regions of varying size [Weigert et al, 1980; Sakano et al, 1981]. V/J and V/D/J joining occur in regions of the light and heavy chains that are directly involved in antigen binding. 2) A hypermutational mechanism appears to operate only during a narrow time span of B-cell differentiation and may well be activated by antigenic stimulation [Crews et al, 1981; Gearhart et al, 1981]. This somatic mutational mechanism is focused only on the rearranged V gene and causes point mutations to occur throughout the V gene and its surrounding flanking sequence regions. To give a specific example, one rearranged V_H gene that recently has been studied appears to have 44 point mutational events in its coding regions and flanking sequences [Kim et al, 1981]. This mutational mechanism expands enormously the variability and hence the potential information of the antibody genes and their corresponding polypeptide products.

Although currently demonstrated only in the antibody gene families, these combinatorial and mutational mechanisms for information expansion may well be employed in other complex eukaryotic systems, such as those seen in the nervous system. In this regard two points are striking. First, the antibody system employs strategies that operate in somatic cells to vastly increase the amount of information that can be expressed from the germline. Certainly some of these strategies for information amplification are directly dependent upon the organization of the antibody genes (eg, combinatorial joining). Second, the existence of enzyme systems for altering DNA in the somatic cells poses intriguing questions as to whether similar mechanisms also exist for the change of genetic information in other somatic systems and, perhaps, in germ cells.

D. Other Eukaryotic Systems Employing DNA Rearrangements

At least two other systems have been described in eukaryotic organisms where there appear to be programmable DNA rearrangements. The mating types of yeast [Hicks et al, 1979] and the alternative expressions of different coat antigens in trypanosomes [Pays et al, 1981] are clearly controlled by DNA rearrangements. However, they appear to be mechanistically different from the rearrangements used to produce antibodies. In both cases, the gene to be expressed is copied and inserted at a different locus, without loss of the original information. Other developmentally interesting systems may well employ DNA rearrangements to express and amplify information.

E. Hierarchical Organization of Developmental Programs

With the increased structural complexity represented by multi-exon genes, multigene families, and DNA rearrangements, the systems regulating gene expression must evolve more sophisticated developmental programs. For example, to construct a millipede one would like to have a developmental program for the construction of body segments and a second developmental program for the construction of legs. Each individual developmental program would include a battery of genes that could be linked or dispersed. These programs could then be employed repeatedly to construct the multisegmental organism. These programs would naturally have to be expressed in a highly controlled manner during development.

The bithorax complex of Drosophila melanogaster provides evidence that there are such developmental programs, exhibiting hierarchical patterns of expression during development. The bithorax complex is a cluster of at least twelve genetic loci that control the developmental fate of the majority of the body segments of the fly. Each different segment is marked morphologically by bristle patterns or structures, like wings and halteres, and can be analyzed in individual mutants. For example, flies containing homozygous mutations at two bithorax loci, bx^3 and pbx, develop two mesothoracic segments and no metathoracic segment; and they therefore have four wings and no halteres [Lewis, 1963]. Since the entire segment phenotype has been altered, it seems likely that a developmental program has itself been changed. There appears to be a developmental hierarchy containing programs for wing formation, bristle structure, wing hairs, etc—all affected simultaneously by two mutations in the bithorax locus.

As we come to study more complex developmental systems, we expect to see a hierarchical organization of numerous batteries of genes that encode the complex phenotypic aspects of higher eukaryotic organisms. Clearly the duplication of a whole battery of genes for an individual developmental program would once again confer enormously enhanced evolutionary potential for the development of new phenotypes in ever evolving and changing organisms.

V. SUMMARY

Most eukaryotic genes are divided into exons and introns, which can be duplicated many times producing complex genes with multiple exons. The exons often appear to encode discrete structural domains with distinct functional potentialities, so their duplication can lead to the rapid, modular construction of extremely sophisticated genes. The resulting polypeptide chains will be composed of multiple domains, capable of acting in a synergistic fashion. Similarly, an individual gene can undergo repeated duplication to generate a multigene family. In some cases, the members of the multigene family will be become quite distinct from one another, each coming to encode a different aspect of a particular phenotypic trait. The individual members of the multigene family also have the potential for precise developmental control or regulation. The multiplicity of genes in such a family can then be assembled into batteries of genes which may either be tandemly linked or dispersed throughout the genome, but together constitute a program for constructing a part of a higher eukaryotic organism such as a leg. Such developmental batteries or programs of genes can be precisely regulated so that they are turned on at appropriate developmental times for the construction of a complex eukaryotic organism.

Thus, three new concepts concerning eukaryotic genetics—introns-exons, multigene families, and dynamic DNA rearrangements—all appear to contribute enormously to the expansion of the information encoded in eukaryotic chromosomes. Indeed, it was at one time a paradox to make the statement that only 50,000 genes were needed in the construction of a man, an organism that employs more than 10^{14} different cells of incredibly diverse and complex phenotype. With the enormous informational expansion that arises from the patterns and strategies of information storage, it is now understandable how only 50,000 genes could code for such complex organisms. The major point is that the linear information of the chromosomes can be assembled in a nonlinear manner, either through exon shuffling, developmental expression of subsets of multigene families, use of DNA rearrangement to reorganize the exons expressed for certain genes, or the evolution of extremely sophisticated developmental programs.

ACKNOWLEDGMENTS

This work has been supported by grants from the National Institutes of Health. Portions of this manuscript were adapted from Hunkapiller et al [1982].

VI. REFERENCES

Abelson J (1979): RNA processing and the intervening sequence problem. Ann Rev Biochem 48:1035–1069.

Brack C, Hirama M, Lenhard-Schuller R, Tonegawa S (1978): A complete immunoglobulin gene is created by somatic recombination. Cell 15:1–14.

Brown DD, Wensink PC, Jordan E (1971): Purification and some characteristics of 5S DNA from Xenopus laevis. Proc Natl Acad Sci USA 68:3175–3179.

Bukhari AI, Shapiro JA, Adhya SL (eds) (1977): "DNA Insertion Elements, Plasmids and Episomes." Cold Spring Harbor, New York: Cold Spring Harbor Laboratory.

Calabretta B, Robberson DL, Barrera-Saldana HA, Lambrou TP, Saunders GF (1982): Genome instability in a region of human DNA enriched in Alu repeat sequences. Nature 296:219–225.

Calame K, Rogers J, Early P, Davis M, Livant D, Wall R, Hood L (1980): Mouse C_μ heavy chain immunoglobulin gene segment contains three intervening sequences separating domains. Nature 284:452–455.

Cameron JR, Loh EY, Davis RW (1979): Evidence for transposition of dispersed repetitive DNA families in yeast. Cell 16:739–751.

Crews S, Griffin J, Huang H, Calame K, Hood L (1981): A single V_H gene segment encodes the immune response to phosphorylcholine: Somatic mutation is correlated with the class of the antibody. Cell 25:59–60.

Davis M, Calame K, Early PW, Livant DL, Joho R, Weissman IL, Hood L (1980a): An immunoglobulin heavy-chain gene is formed by at least two recombinational events. Nature 283:733–739.

Davis MM, Kim SK, Hood LE (1980b): DNA sequences mediating class switching in α immunoglobulins. Science 209:1360–1365.

Doolittle WF (1978): Genes in pieces: Were they ever together? Nature 272:581–582.

Early P, Hood L (1981): Mouse immunoglobulin genes. In Setlow JR, Hollaender A (eds): "Genetic Engineering." New York: Plenum Press, Vol 3, pp 157–188.

Early P, Rogers J, Davis M, Calame K, Bond M, Wall R, Hood L (1980): Two mRNAs can be produced from a single immunoglobulin μ gene by alternative RNA processing pathways. Cell 20:313–319.

Finnegan DJ, Rubin GM, Young MW, Hogness DS (1977): Repeated gene families in Drosophila melanogaster. Cold Spring Harbor Symp Quant Biol 42:1053–1063.

Gally JA, Edelman GM (1972): The genetic control of immunoglobulin synthesis. Ann Rev Genet 6:1–46.

Gearhart P, Johnson N, Douglas R, Hood L (1981): IgG antibodies to phosphorylcholine exhibit more diversity than their IgM counterparts. Nature 291:29–34.

Hagenbüchle O, Bovey R, Young RA (1980): Tissue-specific expression of mouse α-amylase genes: Nucleotide sequence of isoenzyme mRNAs from pancreas and salivary gland. Cell 21:179–187.

Hicks J, Strathern N, Klar AJS (1979): Transposable mating-type genes in Saccharomyces cerevisiae. Nature 282:478–483.

Hood JM, Huang HV, Hood L (1980): A computer simulation of evolutionary forces controlling the site of a multigene family. J Mol Evol 15:181–196.

Hood L, Campbell JH, Elgin SCR (1975): The organization, expression, and evolution of antibody genes and other multigene families. Ann Rev Genet 9:305–353.

Hood L, Campbell J, Crews S, Griffin J, Huang H, Hunkapiller T, Kim S, Kraig E, Perlmutter R (1983): Antibody genes. Science (in preparation).

Houck CM, Rinehart FP, Schmid CW (1979): A ubiquitous family of repeated DNA sequences in the human genome. J Mol Biol 132:289–306.

Hunkapiller T, Huang H, Hood L, Campbell JH (1982): The impact of modern genetics on evolutionary theory. In Milkman R (ed): "Perspectives on Evolution." Sunderland, Mass, USA: Sinauer.

Kehry M, Ewald S, Douglas R, Sibley C, Raschke W, Fambrough D, Hood L (1980): The immunoglobulin μ chains of membrane-bound and secreted IgM molecules differ in their C-terminal segments. Cell 21:393–406.

Kim S, Davis M, Sinn E, Patten P, Hood L (1981): Antibody diversity: Somatic hypermutation of rearranged V_H genes. Cell 27:573–581.

Kleckner N (1977): Translocatable elements in procaryotes. Cell 11:11–23.

Lazowska J, Jacq C, Slonimski PP (1980): Sequence of introns and flanking exons in wild-type and box 3 mutants of cytochrome b reveals an interlaced splicing protein encoded by an intron. Cell 22:333–348.

Lewis EB (1963): Genes and developmental pathways. Am Zool 3:33–56.

Li W, Gojobori T, Nei M (1981): Pseudogenes as a paradigm of neutral evolution. Nature 292:237–239.

Martin MA, Bryan T, Rasheed S, Khan AS (1981): Identification and cloning of endogenous retroviral sequences present in human DNA. Proc Natl Acad Sci USA 78:4892–4896.

McNicholas J, Steinmetz M, Hunkapiller T, Jones P, Hood L (1982): DNA sequence of the gene encoding the BALB/c mouse E_α^d Ia polypeptide. Science (in press).

Moore KW, Rogers J, Hunkapiller T, Early P, Nottenburg C, Weissman I, Bazin H, Wall R, Hood L (1981): Expression of IgD may use both DNA rearrangement and RNA splicing mechanisms. Proc Natl Acad Sci USA 78:1800–1804.

Nevers P, Saedler H (1977): Translocatable genetic elements as agents of gene instability and chromosomal rearrangements. Nature 268:109–115.

Pays E, Van Meirvenne N, Le Ray D, Stewart M (1981): Gene duplication and transposition linked to antigenic variation in Trypanosoma brucei. Proc Natl Acad Sci USA 78:2673–2677.

Proudfoot NJ, Shander MHM, Manley JL, Gefter ML, Maniatis T (1980): Structure and in vitro transcription of human globin genes. Science 209:1329–1336.

Rasmuson B, Green MM, Karlsson B (1974): Genetic instability in Drosophila melanogaster. Mol Gen Genet 133:237–247.

Reitz MS, Poiesz BJ, Ruscetti FW, Gallo RC (1981): Characterization and distribution of nucleic acid sequences of a novel type C retrovirus isolated from neoplastic human T lymphocytes. Proc Natl Acad Sci USA 78:1887–1891.

Rogers J, Early P, Carter C, Calame K, Bond M, Hood L, Wall R (1980): Two mRNAs with different 3' ends encode membrane-bound and secreted forms of immunoglobulin μ chain. Cell 20:303–312.

Sakano H, Rogers JA, Huppi K, Brack C, Traunecker A, Maki R, Wall R, Tonegawa S (1979): Domains and the hinge region of an immunoglobulin heavy chain are encoded in separate DNA fragments. Nature 277:627–634.

Sakano H, Kurosawa Y, Weigert M, Tonegawa S (1981): Identification and nucleotide sequence of a diversity DNA segment (D) of immunoglobulin heavy-chain genes. Nature 290:562–565.

Seidman JG, Leder A, Edgell MH, Polsky F, Tilghman SM, Tiemeier DC, Leder P (1978): Multiple related immunoglobulin variable-region genes identified by cloning and sequence analysis. Proc Natl Acad Sci USA 75:3881–3885.

Smith GP (1976): Evolution of repeated DNA sequences by unequal crossover. Science 191:528–535.

Steinmetz M, Moore KW, Frelinger JG, Sher BT, Shen F-W, Boyse EA, Hood L (1981): A pseudogene homologous to mouse transplantation antigens: Transplantation antigens are encoded by eight exons that correlate with protein domains. Cell 25:683–692.

Weigert M, Perry R, Kelley D, Hunkapiller T, Schilling J, Hood L (1980): The joining of V and J gene segments creates antibody diversity. Nature 283:497–499.

Williams AF (1982): Surface molecules and cell interactions. J Theoret Biol (submitted).

Zambryski P, Holsters M, Kruger K, Depicker A, Schell J, Van Montagu M, Goodman HM (1980): Tumor DNA structure in plant cells transformed by A tumefaciens. Science 209:1385–1391.

Modern Cell Biology 2:329–352
© 1983 Alan R. Liss, Inc., 150 Fifth Avenue, New York, NY 10011

The C-Value Paradox and Genes in Ciliated Protozoa

David M. Prescott

From the Department of Molecular, Cellular and Developmental Biology,
University of Colorado, Boulder, Colorado 80309

I. THE C-VALUE PARADOX

Although it is difficult to give a precise quantitative definition of the structural and functional complexity of an organism, eukaryotes can reasonably be placed in an ordered series based on general estimates of their complexities. As expected, structural and functional complexities of organisms tend to be matched by the complexity of the DNA sequences in their genomes. However, the simplicity of this generalization is confounded on several counts. First, there are pronounced discrepancies in the matching; some organisms—for example, salamanders and plants of the lily family—have far greater sequence complexities in their DNAs than do obviously more complex organisms such as mammals. Second, among closely related species—again for example, among certain salamanders and plants—DNA complexities sometimes differ severalfold within the group whereas structural and functional differences are hardly perceptible. Third, almost all eukaryotes have genomes with far greater DNA complexities than is consistent with the estimated number of genes present or required in the organism. For example, the genomic complexity of Drosophila is equal to $\sim$150,000 kilobase pairs (KBP), at least 10 times more than needed to accommodate the usual estimate of 5,000 genes in this organism. The human genome has a sequence complexity of $\sim$2.0 × 10^6 KBP, an order of magnitude more than believed needed to contain the estimated 50,000 to 100,000 genes in a human. The apparent excess of DNA is usually called the C-value paradox, the C value for an organism being the amount of DNA in its haploid genome [for a review of the C-value paradox, see Gall, 1981]. It is more precise to use the total DNA sequence complexity (sequence value or S value) rather than total amount of DNA (C value) to describe this paradox, since the genomes of most eukaryotes contain considerable amounts of repetitious sequences (including pseudogenes) that increase C values significantly but hardly contribute to the S values.

Two hypothetical solutions to the S-value paradox are that the number of genes in eukaryotic organisms is much greater than supposed or that genes are much larger than supposed—for example, because they contain extremely large amounts of noncoding, intervening sequences. The general occurrence of intervening sequences in eukaryotic genes does account for some noncoding DNA (noncoding in the sense of not coding for a gene product), but not nearly enough to alter the problem of the apparent excess of DNA sequences.

II. THE GENOMES OF HYPOTRICHOUS CILIATES

A different solution to the S-value paradox has emerged from studies of a group of ciliated protozoa called the hypotrichs. Like other ciliates, the hypotrichs have micronuclei and macronuclei (Fig. 1). The particular hypotrich shown in Figure 1 (Oxytricha nova) has four genetically identical micronuclei and two genetically identical macronuclei. A micronucleus contains a diploid set of chromosomes and divides by mitosis; however, none of its DNA is ever transcribed, at least during the vegetative life of an organism. A macronucleus typically contains many times more DNA than a micronucleus, but chromosomes are never observed, and a macronucleus divides by pinching in two (amitosis) (Fig. 14). All the RNA in the cell is produced in macronuclei (except for a small amount synthesized in mitochondria). Micronuclei, in fact, are unnecessary for vegetative life. Organ-

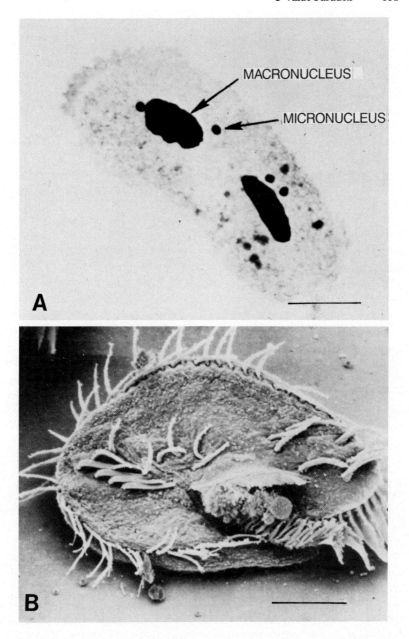

Fig. 1. A) Oxytricha nova stained for DNA by the Feulgen procedure and counterstained with fast green. Four micronuclei and two macronuclei are present. Other stained material is food vacuoles. Bar = 30 μm. B) Scanning electronmicroscope view of Oxytricha nova (by K. G. Murti). Bar = 30 μm.

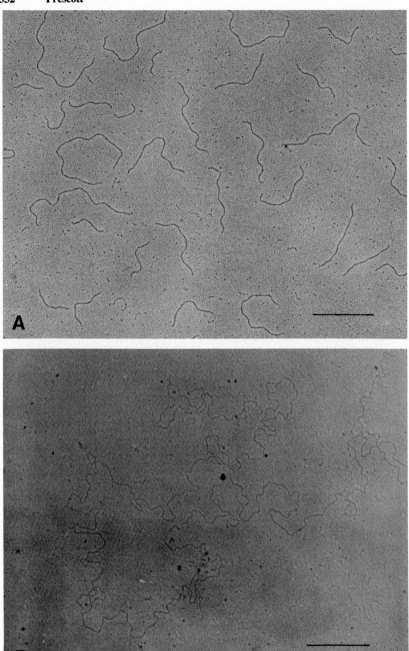

Fig. 2. A) Gene-size, macronuclear DNA molecules from Oxytricha nova (by K. G. Murti). Bar = 1.0 μm. B) A section of a DNA molecule from the micronucleus of Oxytricha nova (by K. G. Murti). Bar = 1.0 μm.

isms that spontaneously lose their micronuclei in laboratory culture and amicronucleated cells isolated from the wild grow and reproduce vigorously.

Another striking difference between a micronucleus and a macronucleus is the molecular weight of their DNAs. The molecular weight of the DNA in the micronucleus is very high—greater than several hundred $\times$ 10^6 daltons. The DNA of the macronucleus occurs only in very small, linear molecules. For example, in the hypotrich, Oxytricha nova, macronuclear DNA ranges in size from 400 base pairs (BP) to $\sim$10,000 BP (10 kilobase pairs or 10 KBP) (0.27 to 6.6 $\times$ 10^6 daltons) with a number average size of 2200 BP ($\sim$1.47 $\times$ 10^6 daltons) [Swanton et al, 1980a]. Figure 2 shows an electron micrograph of undegraded, native macronuclear and micronuclear DNAs [Prescott et al, 1971]. Figure 3 shows macronuclear DNAs of three hypotrich species separated by size by electrophoresis in a 1% agarose gel and stained with ethidium bromide [Swanton et al, 1980a]. Micronuclear DNA does not migrate in such a gel, as expected for DNA of high molecular weight. As shown by electrophoresis in an agarose gel, macronuclear DNA contains a continuum of sizes from $\sim$400 BP to $\sim$10,000 BP; most of the molecules are shorter than 4,000 BP. Superimposed on the size continuum is a series of bands that represents differentially higher copy numbers for molecules of certain sizes. For example, a band occurs at $\sim$7,400 BP; these particular molecules are differentially amplified about 100$\times$ and code for rRNA.

The DNA molecules of the macronucleus are large enough to contain one to a few genes; those analyzed so far contain only single coding functions. In O nova a molecule of $\sim$7,400 BP, for example, codes for a single precursor RNA molecule that contains the 26S, 17S, and 5.8S ribosomal RNAs. The arrangement of leader, trailer, and coding sequences in this gene are shown in Figure 4 [Swanton et al, 1980b]. In O fallax, molecules of 1,511 BP and $\sim$1,400 BP contain single actin genes [Kaine and Spear, 1982]; molecules of 2,030 and 1,880 BP carry single genes for β tubulins; and a molecule of 2,050 BP codes for a single α tubulin gene (Spear, personal communication). A molecule of 690 BP codes for 5S RNA, but whether more than one coding sequence is present is not known [Rae and Spear, 1978]. tRNAs are coded by a series of DNA molecules ranging in size from 400 BP to 3,200 BP (Jahn, unpublished), and histones are coded by another series of molecules [Elsevier et al, 1978; Prescott et al, 1979], although it is not known how many coding sequences are present in the individual molecules in this series. In any case it is clear that the DNA of the micronucleus is in chromosome-size molecules and the macronuclear DNA occurs in gene-size molecules.

Another important difference between micronuclear and macronuclear DNAs concerns their total sequence complexities, as determined by the kinetics of reassociation of their DNAs (after dissociation with heat) (Fig. 5) [Lauth et al, 1976]. Reassociation of macronuclear DNA follows simple second order

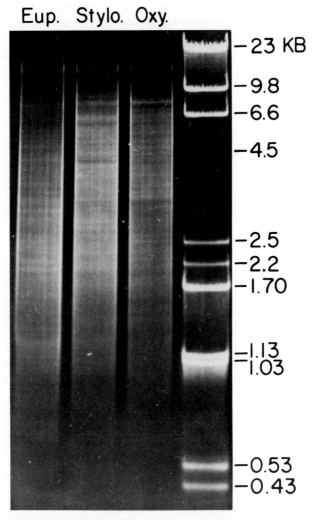

Fig. 3. Macronuclear DNA from three hypotrichs (Stylonychia pustulata, Euplotes aediculatus, and Oxytricha nova) separated by size by electrophoresis on a 1% agarose gel. Size markers are shown in the lane to the right [from Swanton et al, 1980a].

kinetics with half reassociation ($\frac{1}{2}$ Cot) at 24 mole-sec/l. The shape of the curve indicates that most sequences in the macronucleus are present in the same number of copies—ie, a relatively small proportion of the total sequences can be differentially repeated. The $\frac{1}{2}$ Cot value indicates that the total sequence complexity of macronuclear DNA is about 13 times greater than the total sequence complexity of the DNA in E coli.

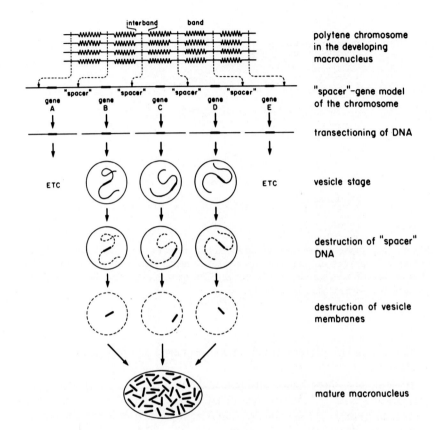

Fig. 4. A map of the coding and noncoding regions of the rDNA molecule from the macronucleus of Oxytricha nova [from Swanton et al, 1980b].

The reassociation of micronuclear DNA follows a more complex course. The first part of the curve (Fig. 5), corresponding to about 30% of the total DNA, represents sequences that are differentially repeated relative to the rest of the DNA. About 70% of the DNA reassociates more slowly; this is made up of unique sequence DNA. The presence of repeated sequences in the micronucleus and their absence from the macronucleus is in agreement with the detection in micronuclear DNA of density satellites in CsCl gradients and the absence of density satellites in macronuclear DNA [Bostock and Prescott, 1972].

Comparison of the $\frac{1}{2}$ Cot value for macronuclear DNA with the $\frac{1}{2}$ Cot values for the repetitious and unique sequences of micronuclear DNA leads to the conclusion that the sequences in the macronucleus contain about 5%

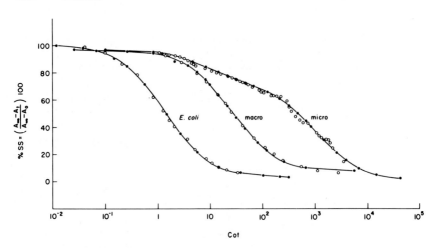

Fig. 5. Reassociation kinetics (Cot curves) for micronuclear and macronuclear DNAs of Oxytricha nova (open circles). The curves were computer-fitted with a program based on idealized second-order kinetics (closed circles). The reassociation curve for E coli DNA is included as a standard [from Lauth et al, 1976].

of the complexity present in the unique sequence component of the micronucleus.

In summary, three major differences distinguish macronuclear from micronuclear DNA: 1) Macronuclear DNA is gene-size; micronuclear DNA is chromosome-size. 2) Macronuclear DNA contains few differentially repeated sequences; micronuclear DNA is made up of ~30% repetitious sequences and 70% unique sequences. 3) Macronuclear DNA has a total complexity equal to 5% or less of micronuclear DNA.

These differences take on extraordinary importance when one considers that the macronucleus forms from a micronucleus following conjugation. In conjugation, which can be induced by starvation, two cells join. This is quickly followed by meiosis of the micronuclei in both cells to produce haploid micronuclei. Each cell donates a haploid micronucleus to the other through a cytoplasmic bridge. The migratory micronucleus fuses with a resident haploid micronucleus to produce a new diploid micronucleus in each cell. All of the unused haploid micronuclei and the macronuclei completely degenerate. The new diploid micronucleus divides mitotically, and one of the daughters develops into a new macronucleus.

The process of forming a macronucleus follows a complex course lasting several days (Fig. 6) which produces the dramatic differences between micronuclear and macronuclear DNAs outlined above. The micronucleus that

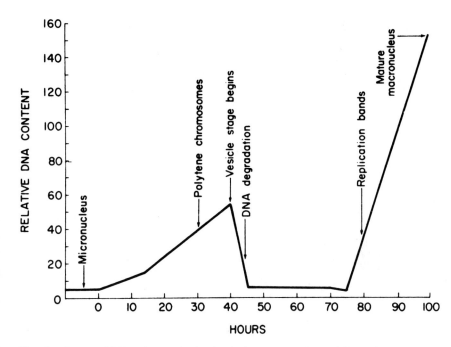

Fig. 6. Course of DNA changes in the developing macronucleus of Oxytricha nova after conjugation [Prescott and Murti, 1974; after Ammermann, 1965].

will give rise to a macronucleus is first identified by an increase in its DNA content and the development of polytene chromosomes (Fig. 7) [Ammermann, 1965]. The polytene chromosomes are probably made up of 128 copies of the DNA, although this varies from one hypotrich species to another. The chromosome number is high, usually around 100, in all hypotrich species examined thus far. Some of the density satellite DNA of the micronucleus is not detected in nuclei with polytene chromosomes and apparently does not replicate [Spear and Lauth, 1976]. This represents the first step in the elimination of sequences.

The polytene chromosomes reach their peak of development at about 48 hours (at 25°C) after fusion of haploid micronuclei in O nova. At this point the polytene chromosomes are rapidly destroyed by formation of septa through all interbands (Fig. 8) [Kloetzel, 1970]. The nucleus is rich in the material that forms the septa, and it has proteinaceous staining properties. The septum subsequently extends to envelope separately each of the bands of all of the chromosomes, forming vesiclelike structures (Fig. 9) [Prescott and Murti, 1974]. In short, the developing macronucleus is converted from a bag of

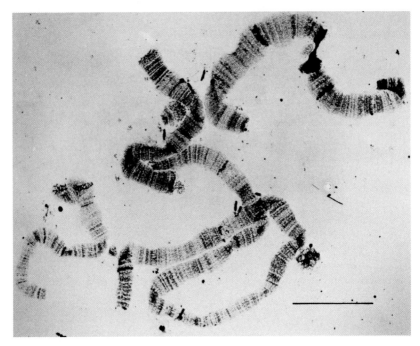

Fig. 7. Four of the many polytene chromosomes in the developing macronucleus of Oxytricha nova (by K. G. Murti). Bar = 10 μm.

polytene chromosomes to a bag of vesicles, each vesicle containing the material formerly present in a band of a polytene chromosome plus material from the two adjacent interbands. If the nucleus of a living cell is broken open at this point, the vesicles quickly float away from each other, showing that the chromosomes have, in fact, been transected band by band.

Immediately after formation of the vesicles most of the DNA in the nucleus is destroyed (Fig. 6) [Ammermann, 1965]. Some destruction occurs in all vesicles [Prescott and Murti, 1974] and is complete in a few hours. Next, the vesicle organization disappears, and the nucleus remains in this DNA-poor stage for ∼20 hours [Ammermann, 1965]. Finally, the remaining DNA, presumably now gene-size molecules, undergoes about five rounds of replication by the replication band mechanism (see below) over ∼24 hours to produce the mature, DNA-rich macronucleus (in O nova the mature, G1 stage macronucleus contains ∼58 pg of DNA compared to ∼1.3 pg in a diploid, G1 stage micronucleus). The cell then resumes vegetative existence with a cell cycle of ∼8 hours under optimum growth conditions.

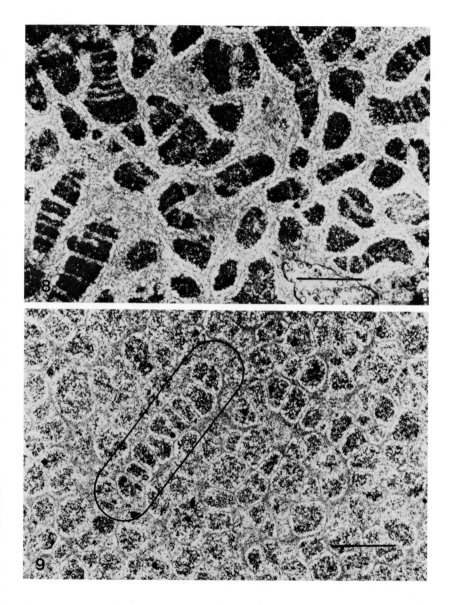

Fig. 8.   Formation of septa through all interbands of the polytene chromosomes of Oxytricha nova [Prescott and Murti, 1974]. Bar = 5 μm.

Fig. 9.   The vesicle stage of macronuclear development. The enclosure of nine vesicles derived from nine consecutive bands of a polytene chromosome [Prescott and Murti, 1974]. Bar = 5 μm.

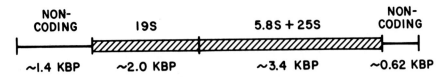

NON-CODING	19S	5.8S + 25S	NON-CODING
~1.4 KBP	~2.0 KBP	~3.4 KBP	~0.62 KBP

Fig. 10. A model showing destruction of chromosomal DNA and release of gene-size molecules during development of the macronucleus in Oxytricha nova, based on cytological changes, DNA changes, and cloning of micro- and macronuclear DNAs [Prescott and Murti, 1974].

We assume that the gene-size molecules are produced in two stages. The first stage consists of the transection of chromosomes to yield subchromosomal molecules. These molecules are then further shortened during the DNA destruction that takes place in vesicles. We have proposed a scheme of chromosome organization in Figure 10 showing separation of one gene from another in micronuclear chromosomes by long spacers of mostly unique sequence DNA. The spacers account for 95% of the sequence complexity of the micronucleus, and when genes are excised from chromosomes, the spacers are destroyed.

III. THE C-VALUE (S-VALUE) PARADOX IN HYPOTRICHS

Whether or not the particular steps in the model by which genes are excised are correct (other models involving reverse transcription of RNA can be envisaged), the facts of macronuclear development and the differences between micronuclear and macronuclear DNAs bear pointedly on the C-value or S-value paradox. First, 5% or less of the DNA sequences in the genome (micronucleus) are ultimately used (in the macronucleus) to code for gene products or to provide any other function during vegetative growth; ie, the macronucleus contains 5% or less of the micronuclear sequences, yet produces all cellular RNA (except mitochondrial). Second, the events of macronuclear development suggest that the eliminated sequences, representing 95% of the unique sequences in the genome, are dispersed between genes throughout the genome. This second point has recently been confirmed by analysis of clones of large fragments of chromosomal DNA that contain sequences corresponding to one to several macronuclear genes [Boswell et al, 1982b]. We found that macronuclear genes sometimes occur singly (as in the model) or in clusters of two to several genes; the genes within a cluster are separated by sequences that are eliminated during macronuclear development. Thus, the model in Figure 10 requires the modification that at least some genes occur in small clusters.

Half of the 20 micronuclear DNA clones examined so far contain an 11 KBP stretch (defined by Eco R1 sites at each end) containing a sequence that is highly repeated in the micronuclear genome. These 11 KBP sequences are adjacent to one or more macronuclear gene sequences in the micronucleus and are all eliminated with other spacer DNA during macronuclear development.

We therefore offer the following explanation of the S-value paradox. First, the apparent excess of DNA sequences in eukaryotic genomes stems from the presence of large spacers between genes and gene clusters. This model of gene organization of eukaryotic chromosomes is not really new; it simply extends to all genes the finding that transcription units for rRNA precursors, 5S rRNA, tRNA precursors, and histone mRNA (all repeated genes) are separated by nontranscribed spacers. Second, the wide deviations from a correspondence between the structural-functional complexity of organisms and the complexity of their genomes derives not from variations in the amount of DNA sequence devoted to genetic coding in the traditional sense, but from wide variations in the amount of DNA sequence present in spacers between genes.

The function (if any) of spacer sequences or why, as proposed, they might vary so widely in size among different species remain to be explained. However, their existence is indisputable.

IV. THE NUMBER OF GENE-SIZE MOLECULES IN THE MACRONUCLEUS

Further understanding of the structure and function of the macronucleus has been gained from studies of the gene-size molecules in the macronucleus. From the rate of reassociation the total complexity of macronuclear DNA of Oxytricha has been estimated to be ~52,800 KBP. The number average size of macronuclear DNA molecules is 2,200 BP. Therefore, the macronucleus contains ~24,000 different kinds of gene-size molecules ((52,800 KBP)/(2,200 BP) = 24,000)). A single macronucleus contains ~58 pg of DNA, which is equivalent to 25×10^6 molecules of 2,200 BP. On the average, therefore, each of the 24,000 different molecules is present in ~1,000 copies per macronucleus. A small proportion of the molecules is present in higher copy numbers, and these molecules give rise to bands in electrophoretic gel separation of DNA (Fig. 3). The rRNA genes are present in ~100,000 copies per macronucleus (Swanton, unpublished) and give rise to the band at ~7,400 BP.

V. REPLICATION OF MACRONUCLEAR DNA

Each of the $\sim 24 \times 10^6$ DNA molecules in the macronucleus must be replicated individually during the cell cycle in hypotrichs; ie, each molecule is a replicon. Replication takes place in a specialized cytological structure

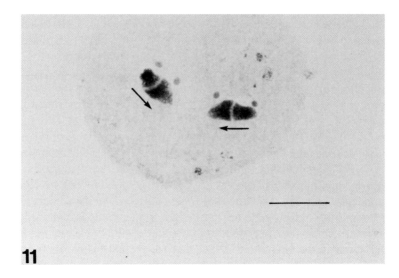

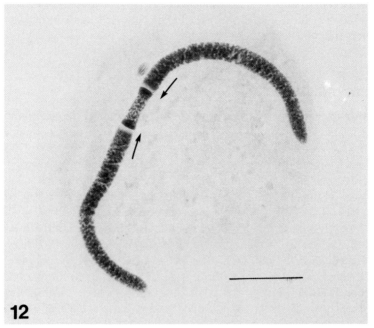

Fig. 11. Oxytricha nova stained with acetocarmine to show replication bands in the macronuclei. Arrows indicate direction of band movement. Bar = 30 μm.

Fig. 12. Euplotes eurystomus stained with acetocarmine to show replication bands. Arrows indicate direction of band movement. The bands originated at the tips of the macronucleus and are approaching each other at the center. The single micronucleus has entered mitosis and will remain arrested in mitosis until cell division takes place. Bar = 30 μm.

called a replication band. In Oxytricha, a replication band originates at the outer end of each of the two oval-shaped macronuclei (Fig. 11) and progresses through the entire nucleus in several hours, replicating all DNA molecules in its path. The phenomenon is particularly striking in hypotrichs of the genus Euplotes, in which the macronucleus is a long, curved rod. Replication bands originate at both tips of the nucleus and move toward each other (Fig. 12). They finally meet in the center of the macronucleus and disappear; the cell then divides. The structure of replication bands has been studied by EM [Murti, 1976; Olins et al, 1981], but little is known about the molecular events in a replication band.

According to electron-microscope observations replication is initiated at or near one or both ends of each molecule. Some molecules show a single replication fork and some show two forks (Fig. 13) (Murti and Prescott, unpublished). We think that every molecule has the potential to initiate replication at both ends, but that the entry of the ends of a molecule into the replication zone of a replication band is random. One end is likely to begin replication before the other, giving rise to some molecules with single forks and other molecules with two forks. In molecules with two forks the forks are almost always at different positions relative to their respective ends, suggesting asynchronous initiation at the two ends.

VI. DISTRIBUTION OF GENES IN NUCLEAR DIVISION

Macronuclei divide by amitosis (Fig. 14). In species with two or more macronuclei, all of the macronuclei fuse together just before division. The initial amitotic division is quickly followed by whatever number of divisions is needed to provide the characteristic number of macronuclei. Amitotic splitting is imprecise; the DNA is distributed between two daughter macronuclei with variable inequality, the inequality commonly ranging up to a 60–40 split in some species.

Despite unequal macronuclear division, variation in DNA content among the macronuclei in a population of cells remains within relatively narrow limits, which implies a mechanism to correct high and low amounts of DNA. When the DNA content of a daughter macronucleus is low, two consecutive rounds of DNA replication, detected as extra replication bands, apparently occur in that nucleus before the next cell division. Daughter nuclei with high contents of DNA apparently skip DNA replication before the next cell division. In this way, DNA contents of macronuclei are constantly adjusted in the direction of the mean value in a population of organisms.

A more perplexing problem of amitosis concerns the distribution of the copies of any particular gene between two daughter macronuclei. In the absence of chromosomes no mechanism is known by which the copies of each gene can be equivalently distributed; it seems likely that such distri-

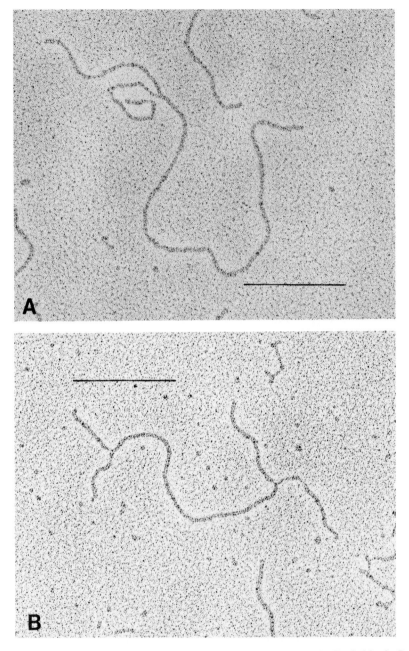

Fig. 13. Replicating molecules of the hypotrich Euplotes eurystomus (by K. G. Murti). Bar = 0.5 μm.

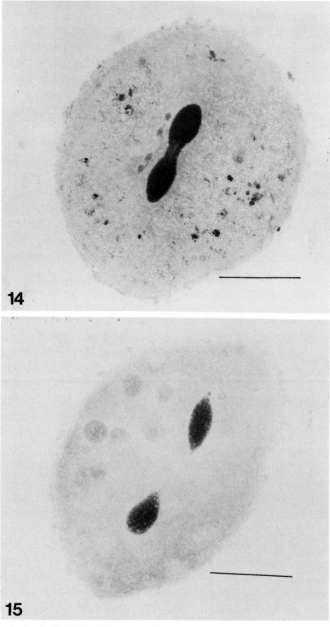

Fig. 14. Amitotic division of the macronucleus of Oxytricha nova. Acetocarmine stain. The four micronuclei are in mitosis. Bar = 30 μm.

Fig. 15. An amicronucleated hypotrich isolated from the wild. Acetocarmine stain. Bar = 30 μm.

butions are random and therefore almost always unequal. Inequalities will be compounded at successive divisions so that gene imbalances severe enough to inhibit cell reproduction should develop. Ultimately, every cell in a population should develop sufficiently severe imbalance to prevent cell reproduction. Clonal lines of some kinds of hypotrichs do senesce and die out, as do some other kinds of nonhypotrich ciliates. Conjugation, however, rejuvenates organisms in aging clones, presumably by replacing the old macronucleus, with its increasing gene imbalances, with a new perfectly balanced macronucleus. According to this hypothesis, amicronucleate hypotrich strains (species ?) of organisms isolated from the wild, which are incapable of conjugation, should senesce and die out. Two types of organisms, one similar in morphology to Oxytricha (Fig. 15) and one belonging to a newly discovered genus, have exceeded 500 generations of clonal reproduction with no perceptible change in reproductive rate. These observations mean either that hypotrichs possess a mechanism to distribute genes equally during amitosis or that they are capable of differential gene replication to correct imbalances created by amitotic division.

VII. STRUCTURE OF GENE-SIZE MOLECULES

The molecules of the macronucleus provide a unique opportunity to study the function and structure of genes. Any gene in a chromosome of any eukaryotic organism can be isolated for study by cutting the chromosomal DNA with restriction nucleases and cloning the piece of DNA containing a given gene in a plasmic or other vehicle, provided a probe is available for selecting and identifying the DNA fragment carrying the particular gene. Although the coding region of a gene can usually be identified if its RNA or protein product is known, it is as yet impossible to identify precisely those sequences adjacent on each side of the coding region that are crucial for gene function.

In this regard the hypotrichs provide a particular advantage because the organism has done the first part of experiments on gene structure and function; the organism itself has excised all of its genes from its chromosomes and eliminated all of its "nongenic" DNA. The organism obviously does this in a way that keeps gene functions intact; it is likely that in excising genes no extraneous sequences are left appended to either end of the DNA molecule. The excision of genes is, in any case, a very precise process since a given coding sequence is always contained in exactly the same size molecule every time a macronucleus develops. In the process of excising its genes from its chromosomes the organism does something extraordinarily important: It defines which sequences in its chromosomes are genes and which are spacers, and allows us, by using the macronuclear DNA molecules, potentially to identify every gene without knowing either its coding sequence or its gene product.

In addition to identifying coding functions of a number of sizes of molecules as described earlier, we have learned the following about the structural properties of macronuclear DNA. When macronuclear DNA is denatured by heating and then rapidly cooled, most of the molecules form single-stranded circles (Fig. 16) [Wesley, 1975]. This means that the ends of a single stranded molecule contain complementary sequences:

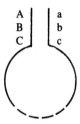

Extrapolating to the original duplex molecules, we can infer the following arrangement:

$$5' \; \text{ABC} \qquad \text{cba} \quad 3'$$
$$\overline{\phantom{5' \; \text{ABC} \qquad \text{cba} \quad 3'}}$$
$$3' \; \text{abc} \qquad \text{CBA} \quad 5'$$

This shows that a molecule has the same sequence at both ends in an inverted order—ie, it possesses inverted terminal repeat sequences. DNA sequencing studies have subsequently confirmed this interpretation and have shown that all of the $\sim 24,000$ gene-size molecules in the macronucleus possess the same inverted terminal repeat sequence [Klobutcher et al, 1981]. Sequencing of total macronuclear DNA from the 5' ends revealed the sequence:

5' CCCCAAAACCCCAAAACCCCXXXXXXXX

We refer to this as the C_4A_4 sequence.

In their classical sequencing paper Maxam and Gilbert [1977] advise the researcher to check sequence determinations by sequencing from both 5' and 3' ends. Sequencing total macronuclear DNA from the 3' ends yielded, as expected, a G_4T_4 sequence to complement the C_4A_4 sequence; unexpectedly, the G_4T_4 sequence was 16 bases longer than the C_4A_4 sequence. Thus, the ends of the gene-size molecules proved to contain a 3' single stretch of 16 bases:

$$5' \; C_4A_4C_4A_4C_4 \ldots\ldots\ldots\ldots\ldots\ldots\ldots\ldots G_4T_4G_4T_4G_4T_4G_4T_4G_4 \; 3'$$
$$3' \; G_4T_4G_4T_4G_4T_4G_4T_4G_4 \ldots\ldots\ldots\ldots\ldots\ldots\ldots\ldots C_4A_4C_4A_4C_4 \; 5'$$

The same inverted terminal repeat sequence and 3' tail have been found in five species of hypotrichs. In a sixth species (Euplotes aediculatus), the

Fig. 16.   Single stranded DNA loops formed by rapid cooling of heat denatured macronuclear DNA of Euplotes aediculatus (by R. Wesley). Bar = 1 μm.

sequence is the same except that it has an additional $\dfrac{C_4A_4}{G_4T_4}$ and that two G's are missing from the 3′ ends [Klobutcher et al, 1981]. The strict conservation of both the length and the specific sequence of the inverted terminal repeat among all the molecules in the macronucleus of a species and among distantly related species of hypotrichs implies that the sequence has a function that is tightly dependent both on the nature of the sequence and its length.

The 3′ tail is probably connected with DNA replication. Viewing it as a 5′ gap instead of a 3′ tail calls attention to the old unsolved 5′ primer problem, or how the 5′ ends of a linear DNA molecule are replicated. A 5′ gap is precisely what one would expect following removal of the last RNA primer of the terminal Okazaki fragment. There is no known way of filling in such a terminal 5′ gap with deoxynucleotides; all nucleic acid polymerases are unable to catalyze nucleotide additions to 5′ ends of nucleic acid chains. Judging from macronuclear DNA the 5′ primer problem may need no solution; instead the deoxynucleotide gaps are left. This raises the problem of

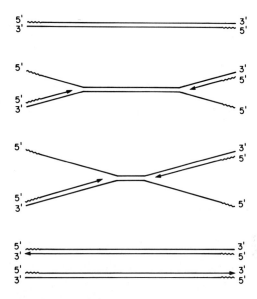

Fig. 17. A model to describe a possible mode of replication of macronuclear DNA. The 5′ ends are made up of 16 ribonucleotides that originally served as primers for DNA polymerase. The 5′ terminal primers remain after replication is completed and must serve as templates for replication of the corresponding regions of the 3′ ends in the next replication round. Replication presumably includes backfilling by means of Okazaki pieces (not shown).

how the 3′ ends are replicated. One possibility is that 5′ gaps are occupied in the cell by a piece of RNA that served as primer for DNA polymerase and then as a template for replicating the 3′ ends of DNA molecules in reverse transcriptase fashion as outlined in Figure 17.

VIII. THE INVERTED TERMINAL REPEAT SEQUENCES IN MACRONUCLEAR DEVELOPMENT

The function or purpose of the inverted terminal repeat sequence is not known, although it may contribute in some way to replication of the ends. Another possibility is that these sequences serve to mark the ends of genes as they exist in integrated form in micronuclear chromosomes and to guide excision of genes in macronuclear development. This idea was tested by examining large fragments of micronuclear DNA (17 KBP and larger) that contained junctions between macronuclear gene sequences and exclusively micronuclear DNA sequences [Boswell et al, 1982a]. None of the nine sequences homologous to macronuclear DNA tested by hybridization with a

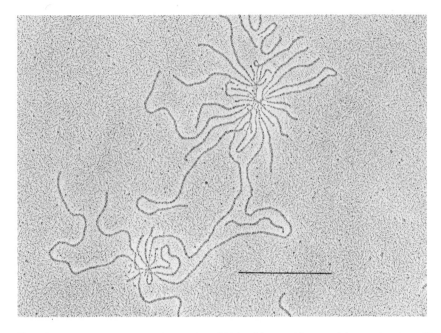

Fig. 18. Aggregated macronuclear DNA of Holosticha sp. Macronuclei were lysed at pH 9.5, treated briefly with protease K, and spread on an aqueous surface. DNA molecules are held together at their ends. Most molecules have one free end; others form loops from an aggregation center. Two molecules connect the centers of the two aggregates, which probably represent the splitting of a single center (Murti and Prescott, unpublished). Bar = 1 μm.

radioactive probe containing the C_4A_4 sequence showed the presence of the inverted terminal repeat. Therefore, the C_4A_4 sequence must be added to a gene during or after its excision from a chromosome.

Micronuclear chromosomes of O nova do contain large amounts of C_4A_4 sequences but not in association with macronuclear genes (Boswell et al, unpublished). These C_4A_4 regions might be the source of the sequences added to genes during macronuclear development.

IX. MACRONUCLEAR CHROMATIN

The DNA molecules in the macronucleus are complexed with histones to form typical nucleosomes [Lawn et al, 1978; Lipps et al, 1974]. In addition, the gene-size molecules are further aggregated, in part at least, by interaction with protein. After a very short digestion of lysed macronuclei with pronase K, discrete aggregates of DNA are observed in the EM (Fig. 18) (Murti and Prescott, unpublished). Longer pronase treatment frees the DNA molecules.

In addition to aggregation with a protein, the DNA molecules apparently interact end to end by means of weak bonds, presumably hydrogen bonds [Lipps et al, 1982], perhaps by formation of triple or tetrastranded regions at molecular ends. This interaction can be mimicked with purified DNA by incubation at a high salt concentration [Lipps, 1980]. The interaction is not abolished by treatment with the single-strand nuclease S1, but is abolished by brief exposure to the double-strand exonuclease Bal 31. This means that the 3' single-stranded tails are not needed for aggregation, but that the inverted terminal repeat sequences probably are. What role aggregation of DNA molecules may serve is not apparent; it does give rise to the chromatin bodies seen in intact macronuclei.

X. FINAL STATEMENT

The foregoing is a brief summary of our major findings with hypotrichous ciliates. As noted, the processing of chromosomes, with the elimination of so much of the genomic sequence complexity, indicates a solution to the S-value (C-value) paradox—ie, most DNA sequences form nongenic spacers between genes, and these spacers may vary widely in size from species to species. We continue to study four basic and related areas with the hypotrichs: 1) the arrangement of macronuclear genes in chromosomes and the sequences between genes (spacers) that are eliminated during chromosome processing; 2) the mechanisms by which genes are excised from chromosomes and by which the C_4A_4 ends are added to genes; 3) the structure and function of macronuclear genes including particularly transcription and chromatin structure; and 4) the mechanism of DNA replication of the gene-size molecules, particularly the role of the C_4A_4 sequences and their possible roles as telomeres on the gene-size molecules.

ACKNOWLEDGMENTS

This work is supported by NIGMS research grant R01 GM19199, and ACS research grant CD-138 to D. M. Prescott.

XI. REFERENCES

Ammermann D (1965): Cytologische und genetische Untersuchungen an dem Ciliaten Stylonychia mytilus Ehrenberg. Arch Protistenk Bd 108:109–152.
Bostock CJ, Prescott DM (1972): Evidence of gene diminution during the formation of the macronucleus in the protozoan, Stylonychia. Proc Natl Acad Sci USA 69:139–142.
Boswell RE, Klobutcher LA, Prescott DM (1982a): Inverted terminal repeats are added to genes during macronuclear development in Oxytricha nova. Proc Natl Acad Sci USA 79:3255–3259.

Boswell RE, Jahn C, Greslin AF, Prescott DM (1982b): Organization of sequences in micronuclear DNA of Oxytricha nova. (In preparation).

Elsevier SM, Lipps HJ, Steinbrück G (1978): Histone genes in macronuclear DNA of the ciliate Stylonychia mytilus. Chromosoma (Berl) 69:291–306.

Gall JG (1981): Chromosome structure and the C-value paradox. J Cell Biol 91:3s–14s.

Kaine BP, Spear BB (1982): Nucleotide sequence of a macronuclear gene for actin in Oxytricha fallax. Nature 295:430–432.

Klobutcher LA, Swanton MT, Donini P, Prescott DM (1981): All gene-sized DNA molecules in four species of hypotrichs have the same terminal sequence and an unusual 3' terminus. Proc Natl Acad Sci USA 78:3015–3019.

Kloetzel JA (1970): Compartmentalization of the developing macronucleus following conjugation in Stylonychia and Euplotes. J Cell Biol 47:395–407.

Lauth MR, Spear BB, Heumann J, Prescott DM (1976): DNA of ciliated protozoa: DNA sequence diminution during macronuclear development of Oxytricha. Cell 7:67–74.

Lawn RM, Heumann JM, Herrick G, Prescott DM (1978): The gene-sized DNA molecules in Oxytricha. Cold Spring Harbor Symp Quant Biol 42:483–492.

Lipps HJ, Sapra GR, Ammermann D (1974): The histones of the ciliated protozoan Stylonychia mytilus. Chromosoma 45:273–280.

Lipps HJ (1980): In vitro aggregation of the gene-sized DNA molecules of the ciliate Stylonychia mytilus. Proc Natl Acad Sci USA 77:4104–4107.

Lipps HJ, Gruissem W, Prescott DM (1982): Higher order DNA structure in macronuclear chromatin of the hypotrichous ciliate Oxytricha nova. Proc Natl Acad Sci USA 79:2495–2499.

Maxam AM, Gilbert W (1977): A new method for sequencing DNA. Proc Natl Acad Sci USA 74:560–564.

Murti KG (1976): Organization of genetic material in the macronucleus of hypotrichous ciliates. In King RC (ed): "Handbook of Genetics." New York: Plenum Press, Vol 5, pp 113–137.

Olins AL, Olins DE, Franke WW, Lipps HJ, Prescott DM (1981): Stereo-electron microscopy of nuclear structure and replication in ciliated protozoa (Hypotricha). Eur J Cell Biol 25:120–130.

Prescott DM, Bostock CJ, Murti KG, Lauth MR, Gamow E (1971): DNA of ciliated protozoa. I. Electron microscopic and sedimentation analyses of macronuclear and micronuclear DNA of Stylonychia mytilus. Chromosoma 34:355–366.

Prescott DM, Murti KG (1974): Chromosome structure in ciliated protozoans. Cold Spring Harbor Symp Quant Biol 38:609–618.

Prescott DM, Heumann JM, Swanton M, Boswell RE (1979): The genome of hypotrichous ciliates. In Engberg J, Klenow H, Leick V (eds): "Specific Eukaryotic Genes, Alfred Benzon Symposium 13." Copenhagen: Munksgaard, pp 85–99.

Rae PMM, Spear BB (1978): Macronuclear DNA of the hypotrichous ciliate Oxytricha fallax. Proc Natl Acad Sci USA 75:4992–4996.

Spear BB, Lauth MR (1976): Polytene chromosomes of Oxytricha: Biochemical and morphological changes during macronuclear development in a ciliated protozoan. Chromosoma 54:1–13.

Swanton MT, Heumann JM, Prescott DM (1980a): Gene-sized DNA molecules of the macronuclei in three species of hypotrichs: Size distributions and absence of nicks. Chromosoma 77:217–227.

Swanton MT, Greslin AF, Prescott DM (1980b): Arrangement of coding and non-coding sequences in the DNA molecules coding for rRNAs in Oxytricha sp. Chromosoma 77:203–215.

Wesley RD (1975): Inverted repetitious sequences in the macronuclear DNA of hypotrichous ciliates. Proc Natl Acad Sci USA 72:678–682.

Modern Cell Biology 2:353–383
© 1983 Alan R. Liss, Inc., 150 Fifth Avenue, New York, NY 10011

The Structure and Expression of Globin Chromatin During Hematopoesis in the Chicken Embryo

Harold Weintraub and Mark Groudine

From the Department of Genetics, Hutchinson Cancer Research Center, Seattle, Washington 98104 (H.W., M.G.), and the Department of Radiation Oncology, University of Washington Hospital, Seattle, Washington 98195 (M.G.)

I. INTRODUCTION

In the developing chicken embryo, hemoglobin (Hb) initially appears in first-generation erythroblasts at approximately 30–35 hours of development.

These cells then serve as the progenitors of six subsequent "maturational" generations of erythroblasts, each of which contains characteristic amounts of Hb and displays unique morphologic characteristics. These cells mature rather synchronously, and by day 6 of embryonic development, most have become erythrocytes. They constitute a "primitive" line of embryonic, erythroid cells. Between days 6 and 7, another lineage of hematopoetic cells, the "definitive" embryonic lineage, enters the circulating red cell population, having arisen from the endothelial lining of the blood vessels. The exact relationship between the precursor cells for the primitive line and those for the definitive line is unknown.

The definitive cell series enters the circulating erythroid compartment between days 6 and 7 as fairly mature erythroblasts, and by days 11–13 matures into erythrocytes that are transcriptionally inactive. These cells are the major erythroid cells in the circulation until hatching, at which time a second definitive line appears. Although minor differences exist, by and large both definitive lines produce similar types of Hb polypeptides [Brown and Brown, 1974], and both differ from the primitive series in the synthesis of specific "β chains," two of the polypeptide subunits of the Hb tetramer.

In this and former studies, we have focused on the switch in β-chain synthesis between the primitive and first definitive series. Given the similarities in β-chain synthesis between the two definitive lines, we have not distinguished between the two types of definitive cells, and, for convenience, refer to both as "adult" red cells since they both produce "adult" β chains. In contrast, given the distinctive embryonic β chains synthesized in the primitive series, we refer to these as "embryonic" β chains and "embryonic" red cells.

While examining the control of embryonic and adult globin gene expression, we previously determined that steady-state globin mRNA is not detectable in the precursor hematocytoblasts present in the embryo before 20–25 hours of development. Furthermore, adult-specific globin RNA is not detectable in steady-state cytoplasmic or nuclear RNA from embryonic red blood cells [Groudine et al, 1974]. As assayed by solution hybridization, globin chromatin is resistant to DNase I before 25 hours, but becomes sensitive to DNase I after overt erythroid differentiation is apparent at 35 hours. By the same assay, adult-specific globin genes are in an "inactive" conformation in embryonic red blood cells, and vice versa [Weintraub and Groudine, 1976].

In the experiments reported here, we use a different assay of DNase I sensitivity, the "blot" hybridization assay, to further probe the structure of globin chromatin in the hematopoetic lineages. Our results will be discussed in the context of the idea that a special round of DNA synthesis is required for a cell to become committed to a particular developmental fate [Holtzer et al, 1975].

II. SENSITIVITY OF GLOBIN GENES AS MEASURED BY LOW-LEVEL DIGESTION OF NUCLEI WITH DNase I

Nuclei were isolated from "embryonic" 4- to 5-day chick red blood cells and from "adult" 14-day red cells. The respective nuclear preparations were digested with very low levels of DNase I [Wu et al, 1979] and the DNA isolated. At the highest level of DNase used in the experiments reported here, the average size of DNA was greater than 10 kilobase pairs (kbp) and in most cases had an average size between 15 and 20 kbp. The purified DNA from these preparations was then cut with the restriction endonuclease Bam H1, separated by electrophoresis on neutral 1% agarose gels, transferred to nitrocellulose filters [Southern, 1975], and hybridized to a ^{32}P-nick-translated adult globin cDNA clone pHb 1001 [Saltser et al, 1979] or to ovalbumin cDNA as a control (Fig. 1A, B). At the level of DNase digestion used, no loss of ovalbumin restriction fragments was observed. As an additional control, nuclei from a T-cell line (MSB) were digested, using our most vigorous DNase I digestion conditions, and this DNA also showed no preferential sensitivity for either the globin probe or the ovalbumin probe. (The ovalbumin Bam H1 pattern is different in MSB cells than in the flock of chickens used for red cell isolation. This probably reflects a genetic polymorphism associated with the ovalbumin gene [Weinstock et al, 1978].)

When the DNaseI-digested DNA blots are hybridized to the adult globin probe pHb 1001, there is a rather steady and gradual loss of the Bam H1 restriction bands that hybridize to the probe as the concentration of DNase increases. This result contrasts with the random and variable intensities displayed by the ovalbumin bands (Fig. 1B), a variability that probably arises from difficulties in loading precise quantities of very viscous DNA onto the gel. In adult nuclei, the 4.5 kbp DNA fragment containing the adult gene is strikingly more sensitive than the 8.5 kbp fragment containing the embryonic gene. The former begins to disappear at about 0.05 μg/ml of DNase, whereas the latter disappears at about 1 μg/ml DNase. These respective values are highly reproducible and are also observed using as probe a λ-genome clone, λCβG1, containing both an embryonic β-globin gene and an adult β-globin gene (β_1, see Fig. 3). For example, Figure 2 is an Eco R1–Hind III double digest, blot-hybridized to λCβG1, showing that in embryonic cells both adult- and embryo-specific fragments are degraded rapidly, but that the embryo-specific fragment becomes much more resistant in adult cells. Similar results are also obtained using a second genomic clone, λCβG2, which carries the 8.5 kbp embryonic gene and a 2 kbp fragment containing β^H globin, (data not shown). Assuming that the initial rates of DNase cutting are roughly proportional to DNase concentration, it is a reasonable first guess that in adult nuclei the adult gene is about 20 times more sensitive to digestion than the embryonic gene, which is in turn more sensitive than the ovalbumin gene.

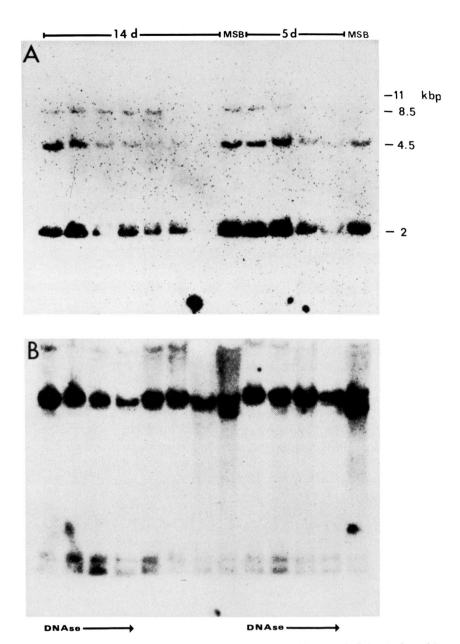

Fig. 1. DNase I sensitivity of the adult and embryonic β-globin genes in 5-day (embryonic) and 14-day (adult) red blood cell nuclei. Control DNA and DNA from 14-day and 5-day nuclei treated with increasing concentrations of very low levels of DNase I were digested with Bam H1, run on 1% agarose gels, blotted, and hybridized to (A) nick-translated pHb 1001 clone DNA and (B) ^{32}P-labeled ovalbumin cDNA. As an additional control, DNA from MSB cell nuclei, incubated with the highest (1.5 μg/ml) DNase I concentration used in the red blood cell nuclear digests was analyzed.

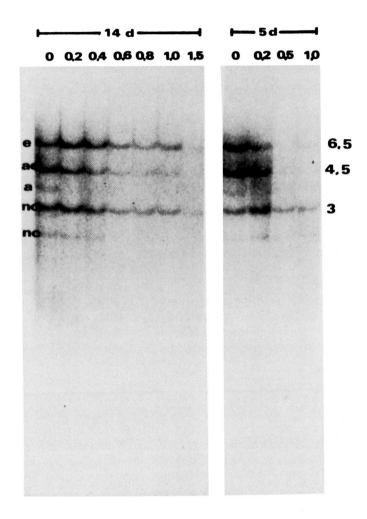

Fig. 2.   Kinetics of DNase I digestion as assayed by hybridization to λCβG1. DNase-treated DNA from 14- and 5-day red cells was digested with R1 and Hind III and the corresponding blots were hybridized to λCβG1. Abbreviations: a, segments coding exclusively for adult genes; e, segments coding exclusively for embryonic genes; ae, segments containing both adult and embryonic genes; nc, segments containing noncoding fragments. Note that these fragments are digested in the highest digest of adult nuclei. We should also point out that the embryonic gene (e) is larger than predicted from restriction analysis of the corresponding genome clone, λCβG1 [Dodgson et al, 1979a, b]. We believe that this result may reflect modified Hind III sites in the nuclear DNA; they are unlikely to represent partially digested fragments because the same digest yields the predicted restriction map when hybridized to different globin gene containing recombinant DNA clones.

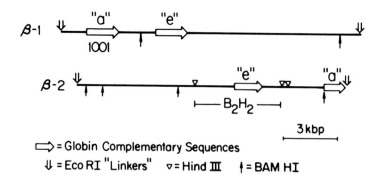

Fig. 3. Restriction maps for two λ Charon 4a recombinant chicken clones carrying globin genes. Abbreviations: a and e, denote adult and embryonic β-globin genes extending over the distance indicated by the thick arrows. The direction of transcription (5′ → 3′) was determined by Dodgson et al [1978] and Dolan et al (unpublished results). The "1001" refers to the cDNA plasmid Hb 1001 [Saltser et al, 1979] which is homologous to the adult β_1-globin gene, as indicated. "B_2H_2" is a subclone of λCβG2, containing the sequences indicated on the diagram of β-2.

Adjusting for the differences in target size, the adult gene may actually be as much as 50 times more sensitive than the embryonic gene. This is clearly a very rough estimate, and the true value might be anywhere between 10 and 100.

A different picture emerges in comparing the relative sensitivity of the embryonic and adult genes in embryonic cells. In embryonic nuclei, both adult and embryonic genes begin to disappear at about 0.1 μg/ml of DNase I. Given the larger target size of the embryonic gene, it is possible that this gene may be slightly more sensitive to DNase digestion; however, this slight increase in sensitivity is marginal when considered against the factor of 20 estimated for the relative sensitivity of the adult gene compared with the embryonic gene in adult cells. Finally, the DNA sizes for the respective DNase concentrations are comparable in 5- and 14-day nuclei. Thus it is also clear that at comparable levels of digestion, the embryonic gene is more sensitive in embryonic nuclei than in adult nuclei, in which approximately ten times more DNase is required for complete digestion. In an absolute sense, the embryonic gene therefore becomes more resistant during the transition from primitive to adult red cells.

III. LINEAGE-DEPENDENT TRANSCRIPTION OF EMBRYONIC AND ADULT β-GLOBIN GENES

In the following experiments, we have employed the two recombinant DNA clones containing the chicken β-globin genes mentioned in the previous section to assay the transcription of these various genes in embryonic and adult red blood cells. The clones (β1 and β2) and the genes they contain are described in the restriction maps illustrated in Figure 3. Purified cloned DNAs were digested with various restriction enzymes, separated on 1% agarose gels, blotted to nitrocellulose filters [Southern, 1975], and hybridized to purified ^{32}P-nuclear RNA (nRNA) synthesized (2.5-minute reactions) by endogenous "run-off" transcription [McKnight and Palmiter, 1979] from embryonic or adult red cell nuclei. Figure 4a shows the ethidium bromide staining patterns and resultant hybridization to specific regons of λCβG-1 after digestion of the clone with Bam H1 and hybridization with ^{32}P-nRNA from either adult or embryonic red cell nuclei. Annealing of "blots" from Bam-digested β1 with ^{32}P-RNA isolated from nuclei of 5-day embryonic red blood cells results in almost exclusive hybridization to the 10.9 kbp embryonic fragment, although a very small signal (< 1% of that observed from 10.9 kbp fragment) can be detected from the 9.3 kbp adult β1 fragment. Conversely, hybridization with ^{32}P-nRNA from 12-day "adult" red cell nuclei results in the almost exclusive labeling of the 9.3 kbp adult fragment, although, again, a small amount of hybridization is observed with the 10.9 kbp embryonic fragment with very long exposures. Given the previous reports of the cross hybridization of the embryonic and adult β-globin genes, we suspect that the small signal observed from the heterologous fragments most likely represents cross hybridization. Additional evidence for thiscomes from the observation that low salt washes preferentially reduce hybridization to the heterologous β-globin gene fragments (data not shown).

Additional data supporting the cell-type-specific transcription of the "adult" β-globin gene is derived from experiments in which we hybridized ^{32}P-nRNA from either embryonic or adult red blood cell nuclei to cDNA plasmid Hb1001, which contains the coding sequences for the adult β-1 globin gene (see Fig. 3). Significant hybridization to this cDNA clone is observed only with ^{32}P-nRNA from the 12-day "adult" red cells, and not from ^{32}P-nRNA from embryonic cells (data not shown).

We next analyzed transcription of the embryonic and adult globin genes located on λCβG-2 (β2). These sequences lie to the immediate 5′ side of β1. Figure 4b shows the results of hybridizing either adult or embryonic ^{32}P-nRNA to blots of Bam III digested β2. As can be seen, when embryonic ^{32}P-nRNA is used, major hybridization is observed to the 7.9 kb embryonic gene-containing fragment, with less, but detectable signal from the 3.9 kbp, noncoding fragment. We believe, however, that the signal obtained from the 3.9 kbp fragment is ambiguous since this fragment is highly repetitive in the chick genome, and therefore the hybridization signal may actually reflect transcription from other regions of the genome rather than transcription from the region in the globin domain. Some evidence for this view comes from

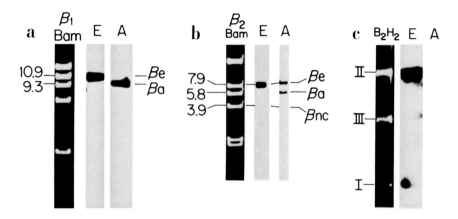

Fig. 4. Hybridization of in vitro synthesized ³²P-RNA from nuclei of 5-day embryonic (E) and 12-day adult (A) red blood cells to a) BAM restriction fragments of λCβG-1, b) BAM restriction fragments of λCβG-2 and β, and c) plasmid B₂H₂. The first lane in each series shows the ethidium bromide stain of DNA from the various clones after restriction and electrophoresis on a 1% agarose gel: βe and βa indicate those fragments containing embryonic globin sequences and "adult" globin genes, respectively.

the fact that this fragment is not digested by DNase I in red blood cells. Lastly, overexpression of the blot reveals a signal from the 5.8 kbp adult β2 gene fragment equal to less than 1% of the signal observed from the embryonic β2 fragment. Again, this may reflect cross hybridization.

When ³²P-nRNA from 12-day, adult, red blood cell nuclei is hybridized with the Bam-digested β2 blot, almost equal signals are observed from the 5.8 kbp (adult) and 7.9 kbp (embryonic) fragments. In addition, as with embryonic ³²P-nRNA, minor signals from the 3.9 kbp, noncoding (and repetitive sequence-containing) fragment is also observed with the adult ³²P-nRNA. Given the large size of the embryonic 7.9 kbp β2 Bam H1 fragment, it is possible that hybridization between this region and adult red blood cell ³²P-nRNA could result from transcription from sequences outside of the embryonic gene. To test this possibility, we hybridized ³²P-nRNA from embryonic or adult red blood cells to a blot containing a PBR subclone of λCβG-2, β2H2, which contains the embryonic β2 gene and its 5' and 3' flanking sequences, defined by the adjacent Hind III sites. As can be seen in Figure 4c, significant hybridization is observed between β2H2 and ³²P-nRNA from 5-day embryonic nuclei, but essentially no hybridization is detected between 12-day adult ³²P-nRNA and β2H2 at similar exposure. Thus, we feel these results indicate that hybridization between ³²P-nRNA from adult red blood cells and the 7.9 kbp Bam embryonic β2 gene-containing fragment is a consequence of transcription from a region outside the embryonic β2 coding sequences, possibly just to the left of the Bam site at the 5′ side of the adult gene in β2. Also, since these bordering regions contain repetitive DNA sequences, it is again possible that the transcription signal actually comes from another region of the genome besides the globin domain. DNase I sensitivity experiments support this possibility. Thus, we conclude that as our probes approach the size of the known coding regions for adult and embryonic β-globin genes, our ability to detect embryonic or adult red cell–specific "run-off" transcription is greatly enhanced. In general, the embryonic genes on both clones are transcribed in embryonic cells, while the adult genes on both clones are transcribed in adult cells. Heterologous transcription is at least 100-fold lower in the respective red cell lineages and may even be zero if the weak signal observed proves to come from cross hybridization.

Although the experiments described above indicate that transcription of adult and embryonic β-globin genes is restricted to their respective developmental compartments, two alternative possibilities could also explain our results: 1) Both embryonic and adult globin genes are transcribed in both adult and embryonic red cells, but the heterologous transcripts are degraded so rapidly as to preclude their detection, even in our 2.5-minute pulses. 2) Polymerases are present on both embryonic and adult globin genes in red cells of both developmental compartments, but are unable to elongate in the heterologous red cells because of "inhibitory" chromosomal proteins. Some evidence against preferential degradation comes from the observation that the hybridization signal obtained with RNA from adult or embryonic nuclei yields the same ratio of adult to embryonic signal, regardless of incubation time from 2.5 to 15 minutes. Nevertheless, we have attempted to test these possibilities by performing the elongation assay in the presence of sarkosyl or heparin. These compounds remove histones and most other chromosomal proteins, but leave initiated RNA polymerases still bound to DNA and capable of elongation, but not new initiation [Gariglio et al, 1974; Green et al, 1975; Gariglio, 1976; Ferencz and Seifart, 1975]; thus, "blocked" RNA polymerases would most likely elongate if histones are usually preventing such elongation in the intact nuclei. In addition, since heparin is a known inhibitor of ribonuclease [Palmiter et al, 1970; Cox, 1976], and since either heparin or sarkosyl would be expected to strip proteins from hnRNP, degradation of specific RNA sequences in the presence of these compounds is less likely.

When either adult or embryonic red blood cell nuclei are incubated in the presence of sarkosyl or heparin, no change in the ratio of transcripts from the adult or embryonic β-globin genes is observed in the presence of these compounds (data not shown). If either sequence-specific posttranscriptional degradation or "blocked" RNA polymerases were major regulatory mechanisms in globin gene switching, we would have expected a significant decrease in the ratio of adult-to-embryonic or embryonic-to-adult globin gene transcription in the adult and embryonic red cell nuclei, respectively. Thus, our results indicate that a major regulatory mechanism in governing hemoglobin switching in the chicken embryo resides at the level of transcription of the respective β-globin genes by RNA polymerase.

We think our current findings may have several implications for the regulation of expression of the globin genes developmentally. We previously reported that the switch from primitive to definitive erythropoiesis is accompanied by a decrease in DNase I sensitivity of the embryonic globin genes [Weintraub and Groudine, 1976]. Our current demonstration that expression of the embryonic β-globin genes is limited to the primitive red blood cells is consistent with the notion that a specific chromosomal conformation is coupled to transcription of this gene. In contrast, as communicated above, by blot hybridization (which measures the first DNase I cut introduced into the gene), both the embryonic and adult β-globin genes are digested at approximately the same rate in primitive erythroid cells. When examined by solution hybridization (which assays the frequency of introducing many cuts into the gene), however, the embryonic β-globin genes are more sensitive than the adult β-globin genes in embryonic cells [Weintraub and Groudine, 1976]. Thus, the "blot" sensitivity of the adult gene in embryonic cells may reflect a "preactivation" chromosomal state. The "run-off" experiments support the conclusion that the adult genes are in a preactivation structure since they show that transcription of the adult genes is limited to the adult cells, and is barely, if at all, detectable in embryonic cells. Thus, though an altered chromosomal state is present prior to overt gene expression and may, in fact, be required for subsequent expression, that altered conformation of chromatin is insufficient by itself to permit transcription. In fact, run-off transcription experiments in the presence of sarkosyl and heparin reveal that few, if any, polymerases are bound to the adult gene in the embryonic cells.

The experiments reported here can be considered in terms of a currently popular model for chromosome structure based upon the organization of lampbrush loops and, to some extent, polytene chromosomes (Fig. 5). We propose that the globin gene cluster is organized into a functional chromosomal domain of about 50–100 kbp, and that such a domain is represented by a chromosomal loop. When the loop is relaxed, a moderately sensitive DNase I domain results. This may be the state of the globin domain, but not of the ovalbumin domain in the erythrocyte stem cell (hematocytoblast). With terminal differentiation to an embryonic red cell, additional elements (ie, HMG 14 and 17, undermethylation, as well as other currently undefined modifications) become associated with the newly activated embryonic gene, resulting in transcription and an increased sensitivity to DNase I. Given the orientation of the embryonic (e) and adult (a) β-globin genes in the chicken

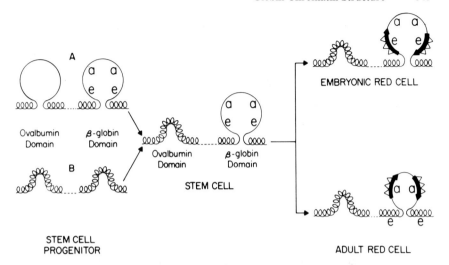

Fig. 5. Interpretation of moderate sensitivity of adult β-globin genes to DNase I in embryonic red cells in terms of "lampbrush" chromosomelike loops or domains. Chromosome domains are either condensed or relaxed. Transcription units become activated in relaxed domains by virtue of their association with HMG 14 and 17, and other modifications (open triangles). The red cell stem cells (hematocytoblasts) are unique in that they are predicted to be defined by a relaxed globin chromatin domain not yet fully activated by HMGs, etc. The hematocytoblast progenitors can have either relaxed (A) or constrained (B) domains; however, whichever proves to be the case, the state of the domains is seen to be the same for ovalbumin and globin.

chromosomes (5′ . . . e . . . a . . . a . . . 6 . . . 3′), overt expression of the embryonic genes requires the "looping out" of the adult β-globin genes as well. Thus, the "blot" DNase I sensitivity of the adult genes in embryonic cells would be a structural consequence of their location in the β-globin domain. As the embryo ages, the hematocytoblasts (or their descendants) switch to adult red cell production, and the adult gene is configured with the modifications mentioned above. However, since the assembly of active adult β-globin genes does not require the concomitant relaxation of the embryonic β-globin gene regions, the embryonic genes are "reeled in," resulting in a greatly reduced sensitivity of these embryonic genes to DNase I in the adult red cells. (Alternatively, the β-globin domain in the immediate precursor cells could be in a completely repressed state, and the whole domain relaxed only in the transition to embryonic red cell. Transition to the adult red cell would be accompanied by relaxation of only the adult genes and surrounding sequences.) Besides focusing one level of control to the base of the loops, Figure 5 also illustrates two extreme states for the progenitor to the hematocytoblast. Ovalbumin and globin domains are both seen to be "open" initially and then differentially "repressed," or repressed initially and then differentially activated.

IV. ASSOCIATION OF HMG 14 AND 17 WITH ACTIVE CHROMATIN

The sensitivity of active genes to DNase I has been shown to depend on the association of two nuclear proteins with these genes [Weisbrod and Weintraub, 1979; Weisbrod et al, 1980; Gazit et al, 1980]. These proteins, HMG 14 and 17 [Goodwin and Johns, 1973], have highly basic and highly acidic domains and can be eluted from nucleosomes with 0.35 M NaCl. After elution, the actively transcribed nucleosomes retain the general features of their structure, but lose their sensitivity to preferential digestion. When, however, pure HMG 14 or HMG 17 is added back to the depleted nucleosomes, the actively transcribed genes regain their preferential sensitivity to DNase I. An example of such a reconstitution is shown in Figure 6. Here chromatin is depleted of HMG 14 and 17 using 0.35 M NaCl. The sample is split in two, and half is reconstituted with pure HMG 14 and 17. Both samples are then digested to equivalent degrees with increasing concentrations of DNase I. The DNA is then isolated, digested with the restriction endonuclease, MSP, and then blot-hybridized to a nick-translated adult β-globin cDNA clone (pHb 1001). Clearly, the globin gene is much more sensitive to DNase I digestion after reconstitution with HMGs. As a control, no preferential digestion of the inactive ovalbumin gene is observed using the same blots (data not shown).

Several very general features of HMG binding to actively transcribed nucleosomes have been established: 1) There is no tissue specificity associated with the HMGs. Thus, HMG 14 and 17 from brain nuclei are capable of restoring DNase I sensitivity when reconstituted with HMG-depleted red cell chromatin. 2) The specificity for HMG binding seems to be in the recipient depleted chromatin. Thus, HMG 14 and 17 from red cell chromatin fail to induce DNase I sensitivity in the globin gene region when reconstituted with HMG-depleted chromatin from brain. 3) Either HMG 14 or HMG 17 can sensitize most actively transcribed genes to DNase I. 4) Genes transcribed at different rates have about the same affinity for HMG 14 and 17. 5) HMG 14 and 17 bind stoichiometrically to actively transcribed nucleosomes. 6) HMG 14 and 17 can restore DNase I sensitivity to isolated, HMG-depleted nucleosome core particles containing 145 base pairs (bp) of DNA. 7) HMGs bind to essentially all of the nucleosomes present in a given transcription unit, and consequently most, if not all, of the nucleosomes in a transcription unit become preferentially sensitive to DNase I. 8) As judged by sensitivity to DNase I, actively transcribed nucleosomes become associated with HMG 14 and 17 within 3 minutes after these genes are replicated.

One of the most interesting features of HMG activation is the specificity of the binding of these proteins to actively transcribed nucleosomes. When

⁻HMG   +HMG

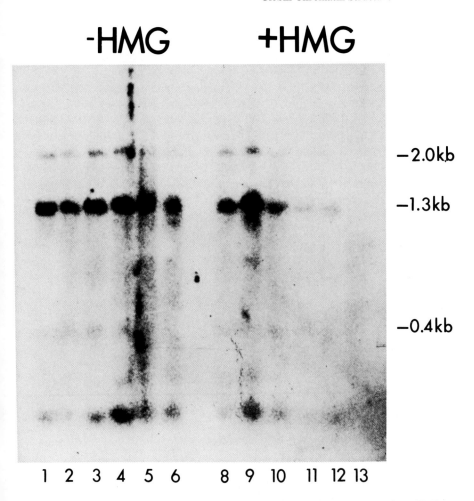

−2.0kb

−1.3kb

−0.4kb

1 2 3 4 5 6   8 9 10 11 12 13

Fig. 6.   Reconstitution of DNase I sensitivity of the globin gene with HMG 14 and 17. HMG-depleted and reconstituted chromatin was digested with increasing concentrations of DNase I and the purified DNA was digested with MSP, separated on 1.4% agarose gels, transferred to nitrocellulose paper, and hybridized to a ^{32}P nick-translated β-globin cDNA clone (2 × 10^3 cpm/μg). Lanes 1–6 represent depleted chromatin digested with 0, 0.01, 0.03, 0.06, 0.1, and 0.3 μg/ml DNase I. Lanes 8–13 represent HMB-reconstituted chromatin digested similarly.

about 1 mole each of pure HMG 14 and 17 was reconstituted onto about 20 moles of HMG-depleted red cell nucleosomes, the DNase I sensitivity of the globin chromatin was restored. The ovalbumin gene used as a control showed no increased sensitivity either before or after reconstitution with HMG 14 and 17. A similar type of result can also be shown by direct measurements of HMG binding. Figure 7 shows a typical result. Labeled HMGs are challenged to bind to a large excess of either pure 200 bp DNA or to nucleosomes. After binding, the mixture is run on sucrose gradients where the 5S DNA is clearly separated from the 11S nucleosomes, and the 2S, unbound HMGs. Figure 7C shows that HMGs prefer depleted nucleosomes to free DNA; however, as shown in Figure 7D, HMGs prefer to bind to pure DNA rather than to undepleted nucleosomes. In separate experiments without addition of the competitor DNA, it is possible to show that the HMGs will bind readily even to undepleted nucleosomes. Thus, taken as a whole, HMGs prefer to bind to actively transcribed HMG-depleted nucleosomes; however, they will also bind to pure DNA and to complete nucleosomes, probably in that order of preference. Given the ability of the HMGs preferentially to induce sensitivity to DNase I in active nucleosomes, there seems to be some feature of these nucleosomes that marks them and allows them to be recognized by HMGs for preferential binding.

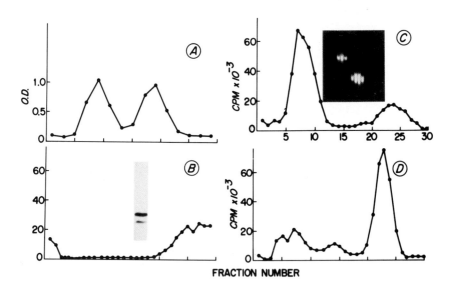

Fig. 7.   Preferential binding of ^{3}H-HMG 14 and 17 to HMG-depleted nucleosome monomers. Each panel shows a 5 → 20% sucrose gradient sedimentation of a mixture of various components as described in the text below. Migration is from right to left. A) Optical density (260 nm) profile of nucleosome monomers and monomer DNA. B) ^{3}H-succinimidyl propionate-labeled HMG 14 and 17 alone. Insert is fluorograph of the ^{3}H-HMG preparation electrophoresed on an 18% SDS-polyacrylamide gel. C) Depleted monomers and monomer DNA were reconstituted with ^{3}H-HMG 14 and 17. Insert is a 4.5% polyacrylamide slab gel profile of each fraction shown in A. D) Undepleted monomers and monomer DNA were reconstituted with ^{3}H-HMG 14 and 17.

In order to gain a better understanding of these additional features of actively transcribed nucleosomes, we have attempted to isolate them using a column composed of HMG 14 and 17 covalently coupled to agarose. As shown in Figure 8, when chromatin is passed over such a column, about 10% of the material binds and is not eluted until NaCl in the elution buffer is raised to 0.3–0.4 M. The material that binds to the column reruns true, as does the material that does not bind. Most important, the material that binds hybridizes to 40%–50% when driven in a hybridization reaction with an excess of nuclear RNA; the material that does not bind is depleted in sequences hybridizing to nuclear RNA (Fig. 9). Assuming asymmetric transcription, it is possible to estimate that the column yields a population of nucleosomes that is between 80% and 100% pure active nucleosomes.

To assay the binding of specific nucleosomes to the HMG column, DNA from red cell nucleosomes prepared from the bound and unbound fractions was spotted onto nitrocellulose and hybridized to a variety of probes. Figure 10 shows that when adult red cells are used, the adult β-globin cDNA clone (pHb 1001) hybridized to DNA in the bound fraction. A flanking region (FL) of this gene that is not transcribed, but displays an intermediate level of DNase I sensitivity, is present in the unbound fraction, as is the ovalbumin gene. A probe that hybridizes predominantly to an embryonic β-globin gene is also found mainly in the unbound fraction. The latter probe also hybridized significantly to DNA in the bound fraction, and we think that this is probably due either to cross hybridization with adult sequences, or more likely, to repetitive DNA sequence elements that have previously been localized to regions contained in this probe.

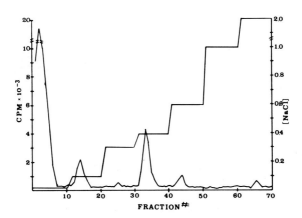

Fig. 8. Binding of chromatin to HMG 14–17 agarose. Chromatin was dialyzed onto the column resin and eluted stepwise with increasing concentrations of NaCl.

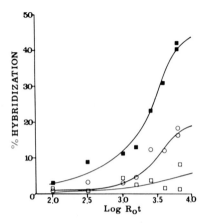

Fig. 9. Sequence analysis of bound and unbound fractions. HMG-agarose affinity chromatography was performed on sheared, ^{3}H-Tdr-labeled chromatin from MSB cells. Hybridization of the purified DNA from each fraction was to an excess of hnRNA. Each reaction contained hnRNA at 20 mgm/ml. ^{3}H-DNA was at 100,000 cpm per reaction (20 μl). ^{3}H-specific activity was about 10^5 cpm/μg. Parallel hybridizations in the absence of hnRNA were also performed and the hybridization due to DNA alone subtracted. This hybridization ranged between 5% and 10%. (■) Fractions 32–36 (Fig. 8), the "bound" chromatin; (□) fractions 1–6 (Fig. 8), the "unbound" chromatin; (○) total ^{3}H-MSB DNA.

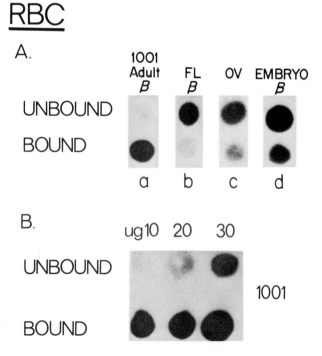

To obtain a more detailed description of the fine structure of active chromatin, we decided to concentrate on the chicken α-globin genes. Restriction maps for this gene cluster are shown in Figure 11. The clones and the restriction maps were kindly provided by Dr J. D. Engel. In particular, we wished to know whether the "spacer" DNA between the two active α genes, α^A and α^D, was DNase I-sensitive and also whether the associated nucleosomes bound HMG 14 and 17. These genes are expressed in both adult and embryonic red cells. Figure 11A (left) shows that after digestion with Bam H1 and Eco R1, and hybridization to αB1.5 (a subclone of α2 in pBR322 that has the 1.5 kb Bam fragment containing both the coding region of α^D and part of the spacer region between α^A and α^D), both α-related fragments are extremely sensitive to DNase I. Most important, they both display approximately equivalent levels of sensitivity. In the same blot a relatively resistant, cross-hybridizing fragment is also observed at about 4 kb. This is probably the inactive U gene present in α5 since a pure U-gene probe gives exactly the same DNase I pattern. The U gene may code for the π-embryonic (α-like) globin [Brown and Ingram, 1980]. This gene is clearly relatively resistant to DNase I and is known not to be expressed in the 14-day definitive adult red cell line used here. It corresponds to a region right at the 5' side of the active α^A–α^D complex.

To look at the sensitivity of the region at the 3' side of α^A, the same blot was hybridized to αB1.2 (a subclone of α2 containing the 1.2 kbp Bam fragment located at the 3' side of α^A). Figure 11A (right) shows that this fragment is as sensitive to DNase I as the coding and the spacer regions. Previously, we showed that the 7.1 and 3 kb Bam fragments more distal to α^A and α^D at their 3' side are relatively resistant to DNase [Stalder et al, 1980a]. In separate experiments (not shown), we have also shown that the 0.8 kb Bam H1 coding fragment located to the right on the 5' side of α^D is DNase I–sensitive. Thus, taken as a whole, the DNase I sensitivity seems to begin very close to the 5' coding region in α^D and extends to about 1 kb beyond the 3' side of α^A, including the spacer region between α^A and α^D.

Using the same probes, we also examined DNA from the bound and unbound fractions from the HMG column for their specific α-related sequences. Figure 11B shows that nucleosomes containing sequences corre-

---

Fig. 10.   Dot hybridization analysis of DNA from 14-day RBC bound and unbound monomers after MG-agarose chromatography. A) Hybridization of 5 μg each of unbound monomer DNA and bound monomer DNA to (a) the adult β-globin cDNA clone (pHb 1001); (b) a 9-kb Bam fragment flanking (FL) the coding sequence for adult β-globin gene in β1 and (c) ovalbumin cDNA; (d) a 2-kb Sac I embryonic globin (EM) fragment from λCβG2. The DNA was covalently linked to DBM paper. B) Hybridization of pHb 1001 to increasing concentration of bound and unbound monomer DNA.

sponding to the spacer region, to the coding region, and to the 3′ side of the α genes bind to the HMG column, but a region 4 kb beyond the coding region (on the 3′ side) does not. Thus, in general, DNase I–sensitive regions correspond to regions that bind to HMG 14 and 17. Moreover, flanking sequences that are located near the DNA coding for the primary stable tran-

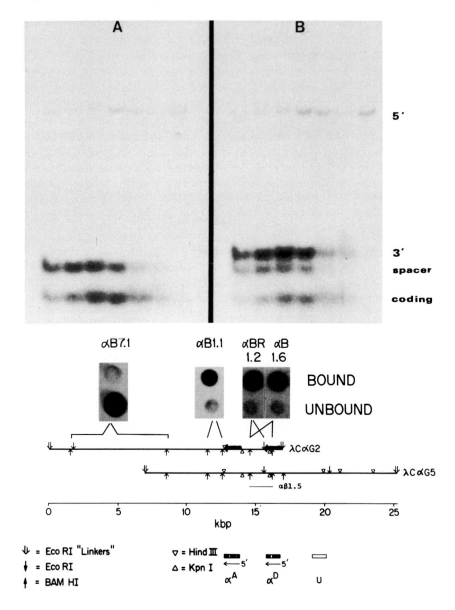

script are also DNase I–sensitive and also bind HMG 14 and 17. Regions of intermediate sensitivity (eg, the 7.1 kilobase pair (kbp) Bam H1 fragment 4 kbp beyond the 3' side of α^A) do not bind to the HMG 14 and 17 column.

The previous results have shown that after depletion and reconstitution, HMG 14 and 17 prefer to bind to actively transcribed nucleosomes. The question arises whether the direct association of these proteins with active nucleosomes can be demonstrated. In the trout testes system, such an association has been shown by Levy and Dixon [1978]. Similar results have also been obtained for the avian globin genes [Albanese and Weintraub, 1980]. Figure 12 shows an electrophoretic separation of monomer, dimer, trimer, etc nucleosomes. Notice that the monomer region is distinctly heterogeneous. By running these particles on a second-dimension protein gel, it is clear from Figure 13B that the monomer heterogeneity is associated with the presence of HMG 14 and 17 which comigrate with the slowest migrating monomers. In this gel system, histones H1 and H5 bind weakly to the nucleosomes and are removed during the electrophoresis (Fig. 13A). In order to study the sequences present in these monomers,

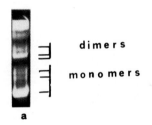

a

Fig. 12. Electrophoretic pattern of nucleosomes. Nuclease-treated chromatin, prepared by micrococcal nuclease digestion of chick erythrocyte nuclei and resuspended in 5 mM Na phosphate, pH 6.8, was diluted with one volume of sample buffer and loaded onto a 5% polyacrylamide slab gel.

Fig. 11. DNase I sensitivity and HMG binding of chromatin associated with the α genes. A) Left: Nuclei from 14-day, adult definitive red cells were digested with increasing concentrations of DNase I (left to right) and the purified DNA restricted with Bam and R1, blotted, and hybridized to αB.16 (a subclone of α2 in PBR322 containing part of the 1.6 kb α^D gene and the spacer region between α^A and α^D (see restriction map below). Right: The blot in A was rehybridized to αB1.2 (see restriction map), a subclone of α2 containing the 3' region adjacent to α^A. The DNase I–resistant fragment probably represents cross hybridization to the gene indicated as "U." This may correspond to the embryonic "π gene" (α-like). As indicated, this fragment would map to the 5' side of α^D. The λCαG2 and λCαG5 were isolated and mapped by Dr. J. D. Engel and J. Dodgson. B) Dot blots of bound and unbound erythrocyte nucleosome fractions were hybridized to the labeled probes as indicated in the restriction map below.

oligo tetra tri di mono   tot

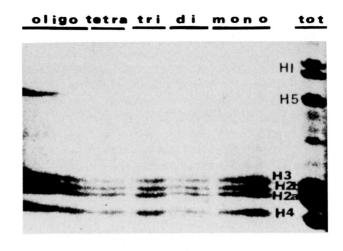

A

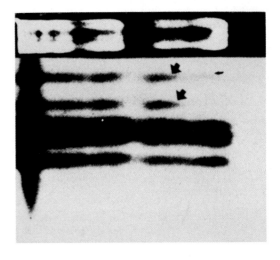

B

Fig. 13.   Second-dimension analysis of the histones associated with electrophoretically fractionated nucleosomes. A) Nuclease-treated chromatin was fractionated on a 5% polyacrylamide gel, in standard conditions. The entire slot of gel corresponding to the sample was then loaded onto a SDS polyacrylamide gel, in order to analyze the protein content of each nucleosome fraction. A total nucleosome sample was also loaded (slot on the right). B) ^{3}H-lysine-labeled nucleosomes were fractionated as in (A) in two dimensions. The thin small arrow indicates A24; the thick arrows, HMG 14 and 17.

the DNA from five fractions across the nucleosome monomer peak was eluted, purified, and run on an agarose gel. The DNA was then transferred to nitrocellulose and hybridized to either a globin probe or an ovalbumin probe (Fig. 14). The data show that the nucleosomes that migrate slowest are enriched in globin sequences, whereas the nucleosomes that migrate

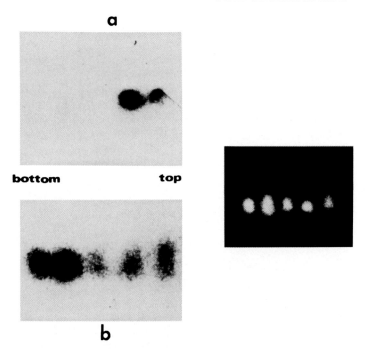

Fig. 14. Blot hybridization of the DNA from different monomer fractions. Twenty micrograms from each of five-monomer fractions was blotted onto nitrocellulose as described in Materials and Methods. The blots were hybridized to either (a) pHb 1001, a β-globin cDNA, or (b) ovalbumin transferred to nitrocellulose and no more than 20% of the transferred DNA was lost during the hybridization and washing procedures; however, when detectable, this loss was the same in each fraction. Blot hybridization to RNA and small DNA fragments immobilized to nitrocellulose has been studied extensively by Dr. Patricia Thomas in this laboratory, and the analysis will be presented elsewhere.

fastest are depleted in globin sequences. Thus, as isolated directly from nuclei, a fraction of nucleosomes that migrate slowest in acrylamide gel electrophoresis contains HMG 14 and 17 and is enriched in globin sequences, but is depleted in ovalbumin sequences.

V. TISSUE-SPECIFIC DNA CLEAVAGES IN THE GLOBIN CHROMATIN DOMAIN INTRODUCED BY DNase I

In assaying the sensitivity of the respective α- and β-globin genes to DNase I using the blot hybridization assay, we noticed that as a consequence of the DNase treatment, specific discrete new bands appeared after hybridization of the blots to the appropriate probes [Stalder et al, 1980b]. These bands contain one end that represents a specific site where DNase I cuts the nuclear chromatin and a second end defined by a restriction endonuclease cleavage. This follows since the discrete new bands do not appear if the purified DNA is not cut with a restriction enzyme. For the most part, these specific cutting sites are not observed in nonerythroid cells, and, surprisingly, the specific cutting sites in the β-globin domain are different, depending on whether the digestion occurs in embryonic or adult nuclei. Figure 15 shows the hybridization of an embryonic β-globin probe to DNase-treated nuclear DNA from embryonic or from adult red cell nuclei. In embryonic cells, a

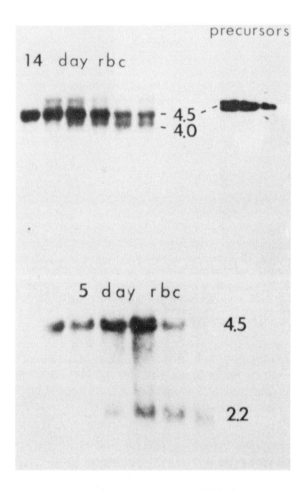

Fig. 15. Absence of DNase I–hypersensitive sites in globin chromatin from precursor cells. Nuclei from primitive erythroblasts (from 5-day embryos) making embryonic Hb, definitive erthroblasts (from 14-day embryos) making adult Hb, and precursor cells (500 embryos) were digested with increasing concentration of DNase I. The DNA was purified, restricted with Hind III and blot-hybridized to an embryonic β-globin gene probe.

discrete new band appears at 2 kbp, whereas in adult cells this hypersensitive region disappears and a different region of hypersensitivity appears, yielding a subfragment of about 4 kbp as compared to the parental fragment of 4.5 kbp. Figure 16 is a summary of the DNase I hypersensitive regions in the β-gene cluster that have been observed to date. There is a definite tendency for these structures to be localized near the 5' and 3' side of genes. In addition, hypersensitive sites are also observed in regions that are thought not to be involved in transcription.

VI. MOLECULAR EVENTS ASSOCIATED WITH THE TRANSITION BETWEEN PRECURSOR RED CELL AND ERYTHROBLAST

In the chick embryo, red cells first appear at about 35 hours of development. There are about 2×10^6 cells in the whole embryo at this time, and of these, some 5×10^5 are red cell precursors [Weintraub et al, 1971]. These cells are localized in a specific region of the developing embryo called the area vasculosa (Fig. 17). It is possible to estimate that at least half of the cells in this region must be red cell precursors in order to account for the number of red cells that are produced during the first few days of development [Weintraub et al, 1971]. To some extent this estimate can be confirmed directly by measuring the percentage of total erythroid colonies obtained from this region when individual cells are plated out in tissue culture (Fig. 18). Our results suggest that although the plating efficiency is very low (about 2–5%), the total precentage of erythroid colonies is very high (80–90%).

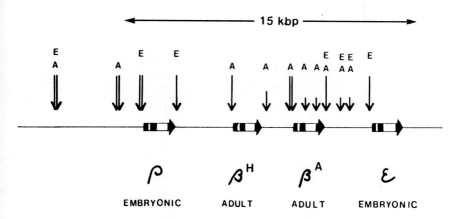

Fig. 16. Summary of known DNase I–hypersensitive sites. "E" represents embryonic cuts observed in 5-day nuclei; "A" represents adult cuts from definitive cells from 12- to 14-day nuclei. We have not probed the region 3' to β¹.

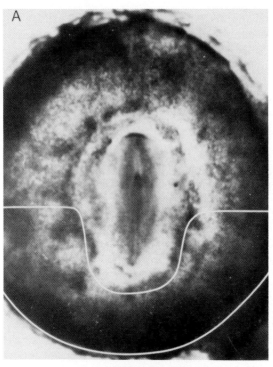

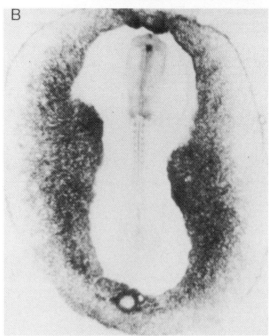

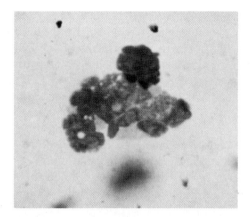

Fig. 18.   In vitro erythroid colonies. The encircled region in Figure 17 was cut out and plated in tissue culture as single cells. The figure shows a high-power field at 10 days. The slide was stained for Hb with benzidine (brownish-yellow cells) and counterstained with Giemsa (blue cells) to detect nonerythroid cells.

Since we have no reason to believe that the culture conditions select against nonerythroid colonies, it is a reasonable first guess that most of the cells in this region of the embryo are in fact erythropoietic, as also indicated by the previous numerological estimate [Weintraub et al, 1971].

The availability of a reasonably pure population of precursor cells allows us to ask several questions about the switch from precursor cell to terminally differentiated cell. We have previously reported [Weintraub and Groudine, 1976] that the globin genes are resistant to DNase I as assayed by solution hybridization in these precursor cells. More recently, we have been able to show that the genes are also resistant as assayed by blot hybridization. Most important, the blot hybridization data also showed that no discrete subfragments appeared after mild DNase I digestion (Fig. 15), indicating the absence of hypersensitive regions before overt differentiation. These experiments suggest that the globin gene is in an inactive chromosomal configuration before the appearance of the first cytologically recognizable erythroid cells at 35 hours of development. This observation has been extended using the "run-off" transcription assay to measure the number of nascent RNA polymerases present on the globin genes in these cells. Finally, we have also investigated the degree of methylation of the globin genes in these precursor cells. There is a growing body of evidence suggesting that the DNA of

---

Fig. 17.   Chicken blastoderms at various stages of incubation. A) Head fold; stage 6; 23 h. B) Stage 11; 40–45 h, containing blood islands. Area outlined in white (A) shows the precursor region that is cut for the biochemical analysis. Most analysis is done with 20- to 22-h embryos.

actively transcribed genes is undermethylated [Bird and Southern, 1978; Bird et al, 1979; McGhee and Ginder, 1979; Mandel and Chambon, 1979; Waalwijk and Flavell, 1978]. The degree of methylation of a gene is currently best assayed by digesting the DNA with methyl-sensitive restriction enzymes. MSP and HPA II both cut DNA at the same site (CCGG); however, HPA II is sensitive to a methyl group at CpG, the most common site of methylation, whereas MSP is not sensitive to methylation in this position. Figure 19 shows MSP and HPA II digests of erythroblast DNA and of erythroid precursor DNA hybridized to either an embryonic β-globin probe or a probe corresponding to the entire α domain (coding and noncoding region). As compared to the precursor DNA, erythroblast DNA is considerably undermethylated, both for the embryonic β-globin gene and for the coding regions of the α genes. In nonerythroid differentiated cells (for example, brain and liver), these genes are highly methylated by this assay. Moreover, in embryonic red cells, the embryonic β-globin genes are unmethylated, but the hatching (β^H) and adult (β^A) genes are methylated; in definitive embryonic cells (synthesizing the adult and the hatching β-globin chains), the embryonic β gene is now methylated, but the others are not; and finally, in adult erythrocytes, the adult β-globin gene is not methylated, but both the hatching and the embryonic genes are methylated [McGhee and Ginder, 1979]. Thus, in the chick β-globin domain, there seems to be a very tight correlation between the expressed gene, the gene that is transcribed, the gene that is undermethylated, and the gene that is sensitive to DNase I and capable of binding to an HMG column. In addition, where it has been analyzed most completely, the coding region of the α-globin domain possesses all these same properties; however, the adjacent, noncoding region on both 5' and 3' sides shows the opposite properties.

Our goal is to understand the events required for activating the globin genes in terms of changes in chromatin structure. It was previously proposed that the assembly of a new chromosome structure about the globin domain required the construction of a new chromosome at the time of DNA replication [Weintraub et al, 1978]. This type of proposal is clearly compatible with the concept of a quantal mitosis previously put forward by Holtzer et al [1975]. In order to test this notion, DNA synthesis was inhibited in 25-hour blastoderms by addition of FUdR to individual embryos. No hemoglobin appeared at 35 hours, and not surprisingly, the embryos died. The principal problem with this type of experiment is the difficulty in distinguishing an effect that is due to a coupling of differentiaton to DNA replication from one that is secondary to the obvious cell death that is occurring in the presence of FUdR.

By lowering the FUdR concentration to inhibit DNA replication by only 60–70% and by reversing the FUdR (by adding thymidine to the embryo) after only 5–8 hours of inhibition, we have been able to define conditions

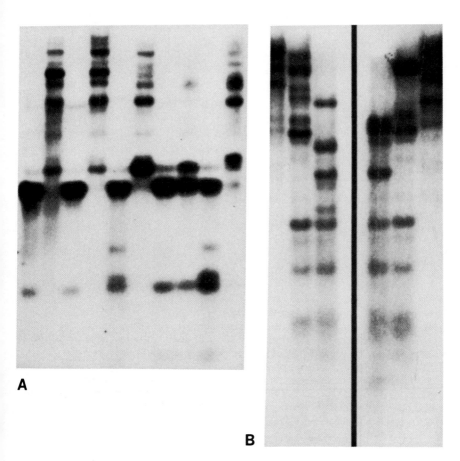

A

B

Fig. 19. Methylation of globin genes in red cell precursors. DNA from red cell precursors or red cells was digested with MSP or HPA II and hybridized to an embryonic β-globin probe (A) or to α5 and α2 (B), an entire α-domain probe (see Fig. 11). A) Left to right: MSP and HPA II (adult erythrocytes); MSP and HPA II (reticulocytes); MSP and HPA II (14-day definitive red cells); MSP and HPA II (5-day primitive embryonic red cells); MSP and HPA II (precursor red cells from 20-hour embryos). B) Left: hybridization to 2. Right: hybridization to 5. Left to right: precursor DNA, HPA II; red cell DNA, HPA II; red cell DNA, MSP; red cell, MSP; red cell DNA, HPA II; precursor DNA, HPA II.

that cause no detectable signs of cell death, but significantly *delay* the appearance of Hb. Figure 20 shows the time course of accumulation of Hb per embryo for the first three days of incubation, as well as the time course for embryos treated with low doses of FUdR (60–70% inhibition of H3-deoxyadenosine incorporation) for increasing periods of time before Hb synthesis begins. The data show that a partial FUdR inhibition of replication, followed

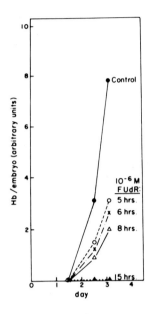

Fig. 20. Delayed appearance of Hb after transient exposure to submaximal doses of FudR. Twenty-five-hour embryos were treated with 10^{-6} M FUdR (3 ml per embryo) for the stated periods of time. Tymidine (1.0 ml at 5 mg/ml) was then added to reverse the FUdR block to DNA synthesis. Hb was determined at various times thereafter. These assays were done on 25–30 embryos for each point according to methods previously described [Weintraub et al, 1971]. The 10^{-6} M FUdR inhibits H3-deoxyadenosine incorporation in OVO into DNA by about 60–70%. Controls were embryos treated simultaneously with FUdR and TdR. Uridine does not reverse the effects of FUdR.

by reversal with thymidine, results in a lag in the accumulation of Hb. Additional experiments have demonstrated that the decreased Hb levels persist through the fifth day of incubation and that even lower concentrations of FUdR (20% inhibition of deoxyadenosine incorporation for 5 hours) produce a significant, though somewhat shorter, delay in the accumulation of Hb.

We do not think that these results can be explained in terms of cell death for the following reasons: 1) Over 80% of the embryos treated with $\leqslant 10^{-6}$ M FUdR for $\leqslant 6$ hours will hatch some 20 days later. 2) During treatment with FUdR, there is no inhibition of H3-leucine incorporation. Autoradiography using labeled leucine or uridine showed all cells to be labeled with either isotope. If "thymidineless" death had been occurring, the characteristic presence of unlabeled cells would have been expected. 3) FUdR treatment several generations earlier, between 5 and 10 hours of incubation, results in no detectable delay in the appearance of Hb (data not shown).

Is the dependence on DNA synthesis a property of Hb synthesis per se or is it only a requirement for its initiation? To answer this question we monitored the effect of FUdR on the synthesis of Hb in cells which had already started to make Hb. Four-day erythroblasts, which make Hb and actively synthesize DNA, were incubated for 10 hours in vitro and in vivo in the presence of doses of FUdR that block DNA synthesis by over 98%. Under these conditions, no inhibition of Hb synthesis was observed. Lower doses of FUdR gave the same results. Longer exposures to high doses of FUdR led to the accumulation of unlabeled cells and obvious signs of cell death. Consequently experiments were not continued beyond 10 hours. In contrast, up to 30 hours of exposure to low doses of FUdR—doses which cause a delay when given for only 5 hours at 25 hours of incubation—result in no signs of cell death and no decrease in Hb accumulation when administered at 4 days of incubation. These experiments show that a partial and transient inhibition of DNA replication inhibits only the initiation of Hb synthesis and not synthesis that has already been initiated. The results are compatible with, but clearly do not prove, the hypothesis that the replication-dependent assembly of a new daughter chromosome is required for the construction of an active chromatin structure and subsequent activation of the globin genes.

In conclusion, our studies to date have defined a variety of biochemical markers that describe the structural state of the α- and β-globin locus before and after the onset of globin gene expression during early erythropoiesis in the chick. The details of how precursor cells make the transition from one state to the next are largely unknown. What molecules define the hematopoietic precursor state? What is the sequence by which all of the features of active chromatin are generated? Is there a defined cascade? Do all properties of active chromatin appear simultaneously? Our current work is designed to approach these types of questions and our approach has been to use precursor cells transformed by a temperature-sensitive AEV (avian erythroblastosis virus) [see Beug et al, 1982]. At 36°C, these cells behave as if they are arrested at various precursor stages; however, at 42°C, the cells activate globin gene transcription and overt erythropoiesis ensues. A preliminary description of the molecular structure of the globin chromatin in these cells has recently been presented [Weintraub et al, 1982]. These results suggest that a defined sequence of chromosomal changes occurs during the activation of globin chromatin; however, some of these features (eg, α-globin DNA demethylation) can exist in the absence of other features (eg, α-globin hypersensitive sites) in cells arrested at 36°C. Thus, we presume that a surprisingly large number of events are required before these genes finally become transcriptionally active. How such a complicated structure is replicated and transmitted to progeny erythroblasts remains a fascinating problem; however, very little is known about this. Although some information has accumulated about the structure of globin chromatin in various states along the activation pathway, it is not obvious how the causal factors regulating this process will be elucidated in the future.

ACKNOWLEDGMENTS

This work was supported by the National Science Foundation and the National Institutes of Health.

VII. REFERENCES

Albanese I, Weintraub (1980): Electrophoretic separation of a class of nucleosomes enriched in HMB 14 and 17 and actively transcribed globin genes. Nucl Acids Res 8:2978–2805.

Beug H, Palmieri S, Freudenstein C, Zentgraf H, Graf T (1982): Hormone-dependent terminal differentiation in vitro of chicken erythroleukemia cells transformed by ts mutants of avian erythroblastosis virus. Cell 28:907–919.

Bird AP, Southern EM (1978): Use of restriction enzymes to study eukaryotic DNA methylation. J Mol Biol 118:27.

Bird AP, Taggart MH, Smith BA (1979): Methylated and unmethylated DNA compartments in sea urchin genome. Cell 17:889.

Brown J, Ingram V (1974): Structural studies on chick embryonic hemoglobins. J Mol Chem 249:3960–3972.

Cox R (1976): Quantitation of elongating form A and B RNA polymerases on chick oviduct nuclei and effects of estradiol. Cell 7:455–465.

Dodgson J, Engel JD, Strommer J (1979a): Organization of globin genes in the chicken chromosome. Abstract 157, Eigth Annual ICN–UCLA Symposium, Molecular and Cellular Biology: 66.

Dodgson J, Strommer J, Engel JD (1979b): Isolation of the chicken β-globin gene and a linked embryonic β-like globin gene from a chicken DNA recombinant library. Cell 17:879–887.

Ferencz A, Seifart K (1975): Comparative effect of heparin on RNA synthesis of isolated rat-liver nucleoli and purified RNA polymerase A. Eur J Biochem 53:605–613.

Gariglio P, Buss J, Green M (1974): Sarkosyl activation of RNA polymerase activity in mitotic mouse cells. FEBS Lett 44:330–333.

Gariglio P (1976): Effect of sarkosyl on chromatin and viral-RNA synthesis isolation of SV40 transcription complex. Differentiation 5 (vol 2–3):179–183.

Gazit B, Panet A, Cedar H (1980): Reconstitution of a DNase I sensitive structure on active genes. Proc Natl Acad Sci USA 77:1787–1790.

Goodwin GH, Johns EW (1973): A new group of chromatin associated proteins with a high content of acidic and basic amino acids. Eur J Biochem 401:215–219.

Green M, Buss J, Gariglio P (1975): Activation of nuclear RNA polymerase by sarkosyl. Eur J Biochem 53:217–225.

Groudine M, Holtzer H, Scherrer K, Therwath A (1974): Lineage-dependent transcription of globin genes. Cell 3:243–247.

Holtzer H, Rubinstein N, Fellini S, Yeoh G, Chi J, Birnbaum J, Okayama M (1975): Quantal and proliferative cell cycles. Q Rev Biophys 8:523–557.

Levy WB, Dixon GH (1978): Partial purification of transcriptionally active nucleosomes from trout testis cells. Nucl Acids Res 5:4155–4163.

Mandel JL, Chambon P (1979): DNA methylation: Organ specific variations within and around the ovalbumin genes. Nucl Acids Res 7:2081.

McGhee JD, Ginder GD (1979): Specific DNA methylation sites in the vicinity of the chicken β-globin genes. Nature 280:419–420.

McKnight GS, Palmiter R (1979): Transcriptional regulation of the ovalbumin and conalbumin genes by steroid hormones in chick oviduct. J Biol Chem 254:9050–9058.

Palmiter R, Christenson A, Schimke R (1970): Organization of polysome from pre-existing ribosomes in chick oviduct by a secondary administration of either estradiol or progesterone. J Biol Chem 245:833–845.

Saltser W, Cummings I, Liv A, Strommer J, Padayatty J, Clarke P (1979): Molecular regulation of hemoglobin switching. In Stamatoyannopoulos G, Neinhuis A (eds). New York: Grune and Stratton, pp 621–645.

Southern EM (1975): Detection of specific sequences among DNA fragments separated by gel electrophoresis. J Mol Biol 98:503–517.

Stalder J, Groudine M, Dodgson J, Engel JD, Weintraub H (1980a): Hb switching in chickens. Cell 19:973–980.

Stalder J, Larsen A, Engel JD, Dolan M, Groudine M, Weintraub H (1980b): Tissue-specific DNA cleavages in the globin chromatin domain introduced by DNase I. Cell 20:451–460.

Waalwijk C, Flavell RA (1978): DNA methylation at a CCGG sequence in the large intron of the rabbit β-globin gene: Tissue-specific variations. Nucl Acids Res 5:4631–4641.

Weinstock R, Sweet R, Weise M, Cedar H, Axel R (1978): Intragenic DNA spacer interrupt the ovalbumin gene. Proc Natl Acad Sci USA 75:1299–1303.

Weintraub H, Cambell G, Holtzer H (1971): Primitive erythropoiesis in early chick embryogenesis. J Cell Biol 50:652–668.

Weintraub H, Groudine M (1976): Chromosomal subunits in active genes have an altered conformation. Science 93:848–853.

Weintraub H, Flint SJ, Leffak M, Groudine M, Grainger R (1978): The generation and propagation of varigated chromosome structures. Cold Spring Harbor Symp Quant Biol 42:401–407.

Weintraub H, Beug H, Groudine M, Graf T (1982): Temperature-sensitive changes in the structure of globin chromatin in lines of red cell precursors transformed by ts-AEV. Cell 28:931–940.

Weisbrod S, Weintraub H (1979): Isolation of a subclass of nuclear proteins responsible for conferring a DNase I sensitive structure on globin chromatin. Proc Natl Acad Sci USA 76:630–634.

Weisbrod S, Groudine M, Weintraub H (1980): Inteaction of HMG 14 and 17 with actively transcribed genes. Cell 19:289–301.

Wu C, Bingham P, Livak K, Holmgren R, Elgin SCR (1979): The chromatin structure of specific genes: I. Evidence for higher order domains of defined DNA sequence. Cell 16:797–806.

Modern Cell Biology 2:385–415
© 1983 Alan R. Liss, Inc., 150 Fifth Avenue, New York, NY 10011

The Three Three-Dimensional Structural Networks of Cytoplasm and Nucleus: Function in Cells and Tissue

Sheldon Penman, David G. Capco, Edward G. Fey, Pradeep Chatterjee, Theresa Reiter, Sabine Ermish, and Katherine Wan

From the Department of Biology, Massachusetts Institute of Technology, Cambridge, Massachusetts 02139

I. INTRODUCTION: WHY THREE DIMENSIONS?

A symposium honoring the contributions of Keith Porter, one of the pioneers of modern cell structure studies, is an appropriate place to reflect on how the views of cell structure and biochemistry in any era are influenced by the morphological techniques then prevalent. Prior to the introduction of the electron microscope, cell science reflected the limitations of the light microscope, leaving us a heritage of "plasms" with their connotation of amorphous, gel-like properties [Wilson, 1925]. The biochemistry of the time dealt with the chemical reactions catalyzed by enzymes freely diffusing in solution. The advent of electron microscopy revealed the cell interior to be composed of an elaborate architecture of membranes and organelles. Many of these were unsuspected prior to thin-section images. The discipline of biochemistry evolved so as to examine the ordered, localized enzymes of subcellular organelles as well as the proteins composing cell structure. Today, we are witness to still another major advance in electron microscope technique which promises to develop further our view of the cell interior. There appears to be a corresponding evolution in the discipline of cellular biochemistry.

For all of its tremendous power, the thin-section technique has afforded us a clouded picture of the cell interior, a picture seemingly supported by the current techniques of cell fractionation and biochemical analysis. There are two major limitations to the thin section. One is quite obvious: A thin section offers us only a two-dimensional slice of three-dimensional objects. Although everyone knows that sectioned material is but a poor representation of reality, few are really aware of how profoundly different the three-dimensional object may be. Serial-section reconstruction requires patience of heroic proprortions, and there are but few practitioners. Their efforts have shown us how surprising the three-dimensional reality can be [a discussion of reconstruction from sections is offered in Ham and Cormack, 1979].

The other major limitation of the embedded thin section is more subtle and more serious. It has to do with the masking of important protein filaments by the electron scattering of the embedding plastic itself. Porter noted this problem with the advent of dense epoxy embedding materials; his words of caution were, alas, unheeded and forgotten [Pease and Porter, 1981].

We now know that the seemingly amorphous appearance of cell cytoplasm and nucleoplasm outside of organelle boundaries is due, in large part, to the failure to image the architectural elements poorly stained by the heavy metals necessary in conventional electron-microscopic procedures. The true extent of this problem became obvious from comparing images of architectural elements of the cells seen in whole mounts with thin-section electron microscopy. Architectural proteins, clearly visible in whole mount, disappear, many times completely, in epoxy-embedded thin sections.

Porter has continued to pioneer in the techniques that serve as the foundation for our concepts in cell biology. His recent research has introduced, or, more correctly, reintroduced electron microscopy of cell whole mounts [Porter et al, 1945; Wolosewick and Porter, 1975, 1976]. The importance of finding a highly structured cytoplasm in both the intact and the detergent-extracted cell interior is indicated by the amount of controversy stirred by these studies in Porter's and other laboratories. Kuhn's view of the "Structure of Scientific Revolutions" [Kuhn, 1962] would suggest we are in the midst of a change in paradigm that results from new experimental procedures brought to bear on previously intractable problems. The attendant Sturm und Drang is probably simply the usual concomitant of a major change in thought.

Our own work, which was for a long time concerned with nucleic acid biochemistry and cell fractionation, wandered or stumbled into the area of three-dimensional electron microscopy by a circuitous and unanticipated route. As a consequence of this history, we have concentrated our studies on the biochemistry and morphology of the filament network and surface lamina remaining after extraction of cultured cells with non ionic detergents.

Our studies of cell architecture began with the seemingly unrelated question of the role of poly (A), the curious homopolymer terminating most, though not all, mRNA molcules. We already knew that the mRNA of membrane-bound polyribosomes is bound to the endoplasmic reticulum and that its poly (A) segment serves as one of the linking elements [Milcarek and Penman, 1974]. Were the so-called "free" polyribosomes bound to some previously undetected structure and could poly (A) play a similar role in the binding of these messages? This question has now been answered in the affirmative (Fey and Penman, in preparation). Its pursuit revealed the presence of the skeletal network and the apparently obligatory binding of message to this framework for translation [Lenk et al, 1977; Lenk and Penman, 1979; Cervera et al, 1981]. We were thus led to study the nuclear and cytoplasmic networks and to reflect on why metazoan cells should be thus organized.

II. THE BASIC EXTRACTION PROCEDURE

These studies depend critically on the method that separates the soluble from the structural proteins and then the separation of structural networks of the cytoplasm from the nucleus. Only in this way can the cell architecture both be viewed microscopically and characterized biochemically. A simple procedure is common to most of the experiments described here. Cultured cells are exposed to a sufficiently strong, nonionic detergent—eg, Triton X-100—in a suitably designed buffer. Most lipids are solubilized, together with about two thirds of cellular proteins; these constitute the "soluble" fraction. Most of the architectural components of the cell remain as an intact entity which retains much of the morphology of the living cell. The complex cytoplasmic network with the polyribosomes attached is now readily visible with transmission electron microscopy. The detailed morphology and protein composition of the extracted structures depend very much on the conditions of extraction. The particular nonionic detergent appears relatively unimportant, provided it is sufficiently lipophilic to effect a complete removal of lipids. Triton X-100 and NP-40 are especially suitable. Some of the various Brijs, Tweens, and Spans can serve, although their rate of lipid solubilization is rather slow. Most important is the ionic composition and pH of the extraction medium.

Our most recent extraction-buffer formulation, designed to optimize the preservation of preexisting structure and, at the same time, effect the complete removal of soluble components, is shown in Table I. Although the exact

TABLE I. Extraction Buffer

Constituent	Concentration
NaCl or KCl	100 mM
MgCl$_2$	3 mM
PIPES (pH 6.8)	10 mM
Sucrose	300 mM
PMSF	1 mM
EGTA	1 mM
Triton X-100	0.5%

concentrations of the constituents are not critical, large departures from these values result in a much reduced skeletal structure. The cation concentrations are close to the activities expected in the living cell interior. A pH of 6.8 appears optimum for preserving filament integrity [Solomon et al, 1979], and 300 mM sucrose has been found empirically to improve the retention of known structure-bound elements. Although the extraction medium appears very hypertonic, this does not result in osmotic stress on cell structure, since, in the presence of Triton, plasma-membrane integrity is breached within seconds and the cell interior rapidly equilibrates with the external medium. The length of extraction must be adequate for soluble components to diffuse completely from the cell interior. We have determined that 3 minutes is adequate for flat, fibroblastic cells, whereas thicker cells, such as those in epithelial sheets, require 10 minutes. Too brief an extraction results in thicker filaments that probably result from adventitious fixation of soluble proteins to skeletal fibers.

Whole mounts of the detergent-extracted cells have proved to be uniquely suited for visualizing cell architecture [Brown et al, 1976; Webster et al, 1978; Fulton et al, 1980, 1981; Schliwa and Van Blerkom, 1981; Schliwa, 1982]. Such whole mounts can afford a far more realistic view than the conventional thin section. In the absence of the embedding plastic required for most thin sections, and without the soluble proteins removed by extraction, the electron scattering by the architectural proteins forms clear images without heavy metal stains. Removal of the soluble proteins with detergent leaves a well-defined skeletal structure and avoids, for now, questions of the organization of the soluble proteins in the cell. Most important, the significance of viewing the cell architecture through its entire thickness cannot be over emphasized. Only in this way can the three-dimensional anastomosing networks be seen: These appear merely as specks or short fibers in thin sections [Wolosewick and Porter, 1979]. It must be noted, however, that all methods of preparation for biological electron microscopy require the transfer of material from an aqueous environment to a vacuum. While we feel that the overall topography of the filaments in our preparations resembles that of the living cell, we cannot tell what alterations have occurred in the protein subunits of the filaments.

III. NETWORKS OF THE CYTOPLASM

The view of cell structure that has emerged from these studies using whole mounts and extractive procedures is summarized schematically in Figure 1. The extracted cell structure is shown here with the three-dimensional anastamosing network with bound polyribosomes filling the cytoplasmic space. The surface is a protein lamina that remains after the removal of lipids by the nonionic detergent. The nucleus is shown as it appears after chromatin is removed; it also has a space-filling three-dimensional network, bounded by the nuclear lamina. The nuclear structure is distinct in organization from the networks of the cytoplasm.

A few typical micrographs of extracted fibroblasts in whole-mount preparations are shown in Figure 2. A chicken embryo fibroblast is shown at low magnification in Figure 2A. Near the edge of the cell is the fine filamentous network rich in microfilaments, characteristic of motile regions of the cell. Closer to the nucleus, the network is organized differently; the filaments are thicker, heterogeneous, and densely studded with polyribosomes. At higher magnification the heterogeneity of the fibers and the complexity of their interconnections becomes apparent as shown in the fibroblasts in Figure 2B and C and the aortic endothelial cell in Figure 2D. There is a considerable range of filament size: many are 5–6 nm in diameter and may well be actin-containing microfilaments, but many are considerably larger. Of particular interest are the numerous structures whose diameter changes significantly throughout their length. These are probably more complicated than simple homopolymers, and are of as yet unknown composition and function. The polyribosomes are bound to these networks and, as Wolosewick and Porter [1976] observed in intact cells, are usually found at the nexus of several filaments.

Polyribosomes appear completely bound to the architectural structures remaining after Triton extraction [Lenk et al, 1977; Fulton et al, 1980; Cervera et al, 1981]. When the polyribosomes are disassembled in vivo by drugs or heat shock, the ribosomes themselves become soluble. The mRNA remains bound to the cells' architecture, or cytostructure, presumably via the proteins of messenger ribonucleoprotein (mRNP). Many different experiments indicate that message binding to the skeleton fibers is not adventitious [Fulton et al, 1980; Cervera et al, 1981]. However, the most compelling is the

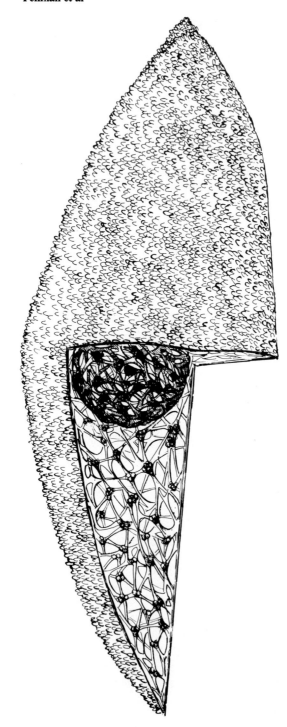

Fig. 1. Schematic representation of an extracted cell whole mount. The diagram represents the fiber networks of the cytoplasm covered by a lamina composed of the plasma membrane protein remaining after detergent extraction. The nuclear lamina shown as the outer boundary of the nuclear network. The nucleus is represented as it appears after the removal of chromatin.

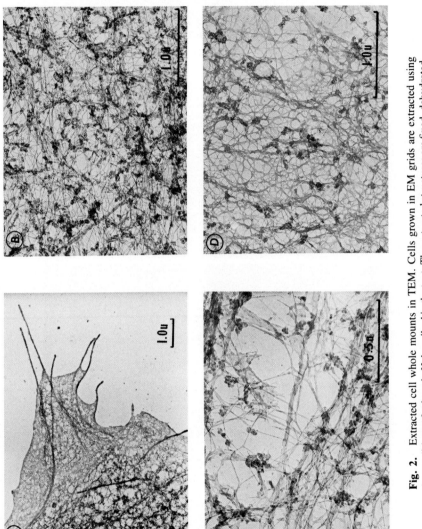

Fig. 2. Extracted cell whole mounts in TEM. Cells grown in EM grids are extracted using the conditions of salt and pH described in the text. The extracted structures are fixed, dehydrated, critical-point dried and viewed in TEM. A) Chicken embryo fibroblast; B) 3T3 fibroblast; C) 3T3 fibroblast; D) bovine aorta endothelial cell.

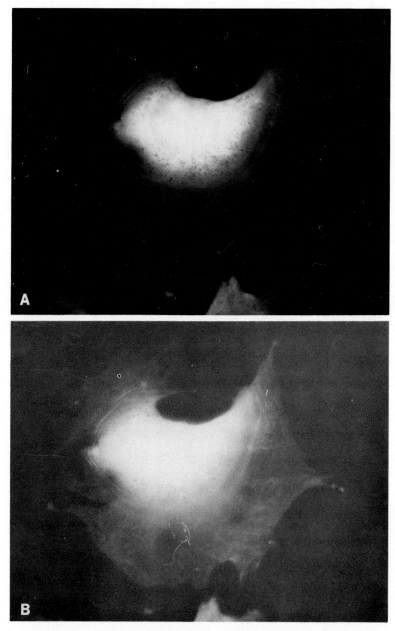

Fig. 3. Polyribosome localization in intact cells shown by comparing polyribosome and total protein fluorescence patterns in the same cell. Endothelial cells were grown on glass microscope slides and, after fixation with formaldehyde, stained for polyribosomes with ethidium (A) and for general protein distribution using thiolyte (B).

cytological evidence which shows that polyribosomes are not free in the living cell. In flat, well-spread cells, polyribosomes are clustered in perinuclear regions and hence are not freely diffusing. Figure 3A shows polyribosomes visualized by ethidium fluorescence in an intact cell. The distribution of cytoplasmic protein in the same cell is shown in Figure 3B by the fluorescence of Thiolyte, a general protein stain. The localized, intense staining in Figure 3A represents polyribosomes, and the faint diffuse fluorescence is due to free ribosomes and t-RNA.

The observation that all polyribosomes were bound to some such cellular structures came as something of a surprise after years of classifying polyribosomes as "free" if they were not bound to membranes. The vigorous procedures developed in earlier experiments to effect a complete separation of cytoplasm from the nucleus actually served to disrupt the then unknown cytoplasmic network and thus yield freely sedimenting polyribosomes. These procedures gave rise to a false impression of a fluid or gel-like cytoplasm which conventional electron microscopy did not dispel. In contrast, the comparatively gentle fractionation described here separates soluble proteins from architectural proteins, leaving the latter associated with the nucleus. When this separation is made, polyribosomes are completely retained on the cytoskeletal framework (CSK) (Fig. 4). This cytostructure can then be selectively solubilized and completely separated from the nucleus by the use of a double detergent method (see below and Holtzman et al [1966]).

IV. THE OUTER BOUNDARY OF THE CYTOPLASMIC NETWORK

Several important aspects of cell morphology are revealed only when lipids are extracted with nonionic detergent. In particular, contrary to intuition, the removal of lipids does not release many proteins of the plasma membrane [Ben-Ze'ev et al, 1979]. Rather, these remain with the cytostructure and form an essentially continuous sheet or lamina, intimately attached to and covering the underlying filament structures. The surface lamina is not readily apparent in most TEM images, such as Figure 2, for reasons that are not clear. The lamina is easily demonstrated either by scanning electron microscopy or by specific radioactive labeling of the cell surface proteins with iodine.

Figure 5 compares scanning microscope images of intact and detergent-extracted cells. The developing myotubes in Figure 5A and B show that the extracted structure is covered by the surface lamina in most places. The lamina is in turn supported by the underlying filaments in approximately the configuration of the plasma membrane of the intact cell. The gross morphology of the intact cells is remarkably unchanged by extraction. Figures 5C and D show a fibroblast surface at higher magnification, before and after

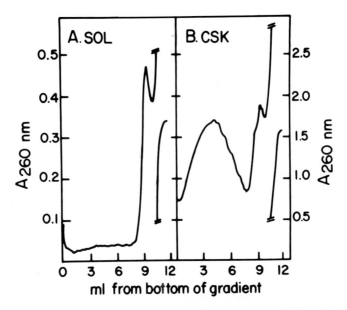

Fig. 4. Polyribosomes in the soluble and cytoskeleton structures of HeLa cells. Cells were extracted with Triton as described in the text and the soluble-fraction analyzed on sucrose density gradients (A); the remaining cytoskeletons were stripped from the nuclei using the double-detergent procedure, the nuclei removed, and the resulting solubilized skeleton analyzed by sucrose density gradient (B).

extraction. The finc structure of the fibroblast surface lamina emerges as a meshwork, presumably composed of clusters of glycoproteins.

Most plasma membrane proteins remain bound to the extracted cell structure, although the detergent does remove a few specific polypeptides. The retained and released proteins are seen by first selectively labeling the plasma membrane proteins by surface iodination of the intact cell and then extracting with Triton [Ben-Ze'ev et al, 1979]. The electropherogram of surface-iodinated proteins in Figure 6 shows the result of such a measurement. The majority of surface proteins remain attached to the cytostructure whereas a few specific polypeptides are solubilized. These latter include a 100-kilodalton protein found to be extractable from most cell types.

There are several other notable examples of selectively extractable surface proteins. The acetylcholine receptor is extractable when it first appears diffusely over the surface of a developing muscle fiber [Prives et al, 1982]. Later, when the receptors patch, it becomes firmly anchored in the surface lamina. Another example, the G protein of vesicular stomatitis virus (in

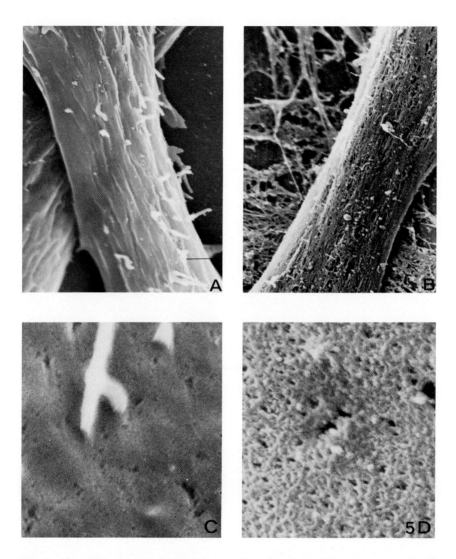

Fig. 5. The surface lamina visualized by scanning microscopy of detergent-extracted cell whole mounts. Cells growing on glass coverslips, both intact and detergent-extracted, were fixed, dehydrated, critical-point dried, sputter-coated with gold, and viewed in the scanning electron microscopy. Chick embryo myotube from cultured myoblasts intact (A); extracted (B) [from Prives et al, 1982]. Human diploid fibroblasts intact (C); extracted (D).

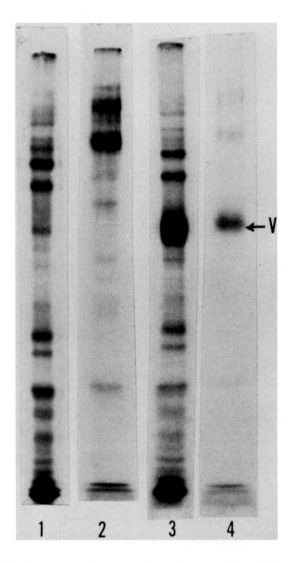

Fig. 6. Retention of plasma membrane proteins on the extracted cytoskeleton. Intact CHO cells were surfaced labeled with ^{125}I and extracted with Triton. Lane 1, proteins retained on the cytoskeleton; lane 2, proteins released with Triton. The retention of a viral plasma membrane protein on the cytoskeleton is shown in lanes 3 and 4. CHO cells were infected with Sindbis virus and then the surface was iodinated and extracted with Triton. Lane 3, cytoskeleton; lane 4, soluble [from Ben-ze'ev et al, 1979].

contrast to that of Sindbis virus) is extractable whether in the plasma membrane or in the mature virion (Chatterjee et al, in preparation). The surface lamina forms the outer boundary of the cytostructure, and it should probably be considered an integral part of the cytoplasmic architecture. Thus, the seemingly anomalous movement of surface receptors [Edidin and Wei, 1977; Webb et al, 1981], which is slow compared to proteins freely diffusing in a lipid bilayer, probably reflects their coupling to the underlying dynamic architectural filaments. The structure of the plasma membrane suggested here is not dissimilar from Singer and Nicolson's [1972] fluid mosaic model in the orientation of the integral membrane proteins and lipids. The major difference is that the structure suggested here implies that protein-protein interactions, both laterally in the plane of the membrane and transversely to the underlying architecture, are fundamental in determining many plasma membrane properties. These relationships are being revealed in elegant studies of a simpler system, the erythrocyte membrane [for review, see Branton et al, 1981]. A schematic summary of our view of the surface lamina is shown in Figure 7. A protein sheet is shown with lateral and perpendicular interprotein connections.

V. THE INNERMOST CELLULAR NETWORK

The cytoplasmic network terminates at the nuclear lamina, which in turn encloses the nuclear interior. The nucleus has presented formidable obstacles to structural studies. The density of chromatin obstructs examination of nuclear architecture and its connections to the cytoplasm unless drastic measures are adopted. The existence of some form of nuclear matrix would seem obvious a priori from the nonrandom distribution of chromatin. Furthermore, the coordinated movements of chromatin into chromosomes indicate the existence of both scaffolding and contractile structures. Indeed, vigorous extraction procedures can remove chromatin and still leave a recognizable nuclear structure which has been partially characterized by biochemical measurements and by thin section microscopy [Aaronson and Blobel, 1974; Berezney and Coffey, 1974; Cook et al, 1976; Keller and Riley, 1976; Herman et al, 1978; Adolph, 1980; Miller et al, 1978; Gerace and Blobel, 1980; McReady et al, 1980; Puvion and Bernhard, 1975]. This matrix has been shown to be the site of chromatin DNA replication [Vogelstein et al, 1980; Berezney and Buchholtz, 1981], hnRNA attachment [Herman et al, 1978; Miller et al, 1978], as well as including the nuclear lamina and nuclear pores [Aronson and Blobel, 1974; Gerace and Blobel, 1980]. However, even more than in the case of cytoplasmic architecture, the three-dimensional anastomosing network of nuclear filaments is very poorly represented in thin section and can be best visualized using a whole-mount technique.

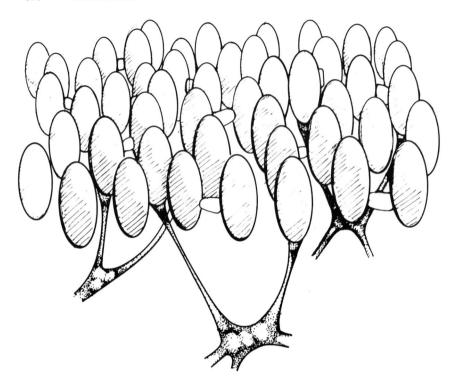

Fig. 7. Schematic of the surface or plasma lamina in detergent-extracted cells. The nearly continuous sheet of proteins remaining after Triton extraction is represented here with suggestions of lateral interaction between transmembrane proteins as well as binding to the underlying cytoskeleton.

The electron opacity of chromatin makes visualization of the nuclear matrix difficult in detergent-extracted cell whole mounts. However, cells with relatively thin, flat nuclei permit visualizing the matrix in a few regions near the edge of the nucleus. An even better view of the nuclear matrix is obtained when the chromatin arrangement is perturbed by virus infection [Capco et al, 1982]. For example, when human diploid fibroblasts are infected with poliovirus, the chromatin condenses revealing regions of nuclear matrix in which the filaments are seen to coalesce with the chromatin (Fig. 8).

The matrix is also suggested by scanning microscopy of dry-fractured whole mounts in stereo in Figure 9. A patch of the nuclear lamina has been torn off, and clumps of chromatin can be observed together with thin, ex-

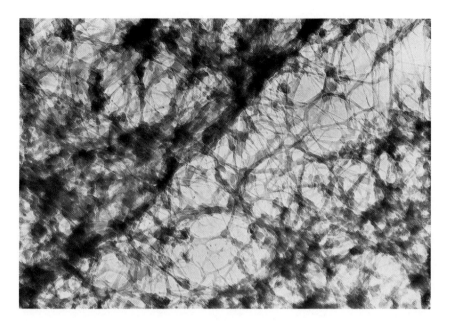

Fig. 8. Nuclear matrix visualized with chromatin still present in a cell whole mount. Human diploid fibroblasts, grown on EM grids, were infected with polio virus and 3 hours later extracted with Triton. Preparation and TEM are as described in text [from Capco et al, 1982].

tended filaments, which are probably the architectural elements of the nucleus.

We have developed a relatively gentle procedure to prepare the nuclear filament network or nuclear matrix for electron microscopy as a whole mount. We do not isolate the nuclei since separation of the nucleus from the cytostructure requires harsh procedures that distort nuclear shape and result in a partial collapse of the filament network. Rather, we prepare the nuclear matrix still bound to the cytoplasmic architecture, or cytoskeleton. These structures are then viewed as cell whole mounts.

A whole mount of the nuclear matrix of a 3T3 mouse fibroblast is shown in Figure 10. The nuclear matrix is to the left of the micrograph; the remnant cytoskeleton with associated polysomes is on the right. A nucleolus can be seen enmeshed in the matrix. Cytoskeletal filaments contact and perhaps enter the nuclear matrix at various points along the nuclear periphery.

The higher magnification TEM micrograph seen in Figure 11 is also obtained from a 3T3 mouse fibroblast. The higher density and complex spatial

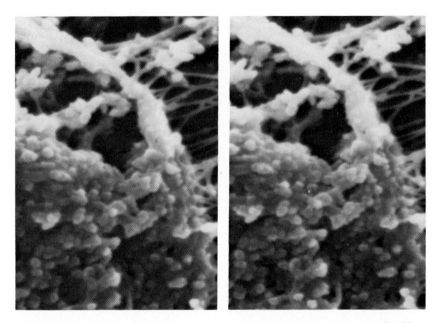

Fig. 9. Fractured whole mount of the nuclear interior by scanning microscopy. Rat fibroblasts were grown on a glass coverslip, extracted with Triton, fixed, dehydrated, and dried through the critical point. Scotch tape was placed on the coverslip and then removed, fracturing most of the skeletons. The samples were then sputter-coated with gold and examined by scanning microscopy at 0° and 8° tilt.

organization of nuclear matrix filaments compared to the remnant cytoplasm can be observed. Enmeshed within the nuclear matrix are residual nucleoli. Figure 12 shows one such nucleolus at high magnification from a DNase I digested HeLa cell nuclear matrix. It appears that many filaments contact or pass through the nucleolus. The whole-mount micrograph shows that this particular nucleolus has at least two holes through it.

Studies of the unique biochemical and structural functions of the nuclear matrix begin with determining the protein composition. This requires removal of the cytoplasmic portion of the filament network. The cells are first extracted to prepare the cytoskeleton and the chromatin-containing nucleus. The "cytoplasmic skeleton" is then stripped from the nucleus using the double-detergent method [Holtzman et al, 1966]. The isolated nucleus is then digested with DNase I and RNase A as described above. Some nuclear components are lost, but at least 60% of matrix components survive this rather harsh fractionation. Whole-mount micrographs of the resulting isolated nuclear matrix show it to be rather collapsed and distorted and some of the

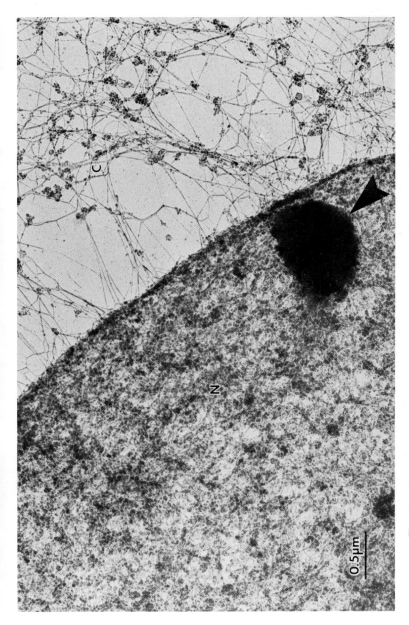

Fig. 10. The nuclear matrix after chromatin removal viewed in whole mount. 3T3 fibroblasts grown on electron microscope grids were extracted with Triton, nuclease, and high salt as described in Capco et al [1982]. Extracted structure is viewed in conventional TEM.

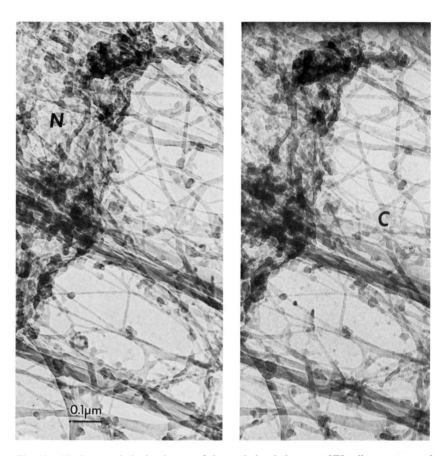

Fig. 11. Nuclear matrix in the absence of chromatin in whole-mount 3T3 cells were prepared as in Figure 10 and viewed stereoscopically in TEM. Specimen is viewed at 0° and 8° tilt.

nuclear lamina may be lost, but the residual nucleoli and much of the internal filament network remain. When the nuclear matrix proteins are analyzed on a two-dimensional gel and compared to those of the cytoskeleton, several proteins, including actin, appear in both preparations. However, most of the proteins are specific to each fraction. Figure 13 shows 2D gel electropherograms of the four basic fractions, soluble, cytoskeleton, chromatin, and nuclear matrix prepared from the epithelial cell line MDCK. Major proteins in the nuclear matrix are actin and the keratins together with many other proteins unique to the matrix fraction.

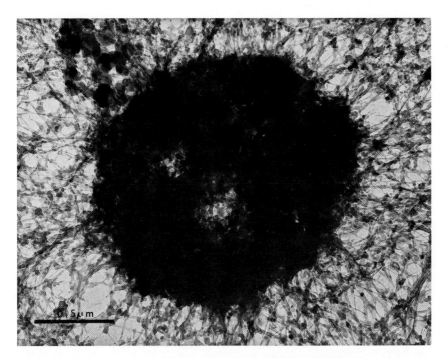

Fig. 12.   Interior of the nuclear matrix viewed in whole mount. A nucleolus is visible at the center of the picture. All around, elements of the nuclear matrix are evident.

The nuclear matrix filaments fill the entire interphase nucleus. An obvious question is; What happens to the nuclear filaments during mitosis? Based on morphological criteria it appears that at least some of the filaments remain associated with the chromatin and form the chromosome scaffolding whereas others, attaching the chromosomes to the cytoplasmic networks, serve in the motile events of mitosis.

In order to study the mitotic cell network it was necessary to surmount two problems. First, cells at mitosis become round whereas whole-mount microscopy requires flat cells. The cell line chosen for our studies is the epithelial cell from the kidney of the rat kangaroo, Potorous tridactylis (PTK$_2$), which remains relatively flat through mitosis. The second problem is that even PTK$_2$ skeletons are too thick to view as a whole mount in the modern short focal length TEMs. The ISI LEM 2000 was used since its long focal length lenses offer a significantly greater depth of field than most modern TEMs.

Soluble

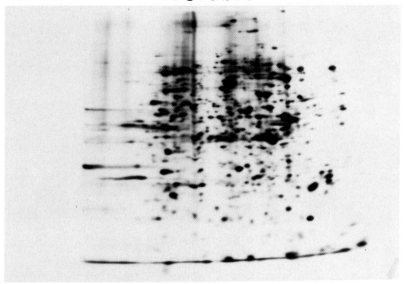

Chromatin

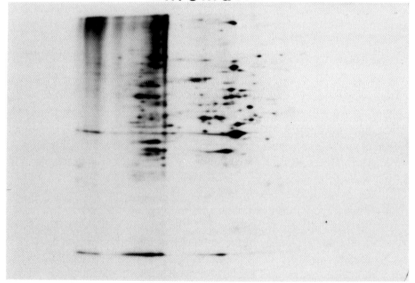

Fig. 13. Proteins of soluble, cytoskeleton, chromatin, and nuclear matrix fractions analyzed by two-dimensional gel electrophoresis. The soluble fraction is obtained by Triton extraction. The cytoskeleton is stripped by double-detergent procedure. The chromatin is obtained by nuclease followed by high salt of the stripped nucleus while the nuclear matrix contains the remaining proteins. The cells used here are the MDCK line of canine kidney epithelium. The major protein spots in the matrix electropherogram are keratins and actin.

Cytoskeleton

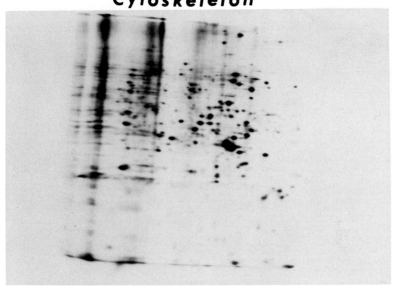

Nuclear Matrix

Figure 13, continued.

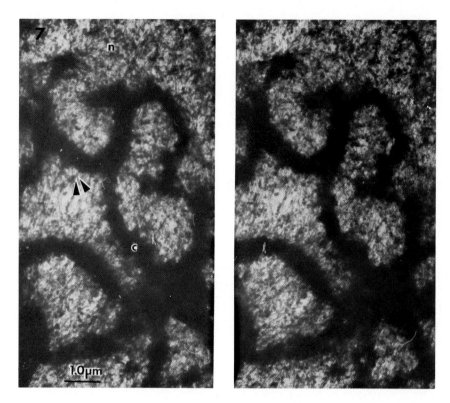

Fig. 14. Whole-mount microscopy of cells in mitosis. PTK cells grown on electron micro-scope grids were synchronized and lysed at the time of mitosis. Cells were extracted as before and viewed by TEM stereo; tilt angle here is 8°. Calcium was added to depolymerize micro-tubules and the spindle apparatus. [from Capco and Penman, 1983].

Once chromatin is compacted into chromosomes, it no longer obscures nuclear structure, and extraction of the nuclei with enzymes and high ionic strength buffers is obviated. Mitotic cells were extracted with our standard skeleton buffer which has an ionic strength close to physiological. When necessary 0.005 M CaCl$_2$ is added to remove the spindle microtubules which can obscure the view of the nonspindle skeletal structures.

Condensation of the chromatin into chromosomes reveals large portions of the nuclear matrix. The stereo pair seen in Figure 14 is a late prophase view of a PTK$_2$ cell. The lamina resulting from the nuclear envelope is still intact. Thus, the filaments within the lamina are nuclear matrix filaments, and those to the exterior are cytoskeletal filaments. Nuclear matrix filaments can be seen contacting the chromosomes along their entire length. The electron-dense chromatin can be seen accumulating around the still forming chromosome associated with the nuclear matrix filaments [Capco and Penman, 1983].

VI. COLLECTIVE ARCHITECTURE AND BIOCHEMISTRY OF CELLS IN TISSUE

The images of complex architectural networks composing the cytoplasm and nucleus of the cell and the first glimpse of their biochemical implications lead to the obvious but difficult question, Why is the cell interior constructed this way? Some general principles are apparent. The cellular networks are space-filling structures possessing considerable mechanical strength. However, they are constructed subject to the constraint that the fibers can withstand compression and tension but not bending movements or shear. Such structures are well known in other contexts. The geodesic dome is a simple example. There are many others which suggest, even more, the networks seen in cells [Pugh, 1976]. In one important regard the cellular networks are unlike anything seen in human scale architecture. They are dynamic structures built from filaments that may undergo cycles of polymerization and with interconnections that may change over time. Much of cellular network behavior is accountable only by its dynamism, such as chromatin movement at mitosis and the saltatory movement of organelles observed in time-lapse cinematography. The frozen images we see here undoubtedly reflect an organization capable of intense and rapid movements. However, the full significance of the structural networks becomes explicable only when cells are considered not as independent entities, but as part of a larger tissue organization.

Many aspects of cell behavior that appear incidental or are unnoticed when cells are studied as individual entities become of paramount importance in the formation of the organized structure of tissue. We have studied two of these. The first is the response of gene activity to cell configuration or shape. This ability of the genome to respond to architecture would seem necessary to regulate the complex patterns adopted by the collective cell organization. The second property is the propensity of the structural fibers in the skeletal frameworks of adjacent cells to organize in response to their neighbors: a quality found only in cells competent to form organized tissue.

The shape dependence of cell behavior, examplified by the anchorage dependence of fibroblasts, is more than a laboratory curiosity. We have studied fibroblasts as a model system for shape modulation of gene expression for many years [Benecke et al, 1978, 1980; Ben-Ze'ev et al, 1980]. Placing anchorage-dependent fibroblast into suspension culture leads to an extensive shut down of all major macromolecular processes and a elaborate program of recovery ensues upon replating cells on a solid substrate. We determined that there are two principal signals when such a cell is placed in suspension. The loss of surface contact results in the shutdown of cytoplasmic protein synthesis, and this can be restored by the cells' simply touching a suitable solid surface. Nuclear metabolism responds to cell shape, however, and is normal only when cells are almost completely spread. An adequate review of these findings would be too lengthy to present here and, very likely, the fibroblast represents but a simple version of the much more complex responses to shape change found in tissue forming cells such as those described below.

One important finding, which we will summarize, is the progressive loss of biochemical responses to shape change as cells become less well growth-regulated or more malignant in their phenotype. Folkman and co-workers made the seminal suggestion that malignant transformation is accompanied by and causally related to the loss of shape regulation of cell growth [Folkman and Greenspan, 1975; Folkman and Moscona, 1978]. We have shown that the decrease in growth regulation is accompanied by a progressive, stepwise loss of response in the basic macromolecular synthetic and processing systems [Wittelsberger et al, 1981]. We examined a series of mouse fibroblasts with varying degrees of growth control ranging from the well-regulated primary diploid fibroblast to the completely unregulated, anchorage-independent, doubly transformed mouse fibroblast line SVPy 3T3. Intermediate in growth regulating properties were the established lines 3T3 and 3T6 and an HDP 3T6 line, developed by our laboratory to be marginally anchorage-dependent. The response of these cell lines to suspension culture is summarized in Figure 15. The cell types are in ascending order with the best regulated at the bottom and the least regulated at the top. The loss of responsiveness to architectural changes is apparent, although the stepwise nature of this loss was a surprise.

The second property of tissue-forming cells, the organization of their skeletal fibers in response to adjacent cells and other external surfaces, has been studied in model tissues such as the MDCK line of canine kidney cells.

The organization of skeletal networks of tissue forming cells is profoundly responsive to the exterior contacts of the cell. We have been studying model tissues formed by the well-differentiated cell line, MDCK, derived from canine kidney. These cells grow as a highly ordered monolayer colonies which closely resemble the epithelial cell structure observed in vitro. They may be plated on electron microscope grids where cell colonies are extracted and viewed as whole mounts, much as single cells. A low-magnification view of such an epithelial cell sheet is shown in Figure 16. The continuity pattern of cytoplasmic filament organization across the cell border is apparent.

The organization of cell skeletons coherent with adjacent cell structure is actually a well-known phenomenon, although its significance has perhaps not been fully appreciated. Tight junctions and especially desmosomes represent structures in which adjacent participating cells must organize colinear

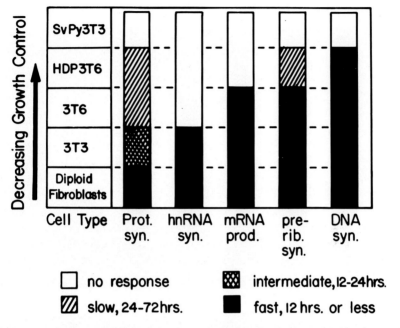

Fig. 15. Progressive loss of shape responsiveness in cell series with increasing growth control. A series of mouse fibroblasts were placed in suspension culture, and the shutdown of major macromolecular systems measured [from Wittelsberger et al, 1981].

filaments which terminate at the same place on the intercell boundary. Actin filaments in cardiac muscle and in epithelial cells appear to be anchored at the cell surface in register with filaments in adjacent cells. Indeed, the coherent organization of the skeletons in a model tissue such as the kidney epithelium shown here suggests that we might view tissue not as a collection of individual cells but, rather, as an organized entity; in a sense we can view tissue as a syncitium or macroskeleton with nuclei embedded throughout. Whether this view of tissue as a supracellular skeleton organization is fruitful will depend on its conceptual utility. For example, the provocative results of Bryant et al [1981] might be approached with this view in mind.

The long-range structural ordering of cytoskeletal structural elements in tissue-forming cells takes on much greater significance in view of the ability of cell architecture to affect genomic activity.

Tumor promoters are agents with an enormous variety of pleotropic effects, including profound mitogenic stimulation. An insight into the mode of action of these remarkable agents is afforded by the electron micrograph in Figure 17 showing the whole-mount skeleton preparation of the MDCK epithelial

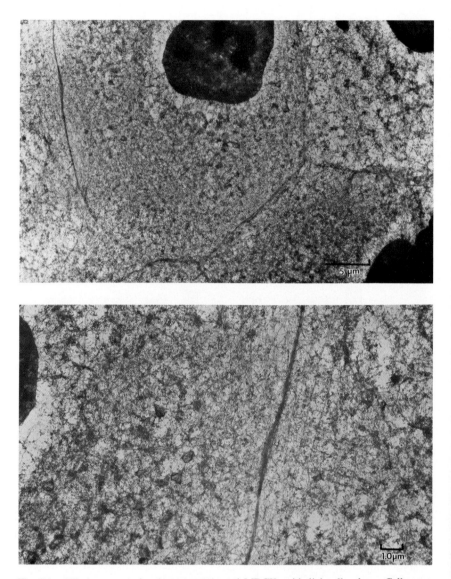

Fig. 16. Whole mount of a detergent-extracted MDCK epithelial cell colony. Cells were grown as colonies on electron microscope grids, extracted with detergent, and viewed as whole mounts by TEM. The higher magnification of the inner-cell border shows the regularity of cytoplasmic filament architecture across the cell border. The empty space about the nuclei is an artifact due to dehydration or beam damage.

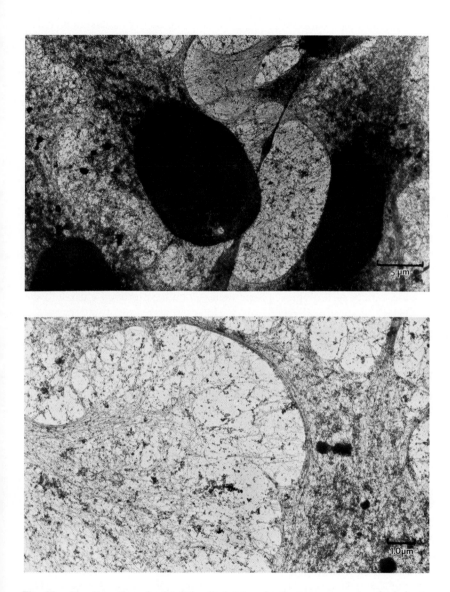

Fig. 17. The disruption of epithelial cell skeletons by the tumor promoter phorbol ester. MDCK cells were prepared as in Figure 16 except for exposure to 10 ng/ml of TPA for 12 hours prior to extraction.

cells [Cereijido et al, 1978] after brief exposure to 5 ng/ml of 12-0-tetra-decanoyl phorbol-13 acetate (TPA). The striking degree of disorganization of the cytoskeleton structures is immediately apparent, and other studies show the nuclear matrix to be similarly affected. The tight junctions have disappeared and no desmosomes remain. Despite these extraordinary changes, the cells remain perfectly viable and actually have an enhanced growth rate, presumably because the architectural control of cells in the epithelial cell sheet is lost. Experiments that will not be described here show by direct biochemical measurement that the shape transduction to the cell nucleus is lost and cells become indifferent to their shape and surroundings. This most likely accounts for the intense mitogenic effect of the tumor promoter. The enhancement of carcinogenesis, however, is more likely due to inability of the aberrant skeleton to organize the mitotic apparatus correctly.

VII. CONCLUSION

The elaborate cell architecture shown here, together with structurally related signals which function at the level of the genome, offer an entree into the wide variety of biological phenomena which have proved elusive to the approaches of traditional biochemistry. Intercell signaling, tissue organization, and pattern formation in development can be seen as manifestations of genome action in setting and controlling architecture. The events involved in tumor promotion may then be seen as a nonlethal disruption of cell architecture and its signaling whereas malignancy results from the permanent disorganization caused by an aberrant karyotype. Only time will establish the validity of this structure-based approach to cell and tissue behavior. Certainly, the promise seems clear.

ACKNOWLEDGMENTS

We would like to thank Dr Alan C. Nelson for access to the ISI LEM 2000. We would especially like to thank Erika A. Hartwieg of our Biology Department electron microscopy facility who facilitated the completion of our work. Laurie J. White is thanked for secretarial assistance. This work was supported by the following grants: NIH CA 08416; NIH CA 12174; NSF 8004696PCM. T. Reiter is supported by a fellowship from the Deutschen Forschungsgemeinschaft; D. G. Capco is an Anna Fuller Fund Fellow (536).

VIII. REFERENCES

Aaronson RP, Blobel G (1974): On the attachment of the nuclear pore complex. J Cell Biol 62:746–754.
Adolph KW (1980): Organization of chromosomes in HeLa cells: Isolation of histone depleted nuclei and nuclear scaffolds. J Cell Sci 42:291–304.

Benecke BJ, Ben-Ze'ev A, Penman S (1978): The control of messenger RNA production, translation, and turnover in suspended and reattached anchorage-dependent fibroblasts. Cell 14:931–939.

Benecke BJ, Ben-Ze'ev A, Penman S (1980): The regulation of RNA synthesis in suspended and reattached 3T6 fibroblasts. J Cell Physiol 103:247–254.

Ben-Ze'ev A, Duerr A, Solomon F, Penman S (1979): The outer boundary of the cytoskeleton: A lamina derived from plasma membrane proteins. Cell 17:859–865.

Ben-Ze'ev A, Farmer S, Penman S (1980): Protein synthesis requires cell-surface contact while nuclear events respond to cell shape in anchorage-dependent fibroblasts. Cell 21:365–372.

Berezney R, Coffey DS (1974): Identification of a nuclear protein matrix. Biochem Biophys Res Commun 60:1410–1417.

Berezney R, Buchholtz LA (1981): Dynamic association of replicating DNA fragments with the nuclear matrix of regenerating liver. Exp Cell Res 132:1–13.

Branton D, Cohen CM, Tyler J (1981): Interaction of cytoskeletal proteins on the human erythrocyte membrane. Cell 24:24–32.

Brown S, Levinson W, Spudich J (1976): Cytoskeletal elements of chick embryo fibroblasts revealed by detergent extraction. J Supramol Struct 5:119–130.

Bryant S, French V, Bryant PJ (1981): Distal regeneration and symmetry. Science 212:993–1002.

Capco DG, Penman S (1983): Mitotic architecture of the cell: The filament networks of the nucleus and cytoplasm. J Cell Biol (in press).

Capco DG, Wan K, Penman S (1982): The nuclear matrix: Three dimensional architecture and protein composition. Cell 29:847–858.

Cereijido M, Robbins ES, Dolan WJ, Rotunno CA, Sabatini DD (1978): Polarized monolayers formed by epithelial cells on a permeable and translucent support. J Cell Biol 77:853–880.

Cervera M, Dreyfuss G, Penman S (1981): Messenger RNA is translated when associated with the cytoskeletal framework in normal and VSV infected HeLa cells. Cell 23:113–120.

Coleman A, Cook PR (1977): Transcription of superhelical DNA from cell nuclei. Eur J Biochem 76:63–78.

Commings DE, Okada TA (1976): Nuclear proteins III. The fibrillar nature of the nuclear matrix. Exp Cell Res 103:341–360.

Cook PR, Brazell IA, Jost E (1976): Characterization of nuclear structures containing superhelical DNA. J Cell Sci 22:303–324.

Cook PR, Brazell IA (1976): Conformational constraints in nuclear DNA. J Cell Sci 22:287–302.

Cook PR, Brazell IA (1978): Spectrofluorometric measurements of the binding of ethidium to superhelical DNA from cell nuclei. Eur J Biochem 84:465–477.

Edidin M, Wei T (1977): Diffusion rates of cell surface antigens of mouse-human heterokaryons. I. Analysis of the population. J Cell Biol 75:475–482.

Folkman J, Greenspan HP (1975): Influence of geometry on control of cell growth. Biochim Biophys Acta 417:211–236.

Folkman J, Moscona A (1978): Role of cell shape in growth control. Nature 273:345–349.

Fulton AB, Wan K, Penman S (1980): The spatial distribution of polysomes in 3T3 cells and the associated assembly of proteins into the skeletal framework. Cell 20:849–857.

Fulton AB, Prives J, Farmer SR, Penman S (1981): Developmental reorganization of the skeletal framework and its surface lamina in fusing muscle cells. J Cell Biol 91:103–112.

Gerace L, Blobel G (1980): The nuclear envelope lamina is reversibly depolymerized during mitosis. Cell 19:277–287.

Ham AW, Cormack DH (1979): "Histology," 8th Ed. Philadelphia: Lippincott.

Herman R, Weymouth L, Penman S (1978): Heterogeneous nuclear RNA-protein fibers in chromatin-depleted nuclei. J Cell Biol 78:663–674.

Holtzman E, Smith I, Penman S (1966): Electron microscope studies of detergent-treated HeLa cell nuclei. J Mol Biol 17:131–135.

Jost E, Johnson RT (1981): Nuclear lamina assembly, synthesis, and disaggregation during the cell cycle in synchronized HeLa cells. J Cell Sci 47:25–53.

Keller JM, Riley DE (1976): Nuclear ghosts: A nonmembranous structural component of mammalian cell nuclei. Science 193:399–401.

Kuhn TS (1962): The structure of scientific revolutions. In Neurath O (ed): "International Encyclopedia of Unified Science," Vol 2, No. 2. Chicago: University Press of Chicago Press.

Lazarides E (1980): Intermediate filaments as mechanical integrators of cellular space. Nature 283:249–256.

Lenk R, Ransom L, Kaufmann Y, Penman S (1977): A cytoskeletal structure with associated polyribosomes obtained from HeLa cells. Cell 10:67–78.

Lenk R, Penman S (1979): The cytoskeletal framework and poliovirus metabolism. Cell 16:289–301.

McCready SJ, Godwin J, Mason DW, Brazell IA, Cook PR (1980): DNA is replicated at the nuclear cage. J Cell Sci 46:365–386.

Milcarek C, Penman S (1974): Membrane-bound polyribosomes in HeLa cells: Association of poly(A) with membranes. J Mol Biol 89:327.

Miller TE, Huang C-Y, Pogo OA (1978): Rat liver nuclear skeleton and ribonucleoprotein complexes containing hnRNA. J Cell Biol 76:675–691.

Pease DC, Porter KR (1981): Electron microscopy and ultramicrotomy. J Cell Biol 91:287s–292s.

Porter KR, Claude A, Fullam EF (1945): J Exp Med 81:233–246.

Prives J, Fulton AB, Penman S, Daniels MP, Christian CN (1982): Interaction of the cytoskeletal framework with Acetylcholine receptor on the surface of embryonic muscle cells in culture. J Cell Biol 92:231–236.

Pugh A (1976): "Introduction to Tensegrity." Berkeley: University of California Press.

Puvion E, Bernhard W (1975): Ribonucleoprotein components in liver cell nuclei as visualized by cryoultramicrotomy. J Cell Biol 67:200–214.

Riley DE, Keller JM (1978): The ultrastructure of non-membranous nuclear ghosts. J Cell Sci 32:249–268.

Robinson SI, Nelkin BD, Vogelstein B (1982): The ovalbumin gene is associated with the nuclear matrix of hen oviduct cells. Cell (in press).

Schliwa M, Van Blerkom J (1981): Structural interactions of cytoskeletal components. J Cell Biol 90:222–235.

Schliwa M (1982): Action of cytochalasin D on cytoskeletal network. J Cell Biol 92:79–91.

Singer SJ, Nicolson GL (1972): The fluid mosaic model of the structure of cell membranes. Science 175:720–731.

Solomon F, Magendantz M, Salzman A (1979): Identification with cytoplasmic microtubules of one of the coassembling microtubule associated proteins. Cell 18:431–438.

Vogelstein B, Pardoll DM, Coffey DS (1980): Supercoiled loops and eucaryotic DNA replication. Cell 22:79–85.

Webb WW, Barak LS, Tank DW, Wu ES (1981): Molecular mobility on the cell surface. Biochem Soc Symp 46:191–205.

Webster RS, Henderson D, Osborn M, Weber K (1978): Three-dimensional electron microscopical visualization of the cytoskeleton of animal cells: Immunoferritin identification of actin- and tubulin-containing structures. Proc Natl Acad Sci USA 75:5511–5515.

Wilson EB (1925): "The Cell in Development and Heredity," 3rd Ed. New York: Macmillan.

Wittelsberger SC, Kleene K, Penman S (1981): Progressive loss of shape-responsive metabolic controls in cells with increasing transformed phenotype. Cell 24:859–866.

Wolosewick JJ, Porter KR (1975): High voltage electron microscopy of WI-38 cells. Anat Rec 181:511–512.

Wolosewick JJ, Porter KR (1976): Stereo high-voltage electron microscopy of whole cells of the human diploid line, WI-38. Am J Anat 147:303–324.

Wolosewick JJ, Porter KR (1979): Microtrabecular lattice of the cytoplasmic ground substance, artifact or reality. J Cell Biol 82:114–139.

Global Organization of Cells and Extracellular Space

Modern Cell Biology 2:419–450
© 1983 Alan R. Liss, Inc., 150 Fifth Avenue, New York, NY 10011

Biogenesis of Epithelial Cell Polarity

David D. Sabatini, Eva B. Griepp, Enrique J. Rodriguez-Boulan,
William J. Dolan, Edith S. Robbins, Stelios Papadopoulos,
Ivan Emanuilov Ivanov, and Michael J. Rindler

From the Department of Cell Biology, New York University Medical Center,
New York, New York 10016 (D.D.S., E.B.G., W.J.D., E.S.R., S.P., I.E.I.,
M.J.R.), and the Department of Pathology, State University of New York,
Downstate Medical Center, Brooklyn, New York 11203 (E.J.R.-B.)

I. INTRODUCTION

Epithelia specialized in transport form continuous monolayers which function as selective permeability barriers between compartments. Within such monolayers, tight junctions, which are beltlike regions of intercellular contact located near the apical surface of the cells [Farquhar and Palade, 1963, 1965], control the movement of substances across the intercellular spaces.

A main attribute of epithelial cells related to their function as permeability barriers is the existence of two distinct plasma membrane domains delimited by the tight junctions. In addition to structural differentiations, such as microvilli in the apical surface and intercellular junctions and membrane infoldings in the basolateral aspects, these domains are characterized by the presence of specific enzymes and transport activities [George and Kenny, 1973; Lojda, 1974; Kinne, 1976]. In general, activities involved in the uptake of substances from the external medium are associated with the apical (or luminal) regions. These include disaccharidases [Sigrist et al, 1975], leucine aminopeptidase [Desnuelle, 1979], alkaline phosphatase [Hugon and Borgers, 1966], and Na^+-dependent sugar [Chesney et al, 1974], and amino acid [Kinne, 1976] transport systems. On the other hand, the Na^+K^+ ATPase [Kyte, 1976; Ernst and Mills, 1977] and the adenylate cyclase activity, as well as hormone receptors [Schwartz et al, 1974; Shlatz et al, 1975] and histocompatibility antigens [Kirby and Parr, 1979] are located in basolateral aspects of the plasma membrane. Tight junctions are thought to be essential for maintaining the functional polarization of epithelial cells by preventing the intermixing of plasma membrane components, which must be segregated exclusively to one or the other cell surface domain. Recently, evidence for a role of tight junctions as barriers to diffusion of membrane components in the plane of the plasma membrane has also been obtained from measurements of the transfer of fluorescent probes from the apical plasma membrane of an epithelial layer to the basolateral surface of the cells [Dragsten et al, 1981].

The biogenesis of epithelial plasma membranes raises interesting questions concerning mechanisms that control the transport of polypeptides to the cell surface and the sorting out processes which effect the segregation of particular polypeptides into the two surface domains. A central question concerns the nature of addressing signals which must be present in newly synthesized polypeptides, and determine their intracellular itinerary and ultimate destination in the plasma membrane [cf Sabatini et al, 1982]. The fact that during membrane recylcing plasma membrane proteins return to their correct site of function [Louvard, 1980] supports the notion that such signals are retained as permanent structural features of the mature proteins.

Cultured epithelial cell lines, such as the kidney-derived MDCK, provide useful model systems for studying the development of epithelial polarity and the role of tight junctions in generating and maintaining the polarized distribution of cell surface components [Misfeldt et al, 1976; Cereijido et al, 1978; Mills et al, 1979; Rindler et al, 1979; Handler et al, 1979; Richardson and Simmons, 1979; Louvard, 1980; Cereijido et al, 1980; Perkins and Hadler, 1981]. Confluent monolayers of MDCK cells exhibit some of the properties of distal convoluted tubules, including the ability to transport fluid and electrolytes vectorially in an apical to basolateral direction [Misfeldt et al, 1976; Cereijido et al, 1978]. It has also been shown that such monolayers acquire a transepithelial resistance which is correlated with the establishment of a complete system of tight junctions between the cells [Misfeldt et al, 1976; Cereijido et al, 1978]. Additional evidence for the functional polarization of MDCK monolayers is provided by the finding that, when infected with enveloped viruses, depending on the virus type, virions assemble asymmetrically at either one or the other plasma membrane domain [Rodriguez-Boulan and Sabatini, 1978].

In this chapter we will review our recent studies utilizing the MDCK cell line to investigate several aspects of epithelial biogenesis including the process of tight junction formation.

II. DEVELOPMENT OF TIGHT JUNCTIONS IN MDCK MONOLAYERS

Freeze-fracture electron microscopy reveals that tight junctions consist of a complex system of anastomosing intramembranous strands involved in maintaining a close contact between adjacent epithelial cells [Kreutziger, 1968; Staehelin et al, 1969; Chalcroft and Bullivant, 1970; Friend and Gilula, 1972; Staehelin, 1974; Wade and Karnovsky, 1974; Bullivant, 1978]. It has been proposed that the strands contain proteins and perhaps phospholipids [Staehelin, 1974; Van Deurs and Luft, 1979; Pinto da Silva and Kachar, 1982], but little direct information is available on the supramolecular organization of junctions and the biochemical nature of their components. It is hoped that information concerning these questions will be gained from studies of the biosynthetic processes necessary for the formation of tight junctions in MDCK cells.

Measurements of the transepithelial resistance are a means of monitoring the assembly of junctions which takes place when MDCK cells obtained from confluent cultures by trypsinization are plated at high density on collagen coated, ion-permeable disks. Typically (Fig. 1a), the resistance of such monolayers rises to a maximum (300 ohm-cm^2) during the first 20–24 hours, and then decreases to steady-state levels (80–150 ohm-cm^2), which may be maintained for many days. Cells dissociated by trypsin require protein synthesis to establish junctional complexes, since no significant resistance developed and no junctions were observed by freeze-fracture, when monolayers were formed by cells plated at high density in the presence of cycloheximide or puromycin (Fig. 1a). These inhibitors did not, however, prevent the settling and spreading of the cells, which are necessary processes for the establishment of intercellular contacts. This suggests that proteins exposed in the cell surface and therefore accessible to trypsin participate in the process of junction formation. The resynthesis of such proteins appears to be completed within the first 8 hours after plating, well before maximal resistance is attained, since protein synthesis inhibitors added after that time did not prevent a normal rise in transepithelial resistance.

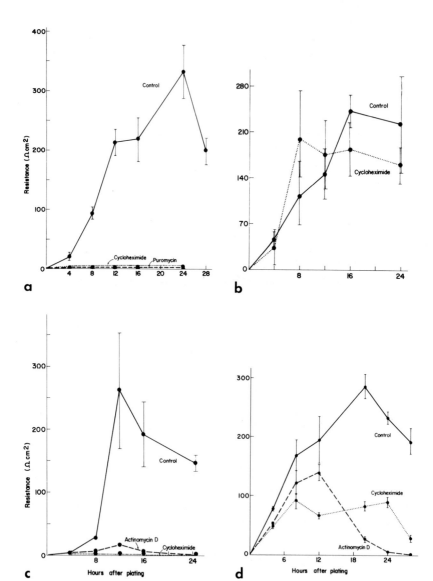

Resistance ($\Omega \cdot cm^2$)

Control
Cycloheximide   Puromycin

a

Control
Cycloheximide

b

Resistance ($\Omega \cdot cm^2$)

Control
Actinomycin D
Cycloheximide

c

Hours after plating

Control
Cycloheximide
Actinomycin D

d

Hours after plating

Glycoproteins do not appear to be involved in junction formation or, if they are, their carbohydrate moieties are not critical for their function, since a transepithelial resistance developed normally when cells obtained from confluent monolayers by trypsinization were plated in the presence of tunicamycin [Griepp et al, 1979]. A normal transepithelial resistance also developed in monolayers formed by mutant cells which synthesize abnormal oligosaccharides [Meiss et al, 1982].

The synthesis of the necessary proteins and the process of junction assembly could be dissociated under several experimental conditions. Cells obtained from confluent monolayers by trypsinization and maintained in spinner culture for 24 hours before plating at high density were capable of reestablishing junctions in the absence of protein synthesis (Fig. 1b). The transepithelial resistance in this case rose faster than it did when monolayers were formed with cells which after trypsinization had not been incubated in spinner culture. The capacity of cells maintained in spinner culture to establish junctions in the presence of cycloheximide was abolished by brief treatment with trypsin before plating, as would be expected if the proteins necessary for junction formation were synthesized and transferred to the surface of the cells during maintenance in spinner medium. Assembly of junctions from preexisting components could also be achieved in monolayers briefly incubated with EGTA [Dolan et al, 1978] or diamide, or exposed to a medium of alkylene pH. These treatments, which have marked effects on cell shape, led to rapid loss of transepithelial resistance, which was, however, restored without the need for protein synthesis after the perturbing agents were removed. Reassembly of junctions in the presence of cycloheximide has also

---

Fig. 1.   Development of a transepithelial resistance in monolayers formed by MDCK cells. MDCK cells which were cultured under various conditions were plated at high density (10^6 cells/ml) onto collagen-coated disks. Disks were transferred to fresh medium after 1.5 hours, and the development of an electrical resistance was monitored essentially as previously described [Cereijido et al, 1978]. a) Protein synthesis is required for the development of resistance in monolayers formed with freshly trypsinized cells obtained from confluent cultures. The inhibitors (10 µg/ml) were added at the time of plating and were present continuously. b) Inhibition of protein synthesis does not prevent the rise of the resistance in monolayers formed with cells which after trypsinization were maintained in spinner culture (5×10^5 cells/ml) for 24 hours prior to plating. c) Not only protein synthesis but also the synthesis of mRNA is required for the development of a resistance in monolayers formed by plating at high density cells obtained by trypsinization from cultures which were sparsely plated 24 hours before (5×10^5 cells per roller bottle). Actinomycin D (2 µg/ml) or cycloheximide (10 µg/ml) was added at the time of plating at high desnity. d) Synthesis of RNA and proteins necessary for junction formation takes place when cells derived from sparse cultures by trypsinization are maintained in spinner medium for 24 hours prior to plating at high density. Sparsely plated cells were obtained as described in (c) and maintained in suspension in spinner culture as in (b). Actinomycin D (2 µg/ml) or cycloheximide (10 µg/ml) was added at the time of high-density plating.

been found to take place after guinea pig pancreatic acinar cells are dissociated with EGTA [Meldolesi et al, 1978].

If proteins are components of tight junctions, as appears likely, they do not turnover with a high rate, since the transepithelial resistance of mature monolayers did not fall during incubation with cycloheximide for several hours. Instead, the resistance increased by 30–40% during treatment with the protein synthesis inhibitor. This suggests that during inhibition of protein synthesis a reorganization takes place which renders the junctions more impermeable to ions. It is possible that junctional elements are recruited from a preexisting reserve pool and incorporated into an expanding junctional system. Alternatively, suppression of the synthesis of a rapdily turning protein which normally inhibits junction formation may account for the increase in resistance.

Several recent reports indicate that a proliferative increase in tight junction strands can be induced in natural epithelia by administration of drugs such as phalloidin [Montesano et al, 1976] or antimycin A [Meldolesi et al, 1978]. Incubation of excised prostatic tissue at 37°C has also been shown to lead to a massive increase in the total length of tight junction strands between columnar cells [Kachar and Pinto da Silva, 1981]. It was observed that this effect was not prevented by protein synthesis inhibitors or metabolic uncouplers, observations thought to be consistent with a role of lipids in junction formation, although the participation of preexisting membrane proteins in this process could not be excluded [Pinto da Silva and Kachar, 1982].

When MDCK cells are plated on collagen disks at a sufficiently high density, formation of tight junctions takes place without cell division. This has enabled us to compare cells obtained from cultures in different growth states as to their RNA and protein synthetic requirements for the formation of confluent monolayers and development of tight junctions. When plated at high density, cells derived by trypsinization from cultures maintained at confluency for 2–5 days formed junctions in the presence of actinomycin D. This shows that these cells already contained adequate levels of mRNA to sustain the synthesis of proteins necessary for junction formation. In contrast, cells that had been sparsely plated and cultured for only 24 hours, and were therefore not confluent at the time of trypsinization, were unable to form junctions when plated at high density in the presence of actinomycin D (Fig. 1c). However, maintenance in suspension with spinner medium for 24 hours prior to plating at high density enabled such cells to form junctions in the presence of actinomycin D (Fig. 1d). Since growth of MDCK cells is arrested in spinner cultures, these observations suggest that expression of genes involved in tight junction formation may be regulated by the growth state of the cells, so that under normal conditions sufficient levels of the required messenger RNAs only accumulate after confluency. It can also be concluded that neither attachment to a substratum nor extensive intercellular contacts are required to trigger the accumulation of the messengers involved in junction formation.

Although the experiments just described demonstrate that some plasma membrane proteins play a role in the process of junction formation, it should be emphasized that a direct participation of those proteins as specific structural components of the junctions is yet to be established. The finding that cells in various states of growth and culture conditions differ in their readiness to establish tight junctions may facilitate the identification of proteins which directly or indirectly participate in the process of junction assembly.

III. ASYMMETRIC BUDDING OF ENVELOPED VIRUSES FROM EPITHELIAL CELLS

Enveloped viruses (rhabdo-, paramyxo-, myxo-, toga-, and retroviridae) are powerful tools for the study of membrane biogenesis. The membrane surrounding the nucleocapsid of these viruses contains only one or a small number of glycoproteins which are incorporated into the host cell plasma membrane and become part of the envelope when the virion buds from the cell surface [cf Katz et al, 1977; Lenard, 1978; Simons et al, 1980].

Viral envelope polypeptides account for a major fraction of the total protein synthetic activity of infected cells. They are synthesized on membrane-bound ribosomes [Hay, 1974; Morrison and Lodish, 1975; Atkinson, 1978; Garoff et al, 1978; Lingappa et al, 1978; Bonatti and Blobel, 1979; Katz and Lodish, 1979], and are inserted cotranslationally into endoplasmic reticulum membranes. They undergo co- and posttranslational modifications such as proteolytic cleavage [Schulman and Palese, 1971], glycosylation [Hunt et al, 1978; Moyer et al, 1978; Nakamura et al, 1979], and addition of fatty acids [Schmidt and Schlesinger, 1979], which are characteristic of many cellular membrane proteins. Clathrin-coated vesicles may be involved in the intracellular transport of the envelope glycoprotein G of vesicular stomatitis virus from the ER to the Golgi apparatus and from this organelle to the cell surface [Rothman et al, 1980; Rothman and Fine, 1980].

It has been shown that when polarized epithelial cells are infected with enveloped viruses [Rodriguez-Boulan and Sabatini, 1978], virions assemble only on specific domains of the plasma membrane. Thus, influenza, or Sendai or simian virus 5 bud exclusively from the apical surface of MDCK or MDBK cells, whereas vesicular stomatitis virions assemble only on basolateral regions of the cell surface (Fig. 2). The asymmetric assembly of enveloped viruses observed in cultured epithelial monolayers also takes place in natural epithelia [Murphy and Bang, 1952; Bang, 1953; Pickett et al, 1975; Hess and Falcon, 1977], and most likely determines the route by which viral

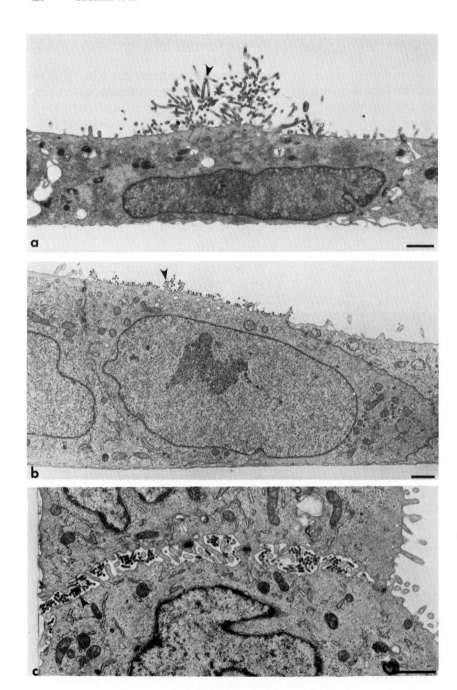

infection spreads within an organism or a population. The plasma membrane domain from which budding takes place is characteristic of each particular virus and is independent of the type of polarized epithelial cell used as host. However, enveloped viruses emerge indiscriminately from all aspects of the plasma membrane of nonpolarized cells such as fibroblasts [Rodriguez-Boulan and Sabatini, 1978].

It has been reported [Rodriguez-Boulan and Pendergast, 1980] that virus budding from an epithelial cell is preceded by the accumulation of viral glycoproteins in the corresponding plasma membrane domain, whereas nucleocapsids appear to be evenly distributed throughout the cytoplasm. These observations suggest that the envelope proteins determine the site of viral budding. Studies with lectin-resistant cells, which synthesize abnormal carbohydrate chains [Green et al, 1981b], and with cells treated with tunicamycin, in which oligosaccharide synthesis is inhibited [Roth et al, 1979; Green et al, 1981b], indicate that the carbohydrate moieties of the envelope glycoproteins are not essential components of the sorting-out signals that direct them to the site of budding. An identification of the polypeptide features which determines the segregation of enveloped glycoproteins in a specific domain of the plasma membrane remains a preeminent goal of research in this field.

It seems reasonable to speculate that the cellular devices that operate on polypeptides synthesized in membrane bound ribosomes and account for their retention in the ER or their selective transfer to other compartments are class-specific and recognize characteristic features common to all polypeptides with the same destination. Simple transmembrane proteins, such as the glycoproteins of enveloped viruses, contain three topologically distinct segments: a middle hydrophobic one, embedded in the bilayer; and two hydrophilic portions, which are exposed on the cytoplasmic and luminal (or exoplasmic) surfaces of the membrane. Segregation of the polypeptide to a specific domain in the plasma membrane could involve interactions between any of these segments and elements of the membrane that are part of the sorting-out mechanism, including peripheral proteins on either side of the membrane. Proteins which, after insertion in the ER membrane, remain with sizeable portions exposed on the cytoplasmic surface could, of course, interact directly with cytoskeletal elements. Because functionally related membrane proteins may form oligomeric complexes soon after they are inserted

---

Fig. 2. Polarized budding of enveloped viruses from monolayers of MDCK cells. SV5 (a) and influenza (b) virions assemble only on the apical plasma membrane domain of cells confluent monolayers, whereas VSV particles (c) bud exclusively from lateral plasma membrane domains and accumulate in the intercellular spaces. The arrows point to the regions of viral budding. Bars represent 1 μm.

in the ER, not all membrane polypetides need contain independent addressing signals.

The complete primary sequence of several viral envelope polypeptides is now available [eg, Rose and Gallione, 1981; Garoff et al, 1980], but it seems likely that recognition of those features which may serve as sorting-out signals will become possible only when sequence information is obtained for whole families of polypeptides which utilize the same subcellular distribution route. Even then, a definitive identification of critical elements involved in sorting out will probably require the application of recombinant DNA techniques to eukaryotic cells for the synthesis of proteins bearing site-specific mutations, or the production of chimeric molecules which will allow the assignment of specific properties to different polypeptide segments.

IV. ROLE OF TIGHT JUNCTIONS IN THE BIOGENESIS OF EPITHELIAL POLARITY

A role of tight junctions in maintaining the characteristic composition of the apical and basolateral domains of the plasma membrane of epithelial cells has been inferred from studies demonstrating a redistribution of cell surface markers after cells are dissociated from native epithelia or cultured mono-layers. Thus, following cell dispersal in Ca^{++}-free medium with EDTA [Pisam and Ripoche, 1976], markers of the apical membrane of toad bladder epithelial cells were found uniformly redistributed over the whole cell surface. A similar observation has been made for the brush-border enzymes leucine aminopeptidase and alkaline phosphatase in dissociated intestinal cells [Ziomek et al, 1980]. It has also been reported that junction disruption caused by EGTA abolishes the difference in concentration of intramembranous particles detected by freeze-fracture between apical and basolateral plasma membrane domains of pancreatic acinar cells [Galli et al, 1976] or MDCK cells within confluent monolayers [Hoi Sang et al, 1979]. It should be noted, however, that experiments of this type do not permit an assessment of the specific contribution of the junctions to the maintenance of the polarized distribution of plasma membrane markers, since the conditions used to dissociate tight junctions also have a general effect on cellular organization and disrupt both cell-cell and cell-substrate interactions. It is known, for example, that Ca^{++} deprivation can markedly affect cell shape [Batten and Anderson, 1981], probably through a reorganization of cytoskeletal elements [Meza et al, 1980], which, it could be speculated, may be involved in controlling the distribution of plasma membrane components (cf Nicolson, 1976).

We have observed that polarized MDCK cells obtained by trypsinization of confluent monolayers lose their structural and functional polarization when maintained overnight in suspension with spinner medium. The cells become rounded and show no morphologically distinct plasma membrane domains. This loss of polarization was clearly reflected in the distribution of budding virions from the surface of infected cells. Thus, when infected with either influenza or VSV, virions emerged indiscriminately from the entire cell surface (Figs. 3, 4), although assembly seemed to take place preferentially within patches of membrane. Other authors have also mentioned [Roth and Compans, 1981] that the polarity of viral budding is abolished when MDCK cells are maintained in spinner culture and when monolayers are treated with EGTA to disrupt tight junctions.

In isolated MDCK cells maintained in suspension, assembly of VSV virions was not always restricted to the cell surface, but frequently also took place on membranes limiting intracellular vacuoles (Fig. 5). This observation suggests that upon cell rounding excess plasma membrane is interiorized by endocytosis and is sequestered intracellularly in a reserve compartment from which it could be reutilized for attachment to a substrate.

Frequently, MDCK cells maintained in suspension formed clusters in which cells were joined by complete tight junctions located near their free surface (Fig. 6). Two plasma membrane domains could be recognized in these cells—a free one with abundant microvilli, and an interior one which frequently showed membrane interdigitations. While influenza virions budded from the free surface of the clusters, VSV particles preferentially emerged from regions of the plasma membrane facing the intercellular spaces, where they accumulated (Fig. 6b). These observations demonstrate that MDCK cells in suspension are capable not only of assembling junctions without attaching to a solid substrate, but also of establishing two functionally differentiated domains on the cell surface. It seems likely that the differentiation of these domains is related to the formation of tight junctions, which is triggered by cell contact.

The orientation of MDCK cells within the clusters is the same as that of thyroid [Mauchamp el al, 1979] or mammary [Emerman et al, 1977] cells, which, in suspension, form spherical vesicles with apical plasma membrane domains delimited by junctions facing the medium. This is the same orientation as that obtained when follicles isolated from the rat thyroid are cultured in a medium with a high concentration of serum, which leads to an inversion of normal polarity [Nitsch and Wollman, 1980]. The opposite polarity, corresponding to that found in native tissues, is attained, however, when dissociated thyroid [Chambard et al, 1981] or mammary [Yang et al, 1979] cells are embedded in a collagen gel. In this case follicles or ductlike structures are formed, respectively, and a mammary tumor virus has been observed to bud into the lumen of the ductlike structure formed by the mammary cells [Yang et al, 1979]. It would therefore be of interest to determine the structural organization of MDCK cells within clusters formed

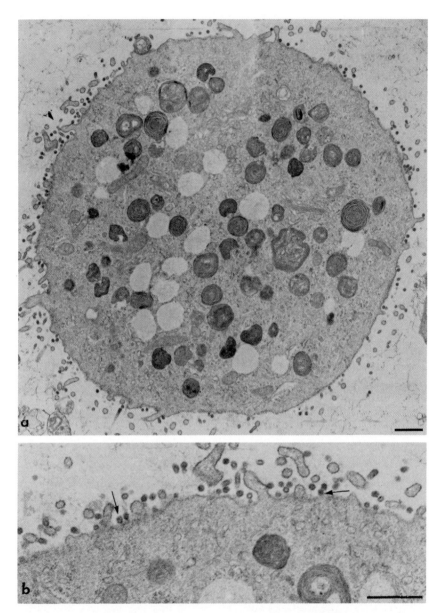

Fig. 3. Nonpolarized budding of influenza virions from an isolated MDCK cell. Cells obtained from confluent monolayers by trypsinization and maintained in spinner culture for 24 hours lose their structural polarity. When such cells are infected, viral budding takes place over the entire cell surface. The region near the arrowhead in (a) is shown at higher magnification in (b). The arrows point to budding influenza virions. Bars represent 1 μm.

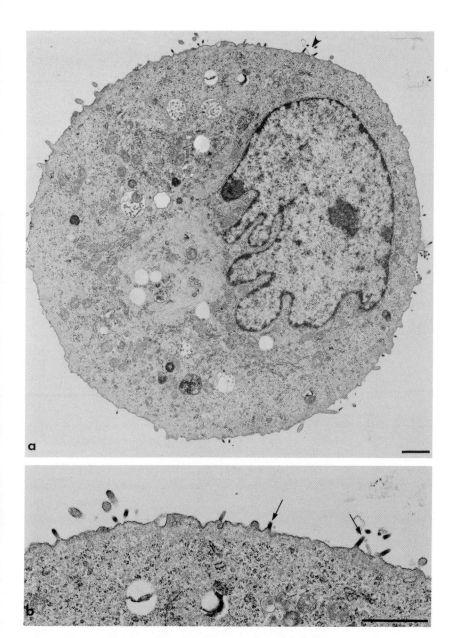

Fig. 4. Nonpolarized budding of VSV from an isolated MDCK cell in suspension. VSV particles bud from the entire surface of an isolated MDCK cell infected in suspension. The region of the cell surface near the arrowhead in (a) is shown at higher magnification in (b). The arrow points to a budding VSV virion. Bars represent 1 μm.

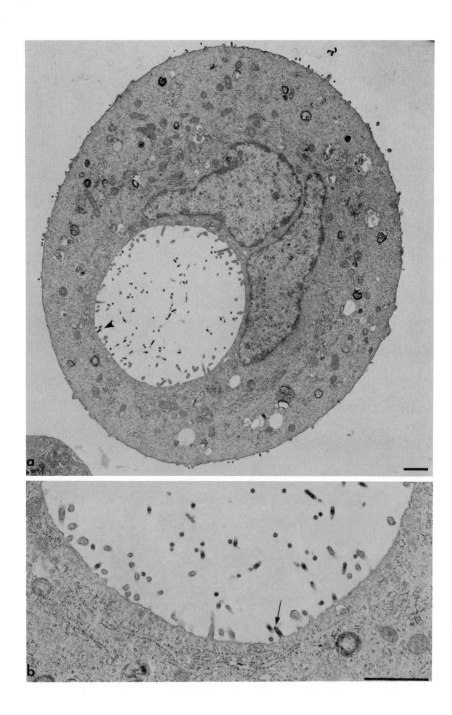

when cells are cultured within a collagen gel and the polarity of viral budding when such cells are infected.

It seems reasonable to presume that attachment to a solid substratum, which may induce important changes in cytoskeletal organization as well as affect directly the distribution of plasma membrane proteins [Michl et al, 1979; Grinnell, 1980; Weigel, 1980], plays a role in determining the polarized organization of cells within epithelial monolayers. Although in some systems the expression of polarity is affected by the nature of the substratum [Emerman et al, 1979; Bennett, 1980], MDCK cells readily form polarized monolayers on a variety of substrates, including glass, plastic, and collagen-coated surfaces. To determine whether attachment to a substratum is sufficent to trigger their functional polarization, MDCK cells were infected in suspension and sparsely plated on a collagen-coated surface. The infected cells attached to the substrate and spread normally. Furthermore, whether single or in contact with others, they became structurally polarized, showing typical microvilli on their free surface and cell adhesion plaques in areas where the basal plasma membrane was closely apposed to the substrate. Cytoplasmic filaments were usually found near the basal region, frequently forming peripheral bundles delimiting the area of attachment. Even though tight junctions were either absent or incomplete in these sparsely plated cells, functional polarization of their luminal and basolateral domains was manifested after viral infection. The polarity of viral budding was that characteristic of cells in confluent monolayers. Influenza virions predominantly emerged from the free surfaces (Fig. 7a) and VSV from regions adjacent to the collagen support (Fig. 7b), although occasionally some VSV virions budded from free surfaces and very infrequently patches of budding influenza virions were found at the adherent face.

The polarized budding from cells that formed clusters in suspension suggests that establishment of complete tight junctions is sufficient to create the conditions necessary for the differentiation of the two surface domains. In contrast, the asymmetric budding of virions from sparsely plated cells demonstrates that such conditions can also be created in the absence of tight junctions or in the presence of incomplete junctions. It is possible, that, in isolated cells, the circumferential region of attachment to the substratum where bundles of filaments approach the cell membrane plays a role equivalent to that of tight junctions in delimiting the two surface domains of cells

---

Fig. 5.   Budding of VSV particles into an intracellular vacuole in an isolated MDCK cell in suspension. Cells infected and maintained in suspension frequently have large cytoplasmic vacuoles. VSV virions bud in a nonpolarized fashion from the entire cell surface, as well as into these vacuoles. The arrowhead in (a) points to a region shown at higher magnification in (b). The arrow points to a budding VSV virion. Bars represent 1 μm.

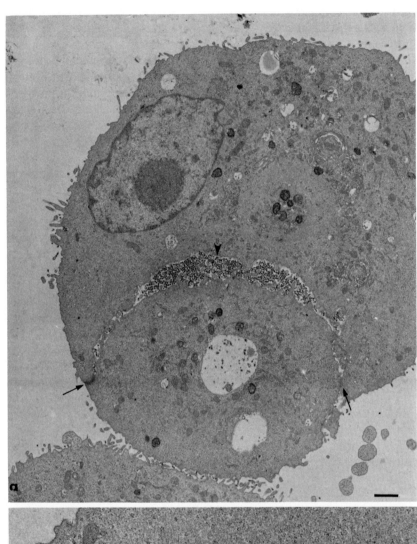

in contact with each other. Alternatively, the segregation of proteins within surface domains of attached single cells may be maintained by direct interactions between specific membrane proteins and cytoskeletal elements. Although fluorescent photobleaching recovery studies have provided strong support for a role of tight junctions in limiting the lateral diffusion of plasma membrane lipids [Dragsten et al, 1981], these studies have shown that binding sites for lectins are immobile in the plasma membrane. Following disruption of the junctions, lectin-binding sites were redistributed over the entire cell surface, but this did not result from their free diffusion in the plane of the membrane [Dragsten et al, 1981]. Thus, the segregation of plasma membrane components to a particular membrane domain need not depend on tight jucntion integrity. Other work has also shown that after disruption of tight junctions, apical enzyme markers of intestinal cells move at a rate lower than expected for free diffusion of proteins in the plane of the membrane [Ziomek et al, 1980].

It has recently been reported [Roth and Compans, 1981] that the polarity of viral budding was completely lost when infected cells, which were dissociated from monolayers by trypsin-EGTA, were replated at low density. The apparent contradiction of these observations with the results reported here may be explained by the fact that, as shown in section II, the development of polarity, after trypsinization, requires cellular protein synthesis, which is suppressed by viral infection. In the experiments described here, however, cells were infected in suspension after they had been incubated in spinner medium for a period suffciently long to complete the resynthesis of tight junctional components damaged by trypsin. As shown in Figure 1b, when plated at high density, such cells are able to carry out the assembly of tight junctions from preexisting components.

The experiments described in this section show that interaction of an epithelial cell with a substrate or with another cell may be a sufficient signal to trigger the segregation of plasma membrane components responsible for polarized viral budding. They do not demonstrate, however, that the two surface domains acquire the level of molecular specificity observed in confluent monolayers. A direct localization of integral plasma membrane proteins in each domain of isolated attached cells will be necessary to elucidate this point.

Fig. 6. Budding of VSV particles into the intercellular space within an aggregate of MDCK cells in suspension. Tight junctions (arrows) delimit the intercellular space (arrowhead) where VSV virions accumulate (a). A higher magnification of a portion of this area is shown in (b). Bars represent 1 μm.

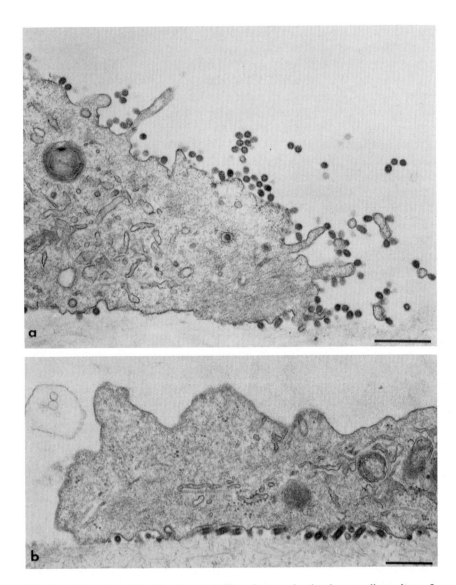

Fig. 7. Polarized viral budding from MDCK cells sparsely plated on a collagen layer. Influenza virions emerge mostly from the free surface near the edge of a single isolated cell attached to a collagen substrate; some virions have been trapped between the basal surface and the substratum (a). VSV virions bud exclusively from the basal surface of a single isolated cell and remain between the cell and the collagen layer (b). Bars represent 0.5 μm.

V. INTRACELLULAR PATHWAYS FOLLOWED BY VIRAL GLYCOPROTEINS IN POLARIZED CELLS

There is now considerable evidence indicating that, following synthesis on membrane-bound ribosomes, proteins destined to the plasma membrane traverse an intracellular route similar to that followed by secretory proteins, which has long been known to include passage through the Golgi apparatus [Caro and Palade, 1964; Jamieson and Palade, 1967]. The presence of retinal opsin [Hall et al, 1969; Papermaster et al, 1975] and the acetylcholine receptor [Fambrough and Devreotes, 1978] in Golgi cisternae of differentiating cells has been demonstrated by autoradiography, and more recently this technique, as well as immunofluorescence and immunoelectron microscopy, has been employed to demonstrate passage of viral envelope proteins through the same organelle [Kääriäinen et al, 1980; Bergmann et al, 1981; Green et al, 1981a; Bergeron et al, 1982; Rindler et al, 1982]. On the other hand, it has been reported [Smilowitz, 1980] that in myotubes the integral membrane protein acetylcholine receptor and the secretory protein acetylcholinesterase are transported to the cell surface through different routes. Although monovalent ionophores, such as monensin, which are known to inhibit secretion and cause morphological changes in the Golgi apparatus [Tartakoff and Vasalli, 1977; Uchida et al, 1979; Ledger et al, 1980; Tajiri et al, 1980], prevented the exocytotic discharge of the esterase which seems to be mediated through noncoated vesicles, the same ionophores did not affect the transfer of the receptor to the cell surface which seems to take place via clathrin-coated vesicles. A recent report [Alonso and Compans, 1981] indicates that monensin equally affects the production of influenza or VSV virions from nonpolarized cells, but has a differential effect in MDCK cells, where it selectively blocks the transport of the G protein of VSV. It was concluded that in the polarized cells the glycoproteins destined to different domains utilize two distinct intracellular pathways to reach the cell surface.

To compare directly the intracellular routes followed by envelope glycoproteins of viruses of opposite budding polarity, and to determine whether they traverse different Golgi apparatus which might serve exclusively one or the other plasma membrane domain, we developed conditions for the double infection of MDCK monolayers with either SV5 and VSV or influenza and VSV. Cells infected with SV5 (1–5 pfu/cell) or influenza (10 pfu/cell) were superinfected with VSV (10–15 pfu/cell) 24–72 hours or 4.5 hours later, respectively. Following these protocols, 5 hours after superinfection more than 95% of the cells could be labeled with specific antibodies against either of the two viruses. Gel electrophoretic analysis of ^{35}S-methionine-labeled cultured monolayers also demonstrated the simultaneous production of proteins of both viruses, although the rate of synthesis of SV5 or influenza proteins decreased with the progress of VSV infection.

The characteristic polarity of viral budding was maintained in doubly infected cells, as long as the junctions remained intact. The distinctive filamentous SV5 (Fig. 8a) or the spherical influenza virions (Fig. 8b) were observed budding from the apical surfaces of cells which, at the same time, showed numerous VSV virions assembling at the basolateral domains (Fig. 8a, b). Only when cells became rounded, tight junctions began to be disrupted, and other cytopathic effects became evident, were patches of assemblying VSV virions seen on the free surfaces of the cells. Regions of VSV budding, however, remained distinct from areas where SV5 or influenza budded, suggesting that, even then, segregation of viral budding was not completely lost.

The intracellular distribution of influenza and VSV viral glycoproteins was examined by immunelectron microscopy in ultrathin frozen sections [Tokuyasu and Singer, 1976; Tokuyasu, 1978] of doubly infected MDCK cells. Sections of cells fixed in glutaraldehyde were incubated with rabbit anti-G or mouse monoclonal anti-HA antibodies, and these primary antibodies were detected by an indirect procedure using affinity purified goat antimouse or antirabbit antibodies, complexed to colloidal gold particles [Horisberger and Rosset, 1977; Geuze et al, 1981].

Antibodies against either HA or G found over the cytoplasm of doubly infected cells were preferentially distributed over the Golgi apparatus (Fig. 9), located in different cisternae as well as adjacent Golgi vesicles and vacuoles. Cisternae throughout the Golgi stacks appeared to be uniformly labeled, and gold particles were found as frequently toward the middle region of the cisternae as toward the rims.

As expected from the changes in the relative ratios of viral glycoprotein synthesis during the course of the double infection, the concentration of anti-HA antibodies over the Golgi complex decreased with the progress of superinfection, as the concentration of anti-G antibodies increased. The latter were first detected in the Golgi apparatus 3 hours after superinfection, when anti-G antibodies were practically absent from the cell surface. Labeling of the Golgi remained prominent at later times, when the G protein was also localized on the plasma membrane. It is worth noting that only low levels of labeling were observed over the rough ER, where synthesis of the poly-

---

Fig. 8.   Simultaneous budding of two viruses of different polarity from doubly infected MDCK cells. The elongated SV5 virions (arrowhead) bud only from the apical surface of a doubly infected cell within a monolayer, whereas the bullet-shaped VSV virions (arrows) bud from the basolateral surfaces of the same cell (a). Influenza virions (arrowhead) assemble at the apical surface of a doubly infected cell, whereas VSV virons emerge from basolateral plasma membrane domains of the same cell and into an intercellular space delimited by tight junctions (arrow) (b). Bars represent 0.5 μm.

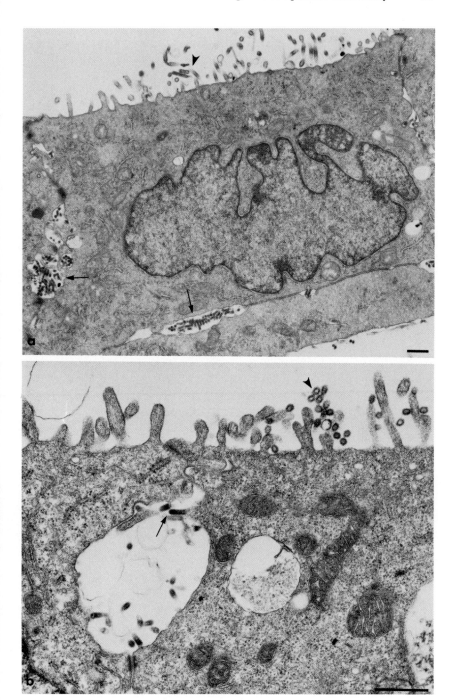

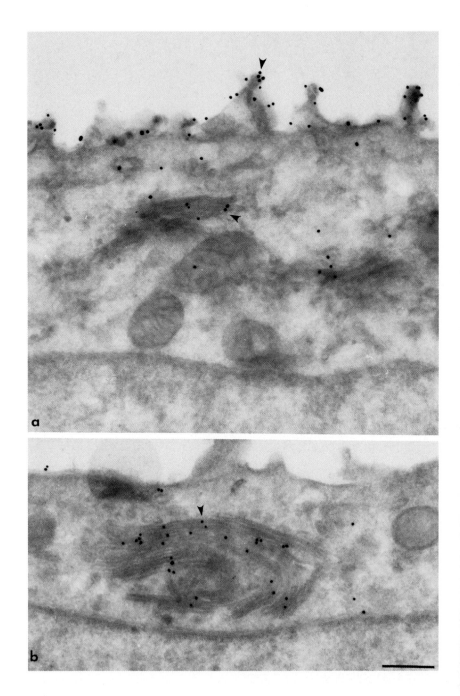

peptides is known to take place. Thus, passage through the Golgi apparatus appears to be a much slower process than exit from the ER, and may be the rate-limiting step in the transport of the newly synthesized glycoproteins to the cell surface.

A simultaneous visualization of the primary mouse anti-HA and rabbit anti-G antibodies within the same cell was achieved using secondary anti-mouse and antirabbit antibodies complexed to colloidal gold particles of different sizes (18 nm and 5 nm in diameter, respectively). In doubly labeled cells, Golgi apparatuses, irrespective of their proximity to one or the other plasma membrane domain, were always labeled with both types of gold particles, which were present either singly or in small clusters (Fig. 10). The two types of gold particles were usually not intermingled within the same aggregates, but often single particles or small aggregates of each kind were found near each other within the same cisterna (Fig. 10b). These findings clearly show that envelope glycoproteins of viruses with opposite polarity traverse the same Golgi apparatus.

Antiviral glycoprotein antibodies were also present in vesicles located between the Golgi apparatus and the plasma membrane, as expected if such vesicles were involved in the transport of the proteins to the cell surface. Although in most cases small vesicles contained only gold particles of one type, larger ones sometimes were labeled with both types of gold particles. Nevertheless, the low level of labeling and the small size of most vesicles makes it difficult to conclude definitively that those vesicles labeled with only one type of antibody did not also carry the other glycoprotein. Moreover, because some labeled vesicles may be involved in the recycling of plasma membrane components [cf Palade, 1982], the direction of movement of individual cytoplasmic vesicles and their role in the segregation of viral glycoproteins cannot be determined unequivocally from these static observations.

An analysis of the distribution of immunolabeled viral envelope proteins in the cell surface showed that more than 80% of the anti-HA antibodies on the cell surface was found throughout the course of infection on the apical

---

Fig. 9.   Frozen thin sections of MDCK cells in monolayer of doubly infected influenza and VSV. Mouse monoclonal antibodies against HA of influenza (a) or rabbit antibodies against the G of VSV (b) were applied to ultrathin frozen sections of a doubly infected monolayer 3 hours after superinfection with VSV. Affinity purified secondary goat antimouse (a) or antirabbit (b) antibodies complexed to colloidal gold particles were employed to visualize the intracellular and cell surface distribution of the primary antibodies bound to the viral glycoproteins. The HA protein of influenza (arrowheads) accumulates both on the apical surface of the cell and intracellularly over the Golgi apparatus (a). In another doubly infected cell (b), the G protein of VSV (arrowheads) is found throughout the cisternae of a Golgi apparatus. Bars represent 0.5 μm.

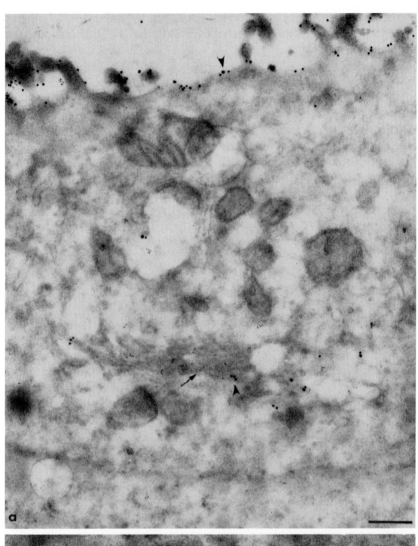

domain of the plasma membrane. In contrast, significant amounts of the G protein were found not only on the basolateral surfaces, but also on the apical domains of the plasma membrane of VSV-infected or superinfected cells. It must be emphasized, however, that in spite of the sometimes intense anti-G apical labeling, budding of VSV was always restricted to the basolateral domains. These observations suggest that the site of VSV assembly may not be determined exclusively by the preferential accumulation of the envelope glycoprotein.

The production of phenotypically mixed virions and pseudotypes from doubly infected polarized cells may provide clues as to the mechanism accounting for the segregation of polypeptides in the two different plasma membrane domains. In these aberrant particles one type of nucleocapsid is surrounded by an envelope composed partly (mixed virions) or totally (pseudotypes) of glycoproteins of the other virus [McSharry et al, 1971]. The production of a mixed virion requires, of course, the presence of both glycoproteins in the same plasma membrane domain. The formation of pseudotypes and mixed virions containing the VSV nucleocapsid and influenza glycoproteins has been monitored in polarized MDCK monolayers infected with both viruses [Roth and Compans, 1981]. The assembly of aberrant particles was first detected after a period (1–2 hrs) during which only wild-type viruses were released. When cytopathic effects became evident, however, the production of aberrant virions increased dramatically, and it was suggested that this reflected an opening of the tight junctions, which allowed the mixing of envelope glycoproteins in the cell surface.

Using the immunolabeling technique we have observed that formation of mixed virions in doubly infected MDCK cells takes place well before there is evidence of junction disruption. Virions with the VSV bullet-shaped nucleocapsid and an envelope labeled with varying amounts of anti-HA were often present in the intercellular spaces and were seen budding from the basolateral domains of the plasma membrane. The frequency of such particles seemed to correlate well with the presence of a small fraction (less than 20%) of the HA glycoprotein on the basolateral aspects of the plasma membrane.

Fig. 10. Simultaneous visualization of the distribution of HA and G glycoproteins in a frozen thin section of doubly infected MDCK cells. a) A frozen thin section of a cell infected with influenza and VSV was treated with both types of primary antibodies (see legend to Fig. 9) in succession, and then with secondary antibodies complexed to gold particles of different sizes (18 nm for the goat antimouse to visualize HA, and 5 nm for the goat antirabbit to visualize G). Gold particles of both sizes are present in the same Golgi apparatus. The apical surface of the cell is mostly labeled with large gold particles, indicating the presence of HA protein of influenza virus. b) Higher magnification of a portion of the Golgi apparatus of another doubly infected cell showing the presence of both HA (18 nm gold particles, arrowhead) and G (5 nm gold particles, arrow) proteins in the same Golgi stack. Bars represent 0.5 μm.

However, mixed-type virions with a rounded influenza nucleocapsid and some anti-G labeling in the envelope were very rarely seen budding from the apical surface, even though the fraction of G protein found on the apical surface was much larger than that of HA on the basolateral region. The relative inability of the apical G protein to be incorporated into virions may reflect a structural state unsuitable for assembly or the requirement for other proteins, such as the M protein or even cellular proteins, which are absent from the apical domain.

VI. CONCLUSIONS

Cultured epithelial cells, such as the kidney-derived MDCK, which form polarized monolayers in vitro, provide useful model systems for studies of the biogenesis and maintenance of epithelial polarity. When such cells form confluent monolayers, the development of a transepithelial electrical resistance provides a means of monitoring the process of junction assembly. We have determined that plasma membrane proteins participate in this process. Such proteins are destroyed when cells are dissociated from confluent monolayers by trypsinization and are resynthesized when cells are plated at high density. Under these conditions synthesis of the necessary proteins is completed well before the resistance rises to a maximum. Cells harvested from sparsely plated cultures do not contain sufficient levels of proteins required for junction formation or of messengers directing the synthesis of these proteins. Cessation of growth, which occurs when cells reach confluency or when cells are maintained in suspension, appears to be a main factor in triggering the expression of genes necessary for tight junction formation.

The polarized nature of MDCK cells is reflected in the asymmetric assembly of virions which takes place when monolayers are infected with enveloped viruses. Myxoviruses (influenza) and paramyxoviruses (Sendai, SV5) bud from the apical surface of the cells, whereas rhabdoviruses (VSV) assemble on the basolateral plasma membrane domains. The characteristic polarity of MDCK cells is abolished when they are maintained in suspension on single cells for 24 hours and enveloped viruses bud indiscriminately from the entire surface of such cells. Single cells, however, become polarized when they are plated sparsely and attach to a substratum without making contact with other cells. Viruses bud from such cells with the same polarity as in established monolayers. A similar polarization is attained when cells in suspension form clusters, in which different plasma membrane domains are delimited by tight junctions.

Virally infected MDCK monolayers provide a convenient system to study the routes followed by newly synthesized glycoproteins to reach specific domains of the plasma membrane. Immunoelectron microscopic observations of doubly infected cells show that glycoproteins of viruses that bud with opposite polarity traverse the same Golgi apparatus. Thus, critical sorting-out steps must take place during or after passage of the glycoproteins through the Golgi cisternae.

ACKNOWLEDGMENTS

We wish to acknowledge the gifts of monoclonal antiserum against HA from Dr R. G. Webster and of rabbit anti-G antiserum from Dr J. Lenard, and to thank H. Plesken, S. Malamet, H. Snitkin, J. Culkin, and B. Zietlow for invaluable technical assistance. We also wish to thank Myrna Chung for the quality of her work and her dedication throughout the typing of the manuscript. This work was supported by National Institutes of Health grants GM 20277, AG 01461, and AG 00378. E.G.B. was supported by a New York Heart Association Investigatorship.

VII. REFERENCES

Alonso FV, Compans RW (1981): Differential effect of monensin on enveloped viruses that form at distinct plasma membrane domains. J Cell Biol 89:700–705.

ATkinson PH (1978): Glycoprotein and protein precursors to plasma membranes in vesicular stomatitis virus infected HeLa cells. J Supramol Struct 8:89–109.

Bang FB (1953): The development of New Castle disease virus in cells of the chorioallantoic membrane as studied by thin sections. Bull Johns Hopkins Hosp 92:309–322.

Batten BE, Anderson E (1981): Effects of Ca^{2+} and Mg^{2+} deprivation on cell shape in cultured ovarian granulosa cells. Am J Anat 161:101–114.

Bennett DC (1980): Morphogenesis of branching tubules in cultures of cloned mammary epithelial cells. Nature (Lond) 285:657–659.

Bergeron JJM, Kotwal GJ, Levine G, Bilan P, Rachubinski R, Hamilton M, Shore GC, Ghosh HP (1982): Intracellular transport of the transmembrane glycoprotein G of vesicular stomatitis virus through the Golgi apparatus as visualized by electron microscope radioautography. J Cell Biol 94:36–41.

Bergmann JE, Tokuyasu KT, Singer SJ (1981): Passage of an integral membrane protein, the vesicular stomatitis virus glycoprotein through the Golgi apparatus en route to the plasma membrane. Proc Natl Acad Sci USA 78:1746–1750.

Bonatti S, Blobel G (1979): Absence of a cleavable signal sequence in Sindbis virus glycoprotein PE_2. J Biol Chem 254:12261–12264.

Bullivant S (1978): "The Structure of Tight Junctions." Toronto: Ninth International Congress on Electron Microscopy, Vol III, pp 659–672.

Caro LG, Palade, GE (1964): Protein synthesis, storage, and discharge in the pancreatic exocrine cell. J Cell Biol 20:473–495.

Cereijido M, Ehrenfeld J, Meza I, Martinez-Palomo A (1980): Structural and functional membrane polarity in cultured monolayers of MDCK cells. J Membr Biol 52:147–159.

Cereijido M, Robbins ES, Dolan WJ, Rotunno CA, Sabatini DD (1978): Polarized monolayers formed by epithelial cells on a permeable and translucent support. J Cell Biol 77:853–880.

Chalcroft JP, Bullivant S (1970): An interpretation of liver cell membrane and junction structure based on the observation of freeze fracture replicas of both sides of fracture. J Cell Biol 47:49–60.

Chambard M, Gabrion J, Mauchamp J (1981): Influence of collagen gel on the orientation of epithelial cell polarity: Follicle formation from isolated thyroid cells and from preformed monolayers. J Cell Biol 91:157–166.

Chesney R, Sacktor B, Kleinzeller A (1974): The binding of phloridzin to the isolated luminal membrane of the renal proximal tubule. Biochim Biophys Acta 332:263–277.

Desnuelle P (1979): The tenth Sir Hans Krebs lecture. Intestinal and renal aminopeptidase: A model of a transmembrane protein. Eur J Biochem 101:1–11.

Dolan W, Griepp EB, Robbins ES, Sabatini DD (1978): Synthesis and assembly of tight junction components of epithelial cells in vitro. J Cell Biol 79:220a.

Dragsten PR, Blumenthal R, Handler JS (1981): Membrane asymmetry in epithelia: Is the tight junction a barrier to diffusion in the plasma membrane? Nature 294:718–722.

Emerman JT, Burwen SJ, Pitelka DR (1979): Substrate properties influencing ultrastructural differentiation of mammary epithelial cells in culture. Tissue Cell 11:109–119.

Emerman JT, Enami J, Pitelka DR, Nandi S (1977): Hormonal effects on intracellular and secreted casein in cultures of mouse mammary epithelial cells on floating collagen membranes. Proc Natl Acad Sci USA 74:4466–4470.

Ernst SA, Mills JW (1977): Basolateral plasma membrane localization of ouabain-sensitive sodium transport sites in the secretory epithelium of the avian salt gland. J Cell Biol 75:74–94.

Fambrough DM, Devreotes PN (197): Newly synthesized acetylcholine receptors are located in the Golgi apparatus. J Cell Biol 76:237–244.

Farquhar MG, Palade G (1963): Junctional complexes in various epithelia. J Cell Biol 17:375–412.

Farquhar MG, Palade G (1965): Cell junctions in amphibian skin. J Cell Biol 26:263–291.

Friend DS, Gilula NB (1972): Variations in tight and gap junctions in mammalian tissues. J Cell Biol 53:758–776.

Galli P, Brenna A, De Camilli P, Meldolesi J (1976): Extracellular calcium and the organization of tight junctions in pancreatic acinar cells. Exp Cell Res 99:178–182.

Garoff H, Simons K, Dobberstein B (1978): Assembly of the Semliki Forest virus membrane glycoproteins in the membrane of the endoplasmic reticulum in vitro. J Mol Biol 124:587–600.

Garoff H, Frischauf AM, Simons K, Lehrach H, Delius H (1980): Nucleotide sequence of cDNA coding for Semliki Forest virus membrane glycoproteins. Nature (Lond) 288:236–241.

George SG, Kenny AJ (1973): Studies on the enzymology of purified preparations of brush border from rabbit kidney. Biochem J 134:43–57.

Geuze HJ, Slot JW, Van der Ley PA, Scheffer RCT (1981): Use of colloidal gold particles in double labelling immunoelectron microscopy of ultrathin frozen sections. J Cell Biol 89:653–665.

Green J, Griffiths G, Louvard D, Quinn P, Warren G (1981a): Passage of viral membrane proteins through the Golgi complex. J Mol Biol 152:663–698.

Green RF, Meiss HK, Rodriguez-Boulan E (1981b): Glycosylation does not determine segregation of viral envelope proteins in the plasma membrane of epithelial cells. J Cell Biol 89:230–239.

Griepp E, Robbins E, Malamet S, Dolan W, Sabatini DD (1979): Studies on the role of glycoproteins in tight junction formation. J Cell Biol 83:88a.

Grinnell F (1980): Fibroblast receptor for cell-substratum adhesion: Studies on the interaction of baby hamster kidney cells with latex beads coated by cold insoluble globulin (plasma fibronectin). J Cell Biol 86:104–112.

Hall MO, Bok D, Bacharach ADE (1969): Biosynthesis and assembly of the rod outer segment membrane system. Formation and fate of visual pigment in the frog retina. Mol Biol 45:397–406.

Handler JS, Steele RF, Wade JB, Peterson AS, Lawson NL, Johnson JP (1979): Toad urinary bladder epithelial cells in culture: Maintenance of epithelial structure, sodium transport, and response to hormones. Proc Natl Acad Sci USA 76:4151–4155.

Hay AJ (1974): Studies on the formation of the influenza virus envelope. Virology 60:398–418.

Hess RT, Falcon LA (1977): Observations on the interaction of baculoviruses with the plasma membrane. J Gen Virol 36:525–530.

Hoi Sang U, Saier MH, Ellisman MH (1979): Tight junction formation is closely linked to the polar redistribution of intramembranous particles in aggregating MDCK epithelia. Exp Cell Res 122:384–391.

Horisberger M, Rosset J (1977): Colloidal gold, a useful marker for transmission and scanning electron microscopy. J Histochem Cytochem 25:295–305.

Hugon J, Borgers M (1966): Ultrastructural localization of alkaline phosphatase activity in the absorbing cells of the duodenum of the mouse. J Histochem Cytochem 14:629–640.

Hunt LA, Etchinson SR, Summers DF (1978): Oligosaccharide chains are trimmed during synthesis of the envelope glycoprotein of vesicular stomatitis virus. Proc Natl Acad Sci USA 75:754–758.

Jamieson JD, Palade GE (1967): Intracellular transport of secretory proteins in the pancreatic exocrine cell. J Cell Biol 34:577–596.

Kääriäinen L, Hashimoto K, Saraste J, Virtanen I, Penttinen K (1980): Monesin and FCCP inhibit the intracellular transport of alphavirus membrane glycoproteins. J Cell Biol 87:783–791.

Kachar B, Pinto da Silva P (1981): Rapid massive assembly of tight junction strands. Science 213:541–543.

Katz FN, Lodish HF (1979): Transmembrane biogenesis of the vesicular stomatitis virus glycoprotein. J Cell Biol 80:416–426.

Katz FN, Rothman JE, Knipe DM, Lodish HF (1977): Membrane assembly: Synthesis and intracellular processing of the vesicular stomatitis viral glycoprotein. J Supramol Struct 7:353–370.

Kinne R (1976): Membrane-molecular aspects of tubular transport. Int Rev Physiol 11:169–210.

Kirby WN, Parr EL (1979): The occurrence and distribution of H-2 antigens on mouse intestinal epithelial cells. J Histochem Cytochem 27:746–750.

Kreutziger GO (1968): Freeze-etching of intercellular junctions of mouse liver. In Arceneaux CJ (ed): "Proceedings of the 26th Annual Meeting of the Electron Microscopic Society of America." Baton Rouge, Louisana: Claitor's Publishing, pp 234–235.

Kyte J (1976): Immunoferritin determination of $(Na^+ + K^+)$ ATPase over the plasma membranes of renal convoluted tubules. I: Distal segment. J Cell Biol 68:287–303.

Ledger PW, Uchida N, Tanzer ML (1980): Immunocytochemical localization of procollagen and fibronectin in human fibroblasts: Effects of the monovalent ionophore, monensin. J Cell Biol 87:663–671.

Lenard J (1978): Virus envelopes and plasma membranes. Annu Rev Biophys Bioeng 7:139–165.

Lingappa VR, Katz FN, Lodish HF, Blobel G (1978): A signal sequence for the insertion of a transmembrane glycoprotein. J Biol Chem 253:8667–8670.

Lojda Z (1974): Cytochemistry of enterocytes and of other cells in the mucous membrane of the small intestine. In Smith DH (ed): "Biomembranes 4A." New York: Plenum, pp 43–122.

Louvard D (1980): Apical membrane aminopeptidase appears at site of cell-cell contact in cultured kidney epithelial cells. Proc Natl Acad Sci USA 77:4132–4136.

Mauchamp J, Margotat A, Chambard M, Verrier B, Remy L, Michel-Béchet M (1979): Polarity of three-dimensional structures derived from isolated hog thyroid cells in primary culture. Cell Tissue Res 204:417–430.

McSharry JJ, Compans RW, Choppin PW (1971): Proteins of vesicular stomatitis virus and

448   Sabatini et al

of phenotypically mixed vesicular stomatitis virus–simian virus 5 virions. J Virol 8:722–729.

Meiss HK, Green RF, Rodriguez-Boulan E (1982): Lectin resistant mutants of polarized epithelial cells. Mol Cell Biol 2:1287–1295.

Meldolesi J, Castiglioni G, Parma R, Nassivera N, De Camilli P (1978): Ca^{++}-dependent disassembly and reassembly of occluding junctions in guinea pancreatic acinar cells. J Cell Biol 79:156–172.

Meza I, Ibarra G, Sabanero M, Martinez-Palomo A, Cereijido M (1980): Occluding junctions and cytoskeletal components in a cultured transporting epithelium. J Cell Biol 87:746–754.

Michl J, Pieczonka MM, Unkeless JC, Silverstein SC (1979): Effects of immobilized immune complexes on Fc- and complement receptor function in resident and thioglycollate-elicited mouse peritoneal macrophages. J Exp Med 150:607–621.

Mills JW, MacKnight ADC, Dayer J-M, Ausiello DA (1979): Localization of ^{3}H-ouabain-sensitive NA$^+$-pump sites in cultured pig kidney cells. Am J Physiol 236:C157–C162.

Misfeldt DS, Hamamoto ST, Pitelka DR (1976): Transepithelial transport in cell culture. Proc Natl Acad Sci USA 73:1212–1216.

Montesano R, Gabbiani G, Perrelet A, Orci L (1976): In vivo induction of tight junction proliferation in rat liver. J Cell Biol 68:793–798.

Morrison TJ, Lodish HF (1975): The site of synthesis of membrane and non-membrane proteins of vesicular stomatitis virus. J Biochem 250:6955–6962.

Moyer SA, Tsang JM, Atkinson PH, Summers DF (1976): Oligosaccharide moieties of the glycoprotein vesicular stomatitis virus. J Virol 18:167–175.

Murphy JS, Bang FB (1952): Observations with the electron microscope on cells of the chorioallantoic membrane infected with influenza virus. J Exp Med 95:259–268.

Nakamura K, Compans RW (1979): Synthesis of the oligosaccharides of influenza virus glycoproteins. Virology 93:31–47.

Nicolson GL (1976): Transmembrane control of the receptor on normal and tumor cells. Biochim Biophys Acta 457:57–108.

Nitsch L, Wollman SH (1980): Ultrastructure of intermediate stages in polarity of thyroid epithelium in follicles in suspension culture. J Cell Biol 86:875–880.

Palade G (1982): Problems in intracellular membrane traffic. In "Membrane Recycling," Ciba Foundation Symposium 92. London: Pitman, pp 1–14.

Papermaster DS, Converse CA, Siu I (1975): Membrane biosynthesis in the frog retina: opsin transport in the photoreceptor cell. Biochemistry 14:1343–1352.

Perkins F, Handler JS (1981): Transport properties of toad kidney epithelia in culture. Am J Physiol 241:C154–C159.

Pickett PB, Pitelka DR, Hamamoto ST, Misfeldt DS (1975): Occluding junctions and cell behavior in primary cultures of normal and neoplastic mammary gland cells. J Cell Biol 66:316–332.

Pinto da Silva P, Kachar B (1982): On tight-junction structure. Cell 28:441–450.

Pisam M, Ripoche P (1976): Redistribution of surface macromolecules in dissociated epithelial cells. J Cell Biol 71:907–920.

Richardson JCW, Simmons NL (1979): Demonstration of protein asymmetries in the plasma membrane of cultured renal (MDCK) epithelial cells by lactoperoxidase-mediated iodination. FEBS Lett 105:201–204.

Rindler MJ, Chuman LM, Shaffer L, Saier MH (1979): Retention of differentiated properties in an established dog kidney epithelial cell line (MDCK). J Cell Biol 81:635–648.

Rindler MJ, Emanuilov Ivanov I, Rodriguez-Boulan E, Sabatini DD (1982): Biogenesis of epithelial cell plasma membrane. In "Membrane Recycling," Ciba Foundation Symposium 92. London: Pitman, pp 184–202.

Rodriguez-Boulan E, Sabatini DD (1978): Asymmetric budding of viruses in epithelial monolayers: A model for the study of epithelial polarity. Proc Natl Acad Sci USA 75:5071–5075.

Rodriguez-Boulan E, Pendergast M (1980): Polarized distribution of viral envelope proteins in the plasma membrane of infected epithelial cells. Cell 20:45–54.

Rose JK, Gallione CJ (1981): Nucleotide sequences of the mRNA's encoding the vesicular stomatitis virus G and M proteins determined from cDNA clones containing the complete coding regions. J Virol 39:519–528.

Roth MG, Compans RW (1981): Delayed appearance of pseudotypes between vesicular stomatitis virus and influenza virus during mixed infection of MDCK cells. J Virol 40:848–860.

Roth MG, Fitzpatrick JP, Compans RW (1979): Polarity of virus maturation in MDCK cells: Lack of a requirement for glycosylation of viral glycoproteins. Proc Natl Acad Sci USA 76:6430–6434.

Rothman JE, Bursztyn-Pettegrew H, Fine RE (1980): Transport of the membrane glycoprotein of vesicular stomatitis virus to the cell surface in two successive stages. J Cell Biol 86:162–171.

Rothman JE, Fine RE (1980): Coated vesicles transport newly synthesized membrane glycoproteins from endoplasmic reticulum to plasma membrane in two successive stages. Proc Natl Acad Sci USA 77:780–784.

Sabatini DD, Kreibich G, Morimoto M, Adesnik M (1982): Mechanisms for the incorporation of proteins into membranes and organelles. J Cell Biol 92:1–22.

Schmidt MFG, Schlesinger MJ (1979): Fatty-acid binding to vesicular stomatitis virus glycoprotein: A new type of posttranslational modification of the viral glycoproteins. Cell 17:813–819.

Schulman JL, Palese P (1977): Virulence factors of influenza A viruses: WSN virus neuraminidase required for plaque production of MDBK cells. J Virol 24:170–176.

Schwartz IL, Shlatz LJ, Kinne-Saffran E, Kinne R (1974): Target cell polarity and membrane phosphorylation in relation to the mechanism of action of antidiuretic hormone. Proc Natl Acad Sci USA 71:2595–2598.

Shlatz LJ, Schwartz IL, Kinne-Saffran E, Kinne R (1975): Distribution of parathyroid hormone-stimulated adenylate cyclase in plasma membrane of cells of the kidney cortex. J Membr Biol 24:131–144.

Sigrist H, Ronner P, Semenza G (1975): A hydrophobic form of the small intestinal sucrase-isomaltase complex. Biochim Biophys Acta 406:433–436.

Simons K, Garoff H, Helenius A (1980): Alphavirus proteins. In Schlesinger RW (ed): "The Togaviruses Biology, Structure, Replication." New York: Academic Press, pp 317–334.

Smilowitz H (1980): Routes of intracellular transport of acetylcholine receptor and esterase are distinct. Cell 19:237–244.

Staehelin LA, Mukherjee TM, Williams AW (1969): Freeze-etch appearance of the tight junction in the epithelium of small and large intestines of mice. Protoplasm 67:165–184.

Staehelin LA (1974): Structure and function of intercellular junctions. Int Rev Cytol 39:191–283.

Tajiri K, Uchida N, Tanzer ML (1980): Undersulfated proteoglycans are secreted by cultured chondrocytes in the presence of the ionophore monensin. J Biol Chem 365:6036–6039.

Tartakoff AM, Vassalli P (1977): Plasma cell immunoglobulin secretion: Arrest is accompanied

450   Sabatini et al

by alterations of the Golgi complex. J Exp Med 146:1332–1345.

Tokuyasu KT, Singer SJ (1976): Improved procedures for immunoferritin labeling of ultrathin frozen sections. J Cell Biol 71:894–906.

Tokuyasu KT (1978): A study of positive staining of ultrathin frozen sections. J Ultrastruct Res 63:287–307.

Uchida N, Smilowitz H, Tanzer ML (1979): Monovalent ionophores inhibit secretion of procollagen and fibronectin from cultured human fibroblasts. Proc Natl Acad Sci USA 76:1868–1872.

Van Deurs B, Luft JH (1979): Effects of glutaraldehyde fixation on the structure of tight junctions: A quantitative freeze-fracture analysis. J Ultrastruc Res 68:160–172.

Wade JB, Karnovsky MJ (1974): The structure of the zonula occludens: A single fibril model based on freeze fracture. J Cell Biol 60:168–180.

Weigel PH (1980): Rat hepatocytes bind to synthetic galactoside surfaces via a patch of asialoglycoprotein receptors. J Cell Biol 87:855–861.

Yang J, Richards J, Bowman P, Guzman R, Enami J, McCormick K, Hamamoto S, Piterka DR, Nandi S (1979): Sustained growth and three-dimensional organization of primary mammary tumor epithelial cells embedded in collagen gells. Proc Natl Acad Sci USA 76:3401–3405.

Ziomek CA, Schulman S, Edidin M (1980): Redistribution of membrane proteins in isolated mouse epithelial cells. J Cell Biol 86:849–857.

Modern Cell Biology 2:451–481
© 1983 Alan R. Liss, Inc., 150 Fifth Avenue, New York, NY 10011

Transcellular Ion Currents:Signals and Effectors of Cell Polarity

Richard Nuccitelli

From the Zoology Department, University of California, Davis, California 95616

I. INTRODUCTION

The plasma membrane, as diagrammed in Figure 1, contains ion channels and pumps which are responsible for generating the membrane potential found in all cells. Based on the current understanding of membrane fluidity and lateral diffusion rates, one might expect these proteins to be randomly distributed over the plasma membrane as diagrammed in Figure 2. Such a random distribution would result in many highly localized current loops, as carried in this example by Ca^{2+} influx and neighboring K^+ efflux. This current pattern would provide no axis of asymmetry and could contribute little to the cell's overall polarity. However, by simply separating these same channel types in space, or equivalently by separating channels and electrogenic pumps, one could generate a transcellular ion flow as shown in Figure 2. This pattern of ion channels *will* result in an asymmetrical current flow through the cell which might in turn influence the cell's polarity. This chapter will summarize the naturally occurring distribution of steady ion currents in cells and will present the evidence that such currents are more than just signals or indicators of the cell's polarity. They may actually be effectors of polarity.

II. THE DETECTION OF STEADY, TRANSCELLULAR ION CURRENTS

A. Estimates of Possible Transcellular Currents From Ion Tracer Data

All cells use either the Na^+-K^+ ATPase or H^+ ATPase to generate gradients of ion concentration and voltage across their plasma membrane [Bonting, 1970]. Ions will tend to flow down their electrochemical gradients, and

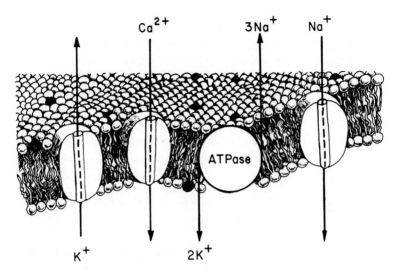

Fig. 1. Sketch of the most common "permeability modifier" proteins in the lipid bilayer plasma membrane.

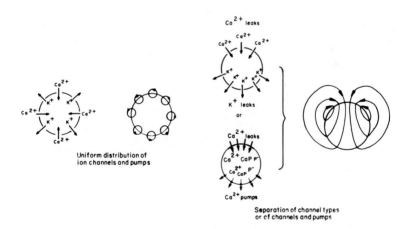

Fig. 2. Possible distributions of ion channels and pumps in a cell's plasma membrane along with the resulting current pattern. Left: A uniform distribution of Ca^{2+} and K^+ channels will result in many highly localized current loops shown on the right. Right: Separation of the Ca^{2+} and K^+ channels (above) or Ca^{2+} channels and pumps (below) will result in the trans-cellular ion current pattern shown on the right.

a steady state will be established in which the number of ions pumped in equals the number leaking out per unit time. These steady-state ion fluxes have been measured for decades by using radioactive ion tracers, and when these measured tracer flux values are combined with the known cell surface area, estimates of the possible transcellular ion currents can be made. For example, in the 100-μm diameter egg of the seaweed Pelvetia, if the total measured K^+ efflux were concentrated in one hemisphere of the egg, one would expect a current density of 6 μA/cm^2 leaving the cell surface [Robinson and Jaffe, 1973]. When the $^{45}Ca^{2+}$ fluxes across this same egg are measured using the ingenious technique of separating the rhizoid and thallus ends of the egg with a nickel screen, a total Ca^{2+} current of 2 pA is found to enter the rhizoid end and leave the thallus end [Robinson and Jaffe, 1975]. This net influx would generate a surface current density at the rhizoid tip of 0.05 μA/cm^2. Therefore, ion tracer data suggests that one might expect to find steady ion currents on the order of 0.1 to 10 μA/cm^2 through single cells. How could such currents be measured directly?

B. The Vibrating Probe Technique for Measuring Steady Transcellular Currents

As diagrammed in Figure 3, such ion currents must flow through the extracellular medium of fixed resistivity. In so doing they will generate a small voltage gradient. By measuring this voltage gradient, one could directly calculate the current density generating it. This can be seen clearly from Ohm's law:

$$I = E/\rho = -(1/\rho)\, \vec{\nabla}\, V$$

$$= -(1/\rho)\, \frac{\partial V}{\partial r}\, \hat{a}_r$$

$$\approx -(1/\rho)\, \frac{\Delta V}{\Delta r}\, \hat{a}_r$$

where I is current density, E is the electric field strength, ρ is the medium resistivity, and ΔV is the voltage gradient measured over the small distance (Δr), and $\hat{a}_r$ is the unit vector in the radial direction. The smaller the Δr used, the better the approximation to the true current density. Therefore, the challenge is to measure extracellular voltage gradients generated by an ion flux of less than 1 μA/cm^2 over very small distances. The magnitude of this challenge can be appreciated by calculating the expected voltage gradient generated over a 20-μm displacement along a radial vector beginning 25 μm outside the cell's plasma membrane: 6 nV in seawater and 30 nV in serum.

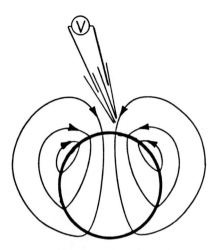

Fig. 3.   Diagram of the transcellular current pattern in a single cell generating an extracellular voltage gradient. Two standard 3-M KCl-filled microelectrodes are positioned outside the cell attempting to measure this voltage gradient over the small displacement necessary to calculate the current density midway between them.

Standard 3M KCl-filled microelectrodes have resistances on the order of 10^6 ohms; their noise places a limit of 10^{-5} volts on their resolution. They are therefore not sensitive enough to detect these steady voltages of 10^{-8} volts. Lionel Jaffe and I overcame this difficulty 9 years ago by developing a much more sensitive electrode system called the vibrating probe [Jaffe and Nuccitelli, 1974]. This was accomplished by lowering the electrode's resistance 1,000-fold (by filling the glass microelectrode with metal and plating a platinum sphere at its tip as shown in Fig. 4) and by signal averaging and filtering using a phase-sensitive lock-in amplifier. The probe impedance is on the order of 10^3 ohms and the system's sensitivity is 10^{-8} volts. These changes resulted in a 100- to 1,000-fold improvement in resolution over the standard microelectrode. This is quite comparable to the increase in resolution obtained when one goes from the light microscope to the electron microscope; with the instrument we can detect the nanovolt gradients generated by the steady ion fluxes flowing through single cells. Moreover, the spatial resolution of the vibrating probe technique is about 10 μm so that the spatial distribution of these membrane ion fluxes can now be studied. A photograph of the vibrating probe is shown in Figure 5. This instrument is now commercially available (Vibrating Probe Co., Davis, California) and is being used in 15 laboratories around the world.

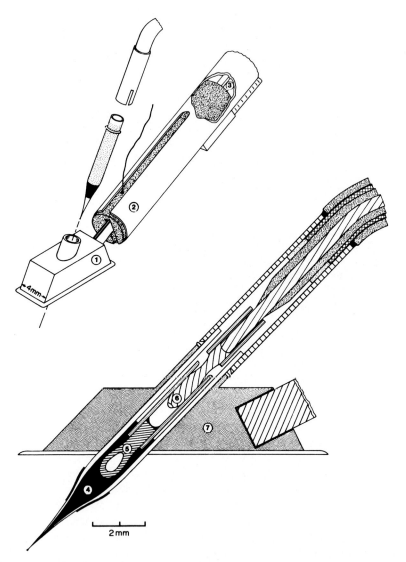

Fig. 4. A scale drawing of the vibrating probe with construction details: 1) lucite boat; 2) 6-gauge stainless steel tube; 3) Bender power cable; 4) Cerrotru solder filling inside of pipette; 5) No. 9 sewing needle eye; 6) gold-plated, braided pin (Malco No. 096-0524-0000, Chicago); 7) lucite boat with coverglass meniscus setter attached to bottom.

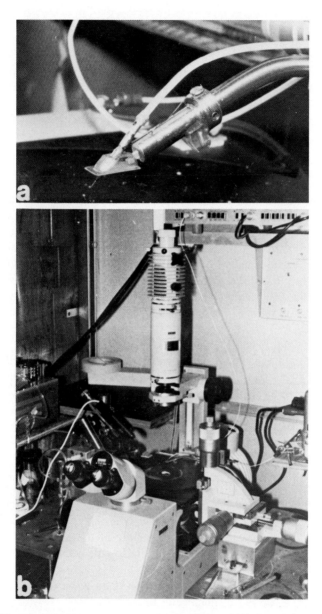

Fig. 5. The vibrating probe. a) Photograph of vibrating probe assembly. A Plezo electric element is housed in the stainless steel tube and attaches to the lucite probe holder to vibrate the probe between the desired two points in the extracellular medium. The coaxial signal cable shown carries the signal from the probe to a lock-in amplifier. b) Photograph of the entire vibrating probe system including the inverted compound microscope, the Line Tool X-Y-Z micropositioner, and the lock-in amplifier used to measure the probe signal.

III. TRANSCELLULAR ION CURRENTS ARE OBSERVED IN CELLS THROUGHOUT THE PLANT AND ANIMAL KINGDOMS

During the past 9 years, a wide variety of cell types in both plants and animals have been studied using the vibrating probe. In every case, steady transcellular ion currents have been found, usually closely correlated with the axis of polarity. It would therefore appear that most cells do not have a uniform distribution of ion channels and pumps, but segregate channel types to varying degrees. Figure 6 summarizes most of the single cell studies to date.

A. Plant Cells

1. Water mold. One of the lowest plant cells to be studied is the water mold, Blastocladiella emersonii [Stump et al, 1980]. In growing cells, positive current on the order of 1 $\mu A/cm^2$ enters the rhizoid and leaves from the thallus. There is some evidence that H^+ carries part of this current. Sporulation is associated with a reversal of this current pattern, and positive current begins to enter the thallus region from which the spores will be released. This cell, therefore, drives a transcellular current through itself along its axis of polarity, and reverses its direction according to the stage of the life cycle.

2. Pelvetia egg. The egg of the brown alga, Pelvetia fastigiata, also drives a current along its axis of polarity [Nuccitelli and Jaffe, 1974, 1975, 1976; Nuccitelli, 1978] and evidence will be presented below that this current, which is carried by Ca^{2+} influx and Cl^- efflux, is a controlling factor in axis determination.

3. Chara internodal cell. The freshwater green alga, Chara corallina, takes in HCO_3^- as its carbon source, fixes the CO_2 and sends out the remaining OH^-. The regions of HCO_3^- influx (outward positive current) and OH^- efflux (inward positive current) are separated to form neighboring bands of current pictured in Figure 6 [Lucas and Nuccitelli, 1980]. Thus, even in this cell with very little growth, the ion channels or transporters are spatially separated. Both the inward and outward currents are light-dependent, falling off within seconds of light removal, and the magnitude of these transport currents is much larger than that of the currents associated with cell elongation, reaching peaks of 80 $\mu A/cm^2$.

4. Barley roots. Growing roots and root hairs of barley seedlings, Hordeum vulgare L, drive steady currents through themselves. Current of about 2 $\mu A/cm^2$ in magnitude enters both the main elongation zone of the root and the growing tips of elongating root hairs [Weisenseel et al, 1979]. Both the inward and outward currents appear to be carried largely by H^+, OH^-, or HCO_3^-.

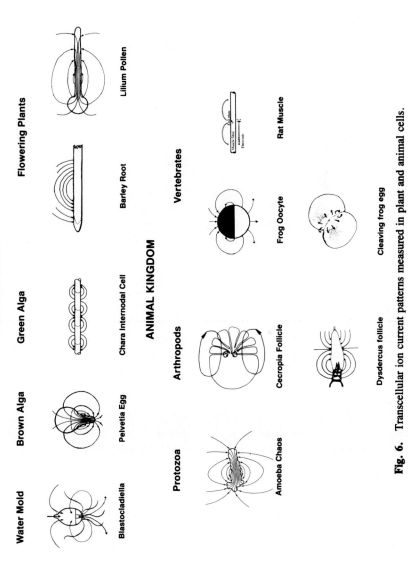

Fig. 6. Transcellular ion current patterns measured in plant and animal cells.

5. Easter lily pollen grain. The higher flowering plant, Lilium longiflorum, exhibits transcellular currents in its germinating pollen grain [Weisenseel et al, 1975; Weisenseel and Jaffe, 1976]. Positive current of about 4 $\mu A/cm^2$ enters the ungerminated grain's prospective growth site and leaves its opposite end. After the grain germinates to form the pollen tube, current enters along most of the tube and leaves the grain as diagrammed in Figure 6. This is another example of a tip-growing plant cell which drives an ion current into the region of vesicle secretion and along its axis of polarity.

B. Animal Cells

Transcellular ion currents have been measured in a wide range of animal cells as well. As in most of the plant cells mentioned, the current pattern is closely correlated with the axis of polarity.

1. Amoebae. The amoeba, Chaos chaos, drives a steady current of 0.1 $\mu A/cm^2$ into its tail or uroid zone and out its pseudopods [Nuccitelli et al, 1977]. This current pattern changes as the amoeba's morphology changes, and there is a striking correlation between the region of highest inward current density and tail formation. Spontaneous cytoplasmic streaming polarity reversals occur frequently in Chaos, and these are always *preceded* by a change in the transcellular current such that the region of largest inward current becomes the new tail. Ca^{2+} influx appears to carry some of the inward current.

2. Silk moth follicle. The nurse cell–oocyte complex of the silk moth, Hyalophora cecropia, drives an ion current into the nurse cell end and out the oocyte end [Woodruff and Telfer, 1974; Jaffe and Woodruff, 1979]. This current is important for the polarized transport of protein along the cytoplasmic bridge connecting the nurse cells with the oocyte [Woodruff and Telfer, 1973], and there is strong evidence discussed in detail below that this current drives the electrophoretic transport of protein from nurse cells to oocyte.

3. Cotton bug ovariole. The morphology of the ovariole of the cotton bug, Dysdercus intermedius, is quite different from the silk moth ovariole because nurse strands between the trophic syncytium and oocytes can measure several millimeters in length. The basic current pattern, however, is analogous to that found in Cecropia with current entering over the tropharium and leaving the small follicles [Dittman et al, 1981]. This current is also about an order of magnitude larger than that measured in Cecropia. In another telotrophic hemipteran, Rhodnius prolixus, an intercellular voltage gradient of up to 10 mV with the oocyte more positive has been measured [Telfer et al, 1981]. The magnitude of this gradient is very sensitive to juvenile hormone. In these three insect systems, the transcellular current pattern is strongly correlated with the axis of polarity and may be directly involved in the polarized transport of protein between cells.

4. Frog oocyte. The immature oocyte of the frog, Xenopus laevis, drives $1 \mu A/cm^2$ into the animal hemisphere and out of the vegetal hemisphere prior to complete maturation [Robinson, 1979]. Here, again, the current pattern is along the main axis of polarity and the inward current exhibits Ca^{2+} influx and Cl^- efflux components. This current pattern is found in all oocytes; however, when progesterone or any of several other maturation-producing agents are applied, the current decreases to nearly zero. Thus, the mature egg appears to be in a more quiescent state waiting for fertilization to reactivate it.

5. Cleaving frog egg. The mature, unfertilized Xenopus egg has no detectable transcellular current until fertilization triggers an inward current pulse which enters the egg at the site of the sperm-egg fusion and spreads over the egg in a ring-shaped wave during about 3 min after fertilization. The fertilized egg then becomes quiescent until first cleavage. About 10 min after first cleavage has begun, new, unpigmented membrane begins to appear in the cleavage furrow. At this same time, outward current carried by K^+ efflux is measured near this unpigmented membrane [Kline et al, 1983]. From this point on through at least the 32-cell stage, each blastomere drives a current through itself, into the old membrane and out of the newly inserted membrane. These blastomeres form an epithelium with the blastocoel on one side and the external medium on the other side, and the outward current region forms the basolateral margins of this epithelial layer. Therefore, an apical-basal transcellular current is associated with the apical-basal polarity of these epithelial cells. Jaffe [1981b] has proposed a model of epithelial cell polarity control by transcellular ion currents.

6. Mammalian muscle fiber. The rat skeletal muscle fiber drives a steady current of $10 \mu A/cm^2$ out of the end plate region and into either side of the end plate [Betz et al, 1980]. This current was initially thought to represent the electrogenic component of the Na^+-K^+ ATPase, but more recent studies suggest that it is mainly due to a Cl^- leak (Betz and Caldwell, personal communication). When acetylcholine is added extracellularly, this outward end plate current immediately reverses and increases dramatically as Na^+ rushes in through the open acetylcholine receptor ion channels. Therefore, even in the fully differentiated muscle fiber, there is a segregation of ion channels leading to a steady transcellular current which may act as a guiding factor in muscle innervation.

These investigations of the steady ion current patterns mainly around single cells have led to two generalizations: 1) Ion channels are generally not uniformly distributed in the cell's plasma membrane, but are spatially separated resulting in transcellular ion currents. 2) These ion current patterns are often closely correlated with the axis of cell polarity in both plant and animal cells. Based on these steady ion current measurements made in a wide variety of cell types from both lower and higher plants and animals, it is clear that these transcellular ion currents are *signals* of cell polarity. However, the far more interesting question is whether the currents can affect, influence, or control cell polarity. Does the plasma membrane directly influence cell function by separating ion channel types and driving these ion currents through cells?

IV. DO TRANSCELLULAR ION CURRENTS PLAY A CAUSAL ROLE AS EFFECTORS OF CELL POLARITY?

Transcellular ion currents can have the most direct effect on the cell by either changing local ion concentrations or by generating voltage gradients. Furthermore, these effects could act on the inside or outside of the cell's plasma membrane. The remainder of this review will concentrate on a few well-studied cases which indicate that the transcellular currents do affect cell polarity. See also Jaffe [1979, 1981a, b].

A. Intracellular Ion Concentration Gradients

Tip-growing plant cells provide the strongest evidence that steady transcellular currents can generate ion concentration gradients which, in turn, can influence the cell's polarity. The most extensively studied system is the germinating egg of the brown alga, Pelvetia, and this current pattern is quite similar to that found in the germinating pollen grain of the higher plant, the Easter lily.

The Pelvetia egg has no predetermined axis of polarity and can germinate a rhizoid at any point on its surface. This process involves a localized secretion of wall-softening enzymes and an increase in turgor pressure to five atmospheres, accomplished by pumping K^+ and Cl^- into the cell. The cell then bulges (germinates) at the weakest region where wall softening has occurred, and the originally apolar zygote becomes differentiated, first into two different regions and then into cells. The establishment of this axis of secretion can be influenced by a variety of environmental vectors including light, temperature, and pH [Jaffe, 1968, Table I]. The response to unilateral light is to germinate on the dark side. This would be useful in nature since the rhizoid outgrowth will form the holdfast for the plant and should form opposite the sun. One can use this light response to orient the germination site while studying the transcellular ion current pattern associated with it using the vibrating probe as shown in Figure 7a [Nuccitelli, 1978]. Currents are detected around this egg as early as 30 min after fertilization and tend to enter at the dark hemisphere. The early spatial current pattern is unstable and shifts position, often with more than one inward current region. However, current enters mainly on the side where germination will occur and is usually

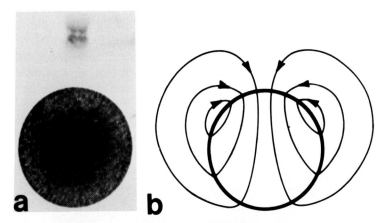

Fig. 7.   Current measurements around the Pelvetia egg. a) Photomicrograph of the vibrating probe near a 3-hour-old Pelvetia egg. The egg's diameter is 100 μm. b) The electrical current pattern generated by the germinating Pelvetia egg just before germination, which will occur at the top of the figure.

largest at the prospective cortical clearing region where the rhizoid forms. The current pattern observed during the 2-hour period prior to germination is more stable and is shown in Figure 7b. The side of inward current always predicts the germination site, even when the axis is reversed by light direction reversal.

In order to understand how this current might be affecting the egg, one must first know what ions are carrying the current. Therefore, these current measurements were next carried out in artificial seawaters in which various ion substitutions were made. These studies indicated that Ca^{2+} influx and Cl^- efflux carried the inward current, with K^+ efflux as the probable outward current carrier. As mentioned earlier, Robinson and Jaffe [1975] directly measured the $^{45}Ca^{2+}$ tracer flux through the egg by separating the rhizoid and thallus ends by fitting eggs tightly into holes in a nickel screen, and using light to define the polarity of development. They measured 2 picoamps of Ca^{2+} current per egg. This means that about 10% of the total transcellular current is carried by Ca^{2+} influx at this stage. While only a small fraction of the current is due to Ca^{2+} influx, the subsequent intracellular free ion concentration change is certainly much larger for Ca^{2+} than for the major current carrier, Cl^-. This is because the free Ca^{2+} concentration in all cells is extremely low (about 10^{-7} M) compared to Cl_i^-, so that even a small Ca^{2+} influx will result in a very large concentration change. A first attempt to look for the Ca^{2+} gradient expected from such local changes was made by John Gilkey while he was in Lionel Jaffe's lab, using low-temperature autoradiography. Figure 8 shows the results of this study in germinated embryos indicating higher Ca^{2+} concentrations at the rhizoid tip than at the thallus end. This technique does not distinguish between free and bound Ca^{2+}, but since the overall Ca^{2+} concentration is higher at the rhizoid end, it is likely that the free concentration is higher there as well. It must also be noted that efforts to detect a Ca^{2+} gradient in the ungerminated egg have so far been unsuccessful.

These results establish the existence of a plasma membrane-driven current, partially carried by Ca^{2+}, entering the prospective germination region and

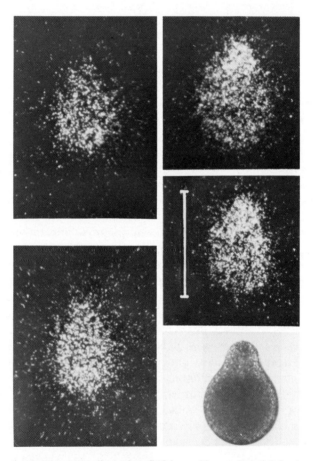

Fig. 8. Low-temperature autoradiographs of 18-hour-old, germinated Pelvetia eggs. Made from deep-frozen median sections of an egg similar to the one shown on the lower right, left column: Grown in seawater labeled with $^{36}Cl^-$. Right column: Grown in seawater labeled with $^{45}Ca^{2+}$. Bar in center autoradiograph: 100 μm. Courtesy of John Gilkey and Lionel Jaffe.

generating a Ca^{2+} gradient within the cytoplasm. If such a Ca^{2+} gradient were important for the establishment of an axis of polarity, manipulations of the gradient should affect cell polarization. The best test for causality is to impose a Ca^{2+} gradient on the ungerminated embryo. This critical experiment was done by Robinson and Cone [1980] using a gradient of the calcium ionophore, A23187. They found that 60–80% of the eggs lying within one egg diameter of a glass fiber coated with ionophore formed a rhizoid on the hemisphere nearer the fiber. This is strong evidence that imposed Ca^{2+} gradients can orient the axis of polarity and supports the hypothesis that the natural transcellular Ca^{2+} current is an effector of cell polarity in the Pelvetia egg. Moreover, a wide variety of other tip-growing plant cells exhibit high Ca^{2+} levels at their growing tips [Jaffe et al, 1975; Reiss and Herth, 1978, 1979] and some drive an ion current into that region [Weisenseel et al, 1975, 1979].

The germinating pollen grain of Lilium longiflorum is the best example of a higher plant cell which exhibits many of the same membrane properties found in Pelvetia. The hydrated, ungerminated pollen grain drives a similar current through itself, the entry region of which accurately predicts the site of tube outgrowth [Weisenseel et al, 1975]. The elongating tube also shows a gradient of total Ca^{2+}, with Ca^{2+} higher in the tip region according to both low temperature autoradiography [Jaffe et al, 1975] and chlorotetracycline fluorescence [Reiss and Herth, 1978]. Therefore, it would appear that in lower and higher plants the plasma membrane is involved in the control of axis establishment by determining the site of Ca^{2+} influx and the transcellular current pattern.

Animal cell polarity can also be influenced by Ca^{2+} gradients. Recent work by William Jeffery [1982] indicates that the "orange crescent" in the egg of the tunicate, Boltenia villosa, forms on the high side of a calcium ionophore gradient formed with the glass fiber technique. This segregation of pigment and mitochondria normally marks the future dorsal side of the embryo. During normal development the two are further segregated into the muscle cells of the tadpole larva. However, when the egg is activated with ionophore by placing it in the appropriate gradient, 82% of the activated eggs formed orange crescents and mitochondria localizations with midpoints 45° or less from the point nearest the ionophore-coated fiber. Moreover, 10% of the eggs positioned between two coated fibers formed two orange crescents, each about half the normal size. This supports the hypothesis that there is no predetermined polarity of cytoplasmic localization in the ascidian egg, and suggests that cytoplasmic Ca^{2+} gradients can strongly influence this axis of polarity.

B. Transcellular Ion Currents Generate Intracellular Voltage Gradients That Can Polarize the Distribution of Cytoplasmic Components

The second way that transcellular ion currents could directly affect cell function is through the intracellular voltage gradient which they might generate as they traverse the cytoplasm. In fact, there is a well-documented example of this very phenomenon in the insect oöcyte-nurse cell syncytium which has been elegantly elucidated by Richard Woodruff and William Telfer (1973, 1974, 1980).

1. The Cecropia moth follicle. In many insects, the oöcyte is connected to several nurse cells by cytoplasmic bridges which are formed by incomplete cytokinesis at cell division. A diagram of this oöcyte-nurse cell complex or follicle is shown in Fig. 9. The main role of the nurse cells is to synthesize protein and RNA and to transport it into the larger oöcyte. This transport occurs through the cytoplasmic bridges and is known to be unidirectional, i.e., in a variety of insect types, RNA and protein is only observed to flow from the nurse cells into the oöcyte and never in the opposite direction (Woodruff and Telfer, 1973). In their study of this polarized intercellular transport, Woodruff and Telfer measured the membrane potential of both the oöcyte and nurse cells and found a surprising 10 mV difference between them with the oöcyte more positive than the nurse cells (Woodruff and Telfer, 1974). This was surprising because these cells have conductive interiors and are joined by a continuous strand of cytoplasm. Such a conductor would be expected to have the same voltage throughout unless there were a steady current flowing through it. Such a current passing across the 35 μm wide intercellular bridges which have a resistance of 13 KΩ could generate a voltage difference between the two cell groups. These investigators studied this possibility using two independent techniques and found that the follicle indeed drives a current through itself. The first method used was to increase

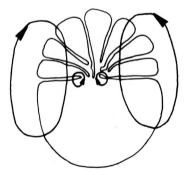

Fig. 9. Diagram of the oocyte–nurse cell syncytium of the Cecropia moth. Three cytoplasmic bridges are evident and the transcellular ion current pattern is drawn in.

the extracellular resistance by drawing the follicle into a tight-fitting capillary so that a larger extracellular voltage would be generated by the same transcellular current. The second method was to map the extracellular current pattern with the vibrating probe in collaboration with Lionel Jaffe [Jaffe and Woodruff, 1979]. Both techniques detected current entering the nurse cell cap and leaving the oocyte's vegetal end as diagrammed in Figure 9. Up to 20 $\mu A/cm^2$ enters the anterior end of the nurse cells, giving a total transfollicular current of about 100 nA. However, the measured intercellular voltage gradient implies a very large back current (on the order of 1,000 nA) across the cytoplasmic bridge from oocyte to nurse cells passing in the direction opposite to that expected from the extracellular current pattern. One possible explanation that has been proposed for this large cytoplasmic bridge back current [Jaffe and Nuccitelli, 1977] is that it is driven by the furrow membranes. This would naturally occur if there were an animal-vegetal polarity in each of the nurse cells and in the oocyte such that the inward current ion channels were located in the animal hemisphere and the outward current channels or pumps were located in the vegetal region of each cell. This would result in the juxtaposition of outward current nurse cell membrane with inward current oocyte membrane at the furrow to drive current across the furrow from nurse cell to oocyte. This current would then leak back through the path of least resistance, which is through the cytoplasmic bridge nearby, as shown by the inner current loops in Figure 9. This furrow membrane battery hypothesis has not yet been tested experimentally.

If an intercellular voltage gradient is involved in the polarized transport of protein, one should be able to demonstrate that the transport is dependent on the protein's net electrical charge. Woodruff and Telfer [1980] showed this by microinjecting fluorescently labeled proteins with different pKs into either the nurse cells or the oocyte and following their movement across the cytoplasmic bridge. Injected electronegative protein, serum globulin, moves only from nurse cell to oocyte, whereas electropositive lysozyme (pK 11.5) only moves from oocyte to nurse cell. Two nearly neutral proteins, hemoglobin and myoglobin, were able to move in both directions across the bridges. These observations suggest that the protein's charge is the main factor determining the direction of intercellular transport. Finally, this conclusion was even more convincingly supported by their demonstration that by simply reversing the charge on lysozyme by methylcarboxylation they could reverse its transport polarity. As shown in Figure 10, essentially the same protein which now has a negative charge will move only from nurse cell to oocyte. These data provide strong evidence in support of an intercellular electrophoresis transport mechanism in the developing Cecropia follicle—ie, a polarized transport of cytoplasmic components which is driven by transcellular ion currents.

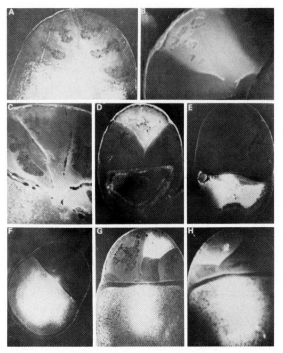

Fig. 10. Fluorescence photomicrographs of follicles injected with proteins labeled with fluorescein isothiocyanate FITC by the method of Nairn. Microinjection was carried out hydrostatically as described earlier. The follicles were incubated in blood from the animal diluted 1:2 with dissecting solution. In dilutions as great as 1:10, blood protein concentration is still sufficient to maintain membrane permeability and steady-state potentials identical to those that occur in whole blood. Following injection and 1–2 h incubation, follicles were fixed either by freeze substitution (A, C) or in Carnoy's acid-alcohol (B, D–H), embedded in paraffin, sectioned, and examined with a fluorescence microscope. Photographs were taken with Kodak Tri-X pan film (ASA 400). A, C) Fluorescent lysozyme (FLy) microinjected into oöcytes; B, D) FLy microinjected into nurse cells. Note in (B) that protein has entered a cobridged nurse cell (see Fig. 1), but failed to cross into the oocyte. E, F) Methylcarboxylated FLy (MCFLy) microinjected into oocytes gave no detectable fluorescence in the nurse cells. G, H) MCFLy microinjected into nurse cells. Magnification: in all cases shown here, the width of the nurse cell cap adjacent to the oocyte was about 450 μm. Courtesy of Woodruff and Telfer [1980], with permission. Copyright © 1980 Macmillan Journals Limited.

2. Elongating fungal hyphae. Twenty years ago Slayman and Slayman [1962] reported a 100 mV voltage gradient along Neurospora hyphae with the elongating tip more positive than the center of the hypha. Since the membrane potential was smaller at the relatively fine cellular tips where electrode penetration injury is more likely, it was difficult to know how much of the measured potential difference was natural and how much was due to a lower membrane potential measurement at the tip where the injury leak could be greater. Nevertheless, recent vibrating probe measurements by Darryl Kropf in Frank Harold's lab indicate that current enters the elongating tip of another fungus, Achlya bisexualis. Perhaps this transcellular current is involved in the generation of the intracellular voltage gradient which could, in turn, electrophorese negatively charged secretory vesicles to the elongating tip where they are secreted.

It should be pointed out here that the intracellular voltage gradient, which a given transcellular current generates, depends very much on the ion species carrying the current. For ions that exhibit little tendency to bind to fixed charges in the cytoplasm (such as K^+ and Cl^-), the voltage generated can be calculated from Ohm's Law. However, ions that bind to fixed charges (such as Ca^{2+} and H^+), will create fixed charge gradients which can generate larger voltage gradients than expected by Ohm's Law [Jaffe et al, 1974]. This is due to the Donnan Potential established by the displacement of K^+ to the end opposite the Ca^{2+} leak. The high K^+ at the low Ca^{2+} end will set up a balancing potential difference across the cytoplasm since it must be at a uniform electrochemical potential. Therefore, the Ca^{2+} leak end of the cytoplasm will become electropositive with respect to the opposite end, and this Donnan Potential can contribute substantially to the intracellular voltage gradient. For example, in the 100-μm diameter Pelvetia egg, 1 μA/cm^2 traversing a 200 Ω-cm cytoplasmic resistivity would generate an ohmic voltage difference of 2 μV across the cytoplasm, but the field due to the fixed charge gradient could be as large as 2 mV [Jaffe et al, 1974].

3. Other intracellular voltage gradients. Three other reports of intracellular voltage gradients have been made on systems that have not all been investigated for transcellular ion currents. The first is a study done on one of Keith Porter's favorite cells, the fish melanophore. This cell's main function is to disperse pigment granules throughout its processes to spread the color, or to aggregate the granules in the centrosphere minimizing the colored area. Using a single microelectrode, Kinosita [1953] found that in the aggregated state the centrosphere was on the average 12 mV more positive than the projections, whereas in the dispersed state it averaged 3 mV more negative. In order to study this apparent internal voltage gradient more directly, Kinosita then inserted two microelectrodes into the melanophore—one into the centrosphere and the other into a thick projection. With this method he could detect only a maximum of 3 mV between the two regions; however, when he stimulated alternating aggregation and dispersion, this voltage difference alternated, changing before the pigment granules moved. The potential differences were smaller than expected from earlier single electrode measurements (probably because of damage caused by impalement with two 2-μm electrodes); nevertheless, the potential changes proceeded granule movement and may be part of the signaling mechanism determining the polarity of granule transport.

The second report of an intracellular voltage gradient is in the epithelial cells of Necturus gallbladder. Zeuthen [1977] has measured a gradient of 0.6 mV/μm across the cell cytoplasm using both single microelectrodes advanced in small steps through the cell and multibarreled electrodes whose tips were staggered in depth. The potential difference across the apical membrane was always about 20 mV smaller than that measured across the basal membrane, and the same potential profile was found when penetrating the epithelium from the serosal side. Furthermore, measurements with multibarreled electrodes whose tips were staggered in depth gave roughly the same internal potential gradient. These epithelial cells are known to drive Na^+ across themselves (Zeuthen [1978] has also measured an intracellular Na^+ gradient), and this ion transport may be involved in generating the intracellular voltage gradient. No extracellular current measurements have been made on this epithelium, but Foskett and Scheffey [1982] have mapped the current pattern outside of a teleost gill epithelium and measure very large negative currents of 400 μA/cm^2 leaving the chloride cells on the apical side. Therefore, it is possible that transepithelial currents may be involved in generating apical-basal intracellular voltage gradients.

A third cell type which very likely generates an intracellular voltage gradient is the retinal rod. Hagins et al [1970] have measured a steady dark current of 70 pA which enters the outer segment, crosses the thin neck connecting the outer and inner segments, and leaves the inner segment. They estimate that this current might generate a 2 mV cytoplasmic voltage drop across the neck region.

C. Transcellular Ion Currents Generate Extracellular Voltage Gradients That Can Polarize the Distribution of Membrane Proteins by Lateral Electrophoresis

1. Redistribution of membrane proteins by weak external electric fields. Many of the externally protruding plasma membrane glycoproteins bear a high negative charge density [Abramson et al, 1942; Ambrose, 1965]. If the lateral mobility and charge heterogeneity among these membrane proteins are sufficient, a steady electric field should redistribute them within the plane of the plasma membrane by lateral electrophoresis. Lionel Jaffe intro-

duced this idea and calculated that rather small extracellular electric fields could polarize the membrane protein distribution [Jaffe, 1977]. For example, a population of 10-μm membrane particles would become one-tenth to one-half polarized by a steady voltage drop of only 0.8–4.0 mV across the cell in 3 hours.

This theoretical prediction was first confirmed by Poo and Robinson [1977] using Xenopus muscle cells in culture. They showed that the Con A receptor distribution could be grossly polarized within 4 hours in a 400 mV/mm field (1.2 mV/cell), independent of cell metabolism. Since then, lateral electrophoresis has also been observed in a wide variety of cells including frog neurons [Patel and Poo, 1982], mouse macrophages [Orida and Feldman, 1982], mouse fibroblasts [Zagyansky and Jard, 1979], and sea urchin eggs [Robinson, 1977]. Other membrane glycoproteins that have been redistributed by the imposed fields include acetylcholine receptors [Orida and Poo, 1978, 1980; Poo et al, 1979], Phaseolus vulgaris lectin receptors [Orida and Feldman, 1982], and ricin receptors [Zagyansky and Jard, 1979; see review by Poo, 1981]. Imposed fields also redistribute freeze-fracture membrane particles in Micrasterias [Brower and Giddings, 1980]. Surprisingly, the Con A and acetylcholine receptors and freeze-fracture particles move in the membrane toward the negative pole, opposite to the expected direction for negatively charged glycoproteins. Jaffe has suggested that the explanation of this important anomaly lay in electro-osmosis. Since the negatively charged membrane proteins will have positive counterions in solution which are in turn surrounded by waters of hydration, the positive counterions may pull the water, which in turn hydrodynamically drags the membrane proteins toward the negative pole. This concept has recently been tested by McLaughlin and Poo [1981], who showed that the direction of Con A receptor migration in an imposed field could be reversed by changing the cell's surface charge, using a positively charged lipid (3,3′ dioctadecylindocarbocyanine iodide) or by cleaving sialic acid residues with neuraminidase. Both of these treatments reduce the net negative surface charge, reduce the zeta potential, and decrease the electro-osmotic flow parallel to the surface. Therefore, it is quite likely that electro-osmosis is responsible for moving the negatively charged glycoproteins to the negative pole.

2. Effects of weak electric fields on cell growth. There is little doubt that weak electric fields on the order of 1 mV per cell diameter can redistribute membrane proteins. However, does this membrane polarization influence cell growth or behavior? The answer to this question is a definite yes. In a previous review, several plant and animal cell types exhibiting morphogenetic responses to weak external fields (0.1–6 mV/cell) were listed [Jaffe and Nuccitelli, 1977, Tables 6 and 7]. Since then, four exciting reports of nerve galvanotropism have appeared [Jaffe and Poo, 1979; Hinkle et al, 1981;

Freeman et al, 1981; Patel and Poo, 1982]. All of these report enhanced growth toward the cathode and reduced growth toward the anode in small steady fields using a variety of neurons (chick dorsal root ganglia, frog neural tube, and goldfish retina). Growing neurites were observed to bend toward the cathode as shown in Figure 11 with an extremely low threshold of 7 mV/ mm (about 0.4 mV/cell) over a 16- to 20-hour period [Hinkle et al, 1981]. The possible targets for such fields could be cytoplasmic, transmembrane, or perimembrane in nature, but the first two are less likely. The cytoplasmic voltage gradient across the interior of a 70 μm growth cone would be 10^{-5}–10^{-3} times the total gradient; since most of the resistance is across the cell membrane [Jaffe and Nuccitelli, 1977], one would expect only 0.004–0.4 μV across the cytoplasm. Given a typical cytoplasmic electrophoretic mobility of 1μm/s per V/cm, cytoplasmic components would move at the most 0.2μ/ h, and would take 50 hours to cross a 10-μm growth cone!

The effect of these low fields on the transmembrane voltage or membrane potential is also very small. The typical neuron membrane potential is −70 to −90 mV, so the threshold field would perturb this by less than 1%. This is not likely to change ion transmembrane driving forces sufficiently to produce the reported polarization response 36% above controls [Hinkle et al, 1982; Jaffe and Nuccitelli, 1977].

The third possibility, perimembrane or lateral electrophoresis of membrane proteins, is the most likely target of these small fields. In fact, Patel and Poo [1982] showed that Con A receptors of Xenopus neurons in culture accumulated on the cathodal side of the cell after 6 hours in a 500 mV/mm field (0.7 mV/cell). Furthermore, incubation in 40 μg/ml Con A during field application blocked Con A receptor migration and also blocked the galvanotropic response completely. This suggests that lateral membrane glycoprotein movements are important for galvanotropism, and supports the contention that the mechanism of action of these low fields is by lateral electrophoresis.

3. Effects of weak electric fields on cell motility. Over the past century, galvanotaxis has been observed in a wide variety of organisms. Recent studies using small, physiological electric fields have revealed a remarkable sensitivity to these fields in three different cell types: macrophages, fibroblasts, and neural crest cells.

Norman Orida [1982] has recently studied pseudopod protrusion in mouse peritoneal macrophages in electric fields. He found that in fields greater than 400 mV/mm (about 10 mV/cell), pseudopodial protrusive activity occurred toward the positive pole of the electric field. This response was dependent on extracellular Ca^{2+} and could be blocked by Ca^{2+} influx inhibitors. He also demonstrated that both Con A and Phaseolus vulgaris lectin receptors were redistributed by the imposed field to opposite ends of the cell. Therefore,

+

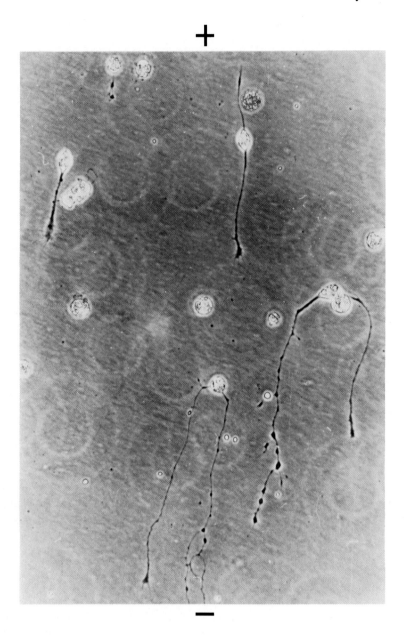

—

Fig. 11. Xenopus neurons in culture in the presence of a steady electrical field of 170 mV/ mm. The poles of the field are indicated by +, −. The neuron at the top, right with a single neurite is 100 μm long. Courtesy of Robinson [1981], with permission. Copyright © 1981, Cambridge University Press.

lateral electrophoresis is certainly occurring at these field strengths and may be involved in the galvanotaxis response.

Carol Erickson and I have been studying galvanotaxis in quail somite fibroblasts and neural crest cells. These cells exhibit a two-stage response to the steady electric field. Within 10 min they begin to align with their long axes perpendicular to the field lines as shown in Figure 12. Then they begin migrating toward the cathode. The threshold voltage for the orientation response appears to be about 150 mV/mm, which would impose 4 to 12 mV across these cells which typically measure 20 μm by 80 μm. Neural crest cells exhibit a similar response and are shown in Figure 13. Two other laboratories have recently found that amphibian neural crest cells also orient perpendicular to the field lines and move toward the cathode (Stump and Robinson, 1982; Cooper and Keller, 1982). Another cell type, the embryonic Xenopus muscle cell, also elongates perpendicular to the field lines [Hinkle et al, 1981]. Therefore, this particular galvanotactic response appears to be relatively common and merits further investigation. There has also been one recent report that chick mesoderm cells display no galvanotaxis [Stern, 1981]. However, the deep chamber used (1 mm) is likely to result in cell heating during field application due to the larger current required and, furthermore, the cells did not appear to be spread and healthy in Figure 4 of that paper.

The important questions remaining are whether or not these various cell types encounter such steady electric fields during development in vivo, and if so, are the fields in the appropriate direction to account for the observed cell migratory behavior?

4. Ion currents and voltages measured in embryos. The electrical fields outside developing chick embryos at the primitive streak stages 3 to 5 have been studied by Jaffe and Stern [1979]. They found that large current densities of the order of 100 μA/cm^2 leave the whole streak and return elsewhere through the epiblast. This current is probably pumped into the intraembryonic space by the epiblast and then leaks out of the streak because it is a zone of junctional disruption. It will certainly generate a voltage gradient across some ingressing cells, and the magnitude of this gradient will depend on the resistivity of the embryonic spaces through which the current flows. The possibility exists that this current might exert a galvanotactic influence on cell migration.

Similar large current densities leave the neural tube in Xenopus embryos (Robinson, personal communication). Again, these currents appear to be driven by a pumping epithelium. The potential difference across the skin of the adult frog has been extensively studied [Lindemann and Voute, 1976], and recently McCaig and Robinson [1982] have shown that this transepi-

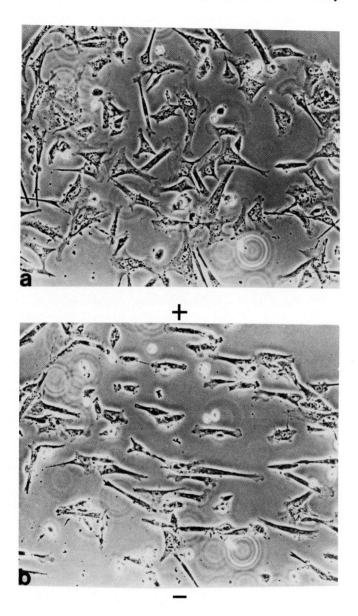

+

−

Fig. 12. Galvanotaxis of quail somite fibroblasts in culture on a glass substratum. a) Fibroblasts that have been in culture for 48 h. b) Same fibroblasts as in (a) after 41 min of 500 mV/mm steady field application. Poles of the field are marked by +, −. The average cell size is 20 μm by 80 μm.

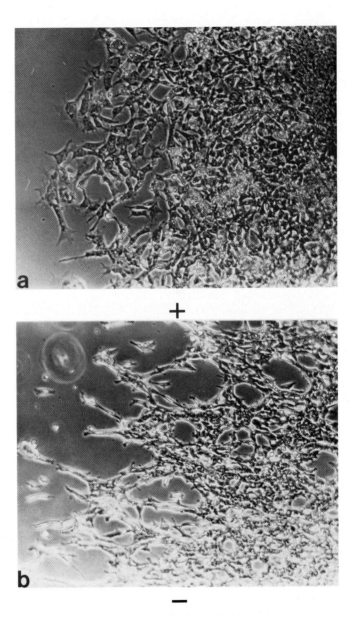

+

—

Fig. 13.   Galvanotaxis of quail neural crest cells in culture on a glass substratum. a) Neural crest cells that have been in culture for 48 h. b) Same cells as in (a) after 90 min of 500 mV/ mm steady electric field application. Poles of the field are indicated by +, −.

dermal potential difference first begins to appear at neurulation. However, no gradients in skin potential within the embryo which might influence nerve or muscle growth have thus far been detected.

5. Ion currents and voltages measured in regenerating systems. The regenerating systems that have been studied also drive currents through themselves using epithelial pumps. Long ago, Monroy [1941] discovered that the distal part of an amphibian limb is positive with respect to the proximal part, and after amputation this potential difference becomes larger. Recent measurements of current densities around regenerating limbs by Borgens et al [1977b, 1979a, b] have confirmed Monroy's observation and indicate that a positive current of 10–100 μA/cm^2 leaves the cut surface of the regenerating limb stump. These large currents persist for up to 2 weeks. They will produce a significant voltage gradient along the inner limb tissue which could exert a galvanotropic influence on the nerve cells in that region. In fact, Borgens et al [1977a, 1979c, d] have shown that when they supplement the natural currents in frog regenerating limbs with an imposed current, they greatly enhance the quantity of nerve which grows into the cathodically stimulated regenerates. Since we know that nerve growth is enhanced toward the cathode in culture, it is quite likely that these imposed fields in vivo are also exerting a direct galvanotropic influence on the limb nerve.

Another system in which imposed voltages enhance nerve growth is the regenerating spinal cord of the lamprey [Borgens et al, 1981]. When 10 μA is imposed across the completely severed spinal cord for 5 days, enhanced regeneration of reticulospinal neurons is observed. As in the frog limb, these imposed currents correspond in direction to the natural extracellular voltage gradient produced by injury-induced current which enters the cut face of the proximal cord segment [Borgens et al, 1980]. Therefore, this constitutes another example of in vivo nerve galvanotropism.

The most direct measurements of a tissue voltage gradient in vivo have recently been made beneath guinea pig skin during wound healing [Barker et al, 1982]. When an incision is made through the glabrous epidermis of the guinea pig, a microampere flows through each millimeter of the cut's edge. These wound currents generate lateral, intraepidermal voltage gradients of about 100–200 mV/mm near the cut. Such gradients are clearly large enough to exert a galvanotactic influence on epidermal cells and may help stimulate wound closure.

In summary, developing and regenerating embryos do exhibit transepidermal currents and voltages which might be used to influence cell movements through galvanotaxis. We do not yet have sufficiently detailed information regarding the spatial and temporal pattern of these embryonic currents to answer the question of whether the embryonic electric fields can account for observed cell migrating behaviors during organogenesis. This is an important area for future study.

V. SUMMARY

This review has focused on the evidence that transcellular ion currents are commonly found in cells, and that the spatial pattern of these currents is intimately correlated with the cell's polarity. With the advent of the vibrating probe technique in 1974, it has become possible to detect the steady transcellular flow resulting from the separation of ion channel types and ion pumps in plant and animal cell plasma membranes. It appears that most cells do not have a uniform distribution of plasma membrane ion channels and pumps, but instead separate these transport sites, generating the transcellular currents. In most cases, these currents are signals or indicators of the cell's polarity, and in some cases there is evidence that these currents play a causal role in determining cell polarity.

There are three ways in which these transcellular ion currents can have the most direct affect on the cell: 1) by changing local intracellular ion concentrations; 2) by generating cytoplasmic voltage gradients; and 3) by generating extracellular voltage gradients. One or two examples of each of these effects is discussed in detail.

The tip-growing plant cells, Pelvetia and lily pollen, provide the strongest evidence that steady transcellular currents can generate ion concentration gradients which, in turn, influence the cell's polarity. Both of these cells drive a current carried partly by Ca^{2+} into their growing tips, and both exhibit intracellular Ca^{2+} concentration gradients. Furthermore, the Pelvetia egg's polarity can be manipulated by imposed Ca^{2+} ionophore gradients, suggesting that the Ca^{2+} gradient plays a causal role in polar axis establishment. The same Ca^{2+} ionophore gradient can also influence tunicate egg polarity, suggesting that animal cells may also use a Ca^{2+} gradient as part of their polarizing mechanism.

The second way that transcellular ion currents can directly affect cell function is by generating cytoplasmic voltage gradients. The best example of this is found in the insect oocyte–nurse cell syncytium which exhibits polarized transport from nurse cell to oocyte across cytoplasmic bridges. These cells drive a steady current through the bridge to generate a 10 mV difference between nurse cells and oocyte which electrophoreses negatively charged proteins and RNA into the oocyte. Protein transport across the cytoplasmic bridges is strictly a function of the protein's net charge, and the same protein can be made to reverse transport direction by simply reversing its net charge.

The third way that transcellular ion currents could directly affect cell function is by generating extracellular voltage gradients to laterally electrophorese membrane proteins. There is much evidence that such membrane protein redistributions can occur at low field strengths on the order of 1 mV/cell, and examples of galvanotropism and galvanotaxis at these low field strengths are discussed. Current measurements in developing and regenerating systems suggest that endogenous embryonic electric fields are present and may be important for guiding cell migration in vivo.

The vibrating probe has opened up a new dimension of cellular organization that is providing many insights into the mechanisms of cell polarity control via transcellular ion currents.

VI. REFERENCES

Abramson HA, Moyer LS, Goun MH (1942): "Electrophoresis of Proteins and the Chemistry of Cell Surfaces." New York: Reinhold.

Ambrose EJ (1965): "Cell Electrophoresis." Boston: Little, Brown.

Barker AT, Jaffe LF, Vanable JW Jr (1982): The glabrous epidermis of cavies contains a powerful battery. Am J Physiol 242:R358–R366.

Betz WJ, Caldwell JH, Ribchester RR, Robinson KR, Stump RF (1980): Endogenous electric field around muscle fibres depends on the Na^+-K^+ pump. Nature 287:235–237.

Bonting SL (1970): Sodium-potassium activated adenosinetriphosphatase and cation transport. In Bittar EE (ed): "Membranes and Ion Transport." New York: John Wiley, Vol 1, p 273.

Borgens RB, Vanable JW Jr, Jaffe LF (1977a): Bioelectricity and regeneration. I. Initiation of frog limb regeneration by minute currents. J Exp Zool 200:403–416.

Borgens RB, Vanable JW Jr, Jaffe LF (1977b): Bioelectricity and regeneration: Large currents leave the stumps of regenerating new limbs. Proc Natl Acad Sci USA 74:4528–4532.

Borgens RB, Vanable JW Jr, Jaffe LF (1979a): Reduction of sodium dependent stump currents disturbs urodele limb regeneration. J Exp Zool 209:377–386.

Borgens RB, Vanable JW Jr, Jaffe LF (1979b): Role of subdermal current shunts in the failure of frogs to regenerate. J Exp Zool 209:49–56.

Borgens RB, Vanable JW Jr, Jaffe LF (1979c): Small artificial currents enhance Xenopus limb regeneration. J Exp Zool 207:217–226.

Borgens RB, Vanable JW Jr, Jaffe LF (1979d): Bioelectricity and regeneration. Bioscience 29:468–474.

Borgens RB, Jaffe LF, Cohen MJ (1980): Large and persistent electrical currents enter the transected lamprey spinal cord. Proc Natl Acad Sci USA 77:1209–1213.

Borgens RB, Roederer E, Cohen MJ (1981): Enhanced spinal cord regeneration in lamprey by applied electric fields. Science 213:611–617.

Brower DL, Giddings TH (1980): The effects of applied electric fields on Micrasterias. II. The distribution of cytoplasmic and plasma membrane components. J Cell Sci 42:279–290.

Cooper MS, Keller RE (1982): Electrical currents induce perpendicular orientation and cathode-directed migration of amphibian neural crest cells in culture. J Cell Biol 95:323a.

Dittman F, Ehni R, Engels W (1981): Bioelectric aspects of the Hemipteran Teleotrophic ovariole (Dysdercus intermedius). Roux's Arch Dev Biol 190:221–225.

Foskett JK, Scheffey C (1982): The chloride cell: Definitive identification as the salt-secretory cell in teleosts. Science 215:164–166.

Freeman JA, Weiss JM, Snipes GJ, Mayes B, Norden JJ (1981): Growth cones of goldfish retinal neurites generate DC currents and orient in an electric field. Soc Neurosci Abstr 7:550.

Hagins WA, Penn RD, Yoshikami S (1970): Dark currents and photocurrent in retinal rods. Biophys J 10:380–412.

Hinkle L, McCaig CD, Robinson KR (1981): The direction of growth of differentiating neurones

480   Nuccitelli

and myoblasts from frog embryos in an applied electric field. J Physiol 314:121–135.

Jaffe LF (1968): Localization in the developing Fucus egg and the general role of localizing currents. Adv Morphogen 7:295–328.

Jaffe LF (1977): Electrophoresis along cell membranes. Nature 265:600–602.

Jaffe LF (1979): Control of development by ionic currents. In Cone RA, Dowling JE (eds): "Membrane Transduction Mechanisms." New York: Raven Press, pp 199–231.

Jaffe LF (1981a): Control of development by steady ionic currents. Fed Proc 40:125–127.

Jaffe LF (1981b): The role of ionic currents in establishing developmental pattern. Phil Trans R Soc Lond B 295:553–566.

Jaffe LF, Nuccitelli R (1974): An ultrasensitive vibrating probe for measuring steady extracellular currents. J Cell Biol 63:614–628.

Jaffe LF, Nuccitelli R (1977): Electrical controls of development. Annu Rev Biophys Bioeng 6:445–476.

Jaffe LF, Poo M-m (1979): Neurites grow faster towards the cathode than the anode in a steady field. J Exp Zool 209:115–128.

Jaffe LF, Stern CD (1979): Strong electrical currents leave the primitive streak of chick embryos. Science 206:569–571.

Jaffe LF, Woodruff RI (1979): Large electrical currents traverse developing Cecropia follicles. Proc Natl Acad Sci USA 76:1328–1332.

Jaffe LF, Robinson KR, Nuccitelli R (1974): Local cation entry and self-electrophoresis as an intracellular localization mechanism. Ann NY Acad Sci 238:372–389.

Jaffe LA, Weisenseel MH, Jaffe LF (1975): Calcium accumulation within the growing tips of pollen tubes. J Cell Biol 67:488–492.

Jeffery WR (1982): Calcium ionophore polarizes ooplasmic segregation in ascidian eggs. Science 216:545–547.

Konosita H (1953): Studies on the mechanism of pigment migration within fish melanophores with special reference to their electric potentials. Annot Zool Jpn 26:115–127.

Kline D, Robinson KR, Nuccitelli R (1983): Ion currents and membrane asymmetries in the cleaving Xenopus egg. J Cell Biol (in press).

Lindemann B, Voute C (1976): Structure and function of the epidermis. In Llinasand R, Precht W (eds): "Frog Neurobiology." New York: Springer-Verlag, pp 169–210.

Lucas WJ, Nuccitelli R: HCO_3^- and OH^- transport across the plasmalemma of Chara. Planta 150:120–131.

McCaig CD, Robinson KR (1982): The ontogeny of the transepidermal potential difference in frog embryos. Dev Biol 90:335–339.

McLaughlin S, Poo M-m (1981): The role of electro-osmosis in the electric-field-induced movement of charged macromolecules on the surfaces of cells. Biophys J 34:85–93.

Monroy A (1941): Ricerche sulle correnti elettriche dalla superficie del corpo di Tritoni adulti normali e durante la rigenarazione degli arti e della coda. Pubbl Stn Zool Napoli 18:265–281.

Nuccitelli R (1978): Ooplasmic segregation and secretion in the Pelvetia egg is accompanied by a membrane-generated electrical current. Dev Biol 62:13–33.

Nuccitelli R, Jaffe LF (1974): Spontaneous current pulses through developing fucoid eggs. Proc Natl Acad Sci USA 71:4855–4859.

Nuccitelli R, Jaffe LF (1975): The pulse current pattern generated by developing fucoid eggs. J Cell Biol 64:636–643.

Nuccitelli R, Jaffe LF (1976): The ionic components of the current pulses generated by developing fucoid eggs. Dev Biol 49:518–531.

Nuccitelli R, Poo M-m, Jaffe LF (1977): Relations between ameboid movement and membrane-controlled electrical currents. J Gen Physiol 69:743–763.

Orida N, Feldman JD (1982): Directional protrusive pseudopodial activity and motility in macrophages induced by extracellular electric fields. Cell Motil 2:243–255.

Orida N, Poo M-m (1978): Electrophoretic movement and localization of acetylcholine receptors in the membrane of embryonic muscle cells. Nature (Lond) 275:31–36.

Orida N, Poo M-m (1980): On the developmental regulation of acetylcholine receptor mobility in the Xenopus embryonic muscle membrane. Exp Cell Res 130:281–290.

Patel N, Poo M-m (1982): Orientation of neurite growth by extracellular electric fields. J Neurosci 2:483–496.

Poo M-m (1981): In situ electrophoresis of membrane components. Annu Rev Biophys Bioeng 10:245–276.

Poo M-m, Robinson KR (1977): Electrophoresis of concanavalin A receptors along embryonic muscle cell membrane. Nature 265:602–605.

Poo M-m, Lam JW, Orida N, Chao AW (1979): Electrophoresis and diffusion in the plane of the cell membrane. Biophys J 26:1–22.

Reiss HD, Herth W (1978): Visualization of the Ca^{2+} gradient in growing pollen tubes of Lilium longiflorum with chlorotetracycline fluorescence. Protoplasma 97:373–377.

Reiss HD, Herth W (1979): Calcium gradients in tip growing plant cells visualized by chlorotetracycline fluorescence. Planta 146:615–621.

Robinson KR (1977): Application of an electric field to sea urchin eggs redistributes components along the plane of the plasma membrane. J Cell Biol 75:25a.

Robinson KR (1979): Electrical currents through full-grown and maturing Xenopus oocytes. Proc Natl Acad Sci USA 76:837–841.

Robinson KR, Cone R (1980): Polarization of fucoid eggs by a calcium ionophore gradient. Science 207:77–78.

Robinson KR, Jaffe LF (1973): Ion movements in a developing fucoid egg. Dev Biol 35:349–361.

Robinson KR, Jaffe LF (1975): Polarizing fucoid eggs drive a calcium current through themselves. Science 187:70–72.

Slayman CL, Slayman CW (1962): Measurement of membrane potentials in Neurospora. Science 136:876–877.

Stern CD (1981): Behaviour and motility of cultured chick mesoderm cells in steady electrical fields. Exp Cell Res 136:343–350.

Stump RF, Robinson KR (1982): Directed movement of Xenopus embryonic cells in a electric field. J Cell Biol 95:331a.

Stump RF, Robinson KR, Harold RL, Harold FM (1980): Endogenous electrical currents in the water mold Blastocladiella emersonii during growth and sporulation. Proc Natl Acad Sci USA 77:6673–6677.

Telfer WJ, Woodruff RI, Huebner E (1981): Electrical polarity and cellular differentiation in Meroistic ovaries. Am Zool 21:675–686.

Weisenseel MH, Jaffe LF (1976): The major growth current through lily pollen tubes enters as K^+ and leaves as H^+. Planta 133:1–7.

Weisenseel MH, Nuccitelli R, Jaffe LF (1975): Large electrical currents traverse growing pollen tubes. J Cell Biol 66:556–567.

Weisenseel MH, Dorn A, Jaffe LF (1979): Natural H^+ currents traverse growing roots and root hairs of Barley (Hordeum vulgare L). Plant Physiol 64:512–518.

Woodruff RI, Telfer WH (1973): Polarized intercellular bridges in ovarian follicles of the cecropia moth. J Cell Biol 58:172–188.

Woodruff RI, Telfer WH (1974): Electrical properties of ovarian cells linked by intercellular bridges. Ann NY Acad Sci 238:408–419.

Woodruff RI, Telfer WH (1980): Electrophoresis of proteins in intercellular bridges. Nature 286:84–86.

Zagyansky YA, Jard S (1979): Does lectin-receptor complex formation produce zones of restricted mobility within the membrane? Nature 280:591–593.

Zeuthen T (1977): Intracellular gradients of electrical potential in the epithelial cells of the Necturus gallbladder. J Membr Biol 33:281–309.

Zeuthen T (1978): Intracellular gradients of ion activities in the epithelial cells of the Necturus gallbladder recorded with ion-sensitive microelectrodes. J Membr Biol 39:185–218.

Modern Cell Biology 2:483–507
© 1983 Alan R. Liss, Inc., 150 Fifth Avenue, New York, NY 10011

Cellular and Pancellular Organization of the Amphibian Embryo

J. Gerhart, S. Black, and S. Scharf

From the Department of Molecular Biology, University of California, Berkeley, California 94720

I. INTRODUCTION: THE EGG AS CELL AND ORGANISM

In higher metazoa, the egg with its precursor, the oocyte, is the only cell specialized for the production of all cell types, including itself, in a precise multicellular array. One would expect that subcellular organization in the egg determines rather directly the pancellular organization of the embryo cleaved from it during the rapid mitotic divisions following fertilization. Over the last century, embryologists have examined egg structure and ultrastructure in an attempt to find unique organelles and cytoplasmic organization responsible for the egg's unique developmental function, but have been for the most part disappointed in this effort [Gerhart, 1980]. Instead, the egg is found to contain only the common organelles and structures, and the problem becomes one of understanding egg function as a subclass of general cell function. That is to say, development is probably generated by ordinary cell structures and functions, perhaps organized temporally and spatially in a unique way and perhaps exaggerated in some respects and suppressed in others in combinations unique to the egg, but wholly by mechanisms within the province of general cell biology. If this view seems to diminish embryology, it at the same time increases our appreciation of cell biology, for which the egg and embryo would be special cases. On the other hand, embryology can enrich the cell biologist's appreciation of the relation of the cell to the organism, the egg being the source and actuality of both at once.

We have been interested in the egg as a cell, and have sought to establish tests for specific functions performed by the egg in the period from fertilization to first cleavage, functions needed for subsequent embryonic development. The amphibian egg is suitable for such studies, since it is well known to reorganize its contents before first cleavage in a way important for dorsal-ventral axis formation by the embryo. We hope to identify the timing and character of these functions, and later relate them to structural aspects of egg organization and reorganization, when we know more precisely what the egg is actually accomplishing. The following sections contain a review of our progress with these tests.

II. NORMAL EVENTS IN DORSAL-VENTRAL AXIS FORMATION IN AMPHIBIA

In the case of an amphibian such as Xenopus laevis, the tadpole is the product of the egg's development; it contains approximately 10^5 cells, and 10^2 different cell types, and slightly less mass than the single celled egg from which it originated. For our purposes, it is worth noting in Figure 1a the dorsal complex of organs in the tadpole, namely the vertebrate's characteristic anterior-posterior alignment of central nervous system, notochord, and musculature as part of its body axis. On the ventral side of the tadpole, the gut structures are the most apparent.

Since there is no net growth or intake of nutrients during this developmental period, each cell of the tadpole is derived directly from constituents of the large (1.3 mm diameter) egg, and in most cases from the constituents cleaved directly from a corresponding region of the egg; there are a few exceptions, such as the tail which grows from a bud by net mass increase at the expense of nutrients brought from the yolky gut cells by the newly developed circulatory system. But for the most part, the egg and tadpole have direct regional correspondences indicated by a "fate map" in which vital dye marks placed on identified positions of the egg are seen to stain with high fidelity specific structures of the tadpole. Interestingly, in the frog egg this is possible to do *after* the time of fertilization, but not *before,* for the regional assignments of developmental fates are made in relation to the point of sperm entry.

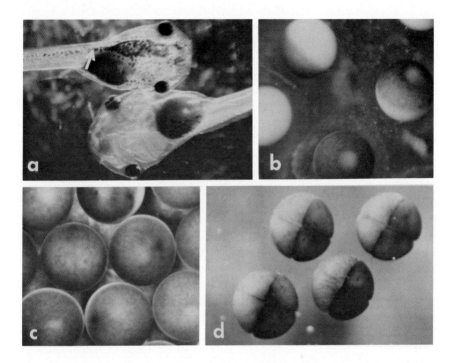

Fig. 1. Developmental stages of Xenopus laevis. a) Tadpole at 4 days postfertilization of the egg, at 22°. Note the dorsal complex of central nervous system, notochord, and segmental muscle blocks in the trunk and tail, indicated by the arrow. b) Unfertilized eggs. Note the pigmented animal hemisphere punctuated by the pigment-free white spot at the animal pole. Note the nonpigmented vegetal hemisphere. Eggs are adsorbed to the vitelline membrane, which is encased in jelly, and are not free to rotate to a vertical position with animal pole at the top. c) Eggs 30 min after fertilization. Note the dark spot of pigment accumulated at the point of sperm entry. The eggs have lifted the vitelline membrane from the egg surface, and are free to rotate to a vertical orientation, with the dense vegetal hemisphere downward. They have been chemically dejellied. d) Four-cell stage, approximately 2 h after fertilization at 22° viewed from the animal pole. The two cleavages have passed vertically through the animal pole. The lighter pair of blastomeres on each egg will provide material for the dorsal embryonic organs, whereas the darker pair of blastomeres will go to ventral organs. The sperm entrance point was on the side of the dark blastomeres.

The unfertilized egg has a dark animal hemisphere and light vegetal hemisphere meeting at an "equator." The egg's contents are arranged in layers with radial symmetry about the animal-vegetal axis, as shown in Figures 1b and 2. These contents include the second meiotic spindle and chromosomes at the externally visible light spot of the animal pole, located in the center of the darkly pigmented animal hemisphere. The pigment granules are located in a "cortical" layer just under the plasma membrane. The animal hemisphere contains a region of nuclear sap near the spindle, and equatorially a layer of organelles, cytosol, and medium-size yolk platelets. The nonpigmented vegetal hemisphere contains close-packed, membrane-bounded yolk platelets and nutrient materials, ending at the vegetal pole geometrically opposite the animal pole. All equatorial positions have the potential to become either dorsal or ventral embryonic structures.

The sperm enters randomly over the animal hemisphere and, as shown in Figure 1c, leaves an easily recognized spot of accumulated pigment, the sperm entry point (SEP). The SEP is rarely if ever at the animal pole, and therefore marks the egg's first departure from radial symmetry. In fact, fate-mapping studies show that as soon as the SEP is visible, that side of the egg can be predicted with high certainty to give rise to embryonic ventral structures whereas the opposite side will generate embryonic dorsal structures. Since the sperm can enter at any position in the animal hemisphere, as shown by localized insemination experiments, it can be concluded that whereas every equatorial region around the egg circumference has the potential to form dorsal, lateral, or ventral structures, the specific selection of one fate or the other is made by the egg on the basis of the locus of sperm entry.

The current state of our understanding of this selection is summarized in Figure 2. The sperm cell fuses with the egg cell and releases its nucleus and microtubule organizing center (MTOC) into the egg cytoplasm. The egg probably does not have a functional MTOC of its own [Heideman and Kirschner, 1975; Maller et al, 1976]. The sperm MTOC generates a large aster which migrates toward the center of the egg, followed by the sperm pronucleus. It is thought that the MTOC and aster signify for the egg the position of sperm entry, whereas the actual darkened spot on the surface does not, since localized nuclear transplantation (Gerhart, unpublished) or injection of sperm constituents [Manes and Barbieri, 1977] obviate the formation of the darkened spot, but do initiate a predictable dorsal-ventral selection. Thus, the SEP is invariably related to the orientation of the dorsal-ventral axis because it is invariably related to the position of the sperm MTOC. The aster enlarges and migrates for approximately 40 min, under conditions where first cleavage will occur at 100 min (19°C). To facilitate comparison of events occuring in different eggs, we use a normalized time scale on which 0 is fertilization and 1 is first cleavage. On this scale the aster migrates until 0.4 and then breaks down. At this time, pronuclear contact occurs and DNA synthesis is just completed [Graham and Morgan, 1966].

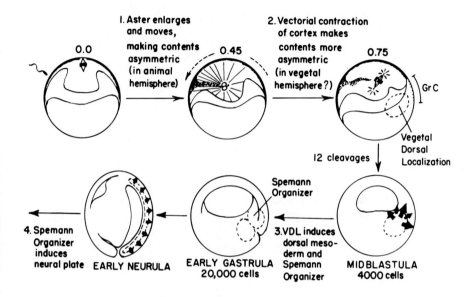

Fig. 2. Schematic diagram of the sequence of events in dorsal-ventral axis formation in anurans. The scheme is largely speculative, based on interpretations of several embryologists, as referenced in the text. In the upper left hand portion of the figure the radially symmetric unfertilized egg is shown at the time of fertilization, with the sperm entering in the animal hemisphere randomly at one point. In the period until 0.45 (45% of the time interval to first cleavage), the sperm MTOC forms a large aster and moves to the center of the egg, coordinating the migration of the sperm and egg pronuclei, which complete DNA synthesis and meet at approximately 0.45. It is suggested that aster formation and movement from one side of the egg causes a slight reorganization of the animal hemisphere contents and perhaps a slight regional difference in the buildup of the animal hemisphere cortex. In the period 0.45–0.75, the cortex shifts relative to the deep contents of the egg, by a rotation or directional contraction toward the side of sperm entry, as revealed by the emergence of the gray crescent. Deep contents adhering to the cortex are drawn up on the gray crescent side. It is suggested that this cortex-driven reorganization of the deep contents activates certain "dorsal determinants" in the vegetal hemisphere, just on one side, to give a "vegetal dorsalizing localization," abbreviated VDL. This region is inherited by vegetal cells arising in the next 12 cleavages, at which time the midblastula transistion begins and vegetal cells induce adjacent animal hemisphere cells to become prospective mesoderm, with the prospective dorsal mesoderm of the Spemann organizer region induced by the VDL. Then, during gastrulation, the prospective dorsal mesoderm migrates up the blastocoel wall and induces the neural plate, which in subsequent steps will form definitive dorsal organs of the nervous systems.

The egg then enters a new phase of the first cell cycle, a G_2-like phase not found in the subsequent 12 cleavage cycles. This phase is devoted to the reorganization of the egg's cortical and endoplasmic contents, particularly at the equatorial level, with the outcome that the egg is further removed from radial symmetry. A new horizontal axis of regional cytoplasmic differences is now visible orthogonal to the old animal-vegetal axis. This new egg axis is used for the subsequent development of the embryonic dorsal-ventral axis.

In more detail, as shown in Figure 3, the cytoplasmic organization of the egg becomes different on its prospective dorsal and ventral sides. The cytology by Klag and Ubbels [1975] of the anuran egg (Discoglossus) documents that its trilayer arrangement of nuclear sap, generalized cytoplasm, and large vegetal yolk platelets is transformed to a skewed apposition of the nuclear sap and vegetal platelets on the prospective dorsal side, whereas the prospective ventral side remains rather similar in organization to the unfertilized egg. Thus, reorganization appears to concern mostly the prospective dorsal side, where a localization or activation of "dorsal determinants" is thought to occur. The external manifestation of this reorganization is seen in the formation of a "gray crescent" which occurs in this period, from 0.45 to approximately 0.80, as measured by Manes and Elinson [1980] in R. pipiens, and illustrated in Figure 4. The gray crescent reflects a cortical change in which pigment granules beneath the plasma membrane, probably embedded in an actin meshwork [Franke et al, 1976], are moved relative to the deeper contents of the egg, by a contraction or rotation of the cortex toward the sperm entrance side, leaving a poorly pigmented zone at the equator of the prospective dorsal side [Elinson, 1980]. It must be made clear that the mechanics of this important reorganization are not yet understood, although some proposals will be made later.

This remarkable period of cytoplasmic reorganization comes to an end at approximately 0.70–0.75 when the egg enters mitosis and the MTOC, which had organized an amphiaster in the G_2-like period, now organizes a spindle. The subsequent cell cycles of the second to thirteenth cleavages consist of S and M periods, without the G_2 delay. Very roughly, the first cycle is lengthened by the amounts of the G_2 period plus the earliest specialized period devoted to the completion of meiosis, whereas the S and M phases of the first cycle are approximately like those of later cycles.

The later effects of the reorganization of the egg's contents along the new axis are seen to be manifold as the intracellular organization of the egg becomes the pancellular organization of the blastula and gastrula. In these stages, the 5,000–20,000 cells cleaved from the egg materials gain motility, adhesive properties, gene expression (for the first time since gametogenesis, Kirschner et al [1981]; Newport and Kirschner [1982]), and developmental autonomy in a patterned way, such that on the dorsal side all these aspects of cell biology are expressed earlier and more strongly than on the ventral side. The dorsal equatorial surface cells expand greatly, the bottle cells appear

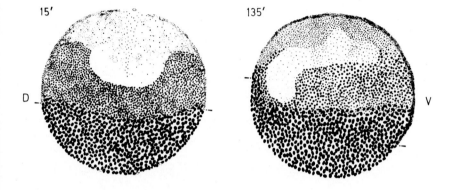

Fig. 3. Internal reorganization of the egg in the period from fertilization to first cleavage, as reported by Klag and Ubbels [1975]. Eggs of the frog Discoglossus pictus were oriented and sectioned at various times. Midsaggital sections are shown, with the animal pole at the top. The section to the left represents an egg 15 min after fertilization; note the approximate radial symmetry of the egg contents around the vertical animal-vegetal axis, and note the trilayer organization of the cytoplasmic contents, with a topmost layer rich in nuclear sap materials from the disrupted germinal vesicle of the oocyte, a middle layer of medium-size yolk platelets interspersed with mitochondria, ribosomes, and familiar cytoplasmic materials, and a bottom layer of large, densely packed yolk platelets. The section on the right represents an egg 135 min after fertilization, with first cleavage due at approximately 150 min. Note the right-hand side of the section where the top and bottom layers have moved into contact, with exclusion of the middle layer. This indicates the reorganization of egg contents on the prospective dorsal side. The gray crescent is formed at the equatorial surface on the right; the dash marks indicating the distance of movement of pigment granules from the crescent area. On the other side of the egg, the contents remain trilayered, as before fertilization; this is the prospective ventral side.

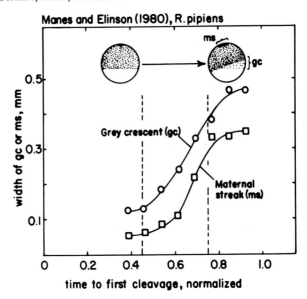

Fig. 4 Time of formation of the gray crescent in Rana pipiens [adapted from Manes and Elinson, 1980]. The width of the gray crescent was measured in approximately 20 eggs at various times, indicated on a normalized decimal scale, with fertilization at 0.0 and first cleavage at 1.0. First cleavage occurred at approximately 150 min. Movement of the polar body spot at the animal pole was also recorded; the spot elongates into a streak, the "maternal streak" (MS). Note that the equatorial movement of the pigment at the gray crescent is approximately equal in magnitude to the movement in the maternal streak, as if the whole cortex moves relative to deeper materials.

first on the dorsal lip, and the internal cells of the prospective dorsal mesoderm begin active migration up the blastocoel wall. These latter cells are thought to comprise the Spemann organizer region of the gastrula, owing to their ability to induce overlying ectoderm cells to form the neural plate, and to induce deep cells to form various mesodermal and endodermal components of the dorsal set of anterior-posterior structures of the body axis, as outlined in Figure 2. It is not known how these dorsal cells of the blastula and gastrula acquire their special properties, although partially the properties are thought to result from their cytoplasmic inheritance of materials of the egg, modified and/or activated in that first G_2-like period of cytoplasmic reorganization. However, this is probably not the whole story, since there is evidence from Nieuwkoop and his colleagues [1973, 1977] for an inductive interaction at the blastula stage, when special dorsal cells located in the *vegetal* hemisphere stimulate neighboring animal hemisphere cells to take on the properties of the Spemann organizer, and of dorsal expansion, migration, and bottle cell formation. In this case, "take on" means at least to begin these activities earlier than the ventral cells, and perhaps also to do them differently, although this is not yet known. These special vegetal cells may be the true recipients of the early "dorsal determinants"—that is, the egg materials localized or activated in the vegetal hemisphere in the first G_2 period of the egg. These cells would then pattern the animal hemisphere by induction at the blastula stage. It is noteworthy that according to the fate map, these vegetal cells themselves never form dorsal embryonic structures, but nonetheless are required to cause neighboring cells to form dorsal structures.

III. EXPERIMENTAL MODIFICATION OF THE NORMAL EVENTS
A. Inhibition of Egg Reorganization

Our first series of experiments involves ways to block the reorganization of the egg's contents in the first G_2 phase, as a test for the developmental importance of this reorganization. There has not previously been an explicit test, although it was certainly assumed that the reorganization is indispensable. Grant and Wacaster [1972] and Malacinski et al [1977] found dramatic inhibitory effects of UV irradiation, applied to the vegetal hemisphere of eggs, on the development of dorsal organs in the embryo, as outlined in Figure 5. In an extension of this observation, we have used brief exposures to low temperature, high pressure, or UV irradiation, and in all cases the egg is sensitive to these treatments before the time of 0.72 but not after, as shown in Figure 6. Despite the basic physical differences of these treatments, their single major effect is to prevent the egg from engaging in dorsal embryonic development. The treated eggs cleave apparently normally, but gastrulate symmetrically instead of starting the lip on the prospective dorsal side, and then never neurulate. They form a so-called "ventral embryo" or "belly piece," lacking central nervous system, notochord, somites, and other organs of the dorsal side. They resemble closely the belly pieces obtained by Spemann [1903] and Fankhauser [1948] from eggs of which a portion containing the gray crescent had been removed by ligation. This ventral development is the extreme effect, found at highest doses, whereas at lower doses, an interesting series of embryos is obtained, in which dorsal structures are completely deleted in a dose-dependent way in anterior to posterior order. One obtains first headless, then headless-trunkless, and finally headless-trunkless-tailless embryos as the strength of the perturbation is increased. Thus, ventral development seems to be the "ground state" of the egg, and we hypothesize that there is a close relationship between the *quantity* of dorsal-type reorganization achieved by the egg in the first G_2 period, and the *quality,* or completeness, of the dorsal anterior-posterior axis of the embryo. We suggest that these treatments block reorganization in the egg and indeed

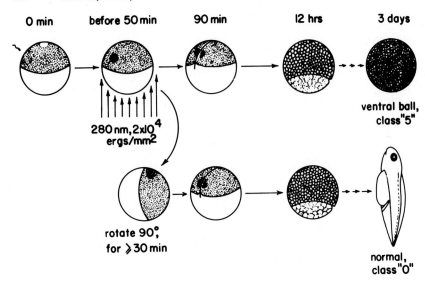

Fig. 5. Experimental interference with the egg's ability to reorganize its contents along the prospective dorsal-ventral axis. Fertilized eggs are irradiated on the vegetal hemisphere with 2.7×10^4 erg/mm^2 ultraviolet light at 280 nm, the maximally effective wavelength [Youn and Malacinski, 1980]. The irradiation must be done before 0.72. The eggs cleave normally, but gastrulate symmetrically and slightly later than normal. As first discovered by Grant and Wacaster [1972], they never neurulate. They form "ventral pieces", so-called "belly pieces", lacking dorsal organs, but possessing a short gut, red blood cells, and a ciliated surface. Irradiated eggs can be rescued by holding them in a horizontal orientation, 90° off axis, for 30–60 min, so that gravity forces the internal contents to reorganize [Scharf and Gerhart, 1980], as indicated in the lower portion of the Figure.

Manes and Elinson [1980] have noted that UV prevents gray crescent formation in R pipiens, and we have noted that it causes reduced and undirected pigment movements in X laevis [Scharf and Gerhart, 1980].

This hypothesis leaves much unsaid because there are probably many ways to block the egg's reorganization of its contents. In general, though, we may expect that the egg has a morphogenetic mechanism to accomplish reorganization, and this must involve both a force-generating agency to cause movement of materials and controls by which the movement is cued to the timing of the cell cycle and to the position of the aster or amphiaster. It is our interpretation that the egg cortex with its actin meshwork generates the force for movement, and that the cortex moves relative to the vegetal endoplasm. In the course of this vectorial movement, the vegetal endoplasm becomes modified, establishing a locus for future dorsalization. If the reorganization relies on the coordination of many components, our treatments may damage different aspects and yet have a common effect.

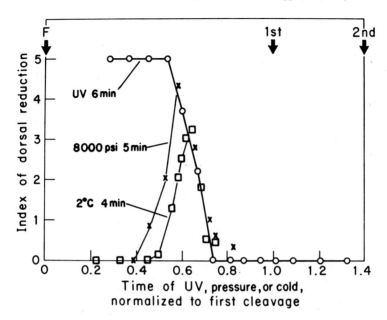

Fig. 6. Time course of sensitivity of the egg to treatments interfering with subsequent dorsal development. Eggs were exposed to UV irradiation on the vegetal hemisphere (2.7 × 10⁴ ergs/mm²), or cold or pressure for a 4- to 6-min interval at the times indicated. Embryos were examined at stage 41, or its equivalent time, for the development of dorsal structures. Data represent averages of eight or more embryos per time point. The index of dorsal reduction (IDR) is a numerical score indicating the severity of the effect of the treatment on dorsal development. A score of 5 means no dorsal structures were formed, the embryo being a radially symmetric "belly piece" possessing a short gut, red blood cells, and a ciliated epidermis. A score of 0 means a normal tadpole. Intermediate scores represent successive deletion of dorsal structures in an anterior-posterior direction, to give headless, then headless-trunkless, and finally headless-trunkless-tailless embryos. See Scharf and Gerhart [1980, 1982] for details. Note the egg's resistance to all of these treatments by the time of 0.72.

With respect to the treatments, cold and pressure are known to depolymerize microtubules, and pressure has been reported to depolymerize actin filaments. Thus, it is plausible that the morphogenetic process employs at least these cytoskeletal elements. There is less information about the effects of UV irradiation on the morphogenetic processes, but Youn and Malacinski [1981] report a 280 nm maximum of the action spectrum, indicating a protein rather than nucleic acid target, and they calculate that the UV penetrates only 5–10 μm into the cortex of the egg, with the vegetal and not animal hemisphere comprising the sensitive region, even in eggs lacking pigment in their animal hemisphere.

B. Control of the Position of the Embryonic Axis

Our second experimental intervention concerns a means to override the egg's control of the direction and position of reorganization, by the use of gravity or centrifugal force to produce an artifactual movement of the egg's contents. As shown in Figure 7, Pasteels [1948] saw clearly the relation of these experimental effects to the normal process of reorganization. The lower portion of the figure shows the directional cortical contraction or rotation pulling a thin layer of vegetal endoplasm upward with the cortex during movement, to give a unique configuration on the prospective dorsal side. The upper portion of the figure shows an egg tipped off axis so that the dense vegetal yolk mass is no longer in a gravitationally stable position. The force of gravity causes the vegetal endoplasm to move *downward* along the cortex, leaving a thin layer of endoplasm stuck to the cortex as it moves. In this latter case, the yolk mass moves relative to the cortex whereas normally the cortex moves relative to the yolk mass, depending on our definition of which egg region undertakes movement, but in both cases the result is the same as far as reorganization of the vegetal endoplasm.

There are three remarkable effects of gravitational or centrifugal treatments on development. First, they are sufficient to control which portion of the equatorial circumference of the egg will be used for dorsal development, irrespective of the location of the point of sperm entry or the sperm MTOC, and in fact even irrespective of the position of the gray crescent in some cases. Born [1885] and later Ancel and Vintemberger [1948, 1949] were the first embryologists to study this effect carefully, using R fusca eggs, and we have modified their procedures to study X laevis eggs. As is seen in Figure 8a, the eggs are tipped on the side so that the equator is vertical, and are deprived of their freedom of rotation by immersing them in 5% Ficoll to dehydrate the perivitelline space, which normally lubricates the egg's return to its usual orientation with dense vegetal yolk mass downward.

These eggs are marked with Nile blue to identify the side of sperm entry and in the illustrated case, the side of sperm entry is directly up. After 20 minutes the eggs are returned to their normal upright position and allowed to develop. At gastrulation, the blastopore normally forms first on the side of the vegetal hemisphere opposite that of sperm entry, as shown in Figure 8b, but in the eggs given the off-axis treatment before first cleavage, the blastopore forms on the *same* side as that of sperm entry, that is, exactly the reverse of the normal axis, as shown in Figure 8c. As it turns out, any part of the equator placed uppermost at this early time becomes dorsal later, irrespective of where the sperm entered [Kirschner et al, 1980; Gerhart et al, 1981] Thus, we conclude that the SEP is not obligatorily linked to the orientation of the axis.

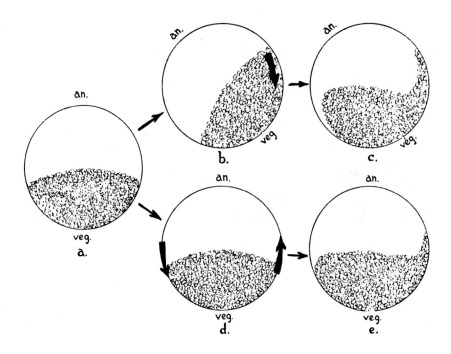

Fig. 7. Interpretation of the effects of oblique orientation of the fertilized, uncleaved egg on alignment of the prospective dorsal-ventral axis. Modified slightly from Pasteels [1948]. In (a) is shown an egg with radially symmetric contents of the animal and vegetal hemispheres. In (d) and (e), the lower portion of the figure, the cortex moves relative to the deep contents, in an upward direction on the right side where the gray crescent is formed; deep contents on the right side are dragged upward near the cortex. This is the prospective dorsal side. In (b) and (c), the egg is tipped 45° off the vertical axis so that the dense vegetal yolk mass is no longer in gravitational equilibrium. These dense materials start to move downward, driven by gravity, leaving a layer of yolk platelets adsorbed to the cortex on the uppermost region of the animal-vegetal interface. This will become the embryonic dorsal region. Notice that the reorganization of the internal materials of the egg looks rather equivalent whether the cortex moves relative to the deep materials (perhaps driven by actomyosin contractility) or whether the deep contents move relative to the cortex (driven by gravity). Thus, the egg contents reorganize irrespective of the motive force, in a way prescribed by the prior organization of the egg, for example the depth to which deep contents adhere to the cortex.

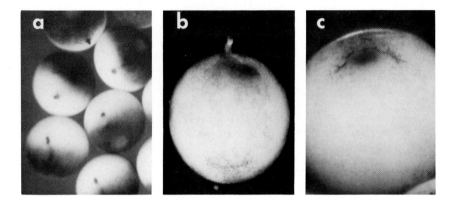

Fig. 8. Experimental reversal of the normal topographic relation between the point of sperm entry and the site of the dorsal blastopore lip in the early gastrula. a) A group of recently fertilized eggs marked at the equator with Nile blue to have a lasting record of the position of sperm entry. In a few eggs, a dark spot of accumulated pigment marking the sperm entry point (SEP) can be seen in the animal hemisphere near the blue spot. Some of the eggs have been left in the normal vertical orientation as controls and others have been turned 90° about a horizontal axis to form the experimental set, where the equator is vertical rather than horizontal. The sperm entry side of the egg is uppermost. The eggs are bathed in 5% Ficoll–20% MMR, as described in Kirschner et al [1980] and Gerhart et al [1981], to dehydrate the perivitelline space and prevent the egg from returning to a vertical orientation, which it normally would do within a few minutes. Eggs are left 20 min, during the period from 0.45 to 0.80, at 22°, and then returned to the original vertical position as indicated by the position of the animal-vegetal hemispheres. b) The vegetal surface of an early gastrula at 10 h, from a control egg not tipped off axis. Notice that the first trace of the blastopore lip appears opposite the Nile blue spot, the indicator of the side of sperm entry. That is, if a circle is drawn through the lip and blue spot, with the vegetal pole as the center, the lip and spot are 180° apart. c) The vegetal surface of an early gastrula from an egg tipped 90° off axis for 20 min. Notice that the Nile blue spot and earliest lip are superimposed, that is, separated by 0°. Thus the relation of the sperm entry point and dorsal lip have been reversed.

We have improved the methodology by using a centrifugal field of 10–30 g to complete the reorganizing treatment in a brief interval of 4 min (Fig. 9) and have studied the cell cycle dependence of these effects. As shown in Figure 10, the egg is very sensitive to gravity or centrifugal force in the 0.4–0.7 period, and rather insensitive afterward as mitosis begins. We think this is really a period of resistance to the force of movement, since the orientation of the axis can be reversed in some eggs when a field of 50g is applied as late as 0.95. This is an important result, since it shows that the position of the gray crescent and of prior reorganization can be unlinked from the orientation of the final axis imposed artificially by centrifugal force [Gerhart et al, 1981; Black and Gerhart, 1982].

CENTRIFUGATION OF X.laevis EGGS, 10 to 50 x g

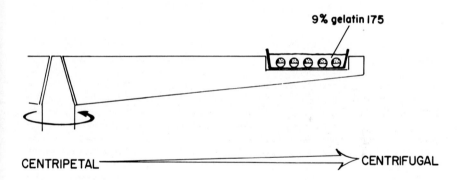

Fig. 9. Centrifugation of eggs of Xenopus laevis to achieve rapid, extensive reorganization of the egg contents. Eggs are fertilized, dejellied, and embedded vertically in 25° molten 9% gelatin–20% MMR (175 Bloom gelatin, Sigma, washed and neutralized) with the sperm entry side oriented toward a marked side of the dish (Falcon, 25 mm, plastic culture dish). The gelatin dehydrates the perivitelline space and hardens quickly to hold the eggs in a fixed orientation. The dish is put in a centrifuge rotor with fitted cups, 10 cm from the center, and centrifuged for 4 min at a selected speed, such as 670 rpm, to achieve 30g. In the orientation shown, the egg contents would be moved across the egg, at 90° to the animal-vegetal axis. A Dayton speed control is used to achieve precise centrifugal fields on a Clinical International centrifuge, and the speed is monitored with a Strobotac. Dishes are removed from the rotor after centrifugation, and placed in a moisture chamber at 18° to allow development to proceed until neurulation, when the position of the neural folds is scored relative to the original sperm entry position. See Black and Gerhart [1982] for details.

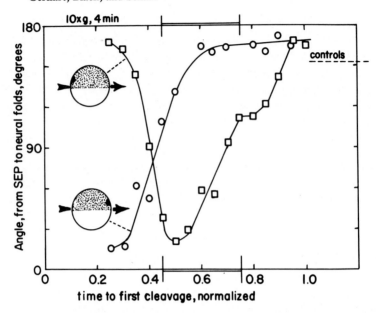

Fig. 10. Time course for the effects of centrifugal force on the direction of the prospective dorsal-ventral axis in the egg X laevis, during the period from fertilization to first cleavage. Eggs were embedded in gelatin, as described in the legend of Figure 9, and centrifuged at 10g for 4 min, with the centrifugal force directed horizontally across the egg, that is, at 90° to the animal-vegetal axis. The sperm entry side was either toward the center of the rotor (□) or away from the center (○). Dishes were removed from the rotor after centrifugation and placed flat in the incubator so that gravity was directed downward on the vertically oriented egg, acting to return the centrifuged contents to their original positions. The positions of neural folds were scored after 24–36 h and assigned angles of 0° to 180° relative to the side of sperm entry. Neural folds opposite the sperm entry point (SEP) were scored at 180°; those through the SEP were scored as 0°. Recall that 180° is the ideal control angle and 0° represents complete reversal of the orientation of the dorsal-ventral axis. See Black and Gerhart [1982] for details.

We are less certain about the period before 0.4 since, as described elsewhere, the effect of gravity at this early stage is exactly opposite in orientation to that found after 0.4 (Black and Gerhart, in preparation). This interesting transition illustrates the occurrence of profound changes at 0.4, the start of the G_2-like phase.

To illuminate some of these effects of centrifugation, we have developed an assay for the extent of movement of vegetal materials at a controlled speed and angle of centrifugation. To do this, we hold the eggs in fixed position by embedding them in gelatin and inject a small amount of Nile blue into the deep vegetal yolk mass, or more shallowly in the cortex, to establish a reference point for later comparison. The dye locally stains the

yolk platelets, oil droplets, and membranous organelles and reveals the movement of the yolk mass when the egg is centrifuged. In these experiments,
the eggs are oriented relative to the centrifugal field so that the contents will
move no more than 90° of arc. As shown in Figure 11, we find that shortly
after fertilization or activation, deep marks do not move extensively when
the egg is centrifuged at 30g. At approximately 0.35–0.40, the marks become
very movable—in fact, maximally so, as quantified by measuring the displacement of the mark from its origin. In the 0.5–0.7 period, the deep marks
become less and less movable, until at 0.8 they are just as resistant to
movement as they were shortly after fertilization. It is to be recalled that the
unfertilized egg is arrested in meiotic metaphase, at which stage the egg's
contents are so structured and adherent that eggs inverted or held obliquely
for hours do not rearrange their contents. This structured state seems to return

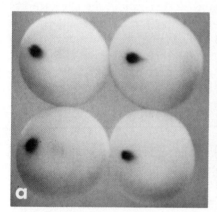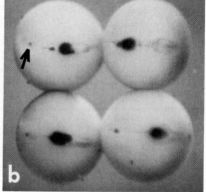

Fig. 11. Cell-cycle dependence of the resistance of the egg's contents to centrifugal force.
a) Eggs embedded in gelatin and marked with a small spot of Nile blue injected in the vegetal
hemisphere. The injection needle was inserted into the vegetal yolk mass and 5 nl of dye (1%
in 10 mM sodium phosphate, pH 7.4) was delivered just near enough to the surface to see the
spot. The eggs were then centrifuged at 30g for 4 min at 0.8 in the interval between fertilization
and first cleavage. Note that the spots moved very little, indicating the resistance of the egg
to centrifugally driven reorganization of its contents at this time of the cell cycle. Equivalent
resistance is found before 0.3. b) Eggs prepared and centrifuged as described for panel (a)
except that centrifugation was done at 0.37. The vegetal hemisphere is shown. The spots moved
extensively, reflecting the displacement of the vegetal yolk mass. The arrow indicates a spot
of residual dye, marking the position of the original injection. The upper right egg was injected
superficially, from the egg surface; note that the mark does not move, indicating the fixed
position of the egg surface layer—ie, the cortex. In each of the other eggs the bulk of the dye
was injected more deeply and has moved in the centrifugal field.

at the first mitotic metaphase. These studies show that the movability of the vegetal endoplasm, relative to the egg cortex, is cell cycle–dependent, and maximal at 0.4. What is assayed here is probably the adherence of the deep vegetal contents to the egg cortex, which is itself essentially unmovable in our assay since it is held in place by the fertilization envelope and the gelatin. Thus, we picture that at 0.4 the egg disengages the connections of its deep cytoplasm to its surface; then in the 0.4–0.8 period the egg normally moves the cortex and endoplasm relative to one another using its cortical contractile machinery. Finally, at 0.8, the egg solidifies and reconnects the cortex and endoplasm, holding the two in the new, reorganized arrangement.

Our most dramatic assay of the effectiveness of these artifically induced reorganizations is found in the rescue of cold, pressure, or UV-irradiated eggs by oblique orientation or centrifugation. As discussed earlier, these eggs have lost the ability to reorganize their contents by the normal cortex-driven morphogenetic process, and would develop as ventral embryos. If, however, the detrimental treatment is followed by off-axis orientation, so that gravity acting as a nonbiological force can cause reorganization, the egg then develops normally with its dorsal structures derived from the equatorial egg regions uppermost in the off-axis period. Examples are shown in Figures 5 and 12. Furthermore, the *extent* of reorganization can be controlled by the length of time the egg is left off axis, and interestingly, the rescue series is the exact reverse of the dose-dependent series of truncations described earlier; here, brief rescue restores tail development, intermediate rescue restores tail and trunk, and full rescue restores tail, trunk, and head [Scharf and Gerhart, 1980].

It is surprising that this rescue works at all, given that gravity or centrifugation provides the morphogenetic force. We infer that the specificity and information for reorganization are not primarily contained in the force, but in the structured response of the egg contents—ie, the endoplasm and cortex—to the force. That is to say, the egg has a certain organization to begin with, and this organization can lead only to certain other organizations, when forced to reorganize. If this is true, it follows that the egg's own force for movement, namely a cortical contraction or rotation, may itself be basically rather crude in spatial detail. The egg responds to this general stimulus in a specific and well structured way because of its inherent organization. Thus, our axis reversal and rescue conditions may be informing us about the existence of developmentally important cellular organization in the egg.

As for timing, it is noteworthy that the rescue can be performed well after the 0.75 period—that is, at times when the egg is normally no longer able to reorganize its contents by its own mechanisms. In fact, rescue can be achieved even after first cleavage. Thus, we conclude that the normal 0.4–0.75 period of axis definition is specified on the one hand by the time during

Fig. 12.   Rescue of UV-irradiated eggs by gravity-driven reorganization of the egg contents. Eggs were irradiated heavily with UV at 0.4 on the vegetal hemisphere, to block dorsal development, as shown by the ventral embryos in the center of the figure. A tadpole from an unirradiated control egg is shown in the upper portion of the figure. Some irradiated eggs were turned 90° off axis and maintained in that orientation for at least 30 min, and then returned to the normal vertical orientation and allowed to develop. This posttreatment completely rescued eggs from the UV effects, and an apparently normal tadpole developed, as shown in the lower portion of the figure. Equivalent rescue was obtained for cold or pressure treated eggs. From Scharf and Gerhart [1980, 1982].

which the egg generates force to move its contents, and on the other hand by the time when the contents are easily moved. However, the 0.4–0.75 period is not the only time when the egg is receptive to the consequences of reorganization. The egg's receptivity to reorganization does not end until well after first cleavage, and, for all we know, may not end until much later, for example, in the blastula stage. If we had the means to reorganize the egg's contents at that time, perhaps dorsal-ventral axis development could still be reoriented.

C. Twinning

Our final impression about the nature of the reorganization process comes from our study of twins. We have said before that reorganization in the 0.4–0.75 period concerns the formation of a special predorsal region in the egg, involving a localization or activation of "determinants" needed for subsequent dorsal embryonic development. And we have said that gravity or centrifugal force can act to provoke the formation of such a region. We find that two sequential centrifugations in opposite directions, or one centrifugation followed by one gravity-driven reorganization, can lead to the formation of Siamese twins at very high frequency. For example, if the first centrifugation is done at 0.4 in a direction to move the egg contents maximally by 90°, and then the egg is left inclined 90° until the eight-cell stage, a single axis embryo results, with its dorsal side derived from the centripetal side of the egg. But if such an egg is returned to its vertical position after centrifugation, so that gravity acts in a direction to move the contents back to the original position, then twins are produced. Figure 13 shows a school of Siamese twins, many of them completely doubled dorsally but fused ventrally, and Figure 14 shows the cell cycle dependence of the frequency of twinning, with 0.4 as the maximum. Similar results can be obtained with UV-irradiated eggs, twice spun, showing that the experimentally controlled reorganizations are responsible for both axes. Several embryologists have studied twinning by gravitational effects on eggs [Penners and Schleip, 1928a, b; Pasteels, 1938; Schultze, 1894; Morgan, 1895], although our conditions give much higher frequency and a clearer indication of the sensitive period.

There are two ways to think about the formation of a predorsal region of determinants during reorganization. One is that the egg has dorsal determinants (whatever they are) evenly distributed around the equatorial cytoplasm and during the reorganization process the egg manages to *sequester* the determinants to one place—namely, the cytoplasmic region near the gray crescent. In this view, the other regions of the egg actually become deficient in dorsal determinants, perhaps losing permanently the *potential* to become dorsal. The alternate view starts with the same assumption that every equatorial position has the potential to become dorsal, there being an ample supply of determinants everywhere but holds that the reorganization process causes the local *activation* of determinants in one region near the gray crescent. Thus, the other regions do not lose the potential; they just don't become activated to use the potential. This latter view, we think, fits more easily our observations on twinning, where a second dorsal center is created opposite and well after the first. Otherwise, we would have to postulate that half the determinants are driven back to the opposite side by the second centrifugation. Rather, it seems more plausible to us that the opposite side was never deficient in the potential to become dorsal, and just becomes activated by the late reorganization of its materials. Such as impression is useful to remember when considering efforts to isolate dorsal determinants; whatever they are, the differences between dorsal and ventral regions of the egg may concern minor modifications of materials found at all positions, and not the presence or absence of specific materials.

Fig. 13. Twins produced by centrifugation of eggs. Fertilized, dejellied eggs were embedded in gelatin as described in the legend of Figure 9 and centrifuged at 30g for 4 min with centrifugal force directed 90° to the animal-vegetal axis. The dish of eggs was removed from the rotor and placed flat in the incubator so that gravity was acting to move the egg contents back to their original positions. Maximum twinning was found for eggs centrifuged at 0.4 in the interval from fertilization to first cleavage. At the late neurula stage, embryos were removed from gelatin by placing the dish in 32° water for 3 min to melt the gelatin, at which time embryos were quickly pipetted into 20% MMR at 22°, in which they were subsequently raised. In this batch, 100% twins was obtained. Note the double axes which join in some cases at the posterior trunk or tail level. All twins share ventral structures. See Black and Gerhart [1982] for details.

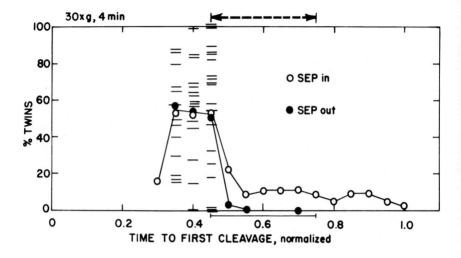

Fig. 14. Time course for twin production by centrifugation of eggs. The procedure follows that described in the legend to Figure 10 except that a force of 30g was used. Some eggs were centrifuged with the sperm entry side toward the center of the rotor (○) and others with that side away from the center (●). Dashes indicate percentages for batches of eggs from different frogs when at least 20 eggs of each batch were tested, and circles indicate averages for all batches tested. See Black and Gerhart [1982] for details. Note that the equatorial orientation of the egg in the centrifugal field is not important for twinning at 0.35–0.45, but is important in the 0.50–0.90 period.

IV. CONCLUSIONS AND FUTURE STUDIES

Our functional tests for developmentally significant egg structures, organization, and morphogenetic events have led us to the conclusion that an unusual G_2-like period during the first cell cycle is devoted by the egg to the differentiation of its cytoplasm along the prospective dorsal-ventral axis, such that a special predorsal region is created on one side of the egg at the equatorial and vegetal levels. We suggest that 1) this region is formed by an activation, not a localization, of components needed for later dorsalization events; 2) the activation results from a reorganization of the cortex and deep cytoplasm (the endoplasm), whereby one moves unidirectionally relative to the other; 3) the G_2-like period begins with the disconnection of structural links between the cortex and endoplasm, continues with the contractile activity of the cortex generating the motive force for reorganization, and ends with the structural reconnection of the cortex and endoplasm as mitosis begins; 4) failure to achieve this reorganization and activation in the uncleaved egg leads to an embryo incapable of developing dorsal organs; 5) the specificity for reorganization in the G_2-like period lies not in the motive force but in the egg's prior cellular organization, since gravity and centrifugation provoke dorsal development as effectively as does the contractile cortex; and 6) the egg can make developmental use of cytoplasmic reorganizations provoked by gravity or centrifugation before, during, or after the G_2-like period, even though it normally conducts its own reorganization only during the G_2-like period.

In the future, we plan to examine the egg cytologically to see what cytoskeletal and organellar organization correlates with our controlled positioning of the predorsal region in the egg, with our ability to block the formation of this region, and with our ability to restore this region artificially in recue experiments. Furthermore, we hope to learn more about the morphogenetic mechanism by which the egg moves its contents and the way this mechanism is tied to the G_2 phase of the first cell cycle and to positional cues from the sperm aster. Thus, we hope to learn which common cell organelles and cell structures serve the uncommon developmental functions of the egg.

But we do not labor under the illusion that this early cytoplasmic differentiation, established before first cleavage, provides the direct and ultimate detailed spatial organization of the 100,000-celled tadpole. Such an expectation was implicit in the venerable double gradient idea of Dalcq and Pasteels [1937]—namely, that the egg at the time of gray crescent formation laid down a quantitative pattern of absolute precision such that cells of the later stages needed only to read the local quantities of morphogens and do what the numbers told them (reviewed by Pasteels [1964]). Although the double gradient idea allowed material simplicity in requiring only two substances, it did require quantitative precision, and made the egg, in a sense, a numerical homunculus as complex as the embryonic product of its development. Instead, our overall impression, with the hindsight of 40 years of molecular and cell biology, is that the egg does not pattern the embryo all at once, but possesses and produces just enough organization (or "pattern") in its contents to meet the developmental demands of the immediate next stage. The next stage will presumably make use of its own set of morphogenetic mechanisms, scheduled for that stage, to transform its organization still further, accomplishing step by step an organized transformation of its prior organization.

In this long sequence, the developmental devices of each stage may differ from those of the first cycle and exploit eventually most of the aspects of the rich repertoire of cell biology.

ACKNOWLEDGMENTS

This research was supported by PHS grant GM 19363 (to J.C.G.). S. D. Black and S. R. Scharf were NIH predoctoral fellows supported by training grants GM 07127 and GM 07232 respectively. The authors thank M. Kirschner, J. Newport, G. Ubbels, K. Hara, and P. Nieuwkoop for beneficial discussion and suggestions in the course of these studies.

V. REFERENCES

Ancel P, Vintemberger P (1948): Recherches sur le déterminisme de la symétrie bilaterale dans l'oeuf des Amphibiens. Bull Biol Fr Belg (Suppl) 31:1–182.

Ancel P, Vintemberger P (1949): La rotation de symétrisation, facteur de la polarisation dorso-ventrale des ébauches primordiales, dans l'oeuf des Amphibiens. Arch Anat Microscop Morphol Exp 38:167–183.

Black SD, Gerhart JC (1982): High frequency twin embryos from eggs of Xenopus laevis centrifuged at precise times, speeds, and angles. (In preparation.)

Born G (1885): Uber den Einfluss der Schwere auf das Froschei. Arch Mikrosk Anat 24:475–545.

Dalcq A, Pasteels J (1937): Une conception nouvelle des bases physiologigues de la morphogénèse. Arch Biol 48:669–710.

Elinson R (1980): The amphibian egg cortex in fertilization and early development. Sympos Soc Dev Biol 38:217–234.

Fankhauser G (1948): The organization of the amphibian egg during fertilization and cleavage. Ann NY Acad Sci 49:684–702.

Franke WW, Rathke PC, Seib E, Trendelenburg MF, Osborn M, Weber K (1976): Distribution and mode of arrangement of the filamentous structures and actin in the cortex of amphibian oocytes. Cytobiologie 14:111–130.

Gerhart JC (1980): Mechanisms regulating pattern formation in the amphibian egg and early embryo. In Goldberger R (ed): "Biological Regulation and Development," Vol 2. New York: Plenum Press, pp 133–316.

Gerhart JC, Ubbels G, Black S, Hara K, Kirschner M (1981): A reinvestigation of the role of the grey crescent in axis formation in Xenopus laevis. Nature 292:511–516.

Graham CF, Morgan RW (1966): Changes in the cell cycle during early amphibian development. Dev Bio 14:439–460.

Grant P, Wacaster JF (1972): The amphibian grey crescent—A site of developmental information? Dev Biol 28:454–471.

Heidemann SR, Kirschner MW (1975): Aster formation in eggs of Xenopus laevis: Induction by isolated basal bodies. J Cell Biol 67:105–117.

Kirschner M, Gerhart JC, Hara K, Ubbels GA (1980): Initiation of the cell cycle and establishment of bilateral symmetry in Xenopus eggs. Symp Soc Dev Biol 38:187–216.

Kirschner MW, Butner KA, Newport JW, Black SD, Scharf SR, Gerhart JC (1981): Spatial and temporal changes in early amphibian development. Neth J Zool 31:50–77.

Klag JJ, Ubbels GA (1975): Regional morphological and cytological differentiation of the fertilized egg of Discoglossus pictus (Anura). Differentiation 3:15–20.

Malacinski GM, Brothers AJ, Chung H-M (1977): Destruction of components of the neural induction system of the amphibian egg with ultraviolet irradiation. Dev Biol 56:24–39.

Maller J, Poccia D, Nishioka D, Kidd P, Gerhart J, Hartman H (1976): Spindle formation and cleavage in Xenopus eggs injected with centriole-containing fractions from sperm. Exp Cell Res 99:285–294.

Manes ME, Barbieri FD (1977): On the possibility of sperm aster involvement in dorso-ventral polarization and pronuclear migration in the amphibian egg. J Embryol Exp Morphol 40:187–197.

Manes M, Elinson R (1980): Ultraviolet light inhibits grey crescent formation in the frog egg. Roux Arch 189:73–76.

Morgan TH (1895): Half embryos and whole embryos from one of the two first blastomeres. Anat Anz 10:623–628.

Newport J, Kirschner M (1982): A major developmental transition in early Xenopus embryos: 1. Characterization and timing of cellular changes at the midblastula stage. Cell 30:675–686.

Nieuwkoop PD (1973): The "organization center" of the amphibian embryo: Its origin, spatial organization, and morphogenetic action. Adv Morphogen 10:1–39.

Nieuwkoop PD (1977): Origin and establishment of embryonic polar axes in amphibian development. Curr Top Dev Biol 11:115–132.

Pasteels J (1938): Recherches sur les facteurs initiaux de la morphogénèse chez les amphibiens anoures. I. Resultats de l'experience de Schultze et leur interpretation. Arch Biol 49:629–667.

Pasteels J (1948): Les bases de la morphogénèse chez vertébrés anamniotes en function de la structure de l'oeuf. Folia Biotheor (Leiden) 3:83–108.

Pasteels JJ (1964): The morphogenetic role of the cortex of the amphibian egg. Adv Morphogen 3:363–388.

Penners A, Schleip W (1928a): Die Entwicklung der Schultzeschen Doppelbildungen aus dem Ei von Rana fusca. Teil I–IV. Z Wiss Zool 130:305–454.

Penners A, Schleip W (1928b): Die Entwicklung der Schultzeschen Doppelbildungen aus dem Ei von Rana fusca. Teil V und VI. Z Wiss Zool 131:1–156.

Scharf SR, Gerhart JC (1980): Determination of the dorsal-ventral axis in eggs of Xenopus laevis: Complete rescue of UV-impaired eggs by oblique orientation before first cleavage. Dev Biol 79:181–198.

Scharf SR, Gerhart JC (1982): Axis determination in eggs of Xenopus laevis: A critical period before first cleavage, identified by the common effects of cold, pressure, and UV-irradiation. (Submitted.)

Schultze O (1894): Die künstliche Erzeugung von Dopplebildungen bei Froschlarven mit Hilfe abnormer Gravitation. Roux Arch 1:160–204.

Spemann H (1903): Entwicklungsphysiologisches Studien am Tritonei. Roux Arch 16:552–631.

Youn BW, Malacinski GM (1980): Action spectrum for ultraviolet irradiation inactivation of a cytoplasmic component(s) required for neural induction in the amphibian egg. J Exp Zool 211:369–378.

Modern Cell Biology 2:509–548
© 1983 Alan R. Liss, Inc., 150 Fifth Avenue, New York, NY 10011

Cell and Extracellular Matrix: Their Organization and Mutual Dependence

Elizabeth D. Hay

From the Department of Anatomy, Harvard Medical School, Boston,
Massachusetts 02115

I. INTRODUCTION

Preceding chapters in this book have considered cell polarity, cytoskeletal organization, and the effects of the cytoskeleton on cell metabolism. Here we shall see that cell polarity, the cytoskeleton, and cell metabolism can be profoundly influenced by molecules of the extracellular matrix that decorate one or another of the cell's surfaces. The cell produces the matrix molecules and controls, from nearby or even from afar, the manner in which they polymerize in the extracellular compartments; the matrix, once deposited, then dictates to the cell on questions of cell organization and cell function. Keith Porter has long been interested in the synthesis, secretion, and assembly of extracellular matrix (ECM), especially of the collagenous components that are so elegantly visualized by electron microscopy [Porter and Vanamee, 1949; Vanamee and Porter, 1951; Porter, 1952, 1956, 1964, 1966; Robbins et al, 1955; Porter and Pappas, 1959; Godman and Porter, 1960; Nadol et al, 1969; Humphreys and Porter, 1976a, b]. In this chapter, we shall have occasion to review a number of Porter's contributions to our understanding of ECM organization, deposition, and interaction with the cell surface. These and the other works discussed here deal mainly with adult and embryonic vertebrates, but it seems clear that wherever multicellular organisms have emerged in evolution, ECM of one kind or another has been created to form scaffoldings. Whether the scaffoldings are merely glycoprotein surface coats or more elaborate fibrous extracellular arrays, they and the cells that created them are always mutually dependent.

II. MATRIX ORGANIZATION

The glycoproteins that coat the apical and lateral surfaces of epithelial cells (Fig. 1), the surface of the oocyte, and so on are in a sense a form of extracellular matrix [Bennett, 1963; Ito, 1965; Rambourg and Leblond, 1967]. We shall limit our discussion, however, to extracellular matrices that contain collagen and glycosaminoglycans (GAG), matrices found usually on the basal surface of epithelia (Fig. 1) or completely surrounding mesenchymal cells (fibroblasts, chondrocytes, osteocytes, myocytes). The basic plan of the vertebrate body provides that the mesenchymal cell is trapped within ECM of this type because of its instinct to invade collagenous matrices. The epithelial cell, by virtue of the distinct programming of its basal surface, will reside on top of ECM [Overton, 1977; Hay, 1982]. Cells that circulate in the bloodstream and move from one compartment to another do not obey these rules, a point that should be considered in generalizing to epithelia and ordinary mesenchymal cells (Fig. 1), the structural relations of plasmalemma and cytoskeleton demonstrated in red and white blood cells.

A. Collagens, GAG, and Structural Glycoproteins

The extracellular matrix proper has been known for some time to be rich in collagen and GAG. Collagen occurs as five or more genetically distinct molecules that polymerize in the form of striated fibrils (types I–III) or in

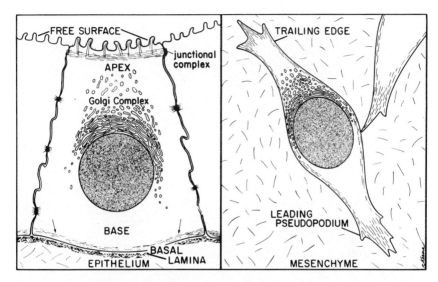

Fig. 1. Diagram showing the relation of epithelial and mesenchymal cells to ECM. The epithelial cell (left) sits on top of ECM, usually on a collagenous basal lamina. The mesenchymal cell (right) moves through extracellular matrix. Arrows, basal cytoskeleton. From Hay [1981b].

the form of basal laminae or so-called basement membranes (type IV, type V?) and other structures [Bornstein and Sage, 1980; Hay, 1981a; Timpl and Martin, 1982]. Proteoglycans (PG), containing GAG (chondroitin sulfate, keratan sulfate, heparan sulfate) linked to protein, occur as monomers associated with basal laminae and collagen fibrils or as aggregates of monomers linked to hyaluronic acid [Hascall and Hascall, 1981] which may influence collagen fibrillogenesis [Seegmiller et al, 1971].

When fixed with ruthenium red, PG monomers appear as small granules in the laminae rarae of the basal lamina (arrows, Fig. 2A), along collagen fibrils (cf Figs. 2A, 3A), and embedded in filamentous ECM meshworks (Fig. 3B) [Bernfield et al, 1972; Trelstad et al, 1974; Hay and Meier, 1974; Hay, 1978; Farquhar, 1981]. The filaments or microfibrils (open arrows, Fig. 3B) in these meshworks are 3–5 nm in diameter and may correspond to hyaluronate cores linking PG granules [Hascall and Hascall, 1981]. PG present in or on the cell membrane (curved arrows, Fig. 3) [Kraemer and Tobey, 1972; Hay and Meier, 1974] may differ structurally from PG secreted into the stroma [Kjellen et al, 1981].

The third class of ECM molecules, the newly discovered structural glycoproteins, consists of fibronectin, laminin, and other glycoproteins with strong binding sites for collagen, GAG, and cells [Yamada, 1981; Hynes,

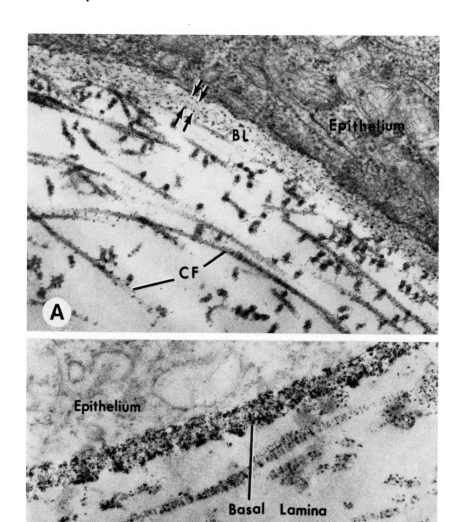

1981; Timpl and Martin, 1982]. Fibronectin is often found in basal laminae (Fig. 2B) and along collagen fibrils (arrows, Fig. 2B), its distribution sometimes resembling that of PG. It also occurs as small nonstriated fibrils (10 nm in diameter) in the embryo [Mayer et al, 1981] and in vitro [Chen et al, 1978; Singer, 1979]. Laminin, however, occurs only in the basal lamina, where it seems to be more concentrated in the laminae rarae than in the lamina densa [Farquhar, 1981].

B. Supramolecular Organization of ECM

Space does not permit consideration of all of the beautiful and useful supramolecular organizations displayed in the vertebrate body by ECM molecules. In bone, the ordered, densely packed collagen fibrils form rigid lamellae that are constantly remodeled by appositional growth in response to outside forces [Bloom and Fawcett, 1975], whereas in cartilage nonstriated collagen fibrils form GAG-rich meshworks of great compressibility that grow interstitially [Revel and Hay, 1963; Hascall and Hascall, 1981; Hendrix et al, 1982]. Type II collagen occurs not only in nonstriated and striated fibrils in cartilage and avian corneal stroma, but also in the broad, nodular subendothelial corneal ECM called Descemet's membrane [Hendrix et al, 1982]. Tendons contain long, parallel arrays of large collagen fibers that obey strict stress-strain rules [Trelstad and Silver, 1981]. Organ parenchyma are often supported by delicate reticular fibers, each composed of only a few collagen fibrils [Bloom and Fawcett, 1975]. The loose arrangement of areolar connective tissue allows the skin of furry mammals to slide over underlying musculoskeletal structures.

One of the most remarkable patterns of collagen polymerization is seen in the acellular dermis (basement lamella) of lower vertebrates and in the cornea of all vertebrates. Here, the collagen fibrils form 20 or more layers arranged at right angles to one another (Fig. 4). The orthogonal arrangement can be distinguished by light microscopy, but is more elegantly resolved by electron microscopy [Weiss and Ferris, 1954; Porter, 1956, 1964, 1966]. The narrowest layers are anchored in the basal lamina (BM, Fig. 4), and both the diameter and number of collagen fibrils in each layer usually increase

Fig. 2. Electron micrographs of the subepithelial ECM of the 10- to 11-day-old embryonic avian cornea. A) Ruthenium red staining reveals proteoglycan granules (arrows) in the lamina rara externa and lamina rara interna of the basal lamina (BL). The lamina densa is not rich in GAG. Collagen fibrils (CF) are coated with proteoglycan granules. B) This cornea was stained with antifibronectin and peroxidase-antiperoxidase as described by Mayer et al [1981]. The laminae rarae internae seem particularly rich in fibronectin. Peroxidase aggregates can also be seen next to collagen fibrils sectioned longitudinally (curved arrows) and transversely (straight arrows). Bar = 300 nm.

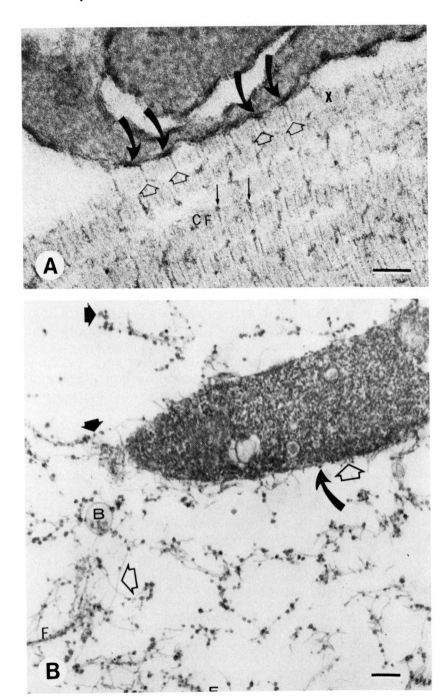

deeper in the dermis, presumably because the juxtaepidermal layers are the most recently formed [Hay and Revel, 1963]. In the avian cornea, the orientation of the anterior orthogonally arranged layers shifts in the same clockwise direction in both eyes and thus is bilaterally asymmetric [Trelstad and Coulombre, 1971]. In the dermis of amphibians and fishes, the layers are displaced in a predictable fashion relative to the overall body pattern [Edds and Sweeny, 1961; Edds, 1964].

Interestingly, the collagenous cuticle on the apical surface of the annelid epidermis also shows an orthogonal pattern that conforms to the curvatures of the animal [Coggeshall, 1966; Humphreys and Porter, 1976a]. The collagen fibrils (Co, Fig. 5) of the cuticle lack electron density and are embedded in a filamentous matrix penetrated by cell processes (P, Fig. 5) that seemingly give rise to membrane-bound corpuscles (C, Fig. 5) on the outer surface of the cuticle [Humphreys and Porter, 1976a]. In the next section, we will consider briefly the possible manner of assembly of such extracellular gridworks in the developing organism.

III. MATRIX DEPOSITION
A. Scaffoldings Associated With Epithelia

Orthogonal collagenous gridworks of the type described above are only laid down in association with epithelium, usually, if not always, on the basal surface subjacent to the basal lamina (BL, Fig. 2A). It is in the subepithelial area (BM, Fig. 4) that new collagen, as detected autoradiographically, is deposited (Fig. 6), and with time the pulse-labeled radioactive collagen fibrils are displaced inward by deposition of unlabeled collagen in the outer layers [Hay and Revel, 1963]. It has been proposed that some sort of switching mechanism exists at the level of the cell or basal lamina that allows each layer to be deposited in "plywood" fashion at right angles to the one previously formed [Weiss, 1957; Edds, 1964].

Fig. 3. Electron micrographs of ruthenium red stained ECM associated with embryonic mesenchymal cells. A) Part of a fibroblast in the compacted zone of a 12-day-old embryonic avian cornea. The inner and outer leaflets of the plasmalemma are particularly well-stained (curved arrows) in regions associated with a repeating band (open arrows) of an adjacent collagen fibril (X). The collagen fibril (CF) in this compacted region is coated with small proteoglycan granules (straight arrows) arranged in a similar repeat pattern. Bar = 75 nm. B) Part of a mesenchymal cell migrating from the sclerotome toward the notochord in a 3-day-old avian embryo. The plasmalemma is coated with small proteoglycan granules (as at the curved arrow). It moves through a hydrated matrix containing proteoglycan granules (solid arrows), small filaments or so-called microfibrils (open arrows), and slightly larger nonstriated fibrils (F). B, bleb. Bar = 200 nm. From Hay [1978].

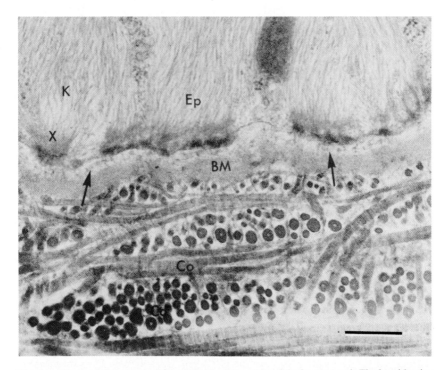

Fig. 4. Electron micrograph of the subepidermal matrix of the lamprey eel. The basal lamina consists of a broad lamina densa (BM) separated from the epithelium (Ep) by a layer of granules (arrows) sturdier than the usual proteoglycan granules in the sense that they survive routine fixation. Adjacent collagen fibrils (Co) are arranged in orthogonal layers, only the most superficial of which are shown here. Cytoplasmic filaments seemingly composed of keratin (K) attach (X) in hemidesmosomes on the basal plasmalemma. B = 200 nm. From Porter [1966].

Porter and his collaborators pointed out the shortcomings of this postulated "plywood" mechanism that requires the information on orthogonality to be dealt out at alternating intervals over long periods of time. They proposed, instead, the "shingle" or "scindulene" theory, which states that all the orthogonal plies insert like shingles into the basal lamina and are growing at the same time [Nadol et al, 1969]. They studied development of the Fundulus dermis, an orthogonal collagenous gridwork which, like that of the dermis of other embryonic and larval vertebrates, is acellular. Since this stroma lacks fibroblasts, its 20 or so layers of collagen fibrils are undoubtedly laid down in part by the overlying epithelium, as in the case of the amphibian dermis and avian cornea [Hay and Revel, 1963; Hay, 1964; Dodson and Hay, 1971, 1974]. Nadol et al [1969] made the rather original observation that one

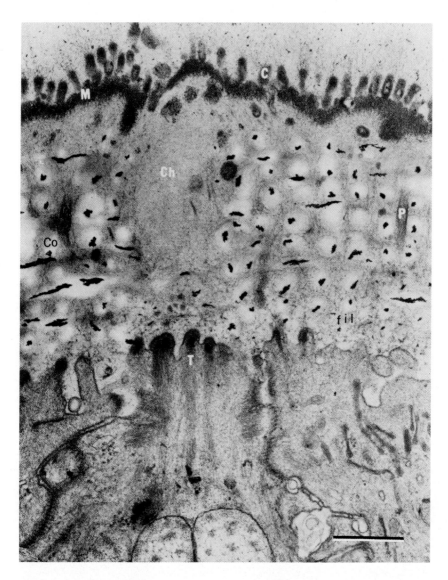

Fig. 5. Electron micrograph of the collagenous cuticle of an earthworm. The cuticle consists of electron lucent collagen fibrils (Co) embedded in filamentous matrix. Fine filaments (fil) separate the collagen fibrils from the apical epithelial surface. The apical cytoplasm is rich in tonofilaments (T) attached to dense cytoplasmic plaques reminiscent of the hemidesmosomes of the basal epithelial surface. Cell processes (P) cross the cuticle and probably give rise to corpuscles (C) on its exterior mat (M). Densities on fibrils are probably staining artifacts. Ch, secretory discharge channel. Bar = 500 nm. From Humphreys and Porter [1976a].

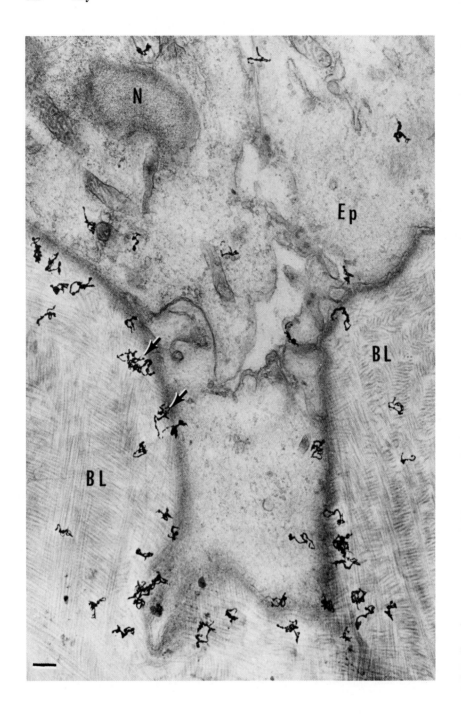

collagen ply (eg, α, Fig. 7) extends for a limited distance along the basal lamina (BM, Fig. 7) before veering off at a slight angle owing to its displacement by the next layer (β, Fig. 7). Tangential sections reveal that the collagen plies are arranged as shingles (Fig. 8C) that are not disposed parallel to the basal lamina, but insert into it at an angle of about 3°. Nadol et al [1969] suggest that "the first fibrils that establish the apparently orthogonal array . . . are all inserted into the basement membrane" (Fig. 8A) and that "with growth more fibrils are inserted into the membrane in each layer" (Fig. 8B).

This hypothesis is compatible with the morphology of other acellular orthogonal stromas. The developing amphibian dermis studied by Hay and Revel [1963], if carefully examined, can be seen to contain subepithelial collagen fibrils that insert into the basal lamina and alternate in direction (eg, α and β, Fig. 9), as would be required by the Nadol et al [1969] theory. Freeze-fractures of developing avian corneal stroma parallel to the long axis reveal shinglelike collections of collagen fibrils (Fig. 10), rather than the plywood sheets formerly predicted from cross sections (Fig. 2). Moreover, collagen fibrils appear to insert into the corneal basal lamina [Figs. 3–9 and 3–11 in Hay and Revel, 1969; Figs. 7–9 in Trelstad and Silver, 1981], as would be predicted by the shingle theory.

The manner in which stromal collagen passes across the basal lamina is unknown. It would probably be "excessive to speculate that collagen fibrils represent a transformation of basement membrane material" [Nadol et al, 1969], because basal lamina collagen seems to be a different type from that found in the stroma [Timpl and Martin, 1982]. How, then, is collagen transported through the lamina meshwork so as to emerge in the form of a striated fibril attached to the filamentous components (Fig. 9) of the lamina rara interna? The possibility that a dense collagenous packet preassembled in a Golgi vacuole is projected intact through the basal lamina into the collagen fibril [Trelstad and Silver, 1981] is not particularly well supported by electron-microscopic observations: The basal lamina itself seems to contain only fine filaments with diameters corresponding to those of individual collagen or procollagen molecules (Fig. 9). Moreover, it is quite clear that the orthogonal stroma also grows interstitially, by addition of collagen to preexisting fibrils

---

Fig. 6. Electron micrograph of an autoradiograph of a section of the subepidermal matrix of a salamander larva. The larva was injected with a single dose of ^{3}H-proline 4 hours before the tissue was fixed. The newly synthesized proline-rich protein (mainly collagen) is secreted into the subepithelial zone of the acellular dermis, the so-called basement lamella (BL). The basal lamina in this region is not well resolved in this micrograph owing to oblique sectioning. Ep, epidermal cytoplasm; N, nucleus. Arrows, silver grains in the overlying emulsion exposed by ^{3}H. Bar = 500 nm. From Hay and Revel [1963].

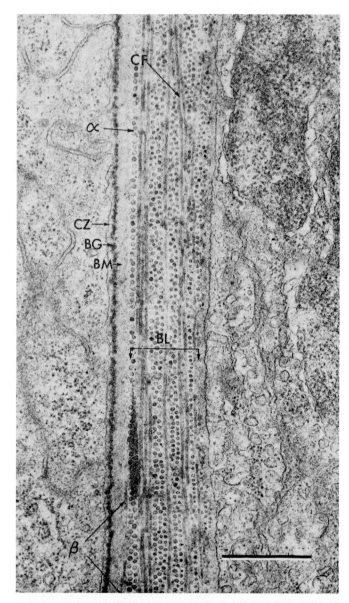

Fig. 7. Electron micrograph of the subepidermal matrix of a hatching fish. Ten orthogonally arranged layers of collagen fibrils (CF) compose this rather narrow basement lamella (BL). The fibrils insert obliquely into the basal lamina (BM). A layer that is cross-sectioned at α, is replaced lower down by a longitudinally sectioned layer, β. BG, basement granules in the lamina rara externa or so-called clear zone (CZ). Bar = 500 nm. From Nadol et al [1969].

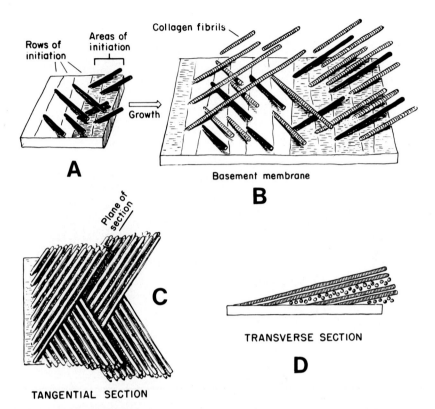

Fig. 8. Diagrams illustrating the scindulene or shingle theory of the development of the orthogonal layers of the fish basement lamella. A) Cell layers are believed to insert into the basal lamina. The number of layers (8–10) remains constant. The basement lamella thickens by an increase in the number of fibrils per layer. B) As the larva grows, the basement lamella expands in diameter by addition of rows of fibrils inserted into the basal lamina or so-called basement membrane. C) The collagen fibrils form shingles inserted into the basal lamina that are best viewed in tangential section. D) In cross section, it can be seen that the layers insert at an angle, exaggerated here. From Nadol et al [1969].

[Hay and Revel, 1963; Nadol et al, 1969]. The role of the epithelium in polymerization of collagen fibrils probably is thus indirect, in the sense that the process is not controlled by direct contact of the cell membrane with each polymerizing molecule.

Oblique sections of the earthworm cuticle also reveal that it consists of a shinglelike array of collagen fibrils [Fib. 12 in Humphreys and Porter, 1976a] that originate at an angle from a finely filamentous material next to the epithelial surface (fil, Fig. 5). The filamentous juxtacellular layer might be

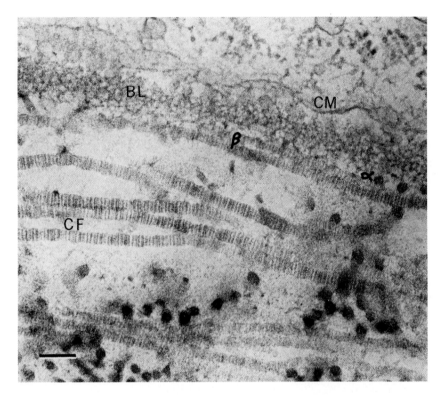

Fig. 9. Electron micrograph of the subepidermal matrix of a salamander larva showing to good advantage the alternating insertion of orthogonally arranged collagen layers (α and β) into the basal lamina (BL). The filamentous component of the basal lamina can be seen to be attached to the basal epithelial cell membrane (CM). Proteoglycan granules are not preserved by O_sO_4 fixation used here. This section of methacrylate embedded tissue was stained with uranyl acetate, which emphasizes the striation pattern of the collagen fibrils (CF). Bar = 100 nm. From Hay [1964], relabeled.

composed of noncollagenous and/or collagenous protein with a "nucleating and organizing capacity" [Humphreys and Porter, 1976b]. There may be a message in the fact that in experimenting with collagen as an exoskeleton in annelids, nature devised an orthogonal array and separated it from the epithelial apical surface by basal laminalike material. The fine filaments that comprise the latter could be "self-ordering, extracellular macromolecules which in turn determine the polymerization of collagen fibrils" [Humphreys and Porter, 1976b].

Fig. 10.   Electron micrographs of a platinum replica of a freeze-fractured and etched 11-day-old embryonic avian cornea to show the shinglelike arrangement of the orthogonal layers of collagen fibrils (arrows). The main figure is a transmission image; the inset is a micrograph of the same specimen taken subsequently with the Jeol 100CX scanning mode. Bar = 2,000 nm.

B. Deposition of Matrix by Mesenchyme

Moving away from epithelium, which is almost invariably separated from collagen fibrils by basal laminalike material, one might ask what is the relation of mesenchymal cells (Fig. 1) to the polymerization of the collagen fibrils that so intimately surround them? Trelstad and Hayashi [1979] suggested that the discharge of collagen aggregates from tendon fibroblasts occurs in deep recesses in the cell surface. They propose that this morphology allows the cell to control the growth of fibrils polymerizing along its plasmalemma. Porter and Pappas [1959], Godman and Porter [1960], and Porter [1964, 1966] proposed some time ago that collagen fibrils might form by a process of ecdysis or shedding from the mesenchymal cell surface (Fig. 11).

Although this process may exist in nature (polymerization of collagen fibrils on or through the plasmalemma), in regenerating connective tissue [Ross and Benditt, 1962, 1965], developing cartilage [Revel and Hay, 1963], and teeth [Weinstock and Leblond, 1971, 1974], it has been shown by

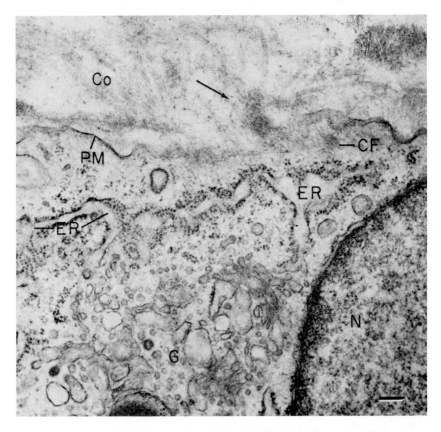

Fig. 11. Electron micrograph of the margin of a fibroblast from the stroma of a hepatoma. At points, the plasmalemma (PM) is transversely sectioned and appears well defined. At other points, presumably because of oblique sectioning, the plasmalemma is not obvious and the dense cortical fibrous material (CF) of the cytoplasm seems continuous with extracellular fibrous material (arrow). Co, collagen fibrils; N, nucleus; ER, endoplasmic reticulum; G, Golgi zone. Bar = 200 nm. From Porter [1964].

autoradiography that newly secreted collagen and other glycoproteins move across previously polymerized ECM before polymerizing at a distance from the cells. Fibroblasts appear to contribute to subepithelial scaffoldings by secreting collagen that diffuses to the subepithelial surface [Hay and Revel, 1963]. Even in bone, which exhibits only appositional growth, the new fibrils are not polymerized directly on the cell surface [Porter, 1966]. Presumably, procollagen is converted to collagen in the extracellular space, and then the molecules polymerize into fibrils [Weinstock and Leblond, 1974; Olsen, 1981].

The very close association of ECM components with the cell surface observed by Porter and his collaborators (see above) and by others may not reflect a direct involvement of the plasmalemma in the polymerization of matrix, but rather may result from a more stable interaction of plasmalemma with already formed matrix that feeds information back to cells. We will examine that possibility in the sections that follow.

IV. MATRIX–CELL SURFACE INTERACTION

A. Fibroblast-ECM Interaction

In electron micrographs published during the 1960s, Porter [1964, 1966] called attention to the very close association of extracellular fibrils, often nonstriated (Co, Fig. 11), with the cell surface of a number of fibroblast type cells. In oblique sections across the plasmalemma, the extracellular fibrils appeared to be continuous with cytoplasmic cortical material (CF Fig. 11) of the same density, leading Porter to conclude that the cortical material was a precursor of the fibrillar extracellular material, presumed to be collagenous. This tendency of extracellular fibrils to coalign with intracellular fibrous components has been rediscovered in recent years by Hynes and Destree [1978], who demonstrated, by double-labeling immunofluorescence, that fibronectin fibrils on the surface of fibroblasts in vitro codistribute with the actin-rich intracellular stress fibers of the cells (Fig. 12).

Singer [1979] confirmed and extended the work of Hynes and others [see Hynes, 1981, for review], providing further evidence for a structural connection between extracellular fibronectin fibrils and intracellular bundles of actin filaments in fibroblasts in vitro. Fibronectin fibrils, identified by ferritin-conjugated antibodies (FN, Fig. 13A), course under the flat cells of planar substrata. Sections cut oblique to the plasmalemma (Fig. 13A–C) show that actin filaments subjacent to the fibrils exhibit a colinear arrangement even when the specimen is tilted through 40° (Fig. 13A, B). Sections cut perpendicular to the cell surface also show that the fibrous components are colinear (Fig. 13D). Singer [1979] termed this junction of extracellular fibrils and actin filaments the "fibronexus," and it is clear that he believed, as did Porter,

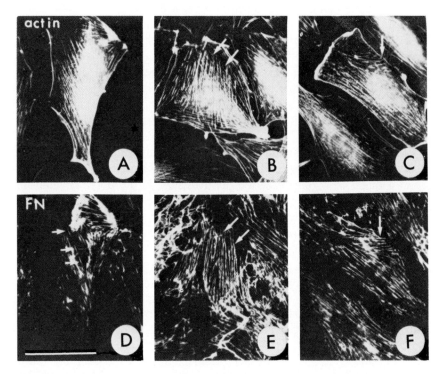

Fig. 12.   Light micrographs of whole mounts of well-spread NIL 8 cells on planar substrata in vitro, viewed by double-label immunofluorescence. A, B, C are stained by antiactin, and D, E, F are photographs of these same cells stained by antifibronectin. In A and D, there are corresponding microfilament bundles and fibronectin fibers in the upper right; arrows indicate interrupted staining with antifibronectin on the under surface of cell. There is good correspondence of microfilament bundles and fibronectin distribution (arrows) in B and E, and C and F. Bar = 50 μ. From Hynes and Destree [1978].

that the extracellular and intracellular components are coaxial—that is, that they connect with each other in the cell membrane. It seems more likely, however, that a binding protein or receptor in the plasmalemma and/or adjacent cytoplasm actually interconnects the two components, as diagrammed (Fig. 14A) by Hynes [1981] and others.

Porter's electron micrographs of sectioned material (Fig. 11) were generally of fibroblasts within extracellular stromas, although Porter and Pappas [1959] also noted very close association of striated and nonstriated fibrils with the surface of fibroblasts grown on planar substrata. It is likely that the nonstriated fibrils (Fig. 11) did contain fibronectin and the cortical fibrous material, actin. Fibroblasts in situ rarely exhibit stress fibers, however. Thus,

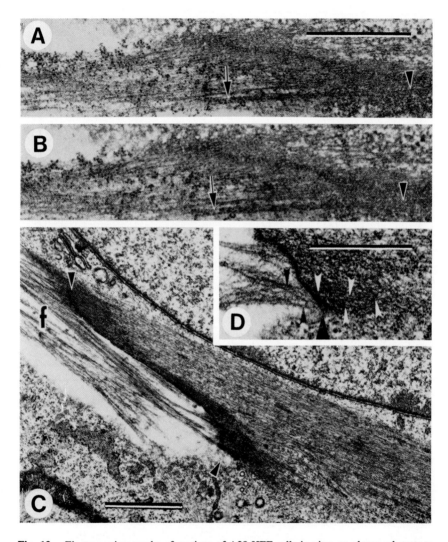

Fig. 13. Electron micrographs of sections of A28-HEF cells in vitro on planar substrata to show the relation between fibronectin (labeled in A and B with ferritin-conjugated antibodies) and the intracellular microfilament bundles. A, B) Micrographs of the same specimen tilted 40°. The section is oblique through the plasmalemma. Note that extracellular fibrils (arrow) and intracellular filaments (arrowheads) remain coaligned through the tilting. C) Oblique en face section of fibronexus stained with tannic acid. Fibronectin fibers (f) insert into a dense membranous zone. Arrowheads indicate obliquely cut cell membrane. D) Transversely sectioned fibronexus. Fibronectin (black arrowheads) appear to enter the plasmalemma to connect with cytoplasmic microfilaments (white arrowheads) embedded in the dense submembranous zone. Bar = 500 nm. From Singer [1979].

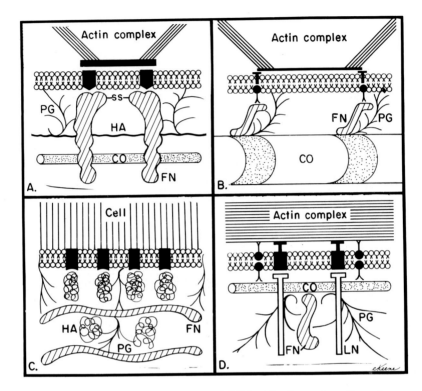

Fig. 14. Diagrams of postulated relations of ECM molecules to the cell membrane and cytoskeleton. A) Based on Hynes [1981a]. B) Based on Kleinman et al [1981]. C) Based on Toole [1981]. D) Based on Sugrue and Hay [1982a]. CO, collagen; FN, fibronectin; HA, hyaluronic acid; LN, laminin; PG, proteoglycan. All models envision receptors for one or more ECM molecules in the plasmalemma rather than direct continuity of ECM with cytoplasmic actin. From Hay [1981b].

a true fibronexus (Fig. 13A) is an in vitro phenomenon reflecting the very strong attachment of the stress fibers of highly flattened cells to planar substrata [Tomasek et al, 1982]. However, fibroblasts in situ or within ECM in vitro do exhibit an apparent capacity for transmembrane interaction of the cytoskeleton with ECM even though they form less striking attachments to fibronectin and collagen (Fig. 11) [Bard and Hay, 1975; Tomasek et al, 1982].

A considerable amount of work has been done recently with cultured fibroblasts and other cell types to identify the proteins responsible for cell attachment to uncoated and collagen coated plastic dishes (planar substrata). Fibroblastic cells adhere to plastic via an "adhesion" protein derived from serum [Grinnell, 1978] that is identical to plasma fibronectin [Table II, Hynes 1981] and that promotes cell spreading as well as attachment. Fibroblasts producing fibronectin do not require exogenous fibronectin to attach to plastic or collagen in vitro, but exogenous fibronectin permits transformed cells with decreased surface fibronectin to attach and spread on planar substrata [Yamada et al, 1976]. Although fibronectin seems to be needed for attachment of fibroblasts to denatured collagen in vitro, it may not always be required for cells to adhere to native collagen fibrils [Grinnell and Bennett, 1981]. Kleinman et al [1981] propose that fibronectin may mediate cell attachment to collagen via a fibronectin receptor, possibly a ganglioside, in the plasmalemma (Fig. 14B). Collagen receptors [Goldberg, 1979] and PG binding sites (Fig. 14C) [Toole, 1981] might also be present in the fibroblast plasmalemma.

B. Epithelium-ECM Interaction

Other adhesion factors have been identified in vitro by this same type of attachment assay. Chondrocytes require a unique glycoprotein, chondronectin, to attach to type II collagen [Hewitt et al, 1980]. Certain epithelial cell types that do not synthesize fibronectin use still another large glycoprotein, laminin, to attach to type IV collagen [Kleinman et al, 1981; Timpl and Martin, 1982]. Not all epithelia require laminin to attach to collagen, however. Some hepatocytes seem to use fibronectin [Berman et al, 1980]; others attach to collagen without helper proteins [Rubin et al, 1981]. Hepatocyte receptors for collagen may recognize repeating sites (eg, Gly-Pro-Pro) in the molecule, and it has been speculated that they are mobile in the cell membrane, moving over a period of time to the base of attached hepatocytes to increase cell-substratum adhesion [Rubin et al, 1981].

The interaction of the epithelial cell surface with ECM molecules has been studied in other ways in vitro. Isolated embryonic avian epidermis [Dodson, 1963], corneal epithelium [Dodson and Hay, 1971, 1974], and spinal cord [Cohen and Hay, 1971] send out cytoplasmic processes or blebs from their basal surfaces and differentiate poorly on Millipore filters; when recombined in culture with frozen-killed dermis or lens capsule, however, these epithelia assume a more normal appearance. Similarly, when mammary gland epithelia are cultured on floating collagen gels, they are able to differentiate remarkably well [reviewed by Bissell, 1981; Bissell et al, 1982]. The cells assume the polarity and the ability to respond to hormones that is characteristic of mammary epithelium in vivo [Emerman et al, 1977].

The basal surface of the embryonic corneal epithelium responds equally well to soluble or polymerized ECM molecules. Even without a supporting substratum under their basal surface, isolated corneal epithelial cells withdraw the blebs (arrows, Fig. 15B) and flatten their basal surface (Fig. 15A) within

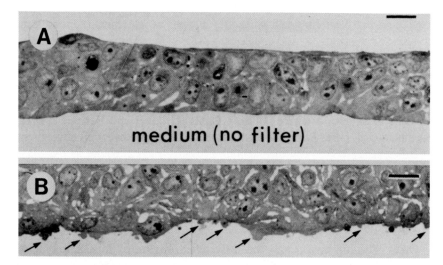

Fig. 15. Light micrographs of sections of isolated embryonic avian corneal epithelial grown in organ culture dishes at the air-medium interface, with no filter support underneath. The epithelium is attached by polylysine to an overlying filter on its free surface. In A, 100 g/ml of solubilized type 1 collagen was added to the medium 6 h before fixation. The basal epithelial surface has flattened. In B, no ECM was added to the medium. The basal surface shows numerous blebs (arrows). Bar = 10 m. From Sugrue and Hay [1981a].

6 hours after soluble collagen (types I–IV), laminin, or fibronectin is added to the culture medium. They continue to bleb in the absence of exogenous ECM [Sugrue and Hay, 1981a], and they resume blebbing if the exogenous ECM molecules are removed from the culture medium [Sugrue and Hay, 1982b].

The flat basal surface configuration induced by ECM seems to be brought about by the assembly of an actin-rich cortical cytoskeleton (mf, Fig. 16A) similar to that observed in the basal cytoplasm of the epithelial cells in vivo. In vivo (Fig. 16B) and in vitro, the cortical filaments exhibit antiparallel orientation when labeled by heavy meromyosin subfragments (HMS). In the blebs, however, actin filaments decorated with HMS exhibit a polarity pointing toward the plasmalemma (Fig. 16C). In microvilli, the polarity is away from the plasmalemma (Tilney, this volume). In ruffling membranes of cells grown on plastic, however, HMS arrowheads point toward the base of the ruffle [Buckley, 1981]. The difference probably lies in the fact that the blebs and ruffles are highly motile. The epithelial blebs extend into filters and can be withdrawn into the cell within a few hours after encountering ECM [Sugrue and Hay, 1981a]. The same actin filaments (Fig. 16C) seem to be used to reconstruct the cortical cytoplasm (Fig. 16A), for the process can occur in the absence of new protein synthesis [Sugrue and Hay, 1982a].

Sugrue and Hay [1982a] postulate that collagen and laminin have different receptors in the cell membrane (Fig. 14D). Either of these molecules is able to flatten the basal epithelial surface in the absence of an endogenous source of ECM molecules. On the contrary, both cycloheximide and LACA (L-azetidine-4-carboxylic acid) abolish the effect of fibronectin, suggesting that fibronectin acts by combining with endogenous collagen secreted by the cells [Sugrue and Hay, 1982a]. The effect on the basal cytoskeleton seems to be mediated by a transmembrane connection of actin and ECM (Fig. 14D), and, moreover, the reorganization of the basal cytoplasm is accompanied by a twofold increase in the amount of collagen and GAG synthesized by the cells [Meier and Hay, 1974a; Sugrue and Hay, 1981b].

The two-cell-thick corneal epithelium studied by Meier and Hay [1974a, b] and Sugrue and Hay [1981a, b; 1982a, b] does not initially exhibit specialized matrix attachments to the basal lamina, but after further development this corneal epithelium stratifies to six or more cells in thickness and develops definitive hemidesmosomes into which basal lamina filaments clearly insert [Figs. 5-13 and 5-14 in Hay and Revel, 1969]. Most stratified epithelia on the outer surfaces of the body develop hemidesmosomes (Fig. 4), suggesting that these cell-ECM junctions enhance the attachment of epithelial cells to the underlying lamina. Other examples of well-developed cell-ECM insertion plaques are seen in smooth and cardiac muscle, whose cells are also attached to basal laminae. In addition to its filtering functions [Farquhar, 1981], then, the basal lamina may be uniquely adapted for the formation of dense cytoplasmic plaques like hemidesmosomes that interact with underlying ECM in various ways [Palade and Farquhar, 1965] to anchor the cells. Although fibroblasts may develop adhesion plaques to collagen in planar substrata [Grinnell and Bennett, 1981], they rarely exhibit prominent adhesion plaques when they lie within collagenous matrices in vivo [Porter, 1964, 1966] or in vitro [Tomasek et al, 1982].

V. MATRIX EFFECTS ON CELL SHAPE
A. Fibroblast Shape and Cytoskeleton

Fibroblasts show a very dramatic change in cell shape in response to collagen. As we have already noted, on planar substrata to which they strongly attach, isolated fibroblasts spread to become highly flattened cells. If, however, they are suspended within collagen gels, they take on a bipolar shape within a few hours (Fig. 17A, B) and over the next 12 hours become quite

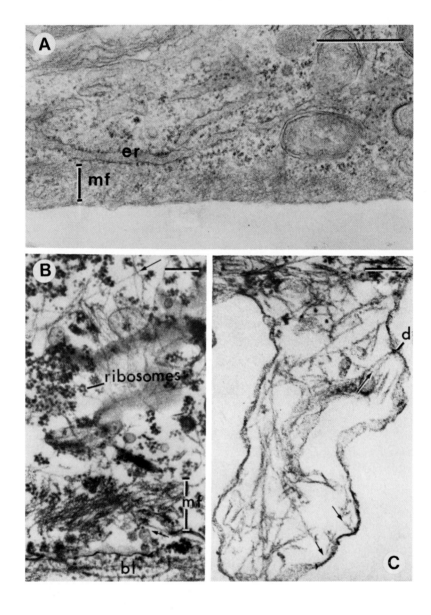

elongate (Fig. 17C). Their ability to become bipolar in the collagen lattice is abolished by treatment with cytochalasin D, suggesting that this component of their elongation (step I, Fig. 17) depends on transmembrane actin-ECM interaction [Tomasek and Hay, 1982]. They are able to become bipolar in the presence of microtubule-disrupting drugs, but are unable to elongate substantially, indicating that the final elongation of the cell (step II, Fig. 17C) is related to tubulin organization [Tomasek and Hay, 1982].

Immunofluorescent studies reveal that the elongated fibroblasts in collagen gels (lower right, Fig. 18) are rich in microtubules that stain with antitubulin and run the length of the cell [Tomasek et al, 1982]. Actin is located in the cell cortex and filopodia, whereas myosin is diffusely distributed in the cytosol (upper right, Fig. 18) [Tomasek et al, 1982]. The flat cells on planar substrata show quite a different cytoskeletal organization. Stress fibers that stain with antiactin and antimyosin are prominent (upper left, Fig. 18), and microtubules form a meshwork in the cytoplasm (lower left, Fig. 18).

Electron micrographs of thin sections (Fig. 19) confirm the impression gained by immunofluorescence (Fig. 18) that stress fibers are not present in corneal and fibroma fibroblasts grown inside collagen gels [Tomasek et al, 1982]. Fibroblasts in collagen gels have an actin-rich cortex (Fig. 19A, B) composed of fine filaments that are closely related to the plasmalemma. The cell membrane may be dense in areas where the cells contact collagen fibrils [Porter, 1966; Hay, 1978], but true adhesion plaques are not observed (Fig. 18B). Filopodia also contain fine actin filaments (Fig. 19C). Fibroblasts in collagen gels are not surrounded by nonstriated, fibronectinlike fibrils, but do produce fibronectin [Sugrue and Hay, 1982a]. We do not yet know whether helper proteins play a role in this cell-ECM interaction (Fig. 19). Interestingly, soluble collagen added to flat fibroblasts on planar substrata does not cause the cells to elongate presumably because the attachment to the substratum is too strong [Tomasek et al, 1982].

Fig. 16. Electron micrographs of routinely fixed (A) and heavy meromysin–treated (B, C) corneal epithelium. A) Micrograph of the flattened basal surface of a corneal epithelium treated with soluble ECM as in Figure 15A. This tissue was fixed by routine methods for electron microscopy, and so the components of the microfilamentous cortical mat (mf) are barely resolvable. Note the close relation of endoplasmic reticulum (er) and ribosomes to the cortical cytoskeleton. Bar = 500 nm. B) Portion of the basal cytoplasm of an epithelial cell fixed in situ in the cornea, extracted with Triton, and stained with heavy meromysin subfragments. The microfilamentous cortical mat (mf) is seen to be composed of decorated actin filaments (arrow). Bar = 200 nm. C) Micrograph of a bleb similar to those depicted in Figure 15B, from isolated epithelium decorated with meromysin as in B. The actin filaments of the cortical cytoskeleton become dispersed in blebs when ECM is removed. Arrows indicate direction of decoration is toward the plasmalemma. d, Dense insertion. Bar = 200 nm. From Sugrue and Hay [1981a, 1982a].

534   Hay

CELL SHAPE IN COLLAGEN LATTICES

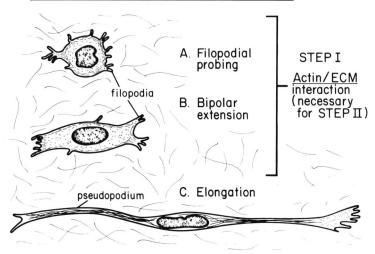

A. Filopodial
   probing

B. Bipolar
   extension

filopodia

pseudopodium   C. Elongation

STEP I

Actin/ECM
interaction
(necessary
for STEP II)

STEP II: <u>Microtubule</u> dependent lengthening of
           pseudopodia (requires STEP I)

Fig. 17. Diagram of steps in the elongation of an isolated corneal fibroblast suspended in a collagen gel. The rounded cell begins to probe the collagen lattice with filopodia (A) and within 6 h becomes bipolar in shape (B). This step is inhibited by cytochalasin and thus is probably an actin/ECM interaction. Further elongation (C) requires another 12 h and is dependent on microtubule integrity. Based on data of Tomasek and Hay [1982].

B. Epithelial Cell Shape and Polarity

Mauchamp and his co-workers [Mauchamp et al, 1979; Chambord et al, 1981] report that when cysts of thyroid epithelium are covered with collagen gels or suspended in solutions of gelatin, they differentiate into thyroid follicles lined with a simple cuboidal epithelium, whose microvillus-lined apical surface faces inward. The surface facing the collagen gel is flat, like the basal surface in situ. Soluble (denatured) collagen in solution also induces this "normal" polarity [Mauchamp et al, 1979], whereas in medium rich in serum, enzyme-isolated thyroid follicles reverse polarity, shifting their apical poles to the outside [Nitsch and Wollman, 1980].

In contrast, when enzyme-isolated corneal or limb epithelia are suspended in collagen gels (upper left, Fig. 20), they do not form follicles, but instead send off mesenchymelike cells into the gel [Greenburg and Hay, 1982]. Greenburg and Hay [1982] also isolated adult or embryonic anterior lens epithelium resting on intact basal lamina by dissection; the lamina (lens capsule) faces the aqueous humor, and there are no mesenchymal cells in

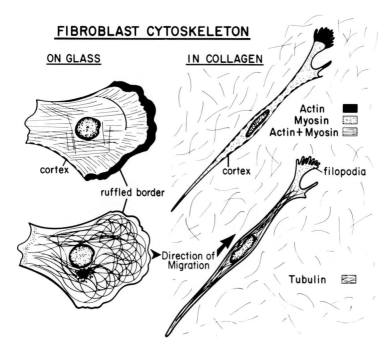

Fig. 18. Diagram of the form of the cytoskeleton of corneal fibroblasts grown on glass (left) or in collagen gels (right). Immunofluorescence shows that actin and myosin occur in stress fibers in cells on glass (upper left). Stress fibers are not present in cells in collagen lattices (upper right). The microtubule pattern also changes when cells are grown in gels (lower left and right). Based on data of Tomasek et al [1982].

the area that would contaminate the preparation (upper right, Fig. 20). The simple cuboidal lens epithelium was then suspended in a gelling solution of collagen (lower right, Fig. 20). After a few days in culture, the epithelium multilayered and sent off mesenchymelike cells from its apical surface (Fig. 21). The individual, bipolar cells migrate for considerable distances through the collagenous matrix, behaving much as if they were fibroblasts.

These results raise a number of questions regarding the control of epithelial and mesenchymal cell shape (Fig. 1). When epidermal and corneal epithelia are placed on top of collagenous matrices, the basal surface flattens and the cells become cuboidal even if the cells are not confluent [Overton, 1977; Hay, 1982]. Isolated kidney (MDCK) cells were shown by Sabatini and co-workers (Sabatini et al, this volume) to reorganize their basal surface on contact with a planar substratum. The isolated cells presumably mix apical and basolateral components of the cell membrane while in suspension, and sort these out on contact with the substratum (Sabatini et al, this volume).

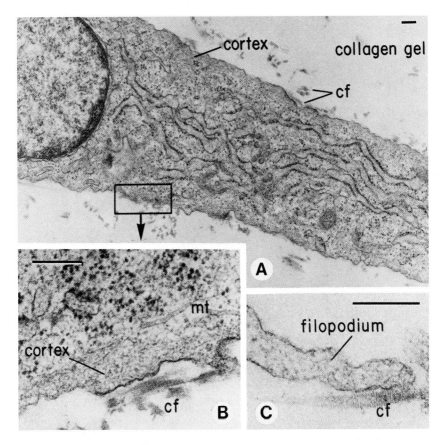

Fig. 19. Electron micrographs of fibroblasts cultured in collagen gels. A) The actin-rich cell cortex contains microfilaments and the adjacent plasmalemma contacts collagen fibrils (cf). B) Higher magnification of a region of the cell cortex similar to that enclosed by the square in A. The close proximity of the plasmalemma to collagen fibrils (cf) is illustrated. mt, Microtubules, C) Filopodia contain an actin-rich poorly defined meshwork of microfilaments. cf, Collagen fibril. Bar = 100 nm. From Tomasek et al [1982].

As we noted earlier, isolated hepatocytes are believed to circulate collagen receptors to the basal surface on contact with collagen [Rubin et al, 1981]. Thus, one mechanism controlling epithelial shape and polarity seemingly resides in the ability of receptors characteristic of the basal surface to attach to an underlying substratum in a way that flattens the basal cytoskeleton.

The isolated lens cells (Figs. 20, 21) did not disobey this rule, for the cells attached to lens capsule by their basal surface remain epithelial [Green-

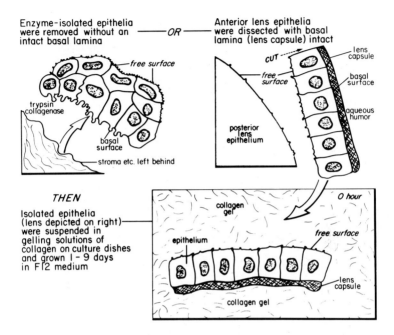

Fig. 20. Diagram of method of isolating epithelia that are then suspended in collagen gels. Based on Greenburg and Hay [1982].

burg and Hay, 1982]. The apical surface, not the basal surface, responded to the immersion in collagen by sending out filopodia and moving into the gel. Enzyme-isolated epithelia suspended in gels also give rise to mesenchymelike cells, but from which surface is unknown [Greenburg and Hay, 1982]. Differences in the response of epithelia to collagenous environments may reflect differences in culture techniques and cell potentialities, rather than disobedience to the basic rules of apical-basal polarity. Mammary epithelial cell line Rama 25 grown on the surface of floating collagen gels sends branching outgrowths into the matrix, but does not form outgrowths when grown on top of nonfloating gels [Ormerod and Rudland, 1982]. Sandwiching the mammary epithelium between two collagen gels results in formation of spindle-shaped cells [Ormerod and Rudland, 1982]. Chambord et al [1981] covered thyroid monolayers with collagen gels, much as did Ormerod and Rudland [1982], but the thyroid cells formed follicles, presumably because they have this particular potential, rather than the potential to form spindle-shaped cells and/or branches.

During embryonic development, basal epithelial surfaces are altered in predictable, seemingly predetermined ways. In the early embryo, somite and

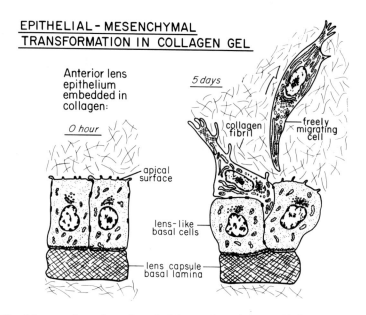

EPITHELIAL - MESENCHYMAL
TRANSFORMATION IN COLLAGEN GEL

Anterior lens
epithelium
embedded in
collagen:

5 days

O hour

collagen
fibril

freely
migrating
cell

apical
surface

lens-like
basal cells

lens capsule
basal lamina

Fig. 21. Diagram of transformation of adult or embryonic lens epithelium to mesenchymal cells. The epithelium was isolated with its basal lamina (lens capsule) intact. After 5 days of culture inside a collagen gel, the epithelium has multilayered and is sending off mesenchymelike cells from its apical surface into the gel. The basal cells in contact with lens capsule remain lenslike. Based on data of Greenburg and Hay [1982].

neural epithelia give rise to mesenchymal cells at a predetermined time. They secrete hyaluronic acid into intercellular spaces before or after the epithelium multilayers, expanding the tissue which now breaks its basal lamina to spew mesenchymelike cells into the surrounding ECM [Solursh et al, 1979; Nichols, 1981, 1982]. Cardiac cushion endothelium seeds underlying hyaluronic acid–rich matrix with mesenchymal cells at a particular time in development [Bernanke and Markwald, 1982]. These secondary mesenchymal cells usually do not form true epithelia again, but primary mesenchymal cells derived from the primitive streak remain coded to redifferentiate into epithelia [Trelstad et al, 1967; Hay, 1968]. When primary mesenchymal cells form kidney tubules, they first begin to secrete laminin in a spotty fashion, then aggregate, and take on epithelial polarities [Ekbloom, 1981]. It has recently been proposed that during regression of the Mullerian duct, the basal lamina is altered and some epithelial cells, instead of dying as we thought, migrate out into the mesenchymal cell compartment [Trelstad et al, 1982]. This possibility does not seem preposterous when the seemingly latent ability of epithelium to form mesenchymal cells (Fig. 21) is taken into account.

VI. MATRIX EFFECTS ON CELL METABOLISM

We have already noted that mammary epithelial cells grown on floating collagen gels alter their shape and their response to hormones [Emerman et al, 1977; Bissell, 1981; Bissell et al, 1982]. Gospodarowicz et al [1978] found that dissociated bovine corneal epithelial cells grown on plastic dishes form flattened monolayers that synthesize DNA in response to fibroblast growth factor (FGF), but that do not respond to epidermal growth factor (EGF). Grown on collagenous matrix, however, corneal epithelial cells become cuboidal and proliferate in response to EGF, but not FGF. Interestingly, another flattened cell type, bovine vascular epithelium, also responds to FGF by proliferating, but if cultured without FGF, becomes immersed in ECM and mesenchymelike in behavior [Greenburg et al, 1980]. Collagenous matrices [Liu and Karasek, 1978] and feeder cells [Rheinwald and Green, 1977] also alter the response of other epithelial cell types to growth factors.

The response of epithelial cell metabolism to culture on collagen gel varies from epithelium to epithelium, but is generally favorable. We saw how the avian embryonic corneal epithelium grown on collagenous matrices steps up its synthesis of corneal stroma components [Meier and Hay, 1974a]. When such epithelia are grown on Millipore filters, they reorganize their basal cytoskeleton in response to soluble collagen molecules just as they did in response to polymerized ECM [Sugrue and Hay, 1981a]. Soluble collagen (100 μg/ml), fibronectin (50 μg/ml), and laminin (2 μg/ml) stimulate such epithelial cells to produce twice as much collagen as they would have in the absence of the exogenous ECM [Sugrue and Hay, 1982b]. GAG does not flatten the basal surface [Sugrue and Hay, 1981a], nor does it stimulate collagen synthesis, although GAG may enhance GAG production by the cells [Meier and Hay, 1974b].

It is tempting to believe that the elongate corneal and fibroma fibroblasts in collagen gels (Fig. 18, right) are more metabolically active than their flattened counterparts on planar substrata (Fig. 18, left) and that the organelle-rich mesenchymal cells derived from lens epithelium (Fig. 21) have started to secrete new proteins. We have, however, only begun to examine this possibility. Certain ECM-derived molecules inhibit ECM synthesis [see Toole, 1981; Horlein et al, 1981], but ECM can enhance collagen synthesis by newly seeded tendon fibroblasts [Bissell et al, 1982]. Nevo and Dorfman [1972] reported that exogenous GAG (eg, chondroitin sulfate) enhances GAG synthesis by epiphyseal chondrocytes, and Kosher and Church [1975] found that both collagens and soluble procollagens stimulate GAG production by somite chondrocytes. These latter findings are of considerable interest when considered in the context of tissue interaction in the embryo, for the inductive effect of the notochord on somite chondrogenesis can be abolished by enzymatic removal of its ECM, and exogenous chondromucoprotein can substitute for notochord in "inducing" somite chondrogenesis in vitro [Kosher et al, 1973; Kosher and Lash, 1975].

Considerable data have accumulated which suggest that in other embryonic tissue interactions, production of ECM by one tissue stabilizes or enhances differentiation of an adjacent tissue. In avian corneal development, the lens appears to be active in the induction of corneal stroma production by corneal epithelium. Since the lens produces considerable ECM early in development [Hay and Meier, 1974] and ECM molecules of this type stimulate corneal epithelium to produce stroma (see above), it is easy to believe that the "inductive" effect of the lens is mediated by the ECM it secretes under the corneal epithelium [Hay, 1980]. The "inductive" effect of fibroblasts on muscle cells can be mimicked by exogenous collagen [Konigsberg and Hauschka, 1965]. The "inductive" effect of matrix on gland differentiation postulated by Grobstein [1955] seems to be indirect, in the sense that it is mesenchymal removal of basal lamina components from the tips of lobules that seems to enhance gland branching and outgrowth [Bernfield and Barnerjee, 1978]. It is possible, however, that collagen assists the branching process by stabilizing epithelial surfaces lining the interlobular clefts; indeed, collagen has been shown to reduce GAG turnover in the ECM that accumulates under mammary gland epithelium in vitro [David and Bernfield, 1981].

In this volume, Penman (Penman et al, this volume) suggests that the organization of the cytoskeleton might affect protein synthesis and other aspects of cell metabolism by dictating the arrangement of polyribosomes associated with cytoplasmic microtrabeculae (Porter-plasm, Porter et al, this volume). Cell shape is known to affect mRNA synthesis in anchorage-dependent fibroblasts [Benecke et al, 1978; Ben-Ze'ev et al, 1980], and to modulate DNA synthesis by endothelial cells, fibroblasts, and several cell types cultivated as cell lines [Folkman and Tucker, 1980]. It is easy to believe that the cytoskeleton can regulate cell shape and thus, directly or indirectly, the metabolism of the cytoplasm and the nucleus.

It seems clear, moreover, that the stimulatory effects of ECM on embryonic corneal epithelial production of stroma are always correlated with cytoskeletal organization [Meier and Hay, 1974a; Sugrue and Hay, 1981a, b, 1982a, b] (Fig. 16). A model for this interaction might be envisioned which states that ECM molecules organize the cortical cytoskeleton by a transmembrane interaction with actin, and that the cytoskeleton, in turn, organizes the secretory organelles (endoplasmic reticulum, Golgi apparatus) involved in collagen production to promote the synthesis and secretion of collagen by the corneal epithelial cells. It will be interesting in the future to investigate the relation of cytoskeletal organization to cell metabolism in other tissue interactions that implicate ECM in tissue morphogenesis. Not only do collagen and structural glycoproteins interact with the cell surface, but also GAG on the plasmalemma (Figs. 3, 22) may affect cell growth and differentiation [Kraemer and Tobey, 1972; Kraemer, 1979; Toole, 1981; Kjellen et al, 1981].

VII. SUMMARY AND CONCLUDING REMARKS

In embryonic and adult vertebrates, the body is composed of epithelial and mesenchymal cells, or their derivatives, which, with the exception of blood cells, form stable interactions with extracellular matrices. The extra-

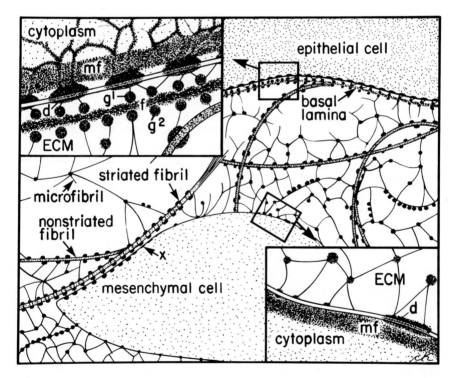

Fig. 22.   Diagram of morphology of basal epithelial surface, mesenchymal cell surface, and ECM as revealed by ruthenium red staining. mf, Microfilaments in cell cortex; g^1, proteoglycan granule in lamina rara interna; g^2, proteoglycan granule in lamina rara externa; f, lamina densa of the basal lamina; d, ruthenium red–staining densities in the plasmalemma; x, close relation of striated collagen fibril to plasmalemma of mesenchymal cell. The nonstriated fibrils in the matrix may contain fibronectin and the microfibrils (3–5 nm diameter) or filaments may contain hyaluronic acid. From Hay [1978].

cellular matrix and the cells form a continuum, in which molecules of the matrix are intimately attached to the cell and the cell to them so that, in a sense, one can say that the cell surface does not actually end at the level of the plasmalemma, but includes the adjacent ECM (Fig. 22). In this chapter, we presented a number of examples of the influence of ECM molecules, including collagen, structural glycoproteins, and GAG, on cytoskeletal organization, cell metabolism, and cell reaction to growth factors and hormones. Cells affect themselves and cells of other tissues by the matrix molecules they manufacture in their endoplasmic reticulum and secrete via Golgi vacuoles and vesicles into the ECM.

We considered the manner in which the secreted ECM molecules might be assembled into structural matrices. The close structural relation of extracellular matrix to the fibroblast cell surface originally emphasized by Porter has been confirmed and extended in recent years. Current interpretation is that this ECM-cell interaction influences mesenchymal cell shape and synthetic activities, but is not a form of ecdysis. Rather, the molecules on or in the cell membrane seem relatively stable once produced. Newly synthesized contents of Golgi vacuoles are revealed by autoradiography to be discharged from the cell to polymerize away in the interstices of existing ECM, at least in cartilage, teeth, and regenerating connective tissue. The mesenchymal cell presumably controls the pattern of polymerization indirectly by the nature of the previously secreted ECM, but other forces may be at work as well (Nuccitelli, this volume).

Epithelia lining the outer surfaces of the body create elegant orthogonal lattices of collagen fibrils in the dermis of aquatic vertebrates and cornea of all vertebrates. Porter has for some time held to the view that the ends of the orthogonal collagen layers insert obliquely into the epithelial basal lamina where their elongation is controlled by the filamentous material there. One might envision the dynamic construction occurring in the cornea and dermis of vertebrate larvae and embryos in the following way. The basal lamina is the first ECM to form under the epithelium, and is, except for expansion to keep pace with body growth, a relatively stable structure that binds basal plasmalemmal ECM receptors and interacts thereby in important ways with the basal epithelial cytoskeleton. Collagen destined for the stromal compartment is of a different type from that in the lamina and thus cannot polymerize in the lamina, nor does it bind to the saturated plasmalemmal receptors. It seems drawn instead into predestined polymerization sites in the lamina rara interna, where the ends of the orthogonal array are elongating, and into the depths of the stroma where existing fibrils are increasing in diameter. What is truly remarkable is that the developing cornea and dermis are completely functional in swimming vertebrate larvae and that the basal lamina is always structurally intact throughout the period of growth and remodeling.

There are many other examples of the mutual dependence of cell and extracellular matrix during embryogenesis. We are currently trying to decipher the rules that underlie the different manner in which the matrix is perceived by epithelial cells as compared to mesenchymal cells. Why do epithelia reside on basal laminae in situ when collagen gels alone support their growth and differentiation in vitro? Why do epithelia give rise to mesenchymal cells at certain times and places during embryogenesis, but appear to be so stable in the adult? Does the secret lie in the nature of the epithelial basal surface? The basal surface normally flattens when epithelia are grown on top of ECM. Epithelia programmed to produce mesenchyme do not follow this rule, however: Basal cells move down into the ECM, presumably after secreting GAG around themselves. Moreover, the epithelia we suspended within collagen gels reacted with collagen via their apical surfaces in much the same way as mesenchymal cells do; the outer cells sent out filopodia, elongated, and entered the collagenous matrix as single cells. Do all epithelia retain an inherent desire to become mesenchymal under appropriate circumstances?

These questions require for answers, not just a knowledge of the chemistry of atoms and molecules, but a complete understanding of how they are put together in cells, in the extracellular matrix, and in the organism.

ACKNOWLEDGMENTS

The research by the author summarized in this chapter was supported by a grant HD-00143 from the United States Public Health Service.

VIII. REFERENCES

Bard JBL, Hay ED (1975): The behavior of fibroblasts from the developing avian cornea: Their morphology and movement in situ and in vitro. J Cell Biol 67:400–418.

Benecke BJ, Ben-Ze'ev A, Penman S (1978): The control of mRNA production, translation, and turnover in suspended and reattached anchorage-dependent fibroblasts. Cell 14:931–939.

Bennett HS (1963): Morphological aspects of extracellular polysaccharides. J Histochem Cytochem 11:14–23.

Ben-Ze'ev A, Farmer SR, Penman S (1980): Protein synthesis requires cell-surface contact while nuclear events respond to cell shape in anchorage dependent fibroblasts. Cell 21:365–372.

Berman MD, Waggoner JG, Foidart JM, Kleinman HK (1980): Attachment to collagen by isolated hepatocytes from rats with induced hepatic fibrosis. J Lab Clin Med 95:660–665.

Bernanke DH, Markwald RR (1982): Migratory behavior of cardiac cushion tissue cells in a collagen-lattice culture system. Dev Biol 91:235–245.

Bernfield MR, Banerjee SD (1978): The basal lamina in epithelial-mesenchymal morphogenetic interactions. In Kefalides N (ed): "Biology and Chemistry of Basement Membranes." New York: Academic Press, pp 137–148.

Bernfield MR, Banerjee SD, Cohn RH (1972): Dependence of salivary epithelial morphology and branching morphogenesis upon acid mucopolysaccharide-protein (proteoglycan) at the epithelial surface. J Cell Biol 52:674–689.

Bissell MJ (1981): The differentiated state of normal and malignant cells or how to define a "normal" cell in culture. Int Rev Cytol 70:27–100.

Bissell MJ, Hall HG, Parry G (1982): How does the extracellular matrix direct gene expression? J Theoret Biol (in press).

Bloom W, Fawcett DW (1975): "A Textbook of Histology." Philadelphia: Saunders.

Bornstein P, Sage H (1980): Structurally distinct collagen types. Ann Rev Biochem 49:957–1003.

Buckley IK (1981): Fine-structural and related aspects of nonmuscle-cell motility. In Dowben RM, Shay JW (eds): "Cell and Muscle Motility," Vol I. New York: Plenum, pp 135–203.

Chambord M, Gabrion J, Mauchamp J (1981): Influence of collagen gel on the orientation of epithelial cell polarity: Follicle formation from isolated thyroid cells and from preformed monolayers. J Cell Biol 91:157–166.

Chen LB, Murray A, Segal RA, Bushnell A, Walsh ML (1978): Studies on intercellular LETS glycoprotein matrices. Cell 14:377–391.

Cohen AM, Hay ED (1971): Secretion of collagen by embryonic neuroepithelium at the time of spinal cord–somite interaction. Dev Biol 26:578–605.

Coggeshall R (1966): A fine structural analysis of the epidermis of the earthworm, Lumbricus terrestris L. J Cell Biol 28:95–108.

David G, Bernfield M (1981): Type I collagen reduces degradation of basal lamina proteoglycan by mammary epithelial cells. J Cell Biol 91:281–286.

Dodson JW (1963): On the nature of tissue interactions in embryonic skin. Exp Cell Res 31:233–235.

Dodson JW, Hay ED (1971): Secretion of collagenous stroma by isolated epithelium grown in vitro. Exp Cell Res 65:215–220.

Dodson JW, Hay ED (1974): Secretion of collagen by corneal epithelium. II. Effect of the underlying substratum on secretion and polymerization of epithelial products. J Exp Zool 189:51–72.

Edds MV (1964): The basement lamella of the developing amphibian skin. In "Small Blood Vessel Involvement in Diabetes Mellitus." Washington: AIBS, pp 245–250.

Edds MV, Sweeny P (1961): Chemical and morphological differentiation of the basement lamella. In Rudnick D (ed): "Molecular and Cellular Synthesis." New York: Ronald Press, pp 111–138.

Ekbloom P (1981): Formation of basement membranes in the embryonic kidney: An immunohistological study. J Cell Biol 911:1–10.

Emerman JT, Enami J, Pitelka DR, Nandi S (1977): Hormonal effects on intercellular and secreted casein in cultures of mouse mammary epithelial cells on floating collagen membranes. Proc Natl Acad Sci USA 74:4466–4470.

Farquhar MG (1981): The glomerular basement membrane. In Hay ED (ed): "Cell Biology of Extracellular Matrix." New York: Plenum, pp 335–378.

Folkman J, Tucker RW (1980): Cell configuration, substratum, and growth control. In Subtelny S, Wessells NK (eds): "The Cell Surface: Mediator of Developmental Processes." New York: Academic Press, pp 259–275.

Godman GC, Porter KR (1960): Chondrogenesis studied with the electron microscope. J Biophys Biochem Cytol 8:719–760.

Goldberg B (1979): Binding of soluble type I collagen molecules to the fibroblast plasma membrane. Cell 16:265–275.

Gospodarowicz D, Greenburg G, Birdwell CR (1978): Determination of cellular shape by extracellular matrix and its correlation with the control of cellular growth. Cancer Res 38:4155–4171.

Greenburg G, Hay ED (1982): The transformation of epithelium to mesenchyme-like cells within a hydrated collagen lattice. Anat Rec 202:69A.

Greenburg G, Vlodavsky I, Foidart J-M, Gospodarowicz D (1980): Conditioned medium from endothelial cell cultures can restore the normal phenotypic expression of vascular endothelium maintained in vitro in the absence of fibroblast growth factor. J Cell Physiol 103:333–347.

Grinnell F (1978): Cellular adhesiveness and extracellular substrata. Int Rev Cytol 53:65–144.

Grinnell F, Bennett MH (1981): Fibroblast adhesion on collagen substrata in the presence and absence of plasma fibronectin. J Cell Sci 48:19–34.

Grobstein C (1955): Tissue interaction in morphogenesis of mouse embryonic rudiments in vitro. In Rudnick D (ed): "Aspects of Synthesis and Order in Growth." Princeton, New Jersey: Princeton University Press, pp 233–256.

Hascall VC, Hascall GK (1981): Proteoglycans. In Hay ED (ed): "Cell Biology of Extracellular Matrix." New York: Plenum, pp 39–63.

Hay ED (1964): Secretion of a connective tissue protein by developing epidermis. In Montagna W, Lobitz WC (eds): "The Epidermis." New York: Academic Press, pp 97–116.

Hay ED: Organization and fine structure of epithelium and mesenchyme in the developing chick embryo. In Fleischmajer R, Billingham RE (eds): "Epithelial-Mesenchymal Interactions." Baltimore: Williams and Wilkins, pp 31–55.

Hay ED (1978): Fine structure of embryonic matrices and their relation to the cell surface in ruthenium red-fixed tissues. Growth 42:399–423.

Hay ED (1980): Development of the vertebrate cornea. Int Rev Cytol 63:263–322.

Hay ED (1981a): Extracellular matrix. In "Discovery in Cell Biology." J Cell Biol 91:205s–223s.

Hay ED (1981b): Collagen and embryonic development. In Hay ED (ed): "Cell Biology of Extracellular Matrix." New York: Plenum, pp 379–409.

Hay ED (1982): Interaction of embryonic cell surface and cytoskeleton with extracellular matrix. Am J Anat 165:1–12.

Hay ED, Meier S (1974): Glycosaminoglycan synthesis by embryonic inductors: Neural tube, notochord and lens. J Cell Biol 62:889–898.

Hay ED, Revel JP (1963): Autoradiographic studies of the origin of the basement lamella in regenerating salamander limbs. Dev Biol 7:152–168.

Hay ED, Revel JP (1969): "Fine Structure of the Developing Avian Cornea." Basel: S. Karger, 144 pp.

Hendrix MJC, Hay ED, Von der Mark K, Linsenmayer TF (1982): Immunohistochemical localization of collagen types I and II in the developing chick cornea by electron microscopy. Invest Ophthalmol Vis Sci 22:359–375.

Hewitt AT, Kleinman HK, Pennypacker JP, Martin GR (1980): Identification of an adhesion factor for chondrocytes. Proc Natl Acad Sci USA 77:385–388.

Horlein D, McPherson J, Goh SH, Bornstein P (1981): Regulation of protein synthesis: Translational control by procollagen-derived fragments. Proc Natl Acad Sci USA 78:6163–6167.

Humphreys S, Porter KR (1976a): Collagenous and other organizations in mature annelid cuticle and epidermis. J Morphol 149:33–52.

Humphreys S, Porter KR (1976b): Collagen deposition on a preformed grid. J Morphol 149:53–72.

Hynes RO (1981): Fibronectin and its relation to cellular structure and behavior. In Hay ED (ed): "Cell Biology of Extracellular Matrix." New York: Plenum, pp 295–327.

Hynes RO, Destree AT (1978): Relationships between fibronectin (LETS protein) and actin. Cell 15:875–886.

Ito S (1965): The enteric surface coat on cat intestinal microvilli. J Cell Biol 27:475–491.

Kjellen L, Pettersson I, Hook M (1981): Cell-surface heparan sulfate: An intercalated membrane glycoprotein. Proc Natl Acad Sci USA 78:5371–5375.

Kleinman HK, Klebe RJ, Martin GR (1981): Role of collagenous matrices in the adhesion and growth of cells. J Cell Biol 88:473–485.

Konigsberg IR, Hauschka SD (1965): Cell and tissue interactions in the reproduction of cell type. In Locke M (ed): "Reproduction: Molecular, Subcellular and Cellular." New York: Academic Press, pp 243–290.

Kosher RA, Church RL (1975): Stimulation of in vitro chondrogenesis by procollagen and collagen. Nature (Lond) 258:327–330.

Kosher RA, Lash JW (1975): Notochordal stimulation of in vitro somite chondrogenesis before and after enzymatic removal of perinotochordal materials. Dev Biol 42:362–378.

Kosher RA, Lash JW, Minor RR (1973): Environmental enhancement of in vitro chondrogenesis. IV. Stimulation of somite chondrogenesis by exogenous chondromucoprotein. Dev Biol 35:210–220.

Kraemer PM (1979): Mucopolysaccharides: Cell biology and malignancy. In Hynes RO (ed): "Surfaces of Normal and Malignant Cells." New York: Wiley, pp 149–198.

Kraemer PM, Tobey RA (1972): Cell-cycle dependent desquamation of heparan sulfate from the cell surface. J Cell Biol 55:713–717.

Liu SC, Karasek M (1978): Isolation and growth of adult human epidermal keratinocytes in cell culture. J Invest Dermatol 71:157–162.

Mauchamp J, Margotat A, Chambard M, Charrier B, Remy L, Michel-Bechet M (1979): Polarity of 3-dimensional structures derived from isolated hog thyroid cells in primary culture. Cell Tissue Res 204:417–430.

Mayer BW Jr, Hay ED, Hynes, RO (1981): Immunocytochemical localization of fibronectin in embryonic chick trunk and area vasculosa. Dev Biol 82:267–286.

Meier S, Hay ED (1974a): Control of corneal differentiation by extracellular materials. Collagen as a promoter and stabilizer of epithelial stroma production. Dev Biol 38:249–270.

Meier S, Hay ED (1974b): Stimulation of extracellular matrix synthesis in the developing cornea by glycosaminoglycans. Proc Natl Acad Sci USA 71:2310–2313.

Nadol JB Jr, Gibbins JR, Porter KR (1969): A reinterpretation of the structure and development of the basement lamella: An ordered array of collagen in fish skin. Dev Biol 20:304–331.

Nevo Z, Dorfman A (1972): Stimulation of chondromucoprotein synthesis in chondrocytes by extracellular chondromucoprotein. Proc Natl Acad Sci USA 69:2069–2072.

Nichols DH (1981): Neural crest formation in the head of the mouse embryo as observed using a new histological technique. J Embryol Exp Morphol 64:105–120.

Nichols DH (1982): Matrix mediated neural crest migration on the head of the mouse embryo; formation of the delaminate gap. Anat Rec 202:136A.

Nitsch L, Wollman SH (1980): Ultrastructure of intermediate stages in polarity reversal of thyroid epithelium in follicles in suspension culture. J Cell Biol 86:875–880.

Olsen BR (1981): Collagen biosynthesis. In Hay ED (ed): "Cell Biology of Extracellular Matrix." New York: Plenum, pp 139–172.

Ormerod EJ, Rudland PS (1982): Mammary gland morphogenesis in vitro: Formation of branched tubules in collagen gels by a cloned rat mammary cell line. Dev Biol 91:360–375.

Overton J (1977): Response of epithelial and mesenchymal cells to culture on basement lamella observed by scanning microscopy. Exp Cell Res 107:313–323.

Palade GE, Farquhar MG (1965): A special fibril of the dermis. J Cell Biol 27:215–224.

Porter KR (1952): Repair processes in connective tissues. In Ragan C (ed): "Connective Tissues, Trans 2nd Conf Josiah Macy Jr Foundation, 1951." New York: Josiah Macy Jr Foundation, pp 126–131.

Porter KR (1956): Observations on the fine structure of animal epidermis: In "Proc Int Conf on Electron Microscopy, London, 1954." London: Royal Microscopical Society, pp 539–546.

Porter KR (1964): Cell fine structure and biosynthesis of intercellular macromolecules. Biophys J 4 Part II:167–196.

Porter KR (1966): Morphogenesis of connective tissue. In Stephens CAL, Stanfield AB (eds): "Cellular Concepts in Rheumatoid Arthritis." Springfield, Illinois: Charles C. Thomas, pp 6–36.

Porter KR, Pappas GD (1959): Collagen formation by fibroblasts of the chick embryo dermis. J Biophys Biochem Cytol 5:153–166.

Porter KR, Vanamee P (1949): Observations on the formation of connective tissue fibers. Proc Soc Exp Biol Med 71:513–516.

Rambourg A, Leblond CP (1967): EM observations on the carbohydrate-rich cell coat present at the surface of cells in the rat. J Cell Biol 32:27–54.

Revel JP, Hay ED (1963): An autoradiographic and electron microscopic study of collagen synthesis in differentiating cartilage. Z Zellforsch 61:110–114.

Rheinwald JC, Green H (1977): Epidermal growth factor and the multiplication of cultured human epidermal keratinocytes. Nature (Lond) 265:421–424.

Robbins WC, Watson RF, Pappas GD, Porter KR (1955): Some effects of anti-collagen serum on collagen formation in tissue culture: A preliminary report. J Biophys Biochem Cytol 1:381–384.

Ross R, Benditt EP (1962): Wound healing and collagen formation. III. A quantitative radioautographic study of the utilization of proline-H^3 in wounds from normal and scorbutic guinea pigs. J Cell Biol 15:99–108.

Ross R, Benditt EP (1965): Wound healing and collagen formation. V. Quantitative electron microscope radioautographic observations of proline-H^3 utilization by fibroblasts. J Cell Biol 27:83–106.

Rubin K, Hook M, Obrink B, Timpl R (1981): Substrate adhesion of rat hepatocytes: Mechanism of attachment to collagen substrates. Cell 24:463–470.

Seegmiller R, Fraser FC, Sheldon H (1971): A new chondrodystrophic mutant in mice. Electron microscopy of normal and abnormal chondrogenesis. J Cell Biol 48:580–593.

Singer II (1979): The fibronexus: A transmembrane association of fibronectin-containing fibers and bundles of 5 nm microfilaments in hamster and human fibroblasts. Cell 16:675–685.

Solursh M, Fisher M, Singley CT (1979): The role of extracellular matrix in the formation of the sclerotome. J Embryol Exp Morphol 54:75–98.

Sugrue SP, Hay ED (1981a): Response of basal epithelial cell surface and cytoskeleton to solubilized extracellular matrix molecules. J Cell Biol 91:45–54.

Sugrue SP, Hay ED (1981b): Effect of fibronectin, laminin, and collagen on the differentiation of embryonic corneal epithelium in vitro. J Cell Biol 91:159a.

Sugrue SP, Hay ED (1982a): Interaction of embryonic corneal epithelium with exogenous collagen, laminin, and fibronectin. Role of endogenous protein synthesis. Dev Biol 92:97–106.

Sugrue SP, Hay ED (1982b): Further studies of the interaction of collagen, laminin, and fibronectin with the basal surface of embryonic corneal epithelium. Anat Rec 202:185A.

Timpl R, Martin GR (1982): Components of basement membranes. In Furthmayr H (ed): "Immunocytochemistry of the Extracellular Matrix," Vol II. Boca Raton, Florida: CRC Press.

Tomasek JJ, Hay ED (1982): Ultrastructure and stability of cytoskeleton of fibroblasts grown in collagen gels. Anat Rec 202:192A.

Tomasek JJ, Hay ED, Fujiwara K (1982): Collagen modulates cell shape and cytoskeleton of embryonic corneal fibroblasts. Distribution of actin, actinin, and myosin. Dev Biol (in press).

Toole BP (1981): Glycosaminoglycans in morphogenesis. In Hay ED (ed): "Cell Biology of Extracellular Matrix." New York: Plenum, pp 259–288.

Trelstad RL, Coulombre AJ (1971): Morphogenesis of the collagenous stroma in the chick cornea. J Cell Biol 50:840–858.

Trelstad RL, Hayashi K (1979): Tendon fibrillogenesis: Intracellular collagen subassemblies and cell surface changes associated with fibril growth. Dev Biol 71:228–242.

Trelstad RL, Silver FH (1981): Matrix assembly. In Hay ED (ed): "Cell Biology of the Extracellular Matrix." New York: Plenum, pp 179–211.

Trelstad RL, Hay ED, Revel JP (1967): Cell contact during early morphogenesis in the chick embryo. Dev Biol 16:78–106.

Trelstad RL, Hayashi R, Toole BP (1974): Epithelial collagens and glycosaminoglycans in the embryonic cornea. Macromolecular order and morphogenesis in the basement membrane. J Cell Biol 62:815–830.

Trelstad RL, Hayashi A, Hayashi K, Donahoe PK (1982): The epithelial-mesenchymal interface of the male rat Mullerian duct: Loss of basement membrane integrity and ductal regression. Dev Biol 92:27–40.

Vanamee P, Porter K (1951): Observations with the electron microscope on the solvation and reconstitution of collagen. J. Exp Med 94:255–266.

Weiss P (1957): Macromolecular fabrics and patterns. J Cell Comp Physiol 49:105–112.

Weiss P, Ferris N (1954): Electron microscopic study of the texture of the basement membrane of larval amphibian skin. Proc Natl Acad Sci USA 40:528–540.

Weinstock A, Leblond CP (1971): Elaboration of the matrix glycoprotein of enamel by the secretory ameloblasts of the rat incisor as revealed by radioautography after galactosc-^{3}H injection. J Cell Biol 51:26–51.

Weinstock M, Leblond CP (1974): Synthesis, migration, and release of precursor collagen by odontoblasts as visualized by radioautography after (^{3}H) proline administration. J Cell Biol 60:92–127.

Yamada KM (1981): Fibronectin and other structural proteins. In Hay ED (ed): "Cell Biology of Extracellular Matrix." New York: Plenum, 95–110.

Yamada KM, Yamada SS, Pastan I (1976): Cell surface protein partially restores morphology, adhesiveness and contact inhibition of movement to transformed fibroblasts. Proc Natl Acad Sci USA 73:1217–1221.

Poster Presentations

Peter Andrews
Department of Anatomy, Georgetown University, Schools of Medicine and Dentistry, 3900 Reservoir Road, NW, Washington, D.C. 20007
"Microfilaments, microtubules and the anionic surface coat in kidney glomerular epithelial cells."

Betty (Mathews) Barbour
San Joaquin Delta College, 5151 Pacific Avenue, Stockton, CA 95207
"Training in electon microscopy."

Gudrun S. Bennett
Department of Anatomy G3, University of Pennsylvania, Medical School, Philadelphia, PA 19104
"Temporal and spatial analysis of the onset of neurofilament expression during neurogenesis."

Mary Bunge and Richard Bunge
Department of Anatomy and Neurobiology, Washington University School of Medicine, 660 South Euclid, St Louis, MO 63110
"Inter-relationship between Schwann cell function and the extracellular matrix."

Robert R. Cardell, Jr.
University of Cincinnati, Cincinnati, OH 45267
"Action of glucocorticoids on smooth endoplasmic reticulum in hepatocytes."

Barry S. Eckert
Department of Anatomy, State University of New York, Buffalo, NY 14214
"Investigation of the intermediate filament organizing center."

Joseph F. Gennaro, Jr.
Department of Biology, New York University, Washington Square, New York, NY 10003
"Patterns of membrane flow in stimulated nerve terminals."

John R. Gibbins
Department of Pathology, University of Sydney, Sydney, Australia
"Modulations of the microtrabecular lattice of motile epithelial cells."

Ursula Goodenough
Department of Biology, Washington University, St Louis, MO 63130
"The subunit structure of the outer dynein arm."

Ed Haskins
Department of Botany, University of Washington, Seattle, WA 98195
Film on "Comparative plasmodial types and sporulation in the Myxomycetes."

J.G. LaFontaine
Biology Department, Laval University, Quebec, Canada G1K 7P4
"A radioautographic and cytochemical study of the nucleolar cycle in the myxomycete, Physarum polycephalum."

Jon C. Lewis, Tilman Prater, and George Wray
Bowman Gray School of Medicine, Department of Pathology, Winston-Salem, NC 27103
"Spatial organization of cytoskeletal elements and cytoplasmic membranes during blood platelet adhesion and release: HVEM and conventional EM studies."

Donald W. Misch
Department of Zoology, Wilson Hall 046A, University of North Carolina, Chapel Hill, NC 27514
"The role of mucus in regulation of mucociliary transport."

Pietro M. Motta
Department of Anatomy, Faculty of Medicine, University of Rome 00161, Rome, Italy
"The liver and the adrenal. A three-dimensional microanatomical parallelism. Results by high resolution scanning electron microscopy."

David H. Robertson
Department of Anatomy, Duke University, Durham, NC 27707
"A new kind of intercellular membrane relationship."
"Studies of urinary bladder hexagonal membranes fractured at near liquid Helium temperatures."

Tom Schroeder
Friday Harbor Lab, Friday Harbor, WA 98250
"Actin in sperm and eggs as visualized with NBD-phallacidin."

Albert W. Sedar
Anatomy Department, Jefferson Medical College, 1020 Locust Street, Philadelphia, PA 19107
"Backscattered electron imaging to visualize arterial endothelial damage in the SEM."

Dorothy Spangenberg
Department of Pathology, Eastern Virginia Medical School, Norfolk, VA 23501
"SEM studies of Aurelia aurita."

George Szabo
Harvard School of Dental Medicine, 188 Longwood Avenue, Boston, MA 02115
"Melanocyte-keratinocyte membrane interactions in vivo and in vitro."

Melvyn Weinstock
Department of Anatomy, McGill University, 3640 University Street, Montreal, Quebec, Canada H3A 2B2
"Gap junctions in odontoblasts from rat incisor teeth."

John P. Wourms, B. Grove, W. Hamlett, F. Knight, and J. Lombardi
Zoology Department, Clemson University, Clemson, SC 29631
"Maternal-fetal nutrient transfer in viviparous fishes."

Summary

Modern Cell Biology 2:553–566
© 1983 Alan R. Liss, Inc., 150 Fifth Avenue, New York, NY 10011

Spatial Organization of the Eukaryotic Cell:
Summary and Perspectives

Peter Satir

From the Department of Anatomy, Albert Einstein College of Medicine, Bronx, New York 10461

I. INTRODUCTION

In the past 30 years our conception of the spatial organization of the eukaryotic cell has changed dramatically. This rapid period of change coincides with the major scientific discoveries of Keith R. Porter and the beginnings of the field now called cell biology. The relationship is a causal and continuing one. Springtime in the Rockies, 1982, has been a time and place for looking back over the years for the threads that make up the tapestry of modern cell biology, and also of looking forward to emerging patterns, where the tapestry remains unfinished.

In 1958, a similar meeting was held in Boulder, and that meeting also produced a book—"Biophysical Science—A Study Program"—which graduate students of the time, including myself, pored over seeking enlightenment on cell organization. In the very first chapter of this book, F. O. Schmitt [1959] summarized the three contrasting theories of the fundamental structure of protoplasm as they came down to that time from the late 19th century. He pointed out that new facts were already available from electron microscopy about the granular, the membranous, and the fibrous structures of the cytoplasm. In retrospect, these facts, although revolutionary and still controversial at that time, were elementary indeed. They consisted of descriptions of the locations and distributions of the main cell organelles, beginning with the endoplasmic reticulum, the mitochondrion, the centriole, the ciliary axoneme, and the then recently described ribosome. Not only microtrabeculae, but also microtubules and microfilaments, were concepts of the future, as were nucleosomes and coated vesicles. The link from structure to biochemistry and cell function was extremely tenuous. Many people still screamed artifact. The pioneers, like Keith Porter, pointed out that, if one were careful enough, the extremely consistent artifacts of electron microscopy had to reflect the form, the chemistry, and the function of the cell to a significant extent [cf Porter, 1957].

From the first electron micrographs of whole mounted cells in 1945, one of Porter's chief preoccupations has been to learn what the microscope can reveal about organization of the ground substance of cytoplasm. As the art of electron microscopy improved, this preoccupation took Porter, and us, from a consideration of the membranous endoplasmic reticulum to studies of a filamentous cytoskeleton, first of microtubules and then of smaller filaments, always in quest of the organizing principles of cytoplasm. It is to Porter's credit that he has persisted in looking for such principles in the face of considerable skepticism; it is further to his credit that he has been right. As the present meeting has indicated, we are just beginning to realize the full subtlety and elegance of the spatial organization of eukaryotic cells. In this sense, we are all, like myself, true students of Keith Porter.

The search for general organizing principles of the nucleus and cytoplasm leads us to the underlying questions addressed at this symposium. These questions are nothing less than the old problems of the role of heredity vs environment considered at the cell biological, and therefore mechanistic, level: what causes specific cells to look and to behave as they do; how do external stimuli produce specific cell responses. Figure 1 is a diagram showing the types of interactions that must form the basis of the answers to these questions. To follow the pathways described in the figure, we need to know the precise spatial, as well as biochemical organization of cell membranes, cytoplasmic and nuclear constituents, and the mechanisms by which feedback occurs between one compartment of the cell and another. You have just read of many specific instances where knowledge of these pathways is being sought. To illustrate the interconnections of the pathways, I want to consider one example of the work presented here as a model of how each contribution to this volume may ramify to generate problems that its neighbors address.

I have chosen the ciliate Oxytricha, as discussed in the article by Prescott (this volume), for my example. Perhaps this is a mite unfair, since ciliates are among the most elaborate of single eukaryotic cells: Both their cytoplasmic and nuclear organizations are unusual and atypical when compared to chick fibroblasts, and, most seriously, they must respond as organisms as well as cells. Nevertheless, Prescott has succeeded in utilizing the ciliates in an interesting way to study the organization of DNA into genes and chromosomes. He has shown that in Oxytricha during the formation of a macronucleus from a micronucleus at the end of conjugation, first the micronuclear chromosomes become polytenic and then they are chopped, band by band, into individual genes. But to get the nuclei to perform their tricks for him, Prescott starves the cells. This simple change provides the environmental stimulus for profound alterations in the organization of cytoplasmic and nuclear function, the end product of which is the generation of a new macronucleus. What are the intervening signals that affect the cell's genetic program and lead to preparation for conjugation (membrane changes?), micronuclear division (microtubules?), macronuclear disappearance, cytoplasmic fusion, exchange of micronuclei (organellar migration), meiosis, DNA synthesis, inhibition of cell division, inhibition of feeding etc?

Every experimental system in this volume generates a similar set of interlocking problems. Because we believe that evolutionarily one eukaryotic cell is similar to another, we need not rely on a single system like Oxytricha for all the answers; instead, the best or most readily studied system for a given problem can be employed, as this symposium often demonstrates. To facilitate summarizing the range of answers presented here, I will search for a few rules, perhaps overly simple, that help to categorize the various contributions of this symposium in terms of general organizing principles applicable to eukaryotic cells and to aspects of the pathways for cellular response, to the extent we understand them.

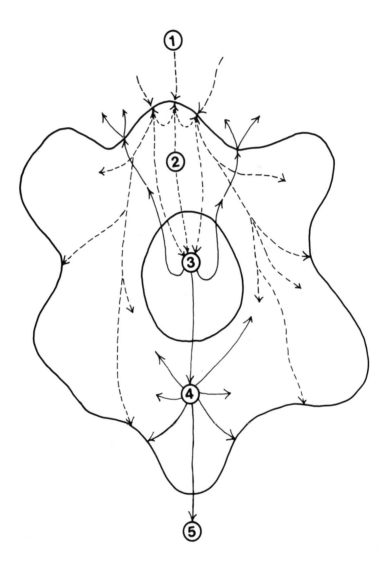

Fig. 1. Cascade of information through the eukaryotic cell following a stimulus. Broken arrows: from environment to genome. Solid arrows: from genome to environment. 1) A localized stimulus acts at the cell surface. 2) Signals are transmitted to the subjacent cytoplasm that affect the cytoskeleton, cytoskeletal-membrane interactions, cytoplasmic organelles, and the nucleus. 3) There is a response in the cell nucleus consisting of turning specific genes on and off. 4) The gene products feedback to influence the cytoplasm, including the cytoskeleton, and the cell membrane. 5) Changes occur in the extracellular environment that act as new stimuli for adjacent cells.

II. OLIGOMERIC, USUALLY HETERO-OLIGOMERIC, PROTEINS ARE REAL CELL STRUCTURES

I begin with this rule because it is one of Keith Porter's crucial contributions to cell biology, and, as such, has pervaded this meeting. You will perhaps say that this rule is obvious, but it is not. Before the invention of the cell biological approach, protein chemists used to think of the cell as a bag of enzymes—and generally water-soluble enzymes at that; physiologists called it a black box. Both were surprised at the complexity of cell compartmentalization. Such structure clearly affected their studies, because enzymatic functions were not randomly distributed throughout the cytoplasm but were localized instead to cell structures.

Penman has indicated in his presentation that a typical eukaryotic cell, after permeabilization to remove soluble components, contains several thousand different proteins, as resolved on 2D gels. Protein molecules are big enough to be visualized individually by electron microscopy. Together with a few other types of macromolecules, they comprise the structures seen in such extracted preparations, including virtually every nonmembranous component of the cell. The locations of specific proteins, their conformations, their binding interactions, and their assembly properties are the essence of what we mean when we talk about cell structure. The cytoskeleton of whole mounted cells that we see throughout this volume is largely the protein residue of the cell after water is removed by the critical-point method. A tremendous effort is now going on to identify the locations, conformations, assembly states, and interactions of a few major components of this residue. Many of the contributions above speak seriously to this theme. Among the best characterized protein families are the following.

A. Microfilaments

Actin and actin-binding proteins comprise this family, which is the subject of the articles by Condeelis, Stossel, and Tilney. This is one of the best-studied families of proteins. By "family," I mean a group of proteins that bind to one another, influencing the structure and assembly of the major constituent of the group, for which the family is named. Although the number of components in the actin family is large, the relevant molecular interactions are being elucidated, and, as Condeelis especially points out, their links to cell structure are beginning to be understood. The actin-binding proteins illustrate the complexity and subtlety of cellular controls. They include myosin, gelsolin, acumentin, α-actinin, spectrin, and profilin, among those singled out for examination. These regulate the polymerization state of actin (eg, profilin), the length and bundling properties of polymerized filaments (gelsolin, acumentin, 95 kd actin binding protein of Dictyostelium), and the capability of relative active movement (myosin). Some of these interactions depend on pH or Ca^{2+} in physiologically significant ways, while others do not. In this way, after a stimulus to the cell, the actin gel comes and goes in local regions of the cytoplasm to contribute to cell form, organelle placement and cell motility.

The relationship of actin to spectrin in red blood cell ghosts is particularly illustrative, since this provides a model for the linkage between the changing actin gel (microfilamentous cytoskeleton) and a membrane through a number of defined accessory proteins: spectrin, ankyrin, and so on. Spectrinlike proteins and similar linkages may occur in other more typical eukaryotic cells. The nature of the specific protein-protein interactions and the lengths and arrangements of particular protein molecules determine spatial relationships between different membrane receptors, and may also determine more general characteristics such as cell shape or membrane vesiculation. Tilney discusses the power of the microfilament systems to mold individual, stable cell organelles, such as the microvilli of the intestinal brush border or the hair cell, and the acrosomal filaments of sperm.

B. Microtubules

Tubulin and microtubule-associated proteins comprise this family, which owes much of its early exploration to Porter. Figure 2, excerpted from the Annual Report of the Rockefeller Institute of 1955–56, shows the composition of the Porter laboratory near the beginning of this exploration and is evidence of the early realization of the importance of cytoplasmic microtubules. The present article by Porter shows that this exploration is not over. The article by McIntosh also probes this aspect of cell organization. Microtubules are so obviously distinctive cell structures that their representative value as proof of the theme that protein assemblies and cytoskeletal elements are largely one and the same thing is usually underplayed. The family is bigger than simply a few kinds of α and β tubulin, and a few MAPs. It includes dynein, the ciliary ATPase, and calmodulin, to name only some of its most visibly important ancillary members. The ciliary axoneme, of which microtubules form the principal component (observed in cross sections by Fawcett and Porter [1954]), consists of several hundred polypeptides other than tubulin. These often form subfamilies of identifiable structure—arms, radial spokes, etc—functionally linked to the microtubules. I shall return to this point below. Moreover, molecules analogous to those regulating actin length, bundling, and assembly probably regulate microtubule length, bundling, and assembly. McIntosh points out that the centrosome (cell center) acts as an overall organizing system in the cell, and perhaps is the locus of proteins that cap the depolymerizing (minus) end of microtubules. The number and distribution of such "tubulin-initiating proteins" would determine the number and distribution of microtubules radiating out from the cell center. This is an important feature in the inheritance of structural pattern.

THE ROCKEFELLER INSTITUTE

BULLETIN

ANNUAL REPORT

1955-1956

CYTOLOGY

Laboratory of Keith R. Porter and George E. Palade
    Montrose J. Moses, P. Siekevitz, M. L. Watson, Jean-Gabriel
    Lafontaine, R. J. Barrnett, J. B. Caulfield, E. Yamada, M. A.
    Rudzinska

Exploratory work on the mitotic apparatus of the cell (Porter) and on the elementary "filaments" of cilia has revealed the importance of fine tubular structures in the organization of such elements and has suggested that such tubular units may possess extensive morphogenetic potentialities.

Fig. 2. Facsimile of portions of the Rockefeller Institute Bulletin 1955–56 pertaining to the laboratory of Porter and Palade, including a brief description of microtubules.

C. Intermediate Filaments

This is also a complex family of proteins, which assemble into a defined cell structure: the 10-nm diameter filament. The five or so general classes of intermediate filament proteins resist solubilization, particularly by nonionic detergents, and they correspond roughly to cell histiotypes. For example, the α keratins are found in epithelial cells, vimentin in fibroblasts, and desmin in smooth muscle. However, more than one of these components can be found in a single cell type. In his article, Lazarides points out that both desmin and vimentin surround the Z disk of chicken myotubes, forming a three-dimensional webbing designed to keep the myofibrils in register. He proposes that the intermediate filaments are important everywhere in determining cell shape and organelle distribution. They are linked via associated proteins to the other cytoskeletal families, particularly to the actin system.

D. Other Protein Families

Other protein families of structural interest have not been as thoroughly addressed at this meeting, but at least two deserve brief mention here. One of these contains self-assembling clathrin, the major protein of the coated vesicle, and its associated proteins, including a family of polypeptides at about 110,000 M_r that may link the system to an appropriate membrane. The proteins in this family are involved in relationships between cytoskeleton and membranes, including receptor-mediated endocytosis, membrane vesiculation, fusion, and recognition. The coated vesicle has a special place in this review, because, once again, Porter's laboratory supplied much of the initial impetus for understanding these relationships [Roth and Porter, 1964].

The second family is less well studied, but I include it for completeness and because of its potential importance. It contains the conformationally responsive fibrous proteins of the cytoskeleton such as spasmin. These proteins are found extensively in the ciliates—for example, Stentor—where they are probably responsible for cell contraction. They were once thought to be widely distributed, but they have suffered recently from general neglect. Ciliary rootlets, another Porter fancy, are structures probably made of such proteins.

Does the sum of all the accessory proteins mentioned in each family total 1,000? Clearly not. Yet even they add an enormous complexity to the major assembled elements of the cytoplasm. Structurally, this complexity is in the form of interlocking side branches of microfilaments, microtubules, etc. Because these are molecules, the lengths, distributions, and binding sites of the side branches are determined. Together with the unnamed proteins that make up the remainder of the permeabilized cell residue, they form the lattice work that Porter and co-workers call the microtrabecular lattice (MTL). The structural elements of the MTL have the disadvantage of being generally irregular and therefore difficult to study. However, one regular element often cited by Porter is the radial spoke of the ciliary axoneme. This side branch connects axonemal microtubules in a defined way; it apparently changes orientation and perhaps attachment during ciliary movement. Its composition is known; there are 12 or so different polypeptides assembled in order to form a spoke. Specific mutations prevent spoke assembly. In other words, here is a perfect example of the properties of the MTL, including physiological response. Look carefully at the MTL as Porter describes it in his article. Isn't this essentially the best proof that hetero-oligomeric families of proteins are real cell structures? The definition of the less regular elements of the MTL, their positions, compositions, and precise responses to local changes in environmental conditions will occupy many of us for some time to come.

III. THE BASIS OF THE FUNCTIONAL MOSAICISM OF CELLULAR MEMBRANES IS THE ARRANGEMENT OF OLIGOMERIC, USUALLY HETERO-OLIGOMERIC, MEMBRANE PROTEINS WITHIN A CONTINUOUS SEMILIQUID BACKGROUND OF LIPID

This is another rule that now seems trivial. However, let us look at some quotes from the published summary [Satir B, 1969] of the Estes Park meeting of 1969 on "Control of Form in Cells," a direct forerunner of this symposium: "Two models of membrane structure were considered: the unit membrane (lipid bilayer) model and a lipoprotein subunit model. Neither model can be applied generally. . . . Only the structure of myelin seems settled." We are a long way from this now. The credit for our sophistication must go to many laboratories but in larger measure to Branton and the freeze-fracture workers, who convincingly showed the distinctive nature of local regions of the membrane within a background made largely of lamellar phase lipids. Singer and Nicolson also deserve much credit for the synthesis of results called the "fluid mosaic model." Many contributions to this volume concern membranes, but hetero-oligomeric membrane proteins and their role in determining structure and function of local regions of a membrane are particularly discussed in the articles by Staehelin and Poyton. Staehelin recalls the freeze-fracture signature of these aggregations: the intramembrane particle (IMP) array. IMPs represent transmembrane proteins or lipoprotein complexes in structures such as the gap junction or the chloroplast thylakoid membrane. Stacking of chloroplast membranes involves rearrangement of the IMPs, segregating components including photosystems I and II into granal and stromal regions of the membrane. The light harvesting complex of the chloroplast membrane can be isolated and added to liposomes. In the presence of high salt, such reconstituted liposomes agglutinate and large IMPs are seen where the membranes are held together. Such manipulations, starting with proteins and lipids in separate test tubes, mimics grana formation in the cell. Poyton discusses the hetero-oligomeric proteins of the mitochondrial membranes. To form cytochrome oxidase for example, three mitochondrial gene products and four nuclear gene products assemble. Apparently, insertion of one subunit into the membrane facilitates insertion and recognition of additional subunit and the assembly of a functional enzyme. I will come back to this in a moment.

Most of the membrane in a eukaryotic cell is found within the various specializations of the endoplasmic reticulum. Each specialization is characterized by a particular enzymatic, and therefore protein, composition. But integral membrane proteins are usually inserted cotranslationally at one point in this complex system, the rough endoplasmic reticulum (RER), and then sorted out to other compartments. The problems in eukaryotic cell membrane organization seem more and more to focus on recognition, signaling, sorting, and, finally, interaction with the cytoskeleton. Cytoskeletal–membrane interaction counteracts the tendency toward randomization of components introduced by the sea of lipids in which the membrane proteins are immersed.

IV. LOCATION AND ASSEMBLY PROPERTIES OF PROTEIN FAMILIES ARE DETERMINED BY SPECIFIC PROTEIN–NUCLEIC ACID OR PROTEIN-PROTEIN INTERACTIONS

With this statement, I have tried to summarize three important general questions of eukaryotic cell biology which, in various guises, permeate the presentations of this symposium: 1) Which proteins will be made in a cell at any given time? This is a question of specificity of synthesis, relating to the sequential activation and sequential reading of the genome, to nucleosome structure and to cell cycle stage. 2) By what mechanisms are newly synthesized proteins segregated into their proper locations in the cytoplasm and to various specialized membranes? This is a question of specificity of recognition and signaling. 3) By what mechanisms are the individual families of proteins assembled properly into functional organelles, membranes, and cytoplasm?

That the overall pattern of cell reproduction rests in a defined genetic blueprint is a truism that is amply illustrated in the article by Pickett-Heaps. In desmid cell wall differentiation, an inbuilt pattern is rebuilt, sometimes with variations, from generation to generation. Morphogenesis is dependent on protein synthesis, and morphological mutants are available. But the intricacy of the final pattern points out how difficult it seems to specify definitive overall morphogenetic mechanisms by identifying patterns of active genes. Gerhart discusses similar problems in relation to the establishment of the embryonic axes after fertilization in Xenopus. Here, the entire organismal blueprint must be linked to cell cycle and finally to differential gene activity.

Identifying active genes is no easy task, but the temporal pattern of gene activation forms a basis for the answer to the first question posed above. Without the proper building blocks, cellular morphogenesis and differentiation is thwarted at the outset. Palade has pointed out that the components of the Chlamydomonas chloroplast membrane are synthesized in a sequence dependent on the cell cycle. Weintraub notes that in erythropoiesis in the chick embryo, precursor blood cells somehow know that the globin gene is to be turned on between the 25-hour embryo when no globin mRNA can be detected and the 35-hour embryo. Moreover, this structural change in the gene is transmitted to and propagated by daughters of the precursor cell.

The electron microscope originally seemed relatively uninformative about questions of nuclear structure and activity, although the uniform dimensions of chromatin were noted quite early on [cf Porter, 1958]. A series of clever and important studies has enlarged and changed this perspective so that, as Weintraub shows, we are now able to examine individual active vs inactive nucleosomes in detail. Higher-order nuclear structures are also being reinvestigated, and molecular interactions between chromatin, the nuclear matrix, and the nuclear membrane are beginning to be understood.

I tread lightly over these important nuclear issues to concentrate on some points raised in this symposium with regard to cytoplasmic questions that on the whole illustrate similar problems. For instance, in determining which proteins are found at a particular location, proper gene transcription and mRNA processing are important, but equally important may be the location of the sites of ribosomal activity. One aspect of this seems relatively clear: Mitochondrial and chloroplast ribosomes translate some proteins, cytoplasmic ribosomes translate others. However, Penman and Porter both pose the possibility that the translation sites for specific cytoplasmic proteins are nonrandom, so that the products can be directly, quickly, and preferentially incorporated at their proper locations. It remains to be shown whether this is a general mechanism for protein segregation during morphopoiesis.

An alternative proposition is that, given appropriate gene activation and message processing, translation occurs more or less randomly in cellular space, and that segregation relies on a variety of signaling mechanisms that permit proteins to follow more and more divergent pathways to their final cellular destinations. In 1969, at the Estes Park meeting referred to above, the predominant signaling mechanism for morphopoiesis involved noncovalent interactions as driving forces for "self-assembly." This thermodynamic recognition principle was originally applied to viruses and by extension to the elements of the cytoskeleton, particularly microfilaments and microtubules.

One of the significant points to emerge from the present volume is that "self-assembly" is not enough. The interactions between various proteins or between proteins and other molecules, particularly nucleic acids and possibly lipids, are more complex than general, noncovalent or hydrophobic interactions. They involve not only sequence information and specific conformational states, but regulatory steps involving posttranslational modification and external control factors. I have already summarized the suggestions of various authors that such specific interactions affect gene activation and the organization of the cytoskeletal families of proteins into a latticework.

Self-assembly ideas have found parallel applications in discussions of membrane biogenesis. The self-assembly principle is embodied in the "signal hypothesis" discussed in the articles of Poyton and Sabatini. In its original form, the signal hypothesis assumed that the presence of a hydrophobic leader sequence on a nascent polypeptide determined whether the peptide inserted into and through the ER. In its new version, the signal hypothesis introduces specific signal recognition proteins and binding proteins that join the growing signal and ER membrane. As mentioned previously, Poyton points out that in the formation of functional enzymes within a growing mitochondrial membrane—eg cytochrome c oxidase—one gene product inserted into the membrane recognizes the next in a progressive, stepwise fashion. These examples suggest that we must look harder for preexisting templates in morphogenic processes, perhaps especially in the differentiation of membranes. To begin to understand membrane differentiation and cell patterning perhaps we must invoke the aphorism, "All membranes from preexisting membranes." The preexisting proteins in a membrane may provide receptor sites for newly synthesized proteins that determine membrane function.

The difference in viewpoints between self-assembly and what I shall call sequence-directed assembly lies in the precision with which morphopoiesis is specified. Sequence-directed assembly is akin to the informational processes that Kellenberger [1967] has discussed in "higher orders" of morphopoiesis. One surprise is that the sequence needed for specificity may be relatively small. We must be concerned not only with the critical concentration of the assembling monomer, but with processes analogous to receptor-ligand or enzyme-substrate interactions within each entire interacting protein family. With this concern, we will be in a position to explain how an assembly event can occur in one but not another part of the same cytoplasm or the same membrane, where no obvious barriers to intermixing are seen.

One of the consequences of this change in viewpoint is that more attention is being paid to triggers of morphopoiesis, especially events that change the structure of proteins to inhibit or facilitate a particular physiological event. An important trigger for assembly is posttranslational modification, especially phosphorylation, as Staehelin notes for chloroplast membrane stacking and IMP migration. Local influx of Ca^{2+}, local pH changes, or cyclic nucleotide concentration changes translate environmental stimuli into such protein modifications via changes in activities of relevant kinases and phosphorylases. Several articles in this symposium present examples of such processes, which are known incompletely for most cellular events. Hepler, for example, discusses the probable significance of a Ca^{2+} signal during chromosome movement. Nuccitelli indicates that signaling via transcellular ion currents may be widespread for cellular activities and morphogenetic events.

The concept of sequence-directed assembly suggests an answer to a serious evolutionary question raised by Hood's article in this volume. The immune system, which Hood discusses, provides one of the glories of modern molecular understanding of the relationship between gene organization and protein structure in eukaryotic cells. The genetic mechanisms of the eukaryotic cell nucleus, including exon shuffling and variable RNA splicing, are capable of producing tremendous protein diversity. Since individual protein molecules are put together on a domain-by-domain basis, secreted vs membrane-bound forms of a single class of proteins, such as the immunoglobulins, can be manufactured. In the face of this easily generated potential diversity, why then are the cytoskeletal families of proteins relatively so conservative? One possibility is that the regulation of interaction and assembly-disassembly within each cytoskeletal family is dependent on complex, carefully defined, and therefore conservative, recognition signals.

V. THE EUKARYOTIC CELL IS STRUCTURALLY INTEGRATED VIA INTERACTIONS AND RECOGNITION BETWEEN PROTEIN FAMILIES

Not surprisingly, we return to structural integration as a final theme of this volume in honor of Keith Porter. I have previously discussed examples of membrane-cytoskeletal linkage. Transmembrane proteins are linked to the peripheral proteins that blend into cytoskeletal latticework on the protoplasmic side of the membrane. Hay reminds us that similar bridges exist between transmembrane proteins and extracellular matrix constituents such as fibronectin on the opposite side of the membrane, and that these also play a role in determination of cell shape. In a simple model for membrane differentiation, some transmembrane proteins would be fixed in position by such exterior and interior bridging. These might serve as sites for the stepwise assembly of originally diffusible subsidiary proteins that would define the subsequent function and type of membrane, as mentioned in the previous section. But this might also have consequences for the cytoplasmic matrix, since binding of say ankyrin to such a fixed transmembrane protein would eventually involve the actin family of proteins and determine the lattice pattern of that region of the cytoplasm, whereas binding of, say, a homologue of MAP2 might organize the microtubule cytoskeleton in this region. Gerhart reminds us that traditional differentiation in the vertebrate egg cell must rely on such mechanisms, since physical contact between the submembranous cytokeleton and some components (originally soluble proteins?) of the vegetal cytoplasm results in gray crescent formation and the establishment of the dorsoventral axis of the future embryo.

In the past 30 years during which the field of cell biology has been invented, as we have seen in this symposium, the basic relationships between stimulus and cellular responses, the basic mechanisms of signaling and integration, and the basic manner in which the gene program is translated into cell structure and cell organization have begun to be elucidated. The complexity and specificity of these mechanisms and relationships are perhaps more than we originally anticipated; the lesson is that nature is subtle and that we have much still to learn.

I have tried to capture some of the flavor not only of the scientific questions probed in this symposium, but also some of the flavor of the event itself: a celebration of the contributions of a pioneer and continuing contributor to the field. I take this opportunity to say for all the contributors to this occasion: It is a pleasure to be part of the house that Porter built.

VI. REFERENCES

Fawcett DW, Porter KR (1954): A study of the fine structure of ciliated epithelia. J Morphol 94:221–281.

Kellenberger E (1967): Control mechanisms in bacteriophage morphopoiesis. In Wolstenholme GEW, O'Connor M (eds): "Principles of Biomolecular Organization." Boston: Little, Brown, pp 192–228.

Porter KR (1957): The submicroscopic morphology of protoplasm. Harvey Lec 51:175–228.

Porter KR (1958): Problems in the study of nuclear fine structure. In Bargmann W et al (eds): "Fourth International Congress on Electron Microscopy." Berlin: Springer-Verlag, pp 186–199.

Roth TF, Porter KR (1964): Yolk protein uptake in the oocyte of the mosquito Aedes aegypti L. J Cell Biol 20:313–332.

Satir B (1969): Control of form in cells. Science 167:307–309.

Schmitt FO (1959): Molecular biology and the physical basis of life processes. In Oncley JL et al (eds): "Biophysical Science—A Study Program." New York: Wiley, pp 5–10.

Index

diagram of postulated relations to cell membrane and cytoskeleton, 528
effects on cell shape, 531–538
cytochalasin D, 533
fibroblast, 531–534, 535, 536
heavy meromyosin subfragments, 533
lens, epithelia, and polarity, 534–538
response to collagen, 531, 536
-epithelial cell surface interaction, 529–531
cycloheximide, 531
heavy meromyosin subfragments, 530
LACA, 531
laminin, 529, 530
type IV collagen, 529, 530
epithelial cf. mesenchymal interaction, 543
-fibroblast surface interaction, 525–529
extracellular fibronectin, 525–529
fibronexus, 525–526, 528
intracellular actin, 525–529
morphology, diagram of, 541
organization, 510–515, 542
avian cornea, 513–515, 523, 530
collagen, GAG, and structural glycoproteins, 510–513
supramolecular, 513–515
scaffoldings, 510
Extrons, 306–310, 326

Fatty acids, ER synthesis, 26
Fiber(s), kinetochore, 95, 99
Fibrils, crow's feet, in membrane-actin filament interactions, 177–180, 196
Fibroblast
cell shape, 531–534, 535, 536
growth factor, 539
3T3, 391, 399
see also Extracellular matrix-epithelial cell interaction
Fibronectin, fibronexus, and extracellular matrix-fibroblast interaction, 525–529

Filaments, cytoplasmic intermediate, 145–146
anisotropic association with plasma membrane, 149, 151, 152
antibody decoration, 152–153
distribution in skeletal muscle, 148
three-dimensional visualization in chicken erythrocytes, 150
see also Transcytoplasmic integration
Fluorescence studies
CTC, in mitosis, 99–103, 105, 106
NPN, in mitosis, 102
Fragmin, 229, 234
Frog oocyte, transcellular ion currents, 460, 466–468
Fundulus, 264, 266–267, 516

Galvanotaxis in quail neural crest cells, 474–476, 478–479
Gelsolin, 557
and cell motility, 229, 232, 234
macrophages, 212, 215, 216, 218
Gene(s)
α-amylase, 309
expression
multigene families, developmental control, 313
in physiologic compartmentation, 16
globin. See Globin genes
μ heavy chain, 310
ovalbumin, probes, 355, 365, 372
products
genome, coding in choroplast, 33
mitochondrial, in compartmentation, 33
nuclear, in compartmentation, 28–33
organellar, 33
see also Mitochondrial gene products
pseudo-, 313
regulation of protein transport pathways, 31–33
split, 306–310
yeast invertase, 31–33
see also Chromosome, Hybridization, Membrane memory

DATE DUE